DEVELOPMENTAL BIOLOGY USING PURIFIED GENES

Academic Press Rapid Manuscript Reproduction

Proceedings of the 1981 ICN–UCLA Symposia
on Developmental Biology Using Purified Genes
Held in Keystone, Colorado, on March 15–20, 1981

ICN–UCLA Symposia on Molecular and Cellular Biology
Volume XXIII, 1981

DEVELOPMENTAL BIOLOGY USING PURIFIED GENES

edited by

DONALD D. BROWN

Department of Embryology
Carnegie Institution of Washington
Baltimore, Maryland

Series Editor
C. FRED FOX
Department of Microbiology and Molecular Biology
University of California, Los Angeles
Los Angeles, California

Managing Editor
FRANCES J. STUSSER
ICN–UCLA Symposia
University of California, Los Angeles
Los Angeles, California

ACADEMIC PRESS 1981
A Subsidiary of Harcourt Brace Jovanovich, Publishers
New York London Toronto Sydney San Francisco

ACADEMIC PRESS, INC.
111 Fifth Avenue, New York, New York 10003

United Kingdom Edition published by
ACADEMIC PRESS, INC. (LONDON) LTD.
24/28 Oval Road, London NW1 7DX

Library of Congress Cataloging in Publication Data
Main entry under title:

Development biology using purified genes.

(ICN-UCLA symposium on molecular and cellular biology; v. 23)
"Compilation of papers is from one of the 1981 ICN-UCLA symposia, co-sponsored by ICN Pharmaceuticals, Inc., and organized through the Molecular Biology Institute of the University of California, Los Angeles"--Pref.
Includes index.
1. Developmental genetics--Congresses. 2. Gene expression--Congresses. I. Brown, Donald D. II. ICN Pharmaceuticals, Inc. III. University of California, Los Angeles. Molecular Biology Institute. IV. Title: Purified genes. V. Series. [DNLM: 1. Genetics, Biochemical--Congresses. W3 I322 v. 23 1981 / QH 430 D489 1981]
QH453.D47 616'.001'575 81-17554
ISBN 0-12-137420-3 AACR2

PRINTED IN THE UNITED STATES OF AMERICA

81 82 83 84 9 8 7 6 5 4 3 2 1

CONTENTS

III. MULTI-GENE FAMILIES AND COMPLEX SYSTEMS

IV. GENE AMPLIFICATION

V. CHROMATIN STRUCTURE

VII. EXPRESSION OF EUKARYOTIC GENES: RNA POLYMERASES FORM III

VIII. VIRUSES, VECTORS, AND CELL TRANSFORMATION

IX. METHODS

CONTRIBUTORS

Numbers in parentheses indicate the chapter numbers.

P. Abrescia (29), *Institut für Molekularbiologie I, Universität Zürich, Honggerberg, 8093 Zürich, Switzerland*

Enrico Avvedimento (4) *Building 37, Room 2D20, National Institutes of Health, 9000 Rockville Pike, Bethesda, Maryland 20205*

Carl C. Baker (30), *The Rockefeller University, 66th Street & York Avenue, New York, New York 10021*

Dennis G. Ballinger (32), *Biology Department 16-717, Massachusetts Institute of Technology, Cambridge, Massachusetts 02139*

Thomas Barnett (10), *Department of Biological Sciences, State University of New York, 1400 Washington Avenue, Albany, New York 12222*

Larry R. Beach (20), *Howard Hughes Medical Institute Laboratory, Department of Biochemistry SJ-70, University of Washington, Seattle, Washington 98195*

Lawrence W. Bergman (22), *Building 6, Room B1-32, National Institutes of Health, 9000 Rockville Pike, Bethesda, Maryland 20205*

Alan Bernstein (42), *The Ontario Cancer Institute and Department of Medical Biophysics, University of Toronto, 500 Sherbourne Street, Toronto, Ontario M4X 1K9, Canada*

P. Andrew Biro (7, 8), *Department of Human Genetics, Yale University School of Medicine, 333 Cedar Street, New Haven, Connecticut 06510*

J. Michael Bishop (40), *Department of Microbiology and Immunology, HSE 469, University of California, San Francisco, California 94143*

Daniel F. Bogenhagen (34), *Department of Embryology, Carnegie Institution of Washington, 115 West University Parkway, Baltimore, Maryland 21210*

Beverley J. Bond (12), *Division of Biology, California Institute of Technology, Pasadena, California 91125*

James Bonner (5), *Division of Biology 156-29, California Institute of Technology, Pasadena, California 91125*

Donald D. Brown (34), *Department of Embryology, Carnegie Institution of Washington, 115 West University Parkway, Baltimore, Maryland 21210*

Peter C. Brown (21), *Department of Biological Sciences, Stanford University, Stanford, California 94305*

Kurt Bürki (48), *Department of Animal Biology, University of Geneva, Geneva, Switzerland*

Randall Cassada (18), *Department of Molecular Biology, Max-Planck-Institute for Experimental Medicine, D-3400 Göttingen, Federal Republic of Germany*

Nancy T. Chang (9), *Department of Biochemistry, Roche Institute of Molecular Biology, Nutley, New Jersey 07110*

Tsu-Rong Chiang (9), *Department of Biochemistry, Roche Institute of Molecular Biology, Nutley, New Jersey 07110*

Su-yun Chung (24), *Biochemical Sciences Department, Princeton University, Princeton, New Jersey 08544*

Stuart G. Clarkson (35, 36), *Department of Microbiology, University of Geneva Medical School, 64 Avenue de la Roseraie, 1205 Geneva, Switzerland*

Janine Corlet (35), *Department of Microbiology, University of Geneva Medical School, 64 Avenue de la Roseraie, 1205 Geneva, Switzerland*

Lawrence Costello (9), *Department of Biochemistry, Roche Institute of Molecular Biology, Nutley, New Jersey 07110*

Norman Davidson (12, 13), *Department of Chemistry 164-30, California Institute of Technology, Pasadena, California 91125*

Benoit De Crombrugghe (4), *Building 37, Room 4B27, National Institutes of Health, 9000 Rockville Pike, Bethesda, Maryland 20205*

Henri De Greve (44), *Genetische Virologie, Free University Brussels, Paardenstraat 65, B-1640 St.-Genesius Rode, Belgium*

Eddy M. De Robertis (37), *Biocenter, University of Basel, CH-4056 Basel, Switzerland*

Jon K. deRiel (7), *Fels Research Institute, Temple University School of Medicine, Philadelphia, Pennsylvania 19140*

Kenneth Denich (18), *Department of Molecular Biology, Max-Planck-Institute for Experimental Medicine, D-3400 Göttingen, Federal Republic of Germany*

Peter Dierks (28, 29), *Institut für Molekularbiologie I, Universität Zürich, Honggerberg, 8093 Zürich, Switzerland*

Gordon H. Dixon (15), *Division of Medical Biochemistry, Faculty of Medicine, The University of Calgary, Calgary, Alberta, Canada T2N 1N4*

Carl Dobkin (29), *Institut für Molekularbiologie I, Universität Zürich, Honggerberg, 8093 Zürich, Switzerland*

R. Keith Dudley (46), *Department of Biology, University of California at San Diego, La Jolla, California 92093*

C. A. Duncan (8), *Department of Human Genetics, Yale University School of Medicine, New Haven, Connecticut 06510*

Diane M. Durnam (20), *Department of Biochemistry SJ-70, University of Washington, Seattle, Washington 98195*

William S. Dynan (31), *Department of Biochemistry, University of California, Berkeley, California 94720*

Thomas H. Eickbush (14), *Department of Cellular and Developmental Biology, Harvard University, 16 Divinity Avenue, Cambridge, Massachusetts 02138*

James T. Elder (7), *Department of Human Genetics I125 SHM, Yale University School of Medicine, 333 Cedar Street, New Haven, Connecticut 06510*

M. Duane Enger (19), *Genetics Group, LS-3 MS 886, Los Alamos National Laboratory, Los Alamos, New Mexico 87545*

Andrew Fire (30), *Center for Cancer Research, Room E17-531, Massachusetts Institute of Technology, Cambridge, Massachusetts 02139*

Richard A. Firtel (11), *Department of Biology B-022, University of California at San Diego, La Jolla, California 92093*

Douglas A. Fisher (16), *Division of Biology 156-29, California Institute of Technology, Pasadena, California 91125*

Bernard G. Forget (7, 8), *Department of Internal Medicine, Yale University School of Medicine, 333 Cedar Street, New Haven, Connecticut 06510*

Dana M. Fowlkes (39), *Laboratory of Pathology, National Cancer Institute, Bethesda, Maryland 20205*

John G. Frelinger (16), *Division of Biology 156-29, California Institute of Technology, Pasadena, California 91125*

James Fristrom (13), *Department of Genetics, University of California, Berkeley, California 94720*

Eric A. Fyrberg (12), *Department of Chemistry 164-30, California Institute of Technology, Pasadena, California 91125*

Lashitew Gedamu (15), *University of Biochemistry Group, Division of Biochemistry, The University of Calgary, Calgary, Alberta, Canada T2N 1N4*

Malcolm L. Gefter (7), *Department of Biology, Massachusetts Institute of Technology, Cambridge, Massachusetts 02139*

Denise Gilbert (2), *Department of Biological Chemistry, Harvard Medical School, Boston, Massachusetts 02115*

Shirley Gillam (36, 53), *Department of Biochemistry, University of British Columbia, Vancouver, British Columbia, Canada V6T 1W5*

Jeffrey K. Griffith (19), *Genetics Group, LS-3 MS 886, Los Alamos National Laboratory, Los Alamos, New Mexico 87545*

Terri Grodzicker (41), *Cold Spring Harbor Laboratory, Box 100, Cold Spring Harbor, New York 11724*

Ingrid Grummt (25), *Institut für Biochemie, Universität Würzburg, Röntgenring 11, D87 Würzburg, FRG*

Peter Gruss (43), *Laboratory of Pathology, National Cancer Institute, Bethesda, Maryland 20205*

Richard Guilfoyle (38), *The Wistar Institute, 36th Street at Spruce, Philadelphia, Pennsylvania 19104*
J. B. Gurdon (1), *Medical Research Council, Laboratory of Molecular Biology, Hills Road, Cambridge CB2 2QH, England*
Otto Hagenbüchle (3), *Swiss Institute for Experimental Cancer Research, 1066 Epalinges, Switzerland*
Benjamin Hall (36), *Department of Genetics SK-50, University of Washington, Seattle, Washington 98195*
Ross C. Hardison (6), Department of Microbiology, Cell Biology, Biochemistry, and Biophysics, 3 Althouse Laboratory, The Pennsylvania State University, University Park, Pennsylvania 16802
Jean-Pierre Hernalsteens (44), *Genetische Virologie, Free University Brussels, Paardenstraat 65, B-1640 St.-Genesius Rode, Belgium*
N. Davis Hershey (12), *Department of Chemistry 164-30, California Institute of Technology, Pasadena, California 91125*
Carl E. Hildebrand (19), *Genetics Group, LS-3 MS 886, Los Alamos National Laboratory, Los Alamos, New Mexico 87545*
Robert A. Hipskind (35), *Department of Microbiology, University of Geneva Medical School, 64 Avenue de la Roseraie, 1205 Geneva, Switzerland*
Tadaaki Hirose (54), *Pharmaceutical Institute School of Medicine, Keio University, 35-Shinanomachi, Shinujuku-ku, Tokyo, Japan 160*
Jay Hirsh (2), *Department of Biological Chemistry, Harvard Medical School, Boston, Massachusetts 02115*
Marcelle Holsters (44), *Laboratory of Genetics, State University Gent, K. L. Ledeganckstraat 35, B-9000 Gent, Belgium*
Barry M. Honda (33), *Department of Biological Sciences, Simon Fraser University, Burnaby, Vancouver, Canada V5A 1S6*
Leroy Hood (16), *Division of Biology 156-29, California Institute of Technology, Pasadena, California 91125*
Peter C. Hoppe (48), *Jackson Laboratory, Bar Harbor, Maine*
Stephen H. Howell (46), *Department of Biology C-016, University of California at San Diego, La Jolla, California 92093*
Peter M. Howley (43), *Laboratory of Pathology, National Cancer Institute, Bethesda, Maryland 20205*
Michael Hunkapiller (13), *Division of Biology, California Institute of Technology, Pasadena, California 91125*
Kostas Iatrou (15), *Biological Laboratories, Harvard University, Cambridge, Massachusetts 02138*
Karl Illmensee (48), *Department of Animal Biology, University of Geneva, Geneva, Switzerland*
Edoardo Isnenghi (18), *Department of Molecular Biology, Max-Planck-Institute for Experimental Medicine, D3400 Göttingen, Federal Republic of Germany*
Keiichi Itakura (54), *Molecular Genetics Section, Division of Biology, City of Hope Research Institute, 1450 East Duarte Road, Duarte, California 91010*

Allan Jacobson (47), *Department of Molecular Genetics and Microbiology, University of Massachusetts Medical School, Worcester, Massachusetts 01605*

Pudur Jagdeeswaran (7, 8), *Department of Human Genetics I125 SHM, Yale University School of Medicine, 333 Cedar Street, New Haven, Connecticut 06510*

Linda L. Jagodzinski (5), *Division of Biology, California Institute of Technology, Pasadena, California 91125*

Merrie Jo Johnson (54), *Molecular Genetics Section, Division of Biology, City of Hope Research Institute, 1450 East Duarte Road, Duarte, California 91010*

C. Weldon Jones (14), *Cellular and Developmental Biology, Harvard University, 16 Divinity Avenue, Cambridge, Massachusetts 02138*

Alexandra Joyner (42), *The Ontario Cancer Institute, and Department of Medical Biophysics, University of Toronto, 500 Sherbourne Street, Toronto, Ontario M4X 1K9, Canada*

Fotis C. Kafatos (14), *Cellular and Developmental Biology, Harvard University, 16 Divinity Avenue, Cambridge, Massachusetts 02138*

Markus Kappeler (29), *Institut für Molekularbiologie I, Universität Zürich, Honggerberg, 8093 Zürich, Switzerland*

Eric H. Kawashima (54), *Biogen, S. A., Department of Chemistry, Route de Troinex 3, 1227 Carouge, Geneva, Switzerland*

George Khoury (43), *Laboratory of Molecular Virology, National Cancer Institute, Bethesda, Maryland 20205*

Alan R. Kimmel (11), *Department of Biology B-022, University of California at San Diego, La Jolla, California 92093*

Lawrence E. Kops (26), *Baylor College of Medicine, 1200 Moursund Avenue, Houston, Texas 77030*

Raymond Koski (35, 36), *Department of Genetics SK-50, University of Washington, Seattle, Washington 98195*

Janet Kurjan (36), *Institute of Molecular Biology, University of Oregon, Eugene, Oregon 97403*

Ulrich K. Laemmli (23), *Department of Molecular Biology, 30 quai Ernest-Ansermet, 1211 Geneva 4, Switzerland*

Pennina R. Langer (51), *Roche Institute of Molecular Biology, Nutley, New Jersey 07110*

Ming-Fan Law (43), *Laboratory of Pathology, National Cancer Institute, Bethesda, Maryland 20205*

Jane S. Lebkowski (23), *Department of Molecular Biology, 30 quai Ernest-Ansermet, 1211 Geneva 4, Switzerland*

David C. Lee (33), *Department of Pharmacology SJ-30, University of Washington, Seattle, Washington 98195*

Jan Leemans (44), *Genetische Virologie, Free University Brussels, Paardenstraat 65, B-1640 St.-Genesius Rode, Belgium*

Warren P. Levy (9), *Department of Biochemistry, Roche Institute of Molecular Biology, Nutley, New Jersey 07110*

Catherine D. Lewis (23), *Department of Molecular Biology, 30 Quai Ernest-Ansermet, 1211 Geneva 4, Switzerland*

E. B. Lewis (17), *Division of Biology 156-29, California Institute of Technology, Pasadena, California 91125*

Shuichiro Maeda (9), *Department of Biochemistry, Roche Institute of Molecular Biology, Nutley, New Jersey 07110*

James L. Manley (7), *Department of Biological Sciences, Columbia University, New York, New York 10027*

Suzanne L. Mansour (41), *Department of Biochemistry, University of California, Berkeley, California 94720*

Daniel Marti (28), *Institut für Molekularbiologie I, Universität Zürich, Honggerberg, 8093 Zürich, Switzerland*

Kelly E. Mayo (20), *Department of Biochemistry SJ-70, University of Washington, Seattle, Washington 98195*

Russell McCandliss (9), *Department of Biochemistry, Roche Institute of Molecular Biology, Nutley, New Jersey 07110*

Melissa S. McCoy (24), *Biochemical Sciences Department, Princeton University, Princeton, New Jersey 08544*

Michael McKeown (11), *Department of Biology, B-022, University of California at San Diego, La Jolla, California 92093*

Steven L. McKnight (27), *Department of Embryology, Carnegie Institution of Washington, 115 West University Parkway, Baltimore, Maryland 21210*

Glen Merlino (4), *Building 37, Room 2D25, National Institutes of Health, 9000 Rockville Pike, Bethesda, Maryland 20205*

Joachim Messing (52), *Department of Biochemistry, University of Minnesota, St. Paul, Minnesota 55108*

Francois Meyer (28, 29), *Institut für Molekularbiologie I, Universität Zürich, Honggerberg, 8093 Zürich, Switzerland*

Bernd Meyhack (29), *Institut für Molekularbiologie I, Universität Zürich, Honggerberg, 8093 Zürich, Switzerland*

Katharine S. Mixter (12), *Division of Biology 156-29, California Institute of Technology, Pasadena, California 91125*

Tetsuo Miyake (54), *Wakunaga Pharmaceutical Company Limited, 1-39 Fukushima 3-CHome, Fukushima-ku, Osaka, Japan*

Kevin W. Moore (16), *Division of Biology 156-29, California Institute of Technology, Pasadena, California 91125*

Joyce Morrison (6), *Department of Microbiology, Cell Biology, Biochemistry, and Biophysics, 3 Althouse Laboratory, The Pennsylvania State University, University Park, Pennsylvania 16802*

Maria Mudryj (4), *Building 37, Room 2D20, National Institutes of Health, 9000 Rockville Pike, Bethesda, Maryland 20205*

Edward E. Mullen (29), *Institut für Molekularbiologie I, Universität Zürich, Honggerberg, 8093 Zürich, Switzerland*

Kazuko Nishikura (37), *Department of Structural Biology, Stanford University School of Medicine, Stanford, California 94305*

Joan T. Odell (46), *Department of Biology, University of California at San Diego, La Jolla, California 92093*

Hiroaki Ohkubo (4), *Building 37, Room 2D20, National Institutes of Health, 9000 Rockville Pike, Bethesda, Maryland 20205*

Bert W. O'Malley (26), *Baylor College of Medicine, 1200 Moursund Avenue, Houston, Texas 77030*

Leon Otten (44), *Max-Planck-Institut für Züchtungsforschung, Erwin-Baur-Institut, D-5000, Köln 30 (Vogelsang), Federal Republic of Germany*

Richard D. Palmiter (20), *Department of Biochemistry SJ-70, University of Washington, Seattle, Washington 98195*

J. Pan (8), *Department of Human Genetics, Yale University School of Medicine, New Haven, Connecticut 06510*

Mary Lou Pardue (32), *Biology Department 16-717, Massachusetts Institute of Technology, Cambridge, Massachusetts 02139*

J. S. Park (24), *Biochemical Sciences Department, Princeton University, Princeton, New Jersey 08544*

Ira Pastan (4), *Building 37, Room 4B27, National Institutes of Health, 9000 Rockville Pike, Bethesda, Maryland 20205*

Hugh R. B. Pelham (34), *Department of Embryology, Carnegie Institution of Washington, 115 West University Parkway, Baltimore, Maryland 21210*

Sidney Pestka (9), *Department of Biochemistry, Roche Institute of Molecular Biology, Nutley, New Jersey 07110*

Gert Pflugfelder (25), *Institut für Biochemie, Universität Würzburg, Röntgenring 11, D87 Würzburg, Federal Republic of Germany*

Khosro Radnia (18), *Department of Molecular Biology, Max-Planck-Institute for Experimental Medicine, D-3400 Göttingen, Federal Republic of Germany*

David I. Ratner (47), *Department of Cellular Biology, Scripps Clinic and Research Foundation, 10666 North Torrey Pines Road, La Jolla, California 92037*

J. Garrett Reilly (50), *Department of Cellular Biology, Scripps Clinic and Research Foundation, 10666 North Torrey Pines Road, La Jolla, California 92037*

J. Reiser (28), *Institut für Molekularbiologie I, Universität Zürich, Honggerberg, 8093 Zürich, Switzerland*

Richard J. Roberts (49), *Cold Spring Harbor Laboratory, Box 100, Cold Spring Harbor, New York 11724*

Robert G. Roeder (33), *Department of Biological Chemistry, Washington University, 660 South Euclid, St. Louis, Missouri 63110*

Mark L. Rohrbaugh (6), *Department of Microbiology, Cell Biology, Biochemistry, and Biophysics, 3 Althouse Laboratory, The Pennsylvania State University, University Park, Pennsylvania 16802*

Shigeru Sakonju (34), *Department of Embryology, Carnegie Institution of Washington, 115 West University Parkway, Baltimore, Maryland 21210*

Thomas D. Sargent (5), *Laboratory of Biochemistry, National Cancer Institute, National Institutes of Health, Bethesda, Maryland 20014*

Nava Sarver (43), *Laboratory of Pathology, National Cancer Institute, National Institutes of Health, Bethesda, Maryland 20205*

Jeff Schell (44), *Max-Planck-Institut für Züchtungsforschung, D-5000, Köln 30 (Vogelsang), Federal Republic of Germany*

Ueli Schibler (3), *Swiss Institute for Experimental Cancer Research, 1066 Epalinges, Switzerland*

Einhard Schierenberg (18), *Department of Molecular Biology, Max-Planck-Institute for Experimental Medicine, D-3400 Göttingen, Federal Republic of Germany*

Robert T. Schimke (21), *Department of Biological Sciences, Stanford University, Stanford, California 94305*

Jo Schröder (44), *Max-Planck-Institut für Züchtungsforschung, Erwin-Baur-Institut, D-5000, Köln 30 (Vogelsang), Federal Republic of Germany*

Tanya Z. Schulz (26), *Baylor College of Medicine, 1200 Moursund Avenue, Houston, Texas 77030*

Matthew P. Scott (32), *Department of Biology, Indiana University, Bloomington, Indiana 47405*

Peter H. Seeburg (52), *Division of Molecular Biology, Genentech Incorporated, South San Francisco, California 94080*

Phillip A. Sharp (30), *Center for Cancer Research, Room E17-529B, Massachusetts Institute of Technology, Cambridge, Massachusetts 02139*

B. S. Shastry (33), *Department of Biological Chemistry, Washington University, 660 South Euclid, St. Louis, Missouri 63110*

Charles Shaw (44), *Max-Planck-Institut für Züchtungsforschung, Erwin-Baur-Institut, D-5000, Köln 30 (Vogelsang), Federal Republic of Germany*

Thomas Shenk (39), *Department of Microbiology, Health Sciences Center, State University of New York, Stony Brook, New York 11794*

Beverly Taylor Sher (16), *Division of Biology 156-29, California Institute of Technology, Pasadena, California 91125*

Don Silvert (13), *Department of Genetics, University of California, Berkeley, California 94720*

Robert B. Simpson (45), *ARCO Plant Research Laboratory, Box 2600, 6905 Sierra Court, Dublin, California 94566*

Robert T. Simpson (22), *Building 6, Room B1-28, National Institutes of Health, 9000 Rockville Pike, Bethesda, Maryland 20205*

Kenneth Smith (18), *Department of Molecular Biology, Max-Planck-Institute for Experimental Medicine, D-3400 Göttingen, Federal Republic of Germany*

Michael Smith (36, 53), *Department of Biochemistry, University of British Columbia, Vancouver, British Columbia, Canada V6T 1W5*

Steven S. Smith (50), *Department of Cellular Biology, Scripps Clinic and Research Foundation, 10666 North Torrey Pines Road, La Jolla, California 92037*

Michael Snyder (13), *Division of Biology, California Institute of Technology, Pasadena, California 91125*

Mark Sobel (4), *Building 30, Room 405, National Institutes of Health, 9000 Rockville Pike, Bethesda, Maryland 20205*

Richard A. Spritz (7, 8), *Department of Human Genetics, University of Wisconsin, Madison, Wisconsin*

Michael Steinmetz (16), *Division of Biology 156-29, California Institute of Technology, Pasadena, California 91125*

Sidney V. Suggs (54), *Molecular Genetics Section, Division of Biology, City of Hope Research Institute, 1450 East Duarte Road, Duarte, California 91010*

Bayar Thimmappaya (39), *Department of Microbiology and Immunology, Northwestern University School of Medicine, Chicago, Illinois 60611*

C. A. Thomas, Jr. (50), *Department of Cellular Biology, Scripps Clinic and Research Foundation, 10666 North Torrey Pines Road, La Jolla, California 92037*

Carl S. Thummel (41), *Department of Biochemistry, University of California, Berkeley, California 94720*

Robert Tjian (31, 41), *Department of Biochemistry, University of California, Berkeley, California 94720*

Mario Tosi (3), *Swiss Institute for Experimental Cancer Research, 1066 Epalinges, Switzerland*

Ming-Jer Tsai (26), *Baylor College of Medicine, 1200 Moursund Avenue, Houston, Texas 77030*

Sophia Y. Tsai (26), *Baylor College of Medicine, 1200 Moursund Avenue, Houston, Texas 77030*

D. Tuan (8), *Department of Medicine, Yale University School of Medicine, New Haven, Connecticut 06510*

Axel Ullrich (48), *Genentech Incorporated, South San Francisco, California 94080*

Marc Van Montagu (44), *Laboratory of Genetics, State University Gent, K. L. Ledeganckstraat 35, B-9000 Gent, Belgium*

A. van Ooyen (28, 29), *Institut für Molekularbiologie I, Universität Zürich, Honggerberg, 8093 Zürich, Switzerland*

Gabriel Vogeli (4), *Building 6, Room 203, National Institutes of Health, 9000 Rockville Pike, Bethesda, Maryland 20205*

Günther von Ehrenstein* (18), *Department of Molecular Biology, Max-Planck-Institute for Experimental Medicine, D-3400 Göttingen, Federal Republic of Germany*

Richard M. Walden (46), *Department of Biology C-016, University of California at San Diego, La Jolla, California 92093*

Linda L. Walker (46), *Department of Biology C-016, University of California at San Diego, La Jolla, California 92093*

R. Bruce Wallace (54), *Molecular Genetics Section, Division of Biology, City of Hope Research Institute, 1450 East Duarte Road, Duarte, California 91010*

Ronald A. Walters (19), *Genetics Group, LS-3 MS 886, Los Alamos National Laboratory, Los Alamos, New Mexico 87545*

David C. Ward (51), *Department of Human Genetics, Yale University School of Medicine, New Haven, Connecticut 06510*

Thomas E. Ward (47), *Department of Molecular Genetics and Microbiology, University of Massachusetts Medical School, Worcester, Massachusetts 01605*

*Deceased.

H. Weber (29), *Institut für Molekularbiologie I, Universität Zürich, Honggerberg, 8093 Zürich, Switzerland*

Cary Weinberger (39), *Department of Microbiology, Health Sciences Center, State University of New York, Stony Brook, New York 11794*

Roberto Weinmann (38), *The Wistar Institute, 36th Street at Spruce, Philadelphia, Pennsylvania 19104*

Charles Weissmann (28, 29), *Institut für Molekularbiologie I, Universität Zürich, Honggerberg, 8093 Zürich, Switzerland*

Sherman M. Weissman (7, 8), *Department of Human Genetics, Yale University School of Medicine, 333 Cedar Street, New Haven, Connecticut 06510*

Peter K. Wellauer (3), *Swiss Institute for Experimental Cancer Research, 1066 Epalinges, Switzerland*

Pieter Wensink (10), *Rosenstiel Basic Medical Sciences Research Center, Brandeis University, Waltham, Massachusetts 02254*

B. Wieringa (28), *Institut für Molekularbiologie I, Universität Zürich, Honggerberg, 8093 Zürich, Switzerland*

Lothar Willmitzer (44), *Max-Planck-Institut für Züchtungsforschung, Erwin-Baur-Institut, D-5000, Köln 30 (Vogelsang), Federal Republic of Germany*

John C. Wooley (24), *Biochemical Sciences Department, Princeton University, Princeton, New Jersey 08544*

W. Michael Wormington (34), *Department of Embryology, Carnegie Institution of Washington, 115 West University Parkway, Baltimore, Maryland 21210*

Michael A. Wosnick (15), *Division of Medical Biochemistry, Faculty of Medicine, The University of Calgary, Calgary, Alberta, Canada T2N 1N4*

Yoshihiko Yamada (4), *Building 37, Room 2D20, National Institutes of Health, 9000 Rockville Pike, Bethesda, Maryland 20205*

Yusei Yamamoto (42), *The Ontario Cancer Institute, and Department of Medical Biophysics, University of Toronto, 500 Sherbourne Street, Toronto, Ontario M4X 1K9, Canada*

Maria Yang (5), *Division of Biology, California Institute of Technology, Pasadena, California 91125*

Richard A. Young (3), *Department of Biochemistry, Stanford University School of Medicine, Stanford, California 94305*

David Yuen (13), *Division of Biology, California Institute of Technology, Pasadena, California 91125*

A. Zeltner (29), *Institut für Molekularbiologie I, Universität Zürich, Honggerberg, 8093 Zürich, Switzerland*

Edward B. Ziff (30), *The Rockefeller University, 66th Street & York Avenue, New York, New York 10021*

PREFACE

This compilation of papers is from one of the 1981 ICN–UCLA Symposia, cosponsored by ICN Pharmaceuticals, Inc., and organized through the Molecular Biology Institute of the University of California, Los Angeles. Its success was due in large part to the efficiency of Fran Stusser and her staff, Sandra and Maureen Malone.

We gratefully acknowledge the financial support from New England Biolabs, Hoffmann-La Roche, Genentech, Merck Sharp and Dohme, Cetus, Burroughs-Wellcome, and Eli Lilly. NIH Contract 263-81-CO157, jointly sponsored by Science and Education Administration, USDA, National Cancer Institute, Department of Energy, National Institute of Allergy and Infectious Diseases, and Fogarty International Center, partially defrayed the cost of speaker travel.

Donald D. Brown

CONCEPTS OF GENE CONTROL IN DEVELOPMENT

J. B. Gurdon

Medical Research Council
Laboratory of Molecular Biology
Cambridge, UK.

Probably the single most interesting current problem in development concerns the mechanism by which gene activity is regionally controlled. Why are keratin genes expressed to a detectable level only in skin cells, globin genes in erythrocytes, etc.? In terms of proteins, erythrocytes contain 10^7 globin molecules in contrast to an undetectable level of globin in other unrelated cell-types. At the level of immediate gene products, an erythrocyte contains about 50,000 globin transcripts, by comparison with an unrelated cell such as a fibroblast which has less than one globin transcript per cell.

This aspect of gene regulation, which I refer to as the cell-type specificity of gene control, is not necessarily connected with the identification of promoter and terminator sequences in DNA. For example, it is possible that the accessibility of a promoter sequence to its RNA polymerase may depend on the composition of chromatin in that region, and this in turn may be determined by a DNA sequence distantly located from the promoter. In this case detailed knowledge of the promoter region might not immediately help the identification of the cell-type specificity control mechanism.

Two aspects of cell-type specific gene control seem to me especially important. One is the identification of the DNA sequences and non-DNA control molecules responsible. The other is the mechanism by which these control molecules come to be distributed in development so as to be located in the

DEVELOPMENTAL BIOLOGY USING PURIFIED GENES

ISBN 0-12-137420-3

appropriate cell type but not in others. This last question must at present be discussed more in terms of concepts than experiments.

DNA Sequences and Non-DNA Control Molecules

The identification of DNA sequences involved in control requires an assay in which cloned segments of purified DNA are expressed in a regulated way. These could then be mutated by genetic manipulation and hence the control region identified. This is the approach that has proved so successful in identifying the promoter region of 5S genes. Initially the functional assay involved the injection of cloned 5S genes into *Xenopus* oocyte nuclei (1); subsequently the work which actually identified the promoter region made use of 5S genes incubated *in vitro* in whole oocyte nuclei (2).

At present there appears to be no equivalent assay for the cell-type specific activity of genes. Purified genes are generally transcribed in all cells into which they are introduced. With the special exception of bacterial and mitochondrial tRNA genes (3), all other genes injected into oocytes are transcribed, including those which should be active only in specialized cells such as globin and oviduct ovalbumin (4). There is no report so far of the cell-type specific transcription of purified genes introduced into adult cells; ideally a globin gene would be transcribed in an erythrocyte but not in a fibroblast. It therefore seems that cell-type specific control is lost when DNA is purified, cloned and introduced into other cells. There are two obvious reasons why this could be so. One is that the region of DNA required for cell-type specific control is not included in the lengths of DNA which can be cloned and re-introduced into living cells. The other is that the conditions which are needed to initiate the regulated state no longer exist in the kinds of cells into which genes are introduced. The length of DNA molecules which can be cloned and introduced into cells is now about 40 kb, and it should soon be possible to test the transcription of two or three independently regulated genes on the same length of DNA (e.g. globin genes). If these do not show cell-type regulation, this will argue against the view that a short length of DNA *per se* prevents cell-type specific control. The possibility that a particular stage in development exists at which genes are put into an active or inactive state, which is then propagated, is being tested by injecting DNA into fertilized eggs. It was shown a few years ago that DNA injected into fertilized Xenopus eggs persists remarkably well for over a week, and is present in tadpoles containing about 500,000 cells. The injected DNA was transcribed for at least the first day (5). These experiments have been pursued by Rusconi and Schaffner (personal communication) who have demonstrated the persistence of injected DNA in advanced tadpoles. It is technically feasible to inject quite large amounts of DNA into the eggs of mammals and Drosophila. If genes injected into fertilized eggs come under developmental regulation, this should make possible a deletion analysis of DNA sequences required for developmental regulation. This in turn could lead to the identification of molecules which bind to these sequences.

An alternative pathway toward the identification of non-DNA molecules involved in regulation aims at making biochemically analysable preparations of genes in their regulated state. If a suspension of nuclei or of chromatin fragments could be obtained in which active or inactive genes remained in these states when tested for transcription, this would permit attempts to expose the nuclei or chromatin to treatments which change the state of gene regulation. The nature of such treatments could eventually elucidate the mechanism of the control. This approach is now discussed.

The in vitro transcription of chromatin has in general given a very low yield of RNA much of which may be run-off transcripts initiated in vivo. However a suspension of nuclei injected into amphibian oocytes yields abundant RNA which continues to be synthesized for many days. This procedure and the application of it to the regulation of 5S genes has been described by Korn and Gurdon (6). 5S genes show an extreme form of cell-type specific regulation, since the oocyte-type 5S genes are inactive in all somatic cells, and active only in oocytes. Relevant to the present discussion is the fact that frog erythrocyte nuclei injected into oocytes show transcription of the constitutive 5S genes (the so-called somatic type), but can retain the unexpressed state of the developmentally inactive oocyte-type 5S genes. To investigate the nature of the gene activation, the suspension of nuclei was treated with 0.35 M NaCl, and then the transcription of the 5S genes tested by injection into oocytes. Analysis of labelled RNAs showed that the oocyte-type 5S genes had been activated. By further analysis, it should be possible to determine the nature of the developmental inactivation of the 5S genes. There are 800 somatic-type 5S genes per diploid chromosome set and the activity of these is readily seen. If the sensitivity of transcription could be improved by 400 times, the same assay system could be applied to genes with typical cell-type specific expression, such as globin, many of which are present as only one copy per genome. To increase the sensitivity of transcription recognition by 400 times would not be difficult but it is not yet clear that globin genes would be transcribed in injected nuclei by polymerase II with the same efficiency as are 5S genes by polymerase III.

The end result of these two experimental procedures would, if successful, be the identification of DNA sequences and gene-control molecules responsible for the activity of genes in certain specialized cell types. Although neither of these kinds of information yet exists, it would be surprising if it were long before they are forthcoming. We will then have still to approach the second major aspect of cell-type specific gene control: this is the mechanism by which the gene-control molecules come to be localized only in the cell-type where they are effective.

The Partitioning of Cytoplasmic Determinants

Already at early stages of development (the blastula or equivalent), cells of most animal embryos differ from each other in developmental potential. Once initial differences between cells exist, it is not hard in principle to imagine how further differences can follow. The key question is how gene-control molecules or other developmental determinants come to be unequally distributed among embryonic cells. Three principal concepts exist. The first believes that determinants are already localized in the fertilized but uncleaved egg. The classical example of this is the pole plasm or germ plasm of insect or amphibian eggs. It is completely established that the pole plasm is localized in eggs and causes the embryonic cells into which it is subsequently included to become eggs or sperm (7). However, there is at present no other case of a precisely localized determinant in uncleaved eggs. Furthermore the fact that separated half, quarter, and in some cases one eighth embryos can give rise to complete individuals rules out prelocalization as a general mechanism for generating cell-type specific gene expression.

The second concept is a progressive partitioning of cytoplasmic determinants, and assumes that these are randomly arranged in fertilized eggs, but that they are gradually moved into appropriate positions during cleavage so as to become localized primarily in those cells which are ancestors by lineage of the specialized cells which they determine. The best example of this type of mechanism comes from the remarkable experiments of Freeman (8,9), who was able to cut portions of cytoplasm from dividing ctenophore cells. By following the developmental fate of a nucleus surrounded by particular parts of a cell about to divide into two lineages, he was able to deduce that cytoplasmic determinants for the formation of ciliated as opposed to light-producing cells become localized at opposite ends of the parental cell as it is about to divide. The principle here is the asymmetric distribution of parental cell cytoplasm into its two daughter cells. The important consequence of this event is that large molecules such as proteins which (if over about 1000 in molecular weight) cannot pass through cell walls, can reach very different concentrations in daughter cells even though they may be synthesized at the same rate in all cells. The phenomenon of asymmetric divisions giving rise to daughter cells which undergo divergent differentiation is widespread among plants and animals. Proof that the divergent differentiation results from cytoplasmic rather than chromosomal segregation was provided by an ingenious manipulation in which mitotic chromosomes were pushed with a needle into the opposite end of the cell to that where they would normally go (10).

The third concept which could explain the origin of differences between cells in development supposes that macromolecular determinants do not exist and that gene control results from the localization of small molecules. It is believed that these become arranged in a gradient at some multicellular stage (e.g. blastula). The most plausible variant of this concept has been proposed by Graham (11). On the basis of experiments in which the relative position of early mouse blastomeres is changed experimentally, it has been suggested that all outer cells in contact with the medium become leaky on their exposed surface, in contrast to all the inner cells which are entirely surrounded by other cells, and which do not become leaky. If gene activity is sensitive to the concentration of ions which leak through external surfaces, this could lead to a difference in gene activity between inner and outer cells. It must be assumed that once this first level of gene control has taken place, a product of one of these genes will determine the way in which subsequent changes in ion concentration will affect the genome. In this way, a sequence of gene activity changes could result from the relative location of cells at different developmental stages. In general, the morphological events in the development of most animals commonly result in the formation of sheets of cells the outer layer of which differs in developmental fate from the others. Although this inside/outside hypothesis is attractive, the events of mouse development appear also to be explicable in terms of cytoplasmic partitioning. Figure 1 compares the same morphological events in terms of these two concepts. So long as a cell division immediately preceeds the appearance of two different cell-types, cytoplasmic partitioning could be involved. Indeed, recent experiments of Johnson et al. (12) have demonstrated directly the asymmetric distribution of cytoplasmic materials at the 8 to 16 cell stage in mouse embryos, though it is not yet known whether these materials have a determinant function.

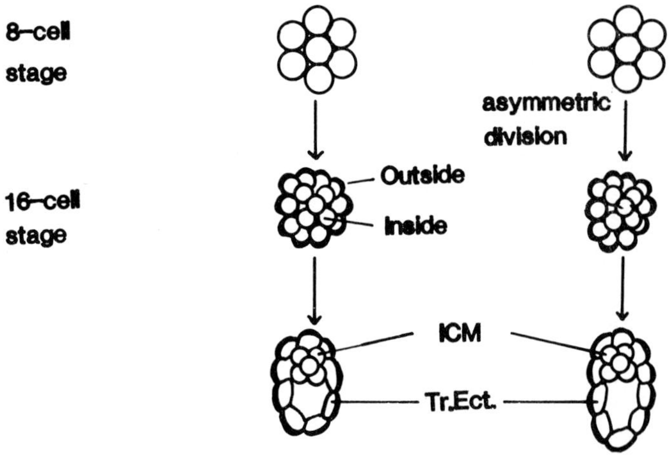

Fig. 1.

Two hypothetical models of early mouse development. Each scheme shows the formation of the inner cell mass (ICM) and trophectoderm (Tr.Ect.), the first two different cell-types to be formed. In the left hand scheme, the cells are presumed to become different on account of their internal or external location at the 16-cell stage. In the right hand scheme, cells are shown becoming different as a result of an asymmetric cell division at the 8-cell stage, which is assumed to segregate cytoplasmic determinants to the cells which subsequently become trophectoderm or inner cell mass.

The purpose of discussing mammalian development in detail is to make the point that the partitioning of cytoplasmic determinants or gene-control substances may be the most widespread mechanism for the origin of cell-type differentiation in development. To subscribe generally to this concept does not require that eggs have a large number of different determinants which are all segregated out during cleavage. Perhaps by the blastula stage, only the ectoderm, mesoderm, and endoderm regions have different determinants, which will activate different genes. The products of these genes could include other determinants which will be asymmetrically distributed during the many cell divisions which take place after the blastula stage. Embryonic inductions involve interactions between cells which become adjacent to each other as a result of the folding of cell layers. Inductions could result in a change in cell type by causing an asymmetric distribution of determinants during division of the responding cells. One objection to the cytoplasmic distribution of determinants as a general mechanism of development is that the few fold differences in concentration that would result from asymmetric cell divisions might be too small to consistently ensure the activity or non-activity of genes. This difficulty would however be overcome if determinants were to show cooperative binding to genes, as has been shown for the λ repressor and T antigen in SV40 (13,14).

The Analysis of Cytoplasmic Localization

The mechanisms involved in asymmetric cell divisions may be subject to genetic analysis in suitable species such as nematodes. A series of asymmetric cell divisions takes place in nematode development, and there is an almost invariant cell lineage by which each specialized cell is derived from early embryonic cells (15). A mutation, named unc86, has been described by Chalfie (16) in which the end product of a certain lineage, a nerve cell, is not formed, but instead dopamine-containing cells of the kind normally formed by the other branches of the nerve-cell lineage. Mutations which alter the lineage of specialized cells in a specific way are what would be expected if a nerve cell determinant were not formed or were unable to be partitioned correctly at cell division. Identification of the gene product affected by this mutation might give a clue about the mechanism responsible for a developmental lineage of this type, and therefore possibly for the partitioning of cytoplasmic determinants. Several "lineage mutations" have now been described in nematode development.

A way of analyzing cytoplasmic localization from a more biochemical point of view may result from the microinjection of purified macromolecules into cells. So far this type of experiment has been applied only to the distribution of molecules between the nucleus and cytoplasm of cells, and not to the asymmetric partitioning of cytoplasmic determinants. It was found some years ago that purified histones become rapidly localized in an oocyte nucleus a few hours after injection into the cytoplasm (17). Subsequently a variety of proteins has been tested by injection into oocytes; of special interest is the result that proteins normally localized in an oocyte nucleus return to the nucleus, and those originally localized in the cytoplasm remain there (18; which includes references to earlier work). In some cases proteins are as much as 100x more concentrated in the nucleus compared to the cytoplasm. Recently Ackerman and Gurdon (unpublished) have found that small nuclear RNAs such as U1 become 20 times concentrated in the nucleus after injection into oocyte cytoplasm, in contrast to tRNA which remains evenly concentrated throughout the cell. A conclusion from these experiments is that molecules can be purified without losing their ability to take up their correct location in a cell. This opens up the possibility of modifying or fragmenting such molecules before injection in order to identify the part of the molecule required for localization. This could suggest which host molecules or mechanisms are responsible. The same type of experimental design may well be applicable to those molecules, when they have been identified, which become localized before cell division, and subsequently partitioned asymmetrically. An interesting possible mechanism for the polarization of molecules, proposed by Jaffe (19), is intracellular electrophoresis. There is evidence for this in newly fertilized sea-weed eggs, and Lepidopteran oocytes (20).

Extracellular Factors

There is still the question of what causes the asymmetrical distribution of intracellular components. This must be some type of external influence on cells. For example, the sperm penetration of an egg, inducers of embryonic induction, or the outside location of some cells as opposed to others (hence the contact of medium with the cell surface). There is no reason to suppose or to require that any of these external stimuli or conditions are specific. The immediate events that follow fertilization can be reproduced by insertion of a glass needle. In sea weed eggs any one of several stimuli will initiate polarization, though light is the usual external stimulus.

The result of an induction event is determined by the properties (or competence) of the responding cells. It therefore seems unlikely that the identification of external stimuli will lead to a further understanding of the intracellular mechanisms involved in cell differentiation and gene control.

Conclusion

I have tried to emphasize what seem to me to be the next generation of problems in understanding gene control in development. The identification of DNA sequences and non-DNA molecules involved in gene regulation will almost certainly be achieved in the reasonably near future by methods and experimental approaches currently available. I have suggested that the intracellular localization of gene control molecules or of other determinants of cell differentiation is certain to be of great, if not universal, importance in the analysis of development. I believe it is still too early to say whether the techniques and procedures currently in use are sufficient to analyze this problem.

REFERENCES

1. Brown, D.D. and Gurdon, J.B. (1978). Proc. Nat. Acad. Sci. US, 75, 2849-2853.
2. Sakonju, S., Bogenhagen, D.F., and Brown, D.D. (1980). Cell, 19, 13-25.
3. Melton, D.A. and Cortese, R. (1979). Cell 18, 1165-1172.
4. Wickens, M.P., Woo, S., O'Malley, B.W., and Gurdon, J.B. (1980). Nature 285, 628-634.
5. Gurdon, J.B. and Brown, D.D. (1977). In The Molecular Biology of the Mammalian Genetic Apparatus, ed. P. T'so, Amsterdam North-Holland Publ. Co., Vol. 2, 111-123.
6. Korn, L.J. and Gurdon, J.B. (1981). Nature 289, 461-465.
7. Illmensee, K. and Mahowald, A. (1974). Proc. Nat. Acad. Sci. US, 71, 1016-1020.
8. Freeman, G. and Reynolds, G.T. (1973). Devel. Biol. 31, 61-100.

9. Freeman, G. (1976). Devel. Biol. 49, 143-177.
10. Carlson, J.G. (1952). Chromosoma, 5, 199-220.
11. Graham, C.F. (1976). Chapter 1.3 in Graham, C.F. and Wareing, P.F. "The Developmental Biology of Plants and Animals", Blackwell.
12. Ziomek, C.A. and Johnson, M.H. (1980). Cell 21, 935-942; Johnson, M.H. and Ziomek, C.A. (1981). Cell 24, in press.
13. Ptashne, M., Jeffrey, A., Johnson, A.D., Maurer, R., Meyer, B.J,, Pabo, C.O., Roberts, T.M., and Sauer, R.T. (1980). Cell 19, 1-11.
14. Tjian, R. (1978). Cell 13, 165-179.
15. Schierenberg, E., Miwa, J., and von Ehrenstein, G. (1980). Devel. Biol. 76, 141-159; Miwa, J., Schierenberg, E., Miwa, S., and von Ehrenstein, G. (1980). Devel. Biol. 76, 160-174.
16. Chalfie, M., Horvitz, H.R., and Sulston, J.E. (1981). Cell 24, in press.
17. Gurdon, J.B. (1970). Proc. Roy. Soc. B. 176, 303-314.
18. De Robertis, E.M., Longthorne, R.F., and Gurdon, J.B. (1978). Nature 272, 254-256.
19. Robinson, K.R. and Jaffe, L.F. (1975). Science 187, 70-72.
20. Woodruff, R.I. and Telfer, W.H. (1980). Nature 286, 84-86.

THE DOPA DECARBOXYLASE GENE LOCUS OF *DROSOPHILA MELANOGASTER*: ORIENTATION OF THE GENE AND PRELIMINARY MAPPING OF GENETIC MARKERS

Denise Gilbert
Jay Hirsh

Department of Biological Chemistry
Harvard Medical School
Boston, Ma. 02115

INTRODUCTION

Few genes have been isolated that produce defined protein products, are subject to striking developmental regulation, and are also amenable to fine structure genetic analysis. The *Drosophila* dopa decarboxylase gene is such a gene. We have recently reported the isolation and preliminary characterization of this gene (1). In this communication, we report the orientation of the gene with respect to the genetic map, and determine a preliminary localization of a genetic breakpoint on the cloned DNA. Our results indicate that a genetic region encompassing five lethal complementation groups adjacent to Ddc is contained in a region of 20 to 30 kb of chromosomal DNA.

Figure 1 summarizes genetic markers and breakpoints of deficiencies adjacent to the structural gene for dopa decarboxylase, Ddc.

ISBN 0-12-137420-3

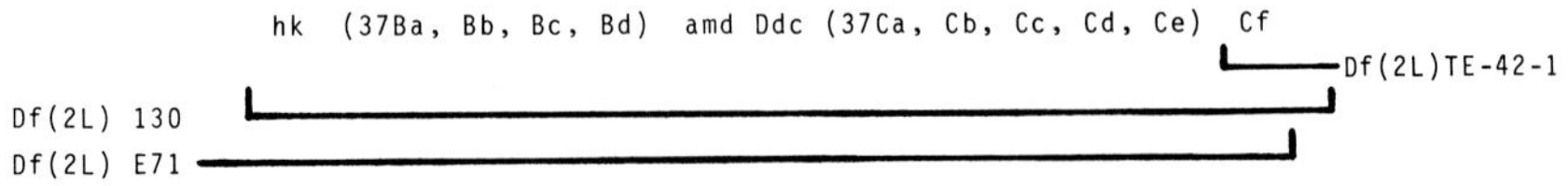

Figure 1. Genetic markers and defiencies near the dopa decarboxylase gene. Redrawn from Wright et al.(2). 37Ba-Bd and 37Ca-Ce are lethal complementation groups which have not been ordered with respect to one another. Note that the names of the lethals do not necessarily imply their precise band localization.

Wright and coworkers(2) have localized Ddc, a visible mutation hook, and nine lethal complementation groups to a region composed of at most seven bands visible on polytene chromosomes. The region is bounded by bands 37B9 and 37C5, on the left arm of chromosome 2. In addition, the lethal complementation group l(2)37Cf has been localized to a region containing at most 4 bands, between bands 37C5 and 37D1. The edges of the region are defined by deficiencies. The centromere distal (left) edge is defined by the distal breakpoint of Df(2L)130. A point separating distal markers from l(2)37Cf is defined by the distal end of Df(2L)TE-42-1. This deficiency was isolated from a stock containing TE-42, a large transposing element containing several bands from the 3C region. The deficiency extends proximally from within the transposing element, removing one of the two major bands carried by the element, and removing w+ function. Since the deficiency removes l(2)37Cf, but leaves all other distal markers, the insertion point of TE-42 must lie between the other markers and l(2)37Cf. The region containing l(2)37Cf is defined by Df(2L)E71, which deletes with a proximal breakpoint between bands 37C5 and 37D1.

RESULTS

We have used the initially isolated Ddc clone to isolate overlapping chromosomal clones extending to presently cover about 100 kb of DNA. The clones including this DNA are shown in Figure 2. This figure shows the gene Ddc, and two other previously described(1) transcribed sequences positioned some

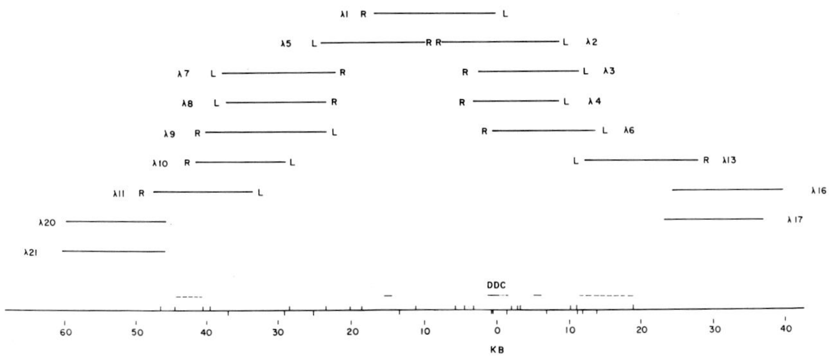

Figure 2. Clones carrying Ddc and flanking regions, and the chromosomal DNA map derived from the clones. The clones were isolated from a random shear library which was inserted into lambda charon-4 with Eco R1 linkers(4). The phage λ-16, λ-17, λ-20, and λ-21 have not been fully mapped. Eco R1 sites; Bam H1 sites.

13 kb to the left, and 5 kb to the right, respectively, from Ddc. Also shown, by dashed lines, are two other regions which show hybridization by cDNA probes synthesized from 3rd instar larval poly(A) RNA. Analysis of transcribed regions within these sequences is not yet complete; some regions have not yet been screened, while other regions have been screened only at low sensitivity with RNA from specific developmental stages.

Ddc lies within or very close to the polytene chromosome band 37C1 (1), as determined by in situ hybridization with λDdc-1 probes. We have used in situ hybridization to determine the orientation of cloned DNA relative to the genetic map. An in situ hybridization using a λDdc-13 probe is shown in Figure 3. Silver grains are observed to lie centromere proximal to band 37C1, in the region between 37C1 and 37D1. The phage λDdc-13 contains chromosomal DNA 13 to 26 kb from Ddc. It is thus likely that Ddc is either within the centromere proximal portion of band 37C1, or lies just adjacent to the band. Consistent with this interpretation, hybridization with probe synthesized from λDdc-11, containing sequences extending 33-46 kb centromere distal from Ddc, is

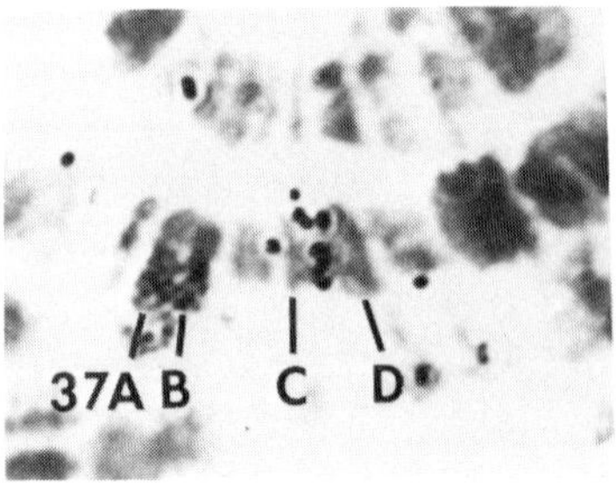

Figure 3. In situ hybridization, λ-13 probe. Hybridization conditions were as described(1).

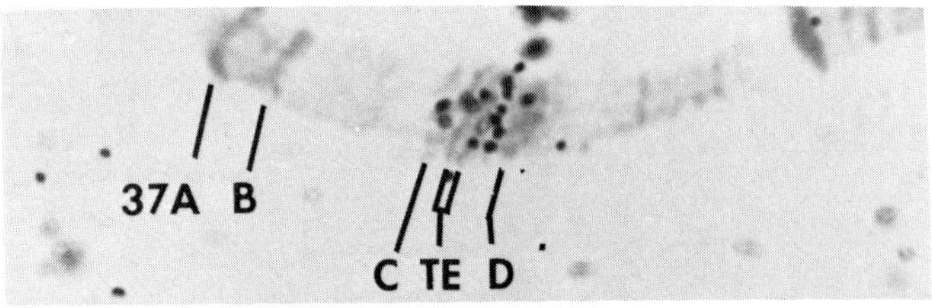

Figure 4. In situ hybridization, λ-16 probe, to chromosomes containing TE-42. The genetic consititution of the flies was w+;Sco TE-42/Cyo. However, the chromosomes on this and several other slides contained well paired, non-inverted 2nd chromosomes. Thus, the chromosome shown is almost certainly homozygous for TE-42.

not clearly separable from 37C1 (not shown).

We have made a low resolution determination of the insertion point of TE-42. Figure 4 shows an in situ hybridization with a probe synthesized from λ-Ddc-16 DNA to chromosomes containing TE-42. Grains are visible to the right of the TE-42 bands, on the far side from 37C1. This shows that λ16-Ddc contains sequences largely or totally past the insertion point of TE-42. It was expected that the clone containing the sequences flanking TE-42 would show parallel lines of grains adjacent to the transposing element when hybridized to TE-42 containing chromosomes. When hybridizations were performed with λDdc-13, we observed one extremely broad band of hybridization covering and obscuring the TE-42 region. We thus think it likely that the

TE-42 insertion occurs within the sequences covered by λDdc-13. The insertion clearly occurs left relative to most of the sequences in λDdc-16, probably near the middle of λDdc-13. This places the insertion point 20-30 kb from Ddc.

We have also examined the centromere distal region. Hybridization with a λDdc-20 probe to heterozygous Df(2L)130 chromosomes indicates that there is little or no hybridization to the deficiency homologue (not shown). Thus, the distal breakpoint of Df(2L)130 must be at least 60 kb from Ddc.

As discussed earlier, and shown in Figure 1, Ddc and the insertion location of TE-42 define a region containing five lethal complementation groups. That this region contains only 20-30 kb of DNA is very intriguing in light of observed functional relations between several of the other lethals and Ddc(2,3). Five of the Df(2L)130 lethals have alleles showing female sterile, or temperature sensitive female sterile phenotypes(2). In addition, Ddc, l(2)amd, hook, and l(2)37Ca, show cuticular phenotypes(2,3). These related phenotypes indicate that this region may consist of a group of developmentally related genes. Our finding that Ddc and five other lethal complementation groups fall within 20-30 kb indicates a rather striking clustering of the lethals. It should be pointed out that most of the lethals in the region have not yet been rigorously shown to represent discrete genes, and that it is possible that some could represent trans-complementing alleles of one genetic locus. Further genetic analysis will clearly be necessary.

ACKNOWLEDGMENTS

This work was supported by an NIH grant to J.H., and a pre-doctoral fellowship to D.G.

REFERENCES

1. Hirsh,J. and Davidson, N.(1981) Molec. and Cell. Biol., in press.

2. Wright, T.R.F., Beerman, W., Marsh, J.L., Bishop, C.P., Steward, R., Black, B.C., Tomsett,A.D., and Wright, E.Y. (1981) Chromosoma, in press.

3. Wright, T.R.F., Bewley, G.C., and Sherald, A.F.(1976) Genetics, 84, 287-310.

4. Maniatis, T., Hardison, R.C., Lacy, E., Lauer, J.,O'Connell, C., Quon, D., Sim, G.K., and Efstratiadis, A.(1978) Cell 15, 687-701.

TISSUE-SPECIFIC EXPRESSION OF MOUSE ALPHA-AMYLASE GENES

Richard A. Young
Department of Biochemistry
Stanford University School of Medicine
Stanford, California 94305

Peter K. Wellauer, Otto Hagenbüchle, Mario Tosi
and Ueli Schibler
Swiss Institute for Experimental Cancer Research
1066 Epalinges, Switzerland

ABSTRACT: Two alpha-amylase genes, $Amy1^A$ and $Amy2^A$, are expressed tissue specifically in the mouse. $Amy1^A$ contains 21 kb of DNA and specifies one alpha amylase mRNA in the salivary gland and a different one in the liver; these two mRNAs contain identical coding and 3'-noncoding regions but differ in their 5'-terminal noncoding segments. The different 5'-terminal CAP sites for these mRNAs reside 3 kb apart in $Amy1^A$. $Amy2^A$ is 11 kb long and contains sequences for the alpha-amylase mRNA found in the pancreas, an mRNA species which is similar in length to its salivary gland and liver counterparts, but differs from them by 11% in nucleotide sequence. A comparison of the two genes reveals that comparable exon sequences are much more highly homologous than adjacent intron residues and that the position of each of the introns is conserved.

INTRODUCTION

We are investigating the expression of mouse alpha-amylase genes to gain insight into the molecular mechanisms which govern tissue-specific gene expression. This is a

ISBN 0-12-137420-3

system of choice for such study since alpha-amylase is produced in at least three mouse tissues - the salivary gland, the liver, and the pancreas - and because there are both quantitative and qualitative differences in the mRNAs from which the amylases are translated in each of these tissues.

The alpha-amylase mRNAs which accumulate in the pancreas, salivary gland and liver account for 25%, 2% and 0.02% of the poly(A) containing RNA, respectively (1). Each of the alpha-amylase mRNAs which have been detected in these three tissues have been sequenced (2,3,4); their general features are summarized in figure 1. The single abundant 1577 nucleotide species found in the pancreas exhibits 11% sequence heterogeneity with comparable portions of its liver and salivary gland counterparts. The major liver and salivary gland alpha-amylase mRNAs (1773 and 1661 nucleotides, respectively) share identical nucleotide sequences except in their extreme 5'-terminal nontranslated regions. Other alpha-amylase mRNAs accumulate, albeit to lower levels, in the liver and in the salivary

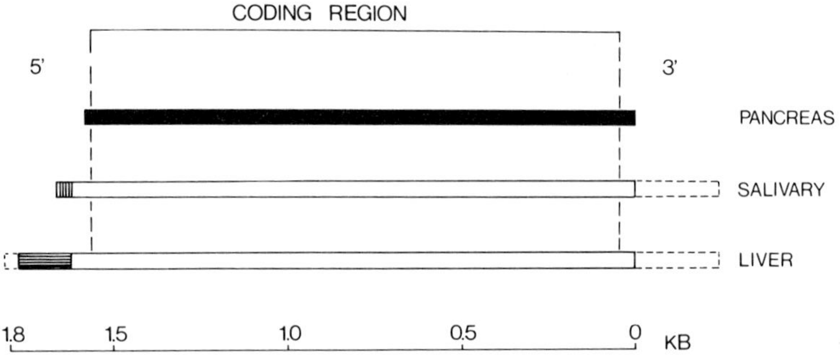

FIGURE 1. The Pancreas, Salivary Gland and Liver Alpha-amylase mRNAs.

The pancreatic alpha-amylase mRNA is shown as a black box to indicate its 11% sequence heterogeneity with comparable portions of the salivary gland and liver species, whose shared sequences are indicated by the open boxes. Tissue-specific leader sequences are shaded vertically for the salivary gland mRNA and horizontally for the liver mRNA. Hyphenated boxes designate residues found in the minor salivary gland and liver alpha-amylase mRNAs.

gland; in each case, these minor species contain major species sequences plus additional residues at their 5'- or 3'-termini (see figure 1). Of the alpha-amylase mRNAs in the liver and salivary gland, 5% contain 237 additional residues preceding the poly(A) tract. Furthermore, 20% of the liver alpha-amylase mRNAs contain 33 additional 5'-terminal residues. Whether these minor species serve some special function or are products of alternate synthesis or processing pathways is not yet clear.

The observation that the major alpha-amylase mRNA species which accumulate in the salivary gland or liver differ solely with respect to their 5'-termini led us to investigate the gene sequences which specify them (5). Through cloning of genomic DNA, we isolated the 5'-terminal one-quarter of $Amy1^A$. Sequence analysis of this DNA revealed that the extreme 5'-terminal noncoding nucleotides of the liver alpha-amylase mRNA are specified by sequences which lie 4.5 kb upstream from those for the common body of the mRNA (figure 2). In contrast, the initial nucleotides for the salivary gland alpha-amylase mRNA reside 7.5 kb from sequences which the two mRNAs share in the genome. DNA titration experiments indicated the presence of a single copy of $Amy1^A$ DNA per haploid genome. Since no gross rearrangement of these gene sequences was observed among liver and salivary gland DNA preparations, it was concluded that the two tissue-specific alpha-amylase mRNAs are differentially transcribed and/or processed from identical sequences in different cell types (5)(Figure 2).

Using recombinant DNA techniques, all the remaining genomic DNA sequences for the alpha-amylase mRNAs which accumulate in the three tissues under study have been isolated. We compare the architecture deduced for $Amy1^A$ and $Amy2^A$ in this report.

MATERIALS AND METHODS

The mouse strain AJ was used in all experiments. Methods have been described elsewhere (1-6).

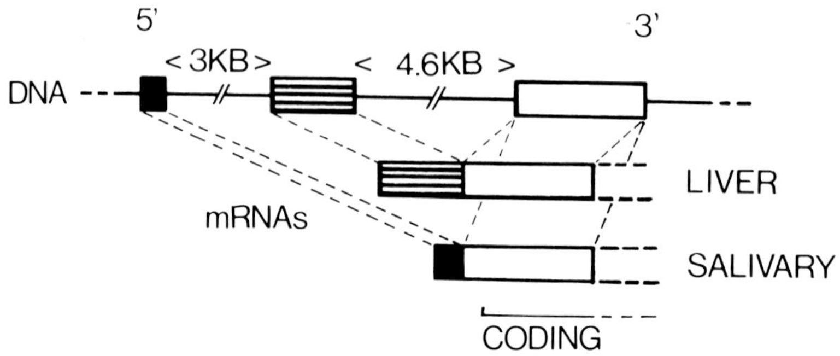

FIGURE 2. Summary of Tissue-Specific mRNA Biosynthesis from $\underline{Amy1}^{A}$.

The pattern by which gene sequences are ultimately incorporated into salivary gland and liver alpha-amylase mRNAs is shown diagrammatically. Tissue-specific sequences are represented by a black box (for the salivary gland mRNA) or by a box shaded with vertical lines (for the liver mRNA). The open box indicates residues shared by both mRNAs. The diagram for the beginning of the gene includes segments for both of the tissue-specific mRNA leaders as well as the first exon containing residues shared by the salivary gland and liver alpha-amylase mRNAs.

RESULTS AND DISCUSSION

A library of cloned mouse parotid DNA was screened for hybridization to salivary gland alpha-amylase cDNA from the clone pMSa104 (1). DNA was purified from several of the clones which gave strong hybridization signals and was analyzed by restriction endonuclease mapping and electron microscopic examination of heteroduplexes formed between the genomic DNA and alpha-amylase cDNA (6). Portions of the DNA from each clone of interest were subjected to sequence analysis. The degree of overlap among the cloned DNA fragments was sufficient to permit construction of a map for the entire $\underline{Amy1}^{A}$ gene. A similar approach was used to isolate and characterize DNA sequences for $\underline{Amy2}^{A}$ using cDNA from the clone pMPa21 (1) as a probe. These data are summarized in figure 3.

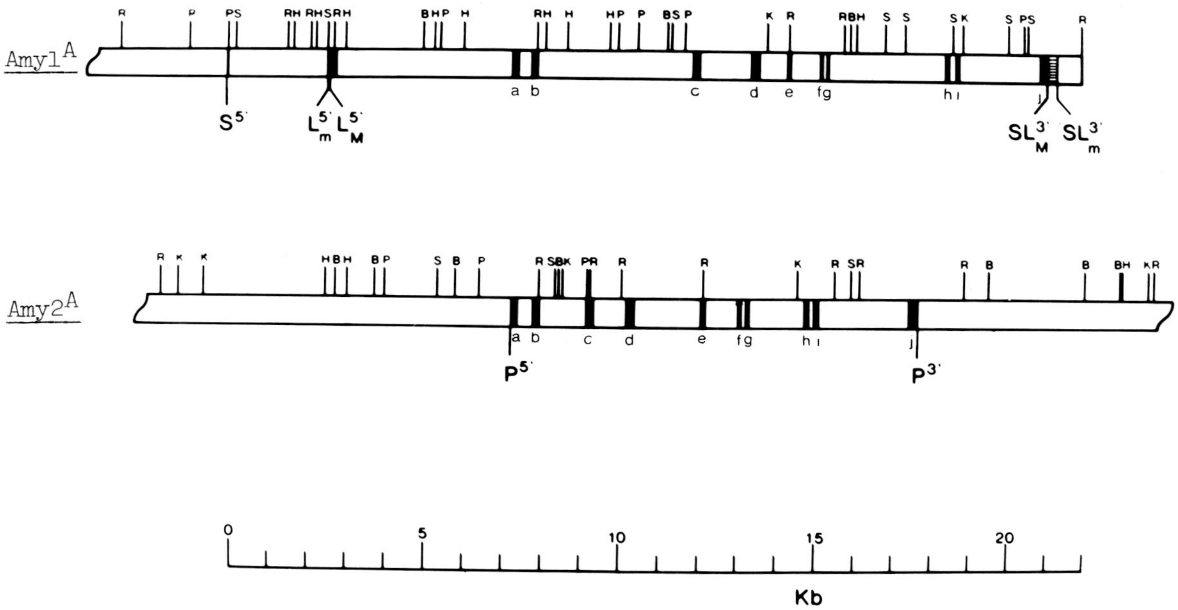

FIGURE 3. Architecture of Amy1^{A} and Amy2^{A}.

Data from a variety of sources (see text and reference 6) are summarized in this map of two segments of the mouse genome. Filled portions designate exons for the salivary gland (S), liver (L) and pancreatic (P) alpha-amylase mRNAs. The CAP and polyadenylation sites for these mRNAs are indicated by the superscripts (5') and (3'), respectively. (M) and (m) subscripts represent sequences for the termini of the major and minor mRNA species, respectively. Restriction endonuclease cleavage sites are shown above each of the two genes; these include EcoRI (R), PstI (P), SacI (S), HindIII (H), BamHI (B) and KpnI (K). Lower case letters designate comparable exons in the two genes.

I. Features of $\underline{Amy1}^A$.

As defined by the distance from the salivary gland mRNA CAP site to the two polyadenylation sites (see below) for both salivary and liver mRNAs, $\underline{Amy1}^A$ is 21,000 base pairs long. The residues which specify the liver and salivary gland mRNAs, which constitute less than 8% of the gene, are separated by at least 10 intervening sequences or introns (very small introns would escape detection in the electron microscope).

Assuming that the CAP site is the site of transcription initiation, as has been demonstrated for a variety of eucaryotic genes in vitro (7,8), transcription initiation occurs at three sites in $\underline{Amy1}^A$ (at the CAP site for the salivary gland mRNA and at the CAP sites for both the major and minor liver mRNAs). This arrangement could provide liver and salivary gland cells with a mechanism to quantitatively regulate alpha-amylase mRNA biosynthesis at the level of transcription initiation. Such a model is consistant with the observation that the relative amounts of alpha-amylase mRNA in salivary gland and liver cytoplasm appear to reflect the relative quantity of alpha-amylase mRNA precursor in nuclear preparations from the two tissues (5).

$\underline{Amy1}^A$ contains multiple polyadenylation sites (4). The 237 additional (relative to the major mRNA) 3'-terminal nucleotides preceding the poly(A) tract in one of the minor alpha-amylase mRNAs (see figure 1) are specified by $\underline{Amy1}^A$ sequences contiguous to residues for the 3'-terminus of the major species. An intriguing feature of the two polyadenylation sites is that they share 14 of 18 of the base pairs which trail the position of poly(A) addition, suggesting that these sequences might be involved in poly adenylation or transcription termination.

II. Features of $\underline{Amy2}^A$.

From the CAP site to the polyadenylation site, $\underline{Amy2}^A$ is approximately 11,000 base pairs long (figure 3). It contains at least 9 introns. We believe that the $\underline{Amy2}^A$ DNA we have cloned specifies the pancreatic alpha-amylase mRNA because all of the sequences we have elucidated for the exon portions of this gene (the region which includes the first two exons, a portion of the fifth exon, and DNA flanking the polyadenylation site have been sequenced) are identical to the residues previously determined for the pancreatic alpha-amylase mRNA (6). However, the number of copies of $\underline{Amy2}^A$ in the mouse genome has not yet been rigorously established as it has for $\underline{Amy1}^A$.

III. Comparison of $\underline{Amy1}^A$ and $\underline{Amy2}^A$.

The data summarized in figure 3 reveal that the initiation of transcription of $\underline{Amy2}^A$ DNA, again assuming that the CAP site is coincident with initiation, occurs within the first exon - that containing alpha-amylase polypeptide coding sequences. In contrast, both of the tissue-specific CAP sites in $\underline{Amy\text{-}1}^A$ are separated from the initial exon containing protein coding sequences by large segments of intervening sequence DNA. As discussed above, one functional reason for the disparity in the position of the transcription initiation site could be related to the large quantitative differences in alpha-amylase mRNA precursor observed in the three tissues examined. Thus, it is tempting to speculate that $\underline{Amy2}^A$ may have dispensed with a long pre-mRNA leader like that specified by $\underline{Amy1}^A$ to accommodate the necessarily efficient synthesis of alpha-amylase mRNA in the pancreas.

Electron microscopic analysis of genomic DNA/alpha-amylase cDNA heteroduplexes indicates, and in every case examined sequence analysis confirms, that intervening sequences occur in homologous positions in the two alpha-amylase genes (see figure 3). Thus, each exon in $\underline{Amy1}^A$ has a counterpart in $\underline{Amy2}^A$ which encodes the same portion of the alpha-amylase polypeptide. The exons exhibit a much greater degree of homology (89%) between $\underline{Amy1}^A$ and $\underline{Amy2}^A$ than do the intervening sequences which flank them. The intron which separates exons a and b (figure 3) in $\underline{Amy1}^A$, for example, bears no apparent sequence relationship to its counterpart in $\underline{Amy2}^A$ (6). Residues adjacent the splice junctions of this intervening sequence in the two genes do exhibit limited homology; however, it is the degree of complementarity to the 5'-terminal sequence of the small nuclear RNA U1 - a posulated component of the splice mechanism (9) - that these particular intron residues share in $\underline{Amy1}^A$ and in $\underline{Amy2}^A$ (6).

ACKNOWLEDGMENTS

We are grateful to A.-C. Pittet and R. Bovey for their excellent technical assistance. This manuscript was improved by criticism from L.J. Korn.

REFERENCES

1. Schibler, U., Tosi, M., Pittet, A.C., and Wellauer, P. (1980) J. Mol. Biol. 142, 93.
2. Hagenbüchle, O., Bovey, R., and Young, R.A. (1980) Cell 21, 179.
3. Hagenbüchle, O., Tosi, M., Schibler, U., Bovey, R., Wellauer, P.K. and Young, R.A. (1981) Nature 289, 643.
4. Tosi, M., Young, R.A., Hagenbüchle, O. and Schibler, U. (1981) Nucleic Acids Res., in press.
5. Young, R. A., Hagenbüchle, O. and Schibler, U. (1981) Cell 23, 451.
6. Schibler, U., Young, R.A., Pittet, A.-C., Hagenbüchle, O., Bovey, R., Tosi, M. and Wellauer, P.K. (1981) manuscript submitted.
7. Hagenbüchle, O. and Schibler, U. (1981) Proc. Natl. Acad. Sci., U.S.A., in press.
8. Contreras, R. and Fiers, W. (1981) Nucleic Acids Res. 9, 215.
9. Lerner, M. R., Boyle, J.A., Mount, S.M., Wolin, S.J. and Steitz, J.A. (1980) Nature 283, 220.

THE COLLAGEN GENE

Hiroaki Ohkubo
Enrico Avvedimento
Yoshihiko Yamada
Gabriel Vogeli
Mark Sobel
Glenn Merlino
Maria Mudryj
Ira Pastan
Benoit de Crombrugghe

Laboratory of Molecular Biology
National Cancer Institute
National Institutes of Health
Bethesda, Maryland

ABSTRACT

The collagens belong to a family of proteins which constitute the principal component of the extracellular matrix of animal tissues. We have used chick embryo fibroblasts in culture as a model system to study the regulation of type I collagen synthesis. Type I collagen, the major collagen species synthesized by these cells, is composed of two alpha 1 subunits and one alpha 2 subunit. When p60$^{\underline{src}}$, the transformation protein encoded by Rous sarcoma virus, is present the synthesis of collagen is severely reduced in these cells. Recent evidence strongly suggests that this regulation, although indirect, is mediated by a transcriptional control mechanism.

The alpha 2 type I collagen gene has a length of about 38 kilobases. Its coding information is subdivided into more than 50 exons which are interrupted by introns of various sizes. Many exons of this gene have an identical length of

ISBN 0-12-137420-3

54 bp although their sequences vary. This strongly suggests that the ancestral gene for collagen arose by amplification of a single genetic unit containing an exon of 54 bp. Later these exon sequences evolved by successive point mutations but also, in rarer instances, by additions or deletions of 9 bp or multiples of 9 bp. To maintain the helicity of the collagen molecule only additions or deletions of 9 bp or multiples thereof were tolerated. Indeed nine bp encode the basic gly-x-y repeat of collagen.

The intron sequences immediately adjacent to the 5' and 3' ends of exons show a higher degree of conservation than in other genes. These intron sequences also exhibit a more marked complementarity with the sequence of the 5' end of a small nuclear RNA, called U_1 RNA. They therefore, strongly support the previously formulated hypothesis (18, 19) that U_1 RNA is involved in splicing. Additionally, splicing sites are found within an intron of the alpha 2 (I) collagen gene. Their location correlates with sequences which are also complementary with the 5' end sequence of U_1 RNA. This correlation further strenghtens the notion that U_1 RNA has a role in splicing.

The structure of the promoter for the alpha 2 (I) collagen gene exhibits three large inverted repeat sequences which overlap each other and hence might form mutually exclusive stem and loop structures. These symmetrical sequences could be potential binding sites for regulatory proteins.

INTRODUCTION

The collagen genes are a family of developmentally regulated genes. They code for a class of proteins which constitute the major protein components of the extracellular matrix in animal tissues. In higher vertebrates there are at least five genetically distinct collagen types with as many as nine different polypeptide chains (1). The differences in primary structure between these collagen chains probably reflect tissue related functional differences.

We are using cultured chick embryo fibroblasts (CEF) as a model system to study the developmental regulation of the genes for type I collagen and their perturbation by oncogenic growth factors. Type I collagen, which consists of two alpha 1 chains and one alpha 2 chain, is the principal collagen made in these cells. The rate of synthesis of type I collagen is severely inhibited by the presence in these cells of p60<u>src</u>, the transforming protein encoded in the genome of Rous sarcoma virus (2). A similar inhibition of type I collagen synthesis is found in rodent fibroblasts transformed by other retroviruses (3). This inhibition is selective because the synthesis of most other cellular proteins is unchanged. Although the effect of p60src on collagen synthesis is almost certainly indirect, there is good evidence that its effect is mediated by a transcriptional control mechanism (4-8) (see below).

In order to study this type of control in reconstituted *in vivo* and *in vitro* systems, we decided to isolate one of the collagen genes which is expressed in chick embryo fibroblasts; we chose to isolate the gene for type I alpha 2 collagen.

STRUCTURE OF THE COLLAGEN PROTEINS

Figure 1 is a model representing type I collagen. The collagen chains are synthesized as precursor molecules and contain, beginning at the NH_2 terminal end: (1) a signal peptide; (2) a N-terminal propeptide; (3) the major helical part of the molecule; (4) a C-terminal propeptide. Like for other secreted proteins, the signal peptide is cleaved off in the rough endoplasmic reticulum, where the chains presumably associate to form the molecule shown in Figure 1 minus the signal peptides. After hydroxylation of many lysine and proline residues and the addition of carbohydrate sidechains, the molecule is secreted. Following the removal of the N

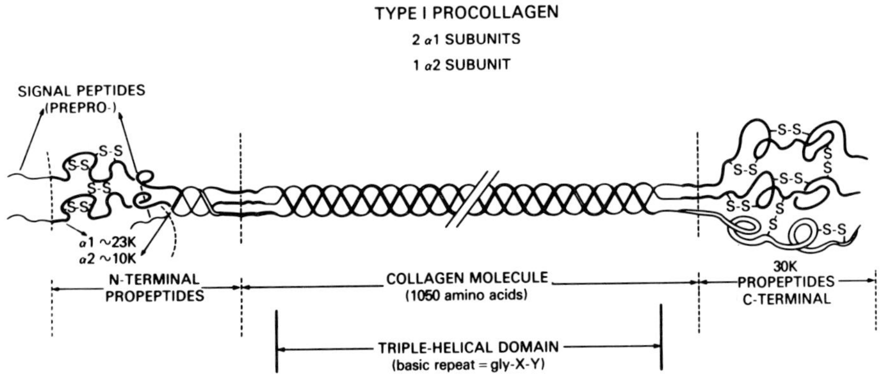

FIGURE 1. Model of type I collagen with its biosynthetic precursors.

and C propeptides by specific enyzmes, the collagen molecules assemble in highly ordered microfibrils and fibrils.

The central helical part of most collagen species have unique structural features. These include (1) the presence of glycine residues every third amino acid; (2) an abundance of proline and lysine residues, many of which are hydroxylated; (3) a characteristic configuration of the molecule composed of three subunits which interact with each other to form a triple helical structure; (4) the presence of many inter- and intra-molecular crosslinks which result in a higher order structural organization. Except for the glycine residues, it is difficult to find extensive sequence homologies between different portions of the same chain or between different collagen chains.

STRUCTURE OF THE ALPHA 2 (I) COLLAGEN GENE: EVOLUTIONARY IMPLICATIONS

We isolated the gene for type I alpha 2 collagen from a library of chick genomic DNA fragments, (generously provided by J. Dodgson, D. Engel and R. Axel) in a series of overlapping clones (9, 10). Figure 2 is a schematic representation of this gene. The vertical bars correspond to exons; the lines between the bars correspond to introns. The gene is 38 Kb in length and contains more introns than any gene reported so far. Its coding information is subdivided in at least 52 exons. The introns which interrrupt these exons have various sizes ranging from about 80 to more than 2,000 bp.

The most attractive hypothesis to explain the multi-exon organization which is a characteristic feature of many eucaryotic genes, postulates that these genes arose by recombination between functional DNA blocks (11). Each functional DNA block corresponds to a specific protein domain. Recombination within introns probably increased the rate of recombination between these functional DNA blocks. This hypothesis is strongly supported by the correlation between the different exons of a given gene with the assignment derived by independent physico-chemical studies of the various domains in the corresponding protein. The immunoglobulin genes are a particularly good example for this type of comparison. During the isolation of the collagen gene we were surprised to find so many exons, and wondered why a molecule as uniform and regular as collagen would be subdivided in so many domains.

To try to answer this question we determined the size of a small number of exons in three different parts of the gene corresponding to three different segments of the helical portion of the collagen molecule (12). Some of these exons were sequenced in collaboration with L. Dickson, B. Olsen, P. Fietzek et al. (13). The results (Fig. 3) show a remarkable conservation of the size of the exons. Nine out of twelve exons which were examined, have an identical length of 54 bp, two exons contain 108 bp (2 x 54) and one has 99 bp. J. Wozney and colleagues (14) have also sequenced exons in the same gene and found in addition to a number of 54 bp and 108 bp exons a few other exons which have a different length but all have lengths that are multiples of 9.

Although most exons of the alpha 2 (I) collagen gene have the same length (54 bp), the sequences within these exons vary. Not surprisingly the sizes of the four exons corresponding to the nonhelical C peptide and that of the promoter

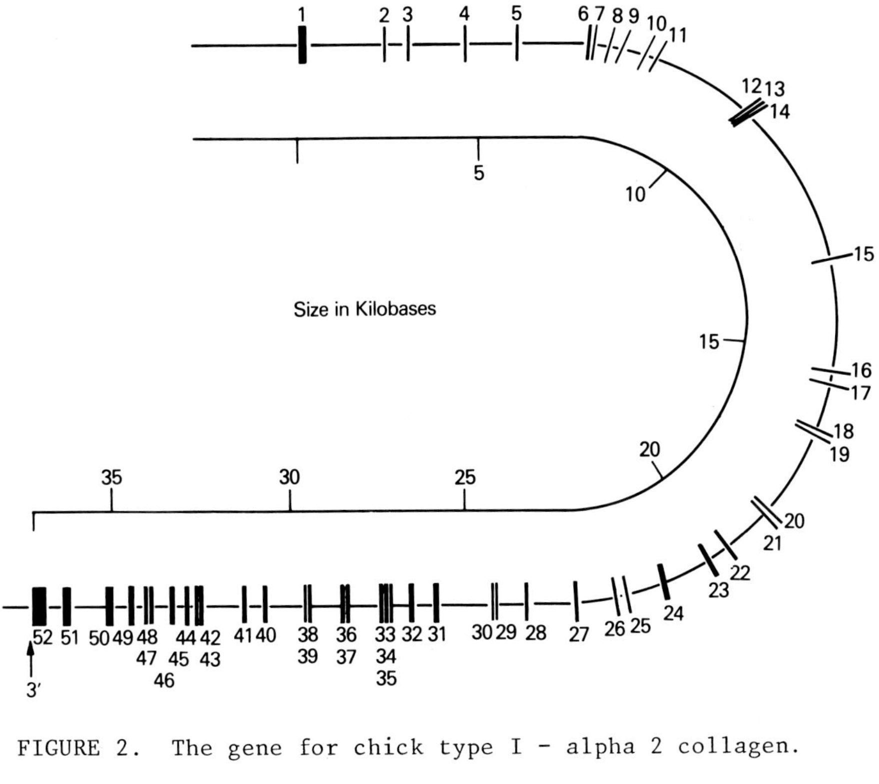

FIGURE 2. The gene for chick type I - alpha 2 collagen.

proximal exon are much larger than 54 bp (between 192 and 400 bp) (10, 13). We know that these exons specify parts of the collagen molecule which are distinctly not helical in structure.

We believe that the 108 bp exons arose by fusion of two adjacent 54 bp exons and resulted from the precise deletion of an intervening intron. A similar finding has been made in one of the rat insulin genes (15). The 99 bp exon shown in Fig.3

```
                              EXONS OF THE ALPHA-2 COLLAGEN GENE

                                                                                                EXON SIZE IN BP

1.   ttgtctag  GGC CCT CCT GGG TTT CAA GGT GTT CCT GGT GAA CCT GGT GAA CCT GGT CAA ACA  gtaagtac     54
               gly pro pro gly phe gln gly val pro gly glu pro gly glu pro gly gln thr
               19                                                                  36

2.   tttaacag  GGT CCC CAA GGT CCT CGT GGT CCC CCT GGT CCT CCA GGA AAG GCT GGT GAA GAT  gtaagtca     54
               gly pro gln gly pro arg gly pro pro gly pro pro gly lys ala gly glu asp
               37                                                                  54

3.   tttttag   GGT CAC CCT GGC AAA CCT GGA AGA CCT GGT GAG AGG GGT GTT GCT GGT CCT CAA  gtaagtaa     54
               gly his pro gly lys pro gly arg pro gly glu arg gly val ala gly pro gln
               55                                                                  72

4.   ttttctag  GGT TTC CCT GGA GCA GAT GGT AGG GTT GGG CCA ATC GGT CCA GCC GGT                       99
               gly phe pro gly ala asp gly arg val gly pro ile gly pro ala gly
               379

                   AAT AGA GGT GAA CCT GGC AAC ATT GGA TTC CCT GGA CCA AAA GGT CCC ACT  gtaagtac
                   asn arg gly glu pro gly asn phe gly phe pro gly pro lys gly pro thr
                                                                                   411

5.   accttcag  GGT GAG CCT GGC AAA CCT GGT GAA AAA GGC AAT GTC GGT CTT GCT GGC CCA CGG  gtactgg      54
               gly glu pro gly lys pro gly glu lys gly asn val gly leu ala gly pro arg
               412                                                                 429

6.   ttcaacag  GGC AAT CCT GGA AAT GAT GGT CCT CCA GGC CGT GAT GGT GCT CCT GGC TTC AAG  gtagactt     54
               gly asn pro gly asn asp gly pro pro gly arg asp gly ala pro gly phe lys
               838                                                                855

7.   gttcacag  GGT GAG CGT GGT GCT CCT GGT AAC CCA GGT CCC ...                                       54
               gly glu arg gly ala pro gly asn pro gly pro
               856

8.                                 ... GGT CCT TCT GGA AAG CCT GGA AAC CGT GGT GAT CCT  gtaagttg     54
                                       gly pro ser gly lys pro gly asn arg gly asn pro
                                                                                   891

9.   tgttccag  GGT CCT GTT GGT CCT GTT GGT CCT GCT GGT GCT TTT GGC CCA AGA GGT CTC GCT  gtaagtct     54
               gly pro val gly pro val gly pro ala gly ala phe gly pro arg gly leu ala
               892                                                                 909

10.  tttcctag  GGC CCA CAA GGT CCA CGT GGT GAG AAA GGT GAA CAT GGT GAT AAG GGA CAT AGA  gtaagtaa    108
               gly pro gln gly pro arg gly glu lys gly glu his gly asp lys gly his arg
               910

               GGT CTG CCT GGC CTG AAC GGA CAC AAT GGG TTG CAG GGT CTT CCT GGT CTT GCT
               gly leu leu gly leu lys gly his asn gly leu gln gly leu pro gly leu ala
                                                                                   945

11.  cccaatag  GGC CAA CAT GGT GAT CAA GGT CCT CCT GGT AAC AAC GGT CCA GCT GGC CCA AGG  gtatgtga     54
               gly gln his gly asp gln gly pro pro gly asn asn gly pro ala gly pro arg
               946                                                                 963

12.  aactttag  GGT CCT CAT GGT CCT TCT GGT CCT CAT GGT AAG GAT GGT CGC AAT GGT CTC CCT
               gly pro his gly pro ser gly pro his gly lys asp gly arg asn gly leu pro
               964                                                                            108
               GGA CCC ATT GGC CCT GCT GGT GTA CGT GGA TCT CAT GGT AGC CAA GGC CCT GCT  gta...
               gly pro ile gly pro ala gly vol arg gly ser his gly ser gln gly pro ala
                                                                                   999
```

FIGURE 3. Sequences of exons in the type I - alpha 2 collagen gene. The column on the right indicates the length of the exons in bp.

could have arisen by a rare recombinational event between two adjacent exons which resulted in the loss of the intervening intron plus 9 bp of exon sequences. Alternatively a deletion in a 108 bp exon could have caused the loss of 9 bp.

The important point is that the length of those exons, which are different from 54 or 108 bp, diverges from their length numbers by 9 or a multiple of 9. For the continuity of the triple helix to be preserved, only deletions of 9 bp or multiples of 9 could have been tolerated. Indeed the basic gly-x-y repeat of the collagen molecule is encoded by 9 bp.

The finding that most exons of the alpha 2 type I collagen gene have an identical length of 54 bp has important implications for the evolutionary assembly of this gene. It suggests that the ancestral gene for collagen was assembled by multiple duplications of a single genetic unit containing an exon of 54 bp (see Fig. 4) (12). Later the exon sequences evolved by successive point mutations but also, in some cases, by deletions or additions of 9 bp or multiples of 9 bp. The 54 bp exon structure is probably common to all collagen genes found in nature. We have recently sequenced two exons in another chick collagen gene; one exon contained 54 bp, the other 108 bp (Y. Yamada and M. Mudryj, personal communication).

Although the exon sequences vary they present enough similarities to suggest that they are derived from each other. We have aligned the sequences of the 54 bp exons (the 108 bp

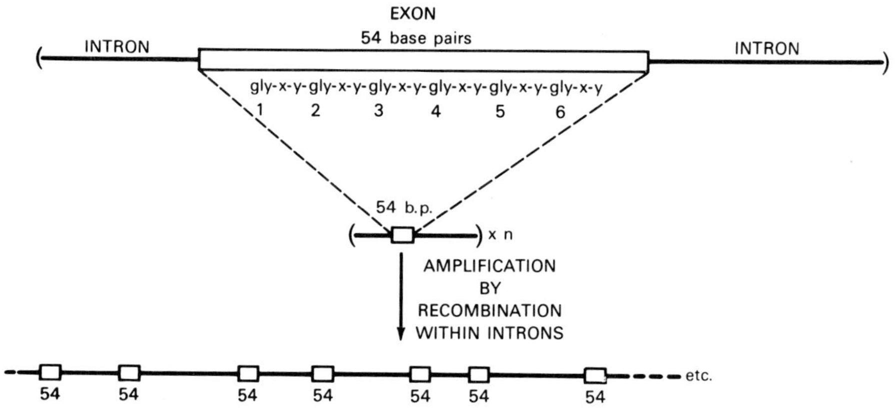

FIGURE 4. Model for the assembly of the primordial collagen gene.

exons were treated as two successive 54 bp exons) and scored the bases which are present at the same location in 50 percent or more of the exons (Fig. 5). In 42 of the 54 positions a given base is found in 50 percent or more of the exons. It is interesting to note that the third base of the glycine codons is very often a T. The data of Fig. 5 also suggest that the primordial exon for the collagens probably encoded a glycine-proline-proline tripeptide repeated six times. It is possible that a glycine-proline peptide of 18 amino acids had the minimal length required to form a stable triple helical structure.

The presence of a triple helical collagen tail in the C_1Q molecule (16) (C_1Q is a component of the complement system) suggests that the original function of the primordial collagen exon peptide may have been different. A short triple helical tail could have helped stablize the tertiary structure of a three-subunit protein. This same primordial exon may later have been utilized to construct the precursor of the collagen gene family.

A number of questions still remain to be answered. Why, for instance, did the amplification process stop at a given stage? What was the function of some of the intermediary genes which were composed of a much smaller number of exons? Did some of the collagen genes branch off during the amplification process? To answer these questions, several other vertebrate and invertebrate collagen genes will need to be characterized.

STRUCTURE OF THE COLLAGEN GENE: IMPLICATIONS FOR SPLICING

The existence of more than 50 introns in the alpha 2 (I) collagen gene (9) implies that the conversion of the primary

GGT	CCX	CXT	GGT	CCT	CXT	GGT	CCT	CCT	GGT	XAX	CAT	GGT	XXX	XCT	GGT	CXX	XCT
1 1 .6	.8 .5 -	.8 - .8	1 1 .5	.5 .5 .7	.6 - .8	1 1 .9	.5 .5 .5	.6 .6 .7	1 1 .6	- .5 -	.5 .5 .6	1 1 .8	- - -	- .6 .8	1 1 .6	.8 - -	- .5 .5

1. 42 OUT OF 54 POSITIONS ARE CONSERVED IN 50% OR MORE OF THE EXONS.
2. T IS HIGHLY FAVORED AS THIRD BASE.
3. ANCESTRAL 54BP EXON WAS PROBABLY COMPOSED OF (GGT-CCT-CCT)6.
   gly-pro-pro

FIGURE 5. Consensus sequence for the collagen exons. The bases which appear at a given position in 50 per cent or more of the exons are indicated.

transcript of the gene to mature translatable collagen mRNA includes more than 50 precise splicing events. This figure is an underestimate because at least some introns are probably removed in more than one step (17).

As in other genes the intron sequences immediately adjacent to the 5' and 3' end of the exons are similar. The finding of such similarities in other genes led Lerner et al. (18) and Rogers and Wahl (19) to postulate that a small nuclear RNA, called U_1 RNA, had a role in the splicing reaction because the sequence at the 5' end of U_1 RNA is complementary to the sequences at both ends of introns. In the gene for type I alpha 2 collagen these "acceptor" and "donor" intron sequences immediately adjacent to the 5' end and 3' ends of exons are conserved to a much higher degree than in any other gene so far studied. As seen in Figure 6, the first 6 bases at the 5' end of the introns are identical in 8 of 12 introns which are examined (12, 13); in three introns there is only one base which is different. The sequences at the 3' end of

EXON

	ACCEPTORS	DONORS
1....	CUUCUUGUCUAG	GUAAGUACAG...
2....	UGUAUUUAACAG	GUAAGUCAUU...
3....	UGAAUUUUUUAG	GUAAGUAAGU...
4....	UUCUUUUUCUAG	GUAAGUACAG...
5....	UUUCACCUUCAG	GUACGUGGAU...
6....	GUACUUCAACAG	GUAGACUUCA...
7....	CCUUGUUCACAG	n.d.
8....	n.d.	GUAAGUUGAU...
9....	UUCCUGUUCCAG	GUAAGUCUAA...
10....	AUCCUUUCCUAG	GUAAGUAAA...
11....	UUUUCCCAAUAG	GUAUGU...
12....	GUUUUCUUCUAG	GUGAGUA...
13....	GUUUCUUGUAG	GUAAGU...
CONSENSUS	5'...UUXXUU(U/C)UX(C/U)AG	GUAAGU3'
	...UAGAGGGAGGUC	CAUUCAUmAmpppGm$_3^{2,2,7}$$U_1$ RNA

FIGURE 6. Sequences adjacent to the 5' and 3' end of exons that code for helical parts of type I - alpha 2 collagen. The sequence of the 5' end of U_1 RNA is given on the last line.

the introns or "acceptor" site, also present considerable homologies. Both these donor and acceptor sequences exhibit a higher degree of complementarity with the 5' end of U_1 RNA than in other genes. They, therefore, clearly support the notion that U_1 RNA participates in the splicing reaction.

Another argument for the participation of U_1 RNA in splicing came from examination of the DNA sequence of a short intron in the collagen gene (16). We found that the sequence at the 5' end of U_1 RNA was complementary not only to sequences at both ends of this intron but also to sequences at one end of the intron and at three locations within the intron. These complementarities suggested the existence of internal splice sites within the intron (17). That these internal splice sites are indeed functional was shown by the existence of discrete RNA species in which defined parts of the intron had been deleted (17). The size of these deletions corresponded precisely to the location of the splice site predicted from the sequence complementarities with U_1 RNA. In addition, once a defined portion of the intron has been removed the sequences at the ends of the newly formed intron are again complementary with the 5' end of U_1 RNA. This suggests that the stepwise removal of this intron, as illustrated in Figure 7, is a functional pathway (17).

Although the participation of U_1 RNA may help provide an essential element of accuracy to the splicing reaction by bringing the bases which need to be spliced in close contact, it does not explain how the appropriate donor and acceptor sequences are spliced. In other words, a mechanism involving U_1 RNA does not explain why splice sites and hence exons are not by-passed. In such cases incomplete RNA molecules would be generated with partial deletions of their coding sequences. The fact that the translatable messenger RNA for alpha 2 (I) collagen is homogeneous indicates that no such errors are made. We know, however, that such "errors" are allowed with other RNAs like SV40 and adeno RNAs or certain species of immunoglobulin RNAs (20, 21, 22, 23). With the latter RNAs a unique precursor RNA can give rise to more than one messenger RNA, because alternate choices are possible in the splicing pattern. These splicing choices are a regulatory mechanism to decide which species of mRNA will be made from a unique precursor RNA. With collagen RNA however, if only a 2 percent possibility of choice existed each time a splice was to be made, no correct messenger RNA would ever be produced. It is therefore likely that for collagen RNA as for many other RNAs a mechanism must exist which prevents alternative RNA splicing patterns. We believe that a unique order in the succession of

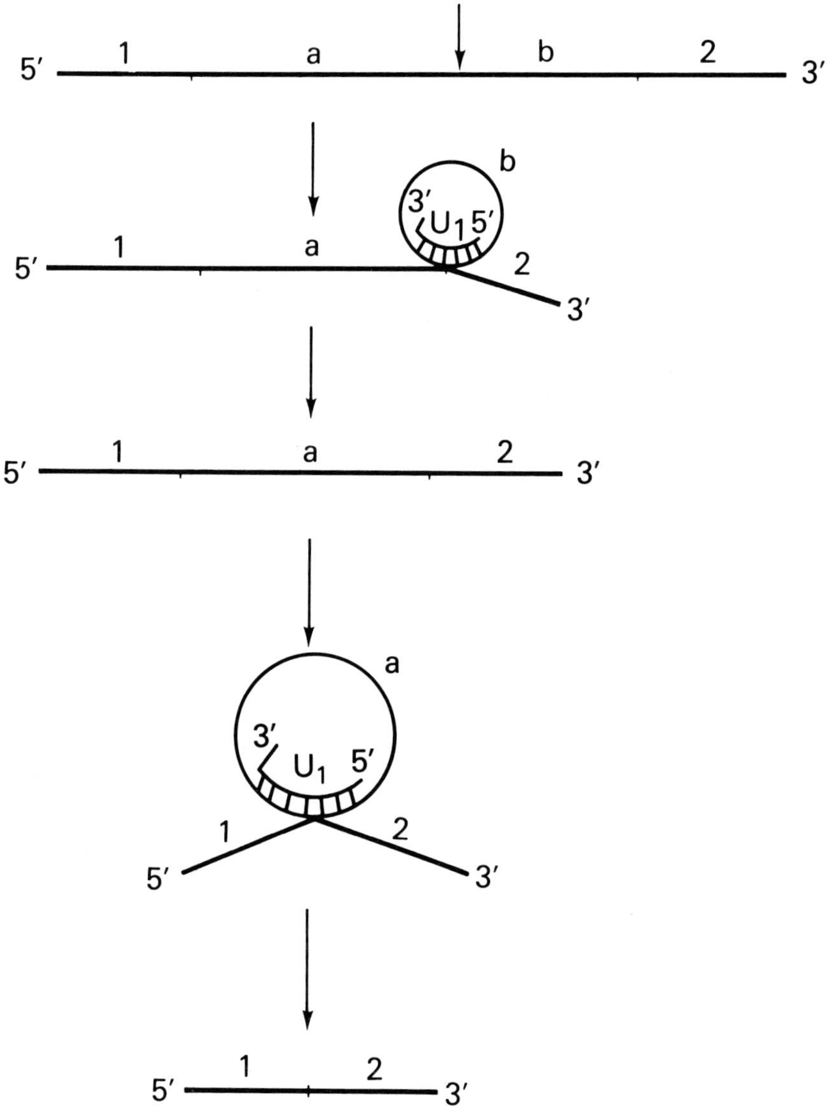

FIGURE 7. Pathway for a two-step removal of an intron.

splices may be essential for this "error"-free processing of a precursor RNA. We have in fact examined a short segment of the alpha 2 (I) collagen gene and have found that within this segment, the introns were removed in a well defined sequence (E. Avvedimento, unpublished results). The secondary structure of the RNA and hence the nucleotide sequences of the exons as well as that of the introns may play an important role in dictating an unique splicing order.

REGULATION OF THE ALPHA 2 (I) COLLAGEN GENE AND STRUCTURE OF ITS PROMOTER

There are a number of arguments which support the notion that the effect of p60<u>src</u> on collagen synthesis is mediated by a transcriptional control mechanism. First, the decrease in type I collagen synthesis in Rous sarcoma virus (RSV) transformed chick embryo fibroblasts is due to a more than 10-fold reduction not only in the levels of translatable cytoplasmic mRNA (4, 5) but also in the levels of nuclear intron specific RNA for alpha 2 (I) collagen (7). Second, Sandmeyer et al. (8) have shown that the synthesis of type I collagen RNA by isolated nuclei is 6- to 8-fold lower with nuclei of CEF transformed by RSV than with nuclei form normal CEF.

In cells infected with a mutant of RSV which is temperature sensitive for transformation, the levels of type I collagen and of type I collagen RNA are low at the permissive temperature and elevated at the non-permissive temperature (6). If the temperature is shifted from a permissive to a non-permissive temperature, a typical induction curve for collagen or collagen RNA is obtained (6). All these experiments strongly suggest that the control of collagen synthesis occurs at the level of transcription. We believe that p60<u>src</u> disrupts normal control mechanisms and, hence, provides us with a tool to study the regulation of this developmentally regulated gene.

Because we assume that these control mechanisms occur at the 5' end of the collagen gene, we have determined the start site for transcription of type I alpha 2 collagen RNA and have characterized the structure of the promoter for this gene (23).

The promoter of the alpha 2 (I) collagen gene shows a typical Goldberg-Hogness sequence between -33 and -26 and a "CAT" box 5' GCCCATT 3' sequence between -89 and -78. Both these conserved sequence are found at approximately the same distance from the start site for transcription (+1) as in other eucaryotic promoters. There are however two more interesting and unique features to the sequences around the start site of this gene (24).

(1) Three translational initiation codons are found in the initial portion of the mRNA. The first two AUG's (+54 to +56 end +117 to +119) are followed by very short coding sequences which could specify a hexapeptide and a tetrapeptide, respectively (23). Only the third AUG which is found at +134 to +136 bases after the start site for transcription, is followed by an open reading frame encoding a sequence which presents considerable homologies with the previously determined amino acid sequence of pre-pro-alpha 1 (type I) collagen (25). The first two AUGs are found within the stem of stem-and-loop structures in the RNA, whereas the third AUG occurs in a loop.

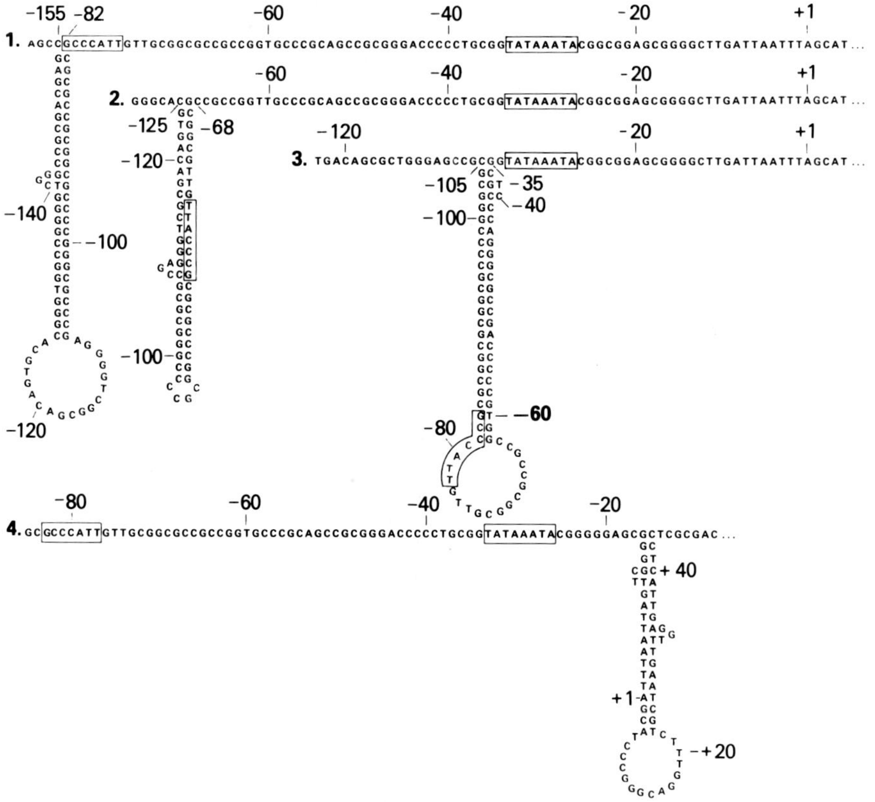

FIGURE 8. Dyads of symmetry in the promoter sequence of the type I - alpha 2 collagen gene.

(2) The promoter sequence also exhibits several large dyads of symmetry. One of these overlaps the start site for transcription (Fig. 8-4). Three other inverted repeats, which precede the start site for transcription, overlap each other and could therefore form mutually exclusive stem and loop structures (Fig. 8-1, 8-2, 8-3). These structures may be binding sites for regulatory proteins and thus play a role in the developmental regulation of this gene (24).

The sequences around the promoter have been used as template for in vitro transcription experiments using crude Hela cell extracts. As with other eucaryotic promoter DNAs in vitro transcription with this system starts at the same place as in intact cells (26).

In summary, the structure of the alpha 2 (I) collagen gene indicates how a single small genetic unit containing an exon of 54 bp, which probably had a different function in a more primitive cell, was utilized to construct the ancestor for a family of genes, encoding a set of abundant extracellular proteins.

The structure of this gene also illustrates that an error-free succession of splicing reactions is essential to ensure the conversion of the primary transcript to translatable messenger RNA. We believe that an unique order in the splicing of introns is a necessary condition for such error-free conversion.

Finally, the examination of the expression of the promoter for the alpha 2 (I) collagen gene in appropriately reconstituted in vivo and in vitro systems will help us investigate the developmental regulation of this gene and its alterations by oncogenic growth factors.

REFERENCES

1. Bornstein, P., and Sage, H. Ann. Biochem. 49, 957 (1980).
2. Levinson, W., Bhatnagar, R. S., and Liu, T.-Z. J. Natl. Cancer Inst. 55, 807 (1975).
3. Hata, R., and Peterkofsky, B. Proc. Natl. Acad. Sci. USA 74, 2933 (1977).
4. Adams, S. L., Sobel, M. E., Howard, B. H., Olden, K., Yamada, K. M., de Crombrugghe, B., and Pastan, I. Proc. Natl. Acad. Sci. USA 74, 3399 (1977).
5. Howard, B. H., Adams, S. L., Sobel, M. E., Pastan, I., and de Crombrugghe, B. J. Biol. Chem. 253, 5869 (1978).
6. Sobel, M., Yamamoto, T., de Crombrugghe, B., and Pastan, I. Biochemistry 20, 2678 (1981).

7. Avvedimento, E., Yamada, Y., Lovelace, E., Vogeli, G., de Crombrugghe, B., and Pastan, I. Nucleic Acids Res. 9, 1123 (1981).
8. Sandmeyer, S., Gallis, B., and Bornstein, P. J. Biol. Chem. 256, 5022 (1981).
9. Ohkubo, H., Vogeli, G., Mudryj, M., Avvedimento, V. E., Sullivan, M., Pastan, I., and de Crombrugghe, B. Proc. Natl. Acad. Sci. USA 77, 7059 (1980).
10. Vogeli, G., Avvedimento, E. V., Sullivan, M., Maizel Jr., J. V., Lozano, G., Adams, S. L., Pastan, I., and de Crombrugghe, B. Nucleic Acids Res. 8, 1823 (1980).
11. Gilbert, W. Nature 271, 501 (1979).
12. Yamada, Y., Avvedimento, V. E., Mudryj, M., Ohkubo, H., Vogeli, G., Irani, M., Pastan, I., and de Crombrugghe, B., Cell 27, 887 (1980).
13. Dickson, L. A., Ninomiya, Y., Bernard, M. P., Pesciottia, D. M., Parsons, J., Green, G., Eikenberry, E. F., de Crombrugghe, B., Vogeli, G., Pastan, I., Fietzek, P. P., Olsen, B. R. J. Biol. Chem. in press (1980).
14. Wozney, J., Hanahan, D., Morimoto, R., Boedtker, H., and Doty, P. Proc. Natl. Acad. Sci. USA 78, 712 (1981).
15. Lomedico, P., Rosenthal, N., Efstratiadis, A., Gilbert, W., Kolodner, R., and Tizard, R. Cell 18, 545 (1979).
16. Porter, R. R., and Reid, K. B. M. Nature 275, 699 (1978).
17. Avvedimento, V. E., Vogeli, G., Yamada, Y., Maizel, J. V., Pastan, I., and de Crombrugghe, B. Cell 21, 689 (1980).
18. Lerner, R. M., Boyle, J. A., Mount, M. S., Wolin, L. S., and Steitz, J. A. Nature 283, 220 (1980).
19. Rogers, J., and Wall, R. Proc. Natl. Acad. Sci. USA 77, 1877 (1980).
20. Griffin, B. E. in "Molecular Biology of Tumor Viruses" 2nd edition, part 2, DNA Tumor Viruses (J. Tooze, ed.) p. 61. Cold Spring Harbor Laboratory (1980).
21. Flint, S. J., and Broker, T. R. in "Molecular Biology of Tumor Viruses" 2nd edition, part 2, DNA Tumor Viruses (J. Tooze, ed.) p. 443. Cold Spring Harbor Laboratory (1980).
22. Maki, R., Roeder, W., Traunecker, A., Sidman, C., Wabl, M., Roschke, W., and Tonegawa, S. Cell 24, 353 (1981).
23. Early, P., Rogers, J., Davis, M., Calame, K., Bond, M., Wall, R., and Hood, L. Cell 20, 313 (1980).
24. Vogeli, G., Ohkubo, H., Sobel, M., Yamada, Y., Pastan, I., and de Crombrugghe, B. Proc. Natl. Acad. Sci. USA in press (1981).
25. Palmiter, R. D., Davidson, J. M., Gagnon, J., Rowe, D. W., and Bornstein, P. J. Biol. Chem. 254, 1433 (1979).
26. Merlino, G., Yamamoto, T., Vogeli, G., de Crombrugghe, B., and Pastan, I. Manuscript submitted (1981).

THE ALBUMIN GENE FAMILY

Linda L. Jagodzinski
Thomas D. Sargent[1]
Maria Yang
James Bonner

Division of Biology
California Institute of Technology
Pasadena, California 91125

I. ABSTRACT

Recombinant DNA clones of the mRNAs and genes which encode rat serum albumin and rat alpha-fetoprotein (AFP) have been isolated and analyzed at the nucleic acid sequence level. Investigations of possible homology between regions within the albumin mRNA sequence and between the albumin and AFP genes revealed that these two genes represent a gene family, which appears to have evolved by a series of at least three intragenic duplications and one intergenic duplication event. Certain features of the evolutionary precursor gene and protein, such as the location of introns and the position of cysteine residues have been rigidly conserved. An evolutionary model is advanced that accounts for the observed patterns of homology within and between these two genes.

II. INTRODUCTION

In the past, most protein-encoding genes have been presumed to be "single copy": that is, to occur only once per haploid genome. Some

[1]*Present address: Laboratory of Biochemistry, National Cancer Institute, NIH, Bethesda, Maryland 20014.*

ISBN 0-12-137420-3

and possibly many of these single copy genes are found to be organized into families of sequences related by homology and sometimes by function. Gene families are thought to arise by a process of intergenic duplication and mutational divergence of common ancestral precursors (globin, actin, ovalbumin, immunoglobulins, histones, keratins, etc.). This process allows evolutionary flexibility and results in gene diversity (9). Duplications can apparently also be intragenic, leading to the expansion of a smaller gene into a larger one with internal redundancy. The genes for collagen, ovomucoid, immunoglobulins appear to have evolved from smaller precursors in this fashion. Thus, intragenic duplication also allows gene diversity and flexibility. Furthermore, the presence of introns in eukaryotic genes may increase the occurrence of intragenic duplication events (3). The albumin gene family, which includes both albumin and alpha-fetoprotein, is an example of both kinds of duplication.

III. RESULTS

A. *Rat Serum Albumin*

The evolutionary history of serum albumin, the major plasma protein in the adult mammal, is reflected in its amino acid sequence. Based upon the observed pattern of internal homology, Brown (1) inferred that serum albumin consists of three weakly homologous structural domains. He further hypothesized that albumin evolved by two intragenic duplications of a smaller ancestor, corresponding to a single domain. The internal homology of the protein is also evident in the nucleotide sequence of rat serum albumin mRNA (12). Recombinant DNA clones complementary to the mRNA sequence of rat serum albumin have been sequenced according to the method of Maxam and Gilbert (7). Statistical analysis of the resultant sequence has demonstrated the existence of homolous regions within the mRNA sequence. These homologies enable the partitioning of the mRNA sequence into three fragments which correspond to the three structural domains of the protein.

The three-fold repeated pattern of serum albumin is clearly present in the structure of the serum albumin gene. The exact size and approximate location of the exons were obtained by sequence determination, restriction mapping, and R-loop analysis (11, 13). A map of the rat serum albumin gene is shown in Figure 1. This gene contains 15 exons and 14 introns and spans 14,900 nucleotides of which 12,900 are intronic. The first exon Z is a leader exon encoding the capping site of the mRNA, the pre-pro albumin signal peptide and the first 2 1/3 amino acids of the mature albumin. The last two exons M and N encode the 3' end of the mRNA. Exon M encodes the

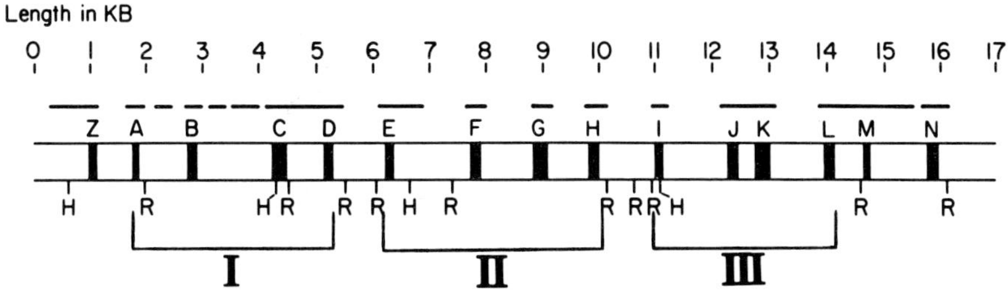

FIGURE 1. Map of the rat serum albumin gene. The black vertical bar denote the exons. The horizontal bars indicate the regions of the cloned gene that have been sequenced to date. H = Hind III, R = Eco RI. The brackets represent the domain regions or subgenes I, II, III.

last 13 carboxyl terminal amino acids and the stop codon (TAA), plus 20 nucleotides of the 3' untranslated region. Exon N is entirely untranslated and probably contains the polyadenylation site of the albumin gene. The remaining 12 exons encode the bulk of the albumin protein. Each of the three structural domains of albumin is encoded by a set of four exons (Fig. 1). These three arrays of exons or subgenes are homologous in nucleotide sequence and with regard to the location of introns within the protein-encoding sequence. Thus, exons ABCD, EFGH, IJKL correspond to the first, second and third structural domains, respectively, of serum albumin (Subgenes I, II, III.)

Each of the 12 internal exons of the albumin gene has been compared to the remaining 11 by use of computer programs (SEQCMP, COMBAT, ROTCMP). The weakest but still statistically significant homology detected is that between the second and fourth exons in each subgene, i.e. between B-D, F-H, and J-L. Each pair is approximately 34% homologous (13). The numerical statistical significance of these and other comparisions is evaluated by calculating an "accidental probability" (Pa) which is the probability that a given level of homology could occur between two sequences purely by accident. The equation is the summation of the Poisson distribution:

$$Pa = \sum_{i=n}^{N} \frac{e^{(-Np)}(Np)^i}{i!}$$

where N is the length of the sequence, n is the number of matches in the comparison, and p is the expected probability of an accidental match between any two nucleotides, or 0.25. The DNA sequence homologies between the second and fourth subgene exons are shown in Table 1. The probabilities that these are accidental matches are 0.014, 0.084, and 0.014, respectively. Individually, these values are not low enough to exclude the possibility of an accidental homology. However, the existence of the observed homology between all three pairs of exons is sufficiently significant to support the conclusion that there was a common ancestor to the second and fourth subgene exons.

TABLE I. Summary of Internal Homology

Comparison	DNA Homology	(%)	Protein homology	(%)
exon B:exon D	47/133	(35)	4/44	(9)
F: H	41/130	(32)	6/43	(14)
J: L	47/133	(35)	7/44	(16)
exon A:exon E	26/58	(45)	3/20	(15)
A: I	22/58	(38)	4/20	(20)
E: I	37/98	(38)	5/33	(15)
exon B:exon F	51/130	(39)	11/43	(26)
B: J	46/133	(35)	6/44	(14)
F: J	55/133	(42)	8/43	(19)
exon C:exon G	93/212	(44)	16/71	(23)
C: K	72/212	(34)	10/71	(14)
G: K	86/215	(40)	11/72	(15)
exon D:exon H	68/133	(51)	18/44	(41)
D: L	51/133	(38)	9/44	(20)
H: L	57/133	(43)	10/44	(23)
subgene 1:subgene 2	238/536	(44)	48/179	(27)
1: 3	191/536	(36)	29/179	(16)
2: 3	235/576	(41)	37/192	(18)
Albumin:AFP	776/1568	(50)	182/523	(34)

The next level of homology is that between the subgenes which encode the three structural domains of the albumin protein. Whereas, there is an overall 20% amino acid homology between the subgenes, the nucleotide sequence homology is 40-50% (Table 1). It is clear that these three subgenes evolved by the triplication of a common ancestor. The sizes of the exons in each subgene has been very rigidly conserved, or in other words, the positions of the introns in the albumin gene are extremely stable. This has been observed in other gene families (globin, etc.; 2, 5, 6, 8, 10). On the other hand, there is no evidence of conservation of the nucleotide sequence or size of albumin introns.

B. *Rat Alpha-Fetoprotein*

Serum albumin has a fetal analog called alpha-fetoprotein. This protein is the dominant plasma protein of the developing fetus. At birth the production of AFP decreases while serum albumin synthesis increases. This inverse relationship in the expression of these two proteins suggests that albumin and AFP may be related. Recombinant DNA clones which encode all but the first 300 nucleotides of rat AFP mRNA were sequenced and the resulting data were compared to the nucleotide sequence of the rat serum albumin mRNA. Computer analysis of these two mRNAs' sequences revealed a single unambiguous alignment of the nucleotide sequences. The highest level of sequence homology is that which exists between the coding regions of the albumin and AFP genes (4). These two mRNAs were found to be highly homologous (50%) over their entire length (Fig. 2). Thus, it can be concluded that AFP and albumin were derived from duplication of a common ancestor gene. Since intron positions are stable in the albumin gene and in many other duplicated genes, presumably the rat AFP gene will be found to have 14 introns located at similar or identical positions as the corresponding introns in the albumin gene. Initial sequencing results of rat AFP genomic clones indicate that this is the case. Eight introns have been found to be located at identical positions in both genes (Fig. 2). These data are further evidence that these two genes evolved from a common ancestral gene.

The amino acid sequences of both proteins were obtained by inference from the nucleotide sequences. The choice of each reading frame was dictated by a lack of premature termination codons and by the assumption that neither mRNA is frameshifted relative to the homologous region in the other mRNA. There is a 34% homology in amino acid sequence between rat AFP and serum albumin. Of particular interest is the conservation of the cysteine residues between these two proteins. All but two are identically positioned in albumin and AFP. The pattern of disulfide crosslinkages defines the secondary folding structure of albumin. Therefore, rat AFP may have a similar secondary protein structure.

C. *Evolution Model*

The sequence homologies within the albumin gene and between albumin and AFP indicate that sequence duplications were responsible for the evolution of this gene family. The recombinational events which led to these duplications are not immediately obvious. There are in fact many potentially correct mechanisms for these evolutionary events. In order to choose one model in preference to all the others, two criteria were applied: 1) translational reading frameshifts must be avoided and 2) the number of discrete recombinational events must be minimized. Any model proposed should account for

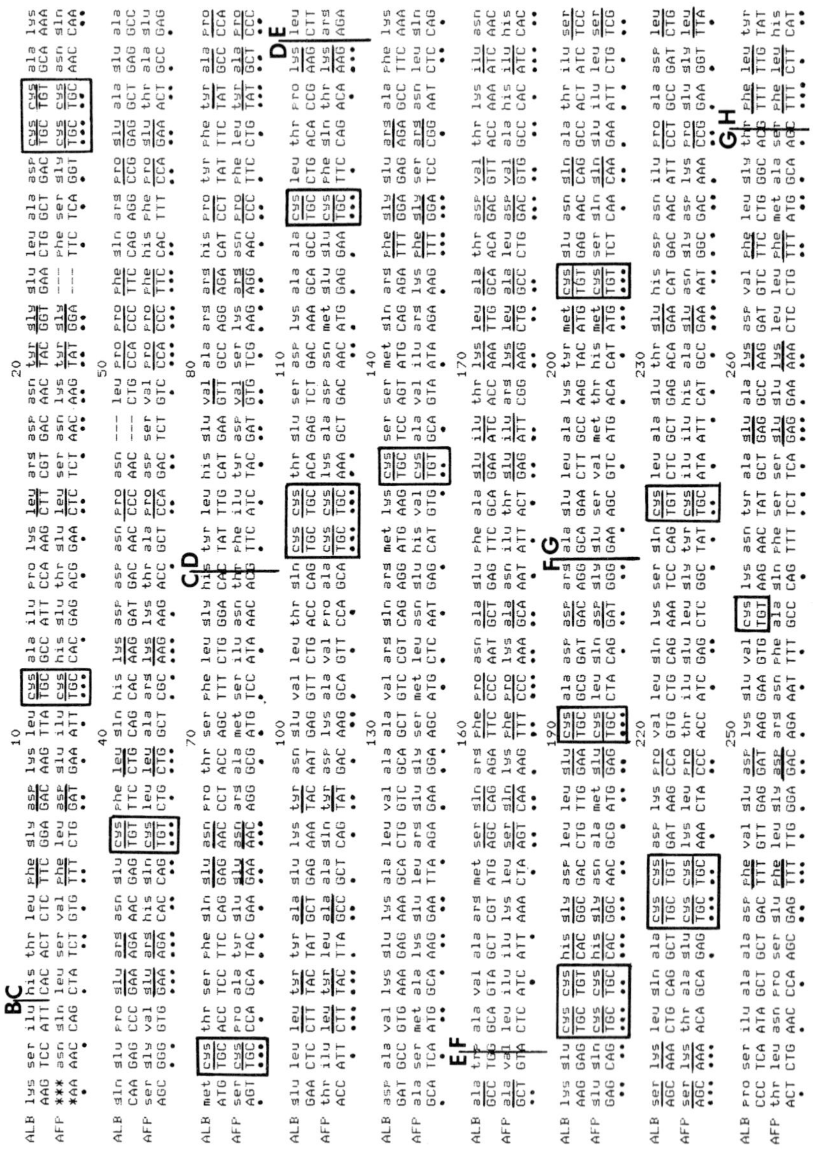

FIGURE 2. Nucleotide sequence comparison. Except for approximately 300 uncloned nucleotides at the 5′ end of the mRNA, the nucleotide sequence of rat AFP mRNA is shown. The inferred amino acid sequence of rat AFP is also indicated, as are the nucleotide and amino acid sequences of rat serum albumin. Matching nucleotides between the two sequences are indicated by the dots. The matching amino acids are underlined. All cysteines are boxed.

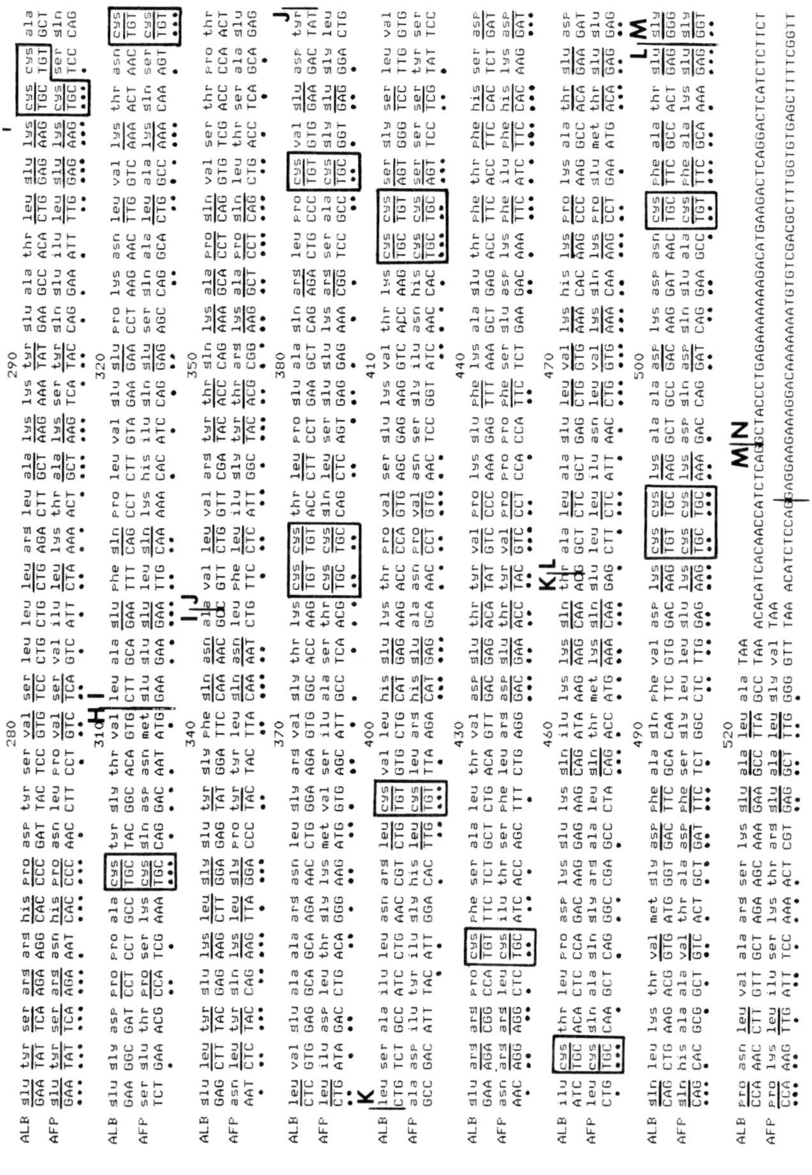

The vertical lines through each sequence represent the location of the introns in the respective genes. The upper case letters designate the exons in the rat serum albumin gene (13). Amino acids are numbered from the 5′ end of the cloned rat AFP mRNA. The dashes in the DNA represent the 3-nucleotide inserts for maximal homology. $) End of rat albumin mRNA.

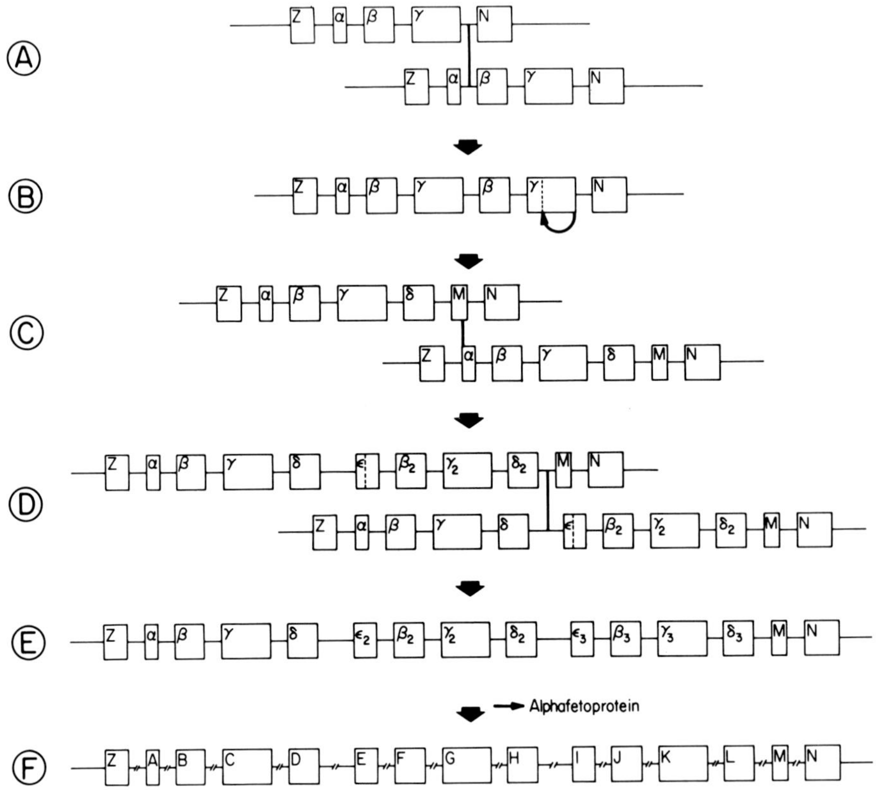

FIGURE 3. Model for the evolution of the albumin gene family. (A) Unequal crossover between two copies of a 5-exon gene duplicates the third and fourth exons. (B) The new second (γ) exon shrinks to 62 nucleotides by a deletion or by evolution of a new splicing signal, creating the "proto-albumin gene". (C) Unequal crossover between two alleles of the "proto-albumin gene". The recombination sites are 40 nucleotides into exon M and the first nucleotide of exon (α). This achieves a simultaneous duplication of most of the protein coding sequences and a frameshift to compensate for the duplication of 3n+2 nucleotides of the first subgene. The second exon (β) also diverges slightly into exon (δ). (D) Second major intragenic duplication, with boundaries within the introns results in three approximately equivalent subgenes encoding three similar protein domains. (E) Divergence of exons to approximately 40% DNA homology and approximately 20% protein homology and extensive divergence of intron sequence and sizes. The alpha-fetoprotein and albumin gene duplication probably occurred during this period. (F) The rat serum albumin gene.

the observed homologies, the greater sizes of exons E and I relative to exon A, and the presence of 3n+2 nucleotides in subgene 1 and 3n nucleotides in subgenes 2 and 3. The series of unequal crossovers diagramed in Figure 3 represent an evolutionary model which satisfies all of these requirements. Of course, it is not possible to prove that any given hypothetical model is the correct one. However, the internal homology within albumin and between albumin and AFP is real and is the result of some sort of duplication series. As depicted in the model the original ancestor gene to the precursor of albumin and AFP consisted of 5 exons which by a series of crossovers expanded and evolved into the complex 15 exon precursor gene. This ancestral gene was duplicated and the resulting gene pair evolved into serum albumin and AFP.

IV. CONCLUSION

Serum albumin consists of three weakly homologous structural domains. This pattern of internal homology is even more evident in the nucleotide sequence of the mRNA and in the structure of the albumin gene, and indicates that the albumin gene evolved from a relatively simple precursor into a complex structural gene by intragenic duplication. Comparative analysis of the rat albumin and AFP mRNAs and gene sequences has also confirmed that these two plasma proteins are related through a common ancestor gene. Thus, the processes of intragenic and intergenic duplication have been responsible for the expansion of a small simple gene of 5 or fewer exons into a complex family of at least two complex highly active genes. This is the most elaborate and detailed evolutionary history yet adduced for an eukaryotic gene family, and it may serve as a general paradigm for the rapid evolution of eukaryotic genomes.

ACKNOWLEDGMENTS

We wish to acknowledge Drs. Jose Sala-Trepat, Bruce Wallace, and Walter Rowekant for their work in the initial phases of this project. Furthermore, we wish to thank Barbara Hough-Evans for her critical comments concerning the preparation of this manuscript. This project was supported in part by Grant No. 5T32 GM 07616, awarded by the National Institute of General Medical Sciences, N.I.H. and Grant No. GM 13762 awarded by the U.S. Public Health Service, N.I.H.

REFERENCES

1. Brown, J. R. (1976). *Fed. Proc. 35,* 2141-2144.
2. Efstratiadis, A., Posakony, J.W., Maniatus, T., Lawn, R.M., O'Connell, C., Spritz, R.A., DeRiel, J.K., Forget, B.G., Weissman, S.M., Slighton, J.L., Blechi, A.E., Smithies, O., Baralle, F.E., Shoulder, C.C., Proudfoot, N.J. (1980). *Cell 21,* 653-668.
3. Gilbert, W. (1978). *Nature 271,* 501.
4. Jagodzinski, L.L., Sargent, T.D., Yang, M., Glackin, C., Bonner, J. (1981). *Proc. Natl. Acad. Sci. USA,* in press.
5. Konkel, D.A., Maizel, J.V., Jr., Leder, P. (1979). *Cell 18,* 865-873.
6. Leder, A., Miller, H.I., Hamer, D.H., Seidman, J.G., Norman, B., Sullivan, M., Leder, P. (1978). *Proc. Natl. Acad. Sci. USA 75,* 6187-6191.
7. Maxam, A.M. and Gilbert, W. (1980). *In* "Methods in Enzymology", eds. L. Grossman and K. Moldave, (Academic Press, New York), Vol. 65: *Nucleic Acids Part I,* pp. 499-560.
8. Nishioka, Y., Leder, P. (1979) *Cell 18,* 875-882.
9. Ohno, S. (1970). *Evolution by Gene Duplication,* Springer-Verlag New York Inc.
10. Perler, F., Efstradiatis, A., Lomedico, P., Gilbert, W., Kolodner, R., Dodgson, J. (1980). *Cell 20,* 555-565.
11. Sargent, T. D., Wu, J.-R., Sala-Trepat, J.M., Wallace, R.B., Reyes, A.A., Bonner, J. (1979). *Proc. Nat. Acad. Sci. USA 76,* 3256-3260.
12. Sargent, T. D., Yang, M., Bonner, J. (1981). *Proc. Natl. Acad. Sci. USA 78,* 246-253.
13. Sargent, T. D., Jagodzinski, L.J., Yang, M., Bonner, J. (1981). *Molecular and Cell. Biol.,* in press.

Expression of β-like Globin Genes During Rabbit Embryogenesis

Ross C. Hardison
Mark L. Rohrbaugh
Joyce Morrison

Department of Microbiology, Cell Biology,
Biochemistry and Biophysics
The Pennsylvania State University
University Park, PA 16802

I. ABSTRACT

The switch from embryonic to adult hemoglobins during rabbit embryogenesis coincides with the shift in site of erythropoiesis from the embryonic yolk sac to the fetal liver. The nucleated, embryonic erythrocytes are replaced by anucleated, fetal erythrocytes from days 13 to 18 of the 30 day gestational period. The levels of precursor RNAs, mature mRNAs and presumptive polypeptides from the embryonic globin genes β3 and β4 decline prior to the onset of fetal liver erythropoiesis, with β3 turning off before β4. These gene products are replaced by the products of adult globin gene β1, whose RNAs and polypeptide increase abruptly at day 15. The coordinate appearance of precursor RNAs, mature mRNA, and polypeptide for each expressed β-like globin gene suggests that the primary level of regulation is either transcription or very rapid RNA turnover. Several RNA species much larger than the expected precursor (cap to poly A, including the intervening sequences) are seen for all three β-like globin genes in either embryonic or fetal erythroid tissue. We do not yet know whether these represent transcription of additional 5' or 3' flanking regions, but they do indicate that the transcription unit for rabbit β-like globins exceeds cap to poly A.

ISBN 0-12-137420-3

II. INTRODUCTION

Like a variety of other organisms, rabbits produce different types of hemoglobin during their embryonic development (Kitchen and Brett, 1974). Six chromatographically and electrophoretically distinct species are seen at day 14 which disappear by day 18 of the 30 day gestational period (Steinheider et al., 1972; Jelkmann and Bauer, 1977). In an effort to understand the molecular basis of these developmental changes in gene expression, the β-like globin gene family of rabbits has been isolated and analyzed (Maniatis et al., 1978; Lacy et al., 1979). As illustrated in Figure 1, the family consists of four related genes arranged in the order 5'-β4-8kb-β3-5kb-β2-7kb-β1-3' (kb=kilobase pairs). All four genes have the same transcriptional orientation and all have similar structures. The mRNA coding sequences are interrupted by two intervening sequences (IVSs) between codons 30 and 31 (IVS I) and 104 and 105 (IVS II) as has been seen for every β-like globin genes so far analyzed. Genes β3 and β4 are expressed in embryonic erythrocytes, gene β2 is an unexpressed pseudogene, and gene β1 is expressed in adult bone marrow (Hardison et al., 1979; Lacy and Maniatis, 1980). Several families of repeated sequences are interspersed within the gene cluster (Shen and Maniatis, 1980), some of which are homologous to prevalent transcripts in the rabbit (Fritsch et al., 1981). Many of the salient features of gene structure and sequence organization in the rabbit β-like globin gene family have been reviewed recently (Hardison et al., 1981).

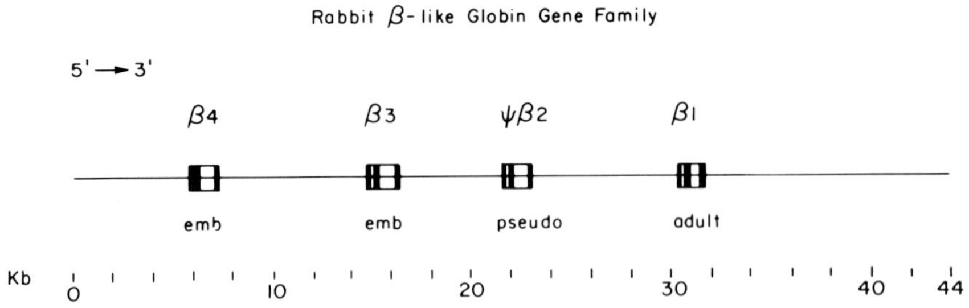

FIGURE 1. The β-like gene family of rabbits. The black boxes are mRNA-coding regions and white boxes are intervening sequences.

We have now analyzed the levels of presumptive precursor RNAs, mature mRNAs and polypeptides for genes β1, β3 and β4 during embryonic and fetal development. The results argue against a major role in regulation being exerted during RNA processing or translation; the simplest interpretation is that of transcriptional control. These analyses also reveal very large RNA species from the three genes, larger than the distance from the cap to the poly-A addition sites.

III. RESULTS

A. Switches in erythroid cell type and pattern of hemoglobin production during development.

Primitive erythrocytes produced in the yolk sac blood islands are large and nucleated. They are replaced about halfway through gestation by anucleated erythrocytes derived from the fetal liver. One can, therefore, monitor the switch in site of erythropoiesis by counting the numbers of nucleated and anucleated erythrocytes circulating in the embryo. Rabbit embryos or fetuses from days 11 (four days after implantation of the blastocyst) through day 22 were dissected free of the placenta and extraembryonic membranes. A small sample of embryonic blood was spread on a microscope slide, stained with benzidine, counterstained with hematoxylin, and examined in the light microscope. Greater than 90% of the benzidine positive (heme-containing) cells were nucleated through day 13 of gestation, after which the fraction of nucleated erythrocytes declined slowly to a level of 10% at day 20 of gestation. Thus, the fetal liver begins erythropoesis at day 14, increasing its contribution to the erythroid cell population through day 20. The bone marrow becomes a major source of erythroid cells around day 20 (Sorenson, 1963), and it continues to provide erythrocytes after birth at day 30.

We also examined the pattern of hemoglobins synthesized during gestation. A set of five embryo-specific globins can be resolved on Triton-acid-urea polyacrylamide gels (Rovera <u>et al</u>., 1978). We do not yet know which are ε^{y}, ε^{z} (the embryonic β-like globins) or χ (embryonic α-like globin), or if some heterogeneity is introduced by post-translational modification. All of the embryo-specific globins migrate slower than adult α- and β-globins, which is also observed for the human $^{G}\gamma$, $^{A}\gamma$, ε and ζ-globins (Rutherford <u>et al</u>., 1981). As shown in Figure 2, synthesis of the rabbit embryonic globins declines from days 13 to 17, whereas adult β-globin production begins at day 15. The adult α-globin is seen as early as we have looked (day 11), and its synthesis reaches a plateau at day 14. We conclude that the developmental switches in rabbit hemoglobin synthesis occur before or during the time that the site of erythropoiesis changes from the yolk sac to the fetal liver. No major globin polypeptide is produced exclusively in the fetal liver, which confirms the observation that rabbits do not have a temporal equivalent of γ-globin (Kitchen and Brett, 1974). Synthesis of the adult α-globin begins considerably earlier than synthesis of adult β-globin.

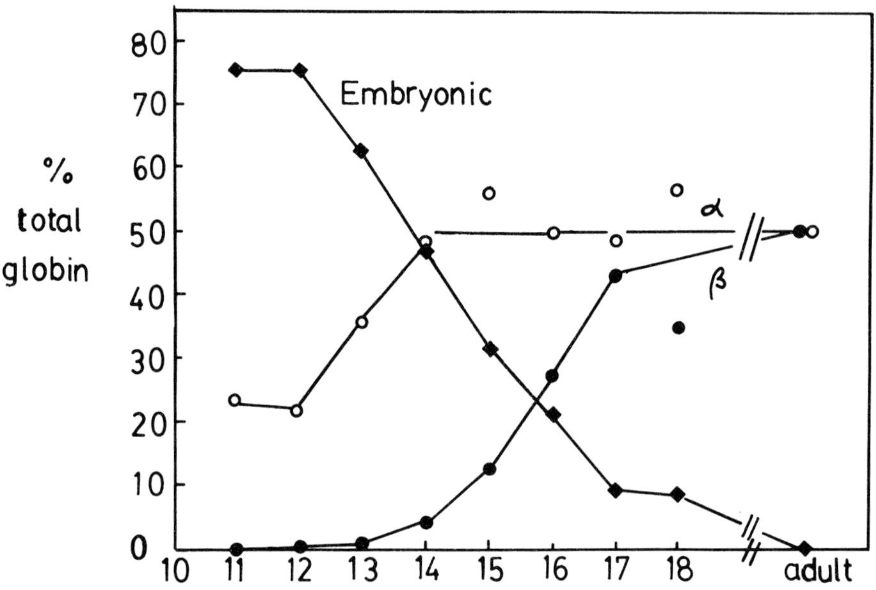

FIGURE 2. Switches in the pattern of globin polypeptide synthesis during development. The amount of each polypeptide was determined by scanning a Triton-acid-urea polyacrylamide gel of globins from the indicated days of gestation (abscissa). The sum of all the embryo-specific globins is plotted on the "embryonic" line.

B. Production of presumptive precursor and mature mRNAs through development.

With this background on the temporal sequence of the switches in erythroid cell type and hemoglobin production, we proceeded to analyze the production of RNAs from the embryonic genes β3 and β4 and the adult gene β1. Previous work

(Hardison et al., 1979) had shown that genes β3 and β4 were transcribed at one stage of embryonic development, but we needed to establish the time of maximal expression. In order to determine the sizes of different transcripts, we used the blot-hybridization assay of Southern (1975) modified for transfer of RNA to nitrocellulose (Goldberg, 1980; Thomas, 1980).

Total RNA from embryos or fetal livers of increasing maturity was separated on a denaturing agarose gel (formaldehyde-agarose, Lehrach et al., 1977), transferred to nitrocellulose, and hybridized with cDNA plasmids for genes β3, β4 (Lacy et al., 1979) or β1 (Maniatis et al., 1976). Figure 3 shows the major band of hybridization in each case is about 700 to 750 nucleotides (nt), the size expected for mature, polyadenylated globin mRNA (Maniatis et al., 1976). The transcripts from gene β3 are maximal at days 11 and 12, and fall off dramatically afterwards. Transcripts from β4 are also maximal at days 11 and 12, but substantial RNA is detected through day 16. Although we do not know which gene is turned on first (both are expressed abundantly as early as we have looked), it is clear that gene β3 turns off before gene β4. In contrast, transcripts from gene β1 appear at day 14, increase dramatically and reach a plateau at day 18. Thus, the switch in mRNA production from β3 and β4 to β1 occurs around days 14-15, at approximately the same time as the switch in polypeptide synthesis or the switch in cell type.

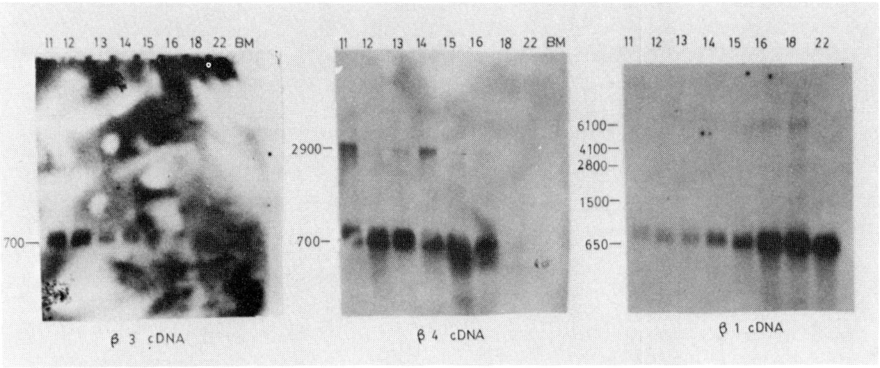

FIGURE 3. Blots of electrophoretically separated RNAs from different days of gestation probed with labeled cDNA plasmids from genes β3, β4 or β1. Sizes are in nucleotides; BM is adult bone marrow.

The cDNA plasmid probes hybridize to RNA species as large as 2900 nt for β4 and 6100 nt for β1 (Figure 3); a 2900 nt RNA is also seen with the β3 cDNA probe on a longer exposure of the autoradiogram. The observation of these large RNAs is not an artifact of nonspecific binding of the probe to the rRNA on the filter, since the same amount of rRNA is in each lane but bands are detected in specific lanes. Also, the observed globin RNAs are not the same size as 18S or 28S rRNA. Large precursor RNAs containing transcripts of the IVSs are expected by analogy with the observations for mouse β-globin (Tilghman et al., 1978; Kinniburgh et al., 1978; Smith and Lingrel, 1978). The expected sizes for precursor RNAs for genes β3 and β4 are about 1800 nt (1540 to 1600 nt from cap to poly A plus about 200 nt poly A) and about 1500 nt for β1 (the large IVS in β1 is about 300 nt smaller than in β3 or β4). In contrast to the globin precursor RNAs reported previously (Curtis and Weissmann, 1976; Ross, 1976), the RNAs in Figure 3 are 1000 to 4600 nt larger than would be found for a transcript of each gene from the cap to the poly A addition site.

In order to determine if the large RNAs contain sequences from the IVSs, as expected for precursor RNAs, we hybridized similar RNA blots with labeled, cloned restriction fragments containing the large IVS from genes β3, β4 or β1. The large IVSs in these genes do not cross-hybridize, and are, therefore, specific probes for transcripts from each gene. The subcloned fragments also contain 74 nt homologous to the mRNA, which accounts for the hybridization to mature message-size RNA (700 nt) seen in Figure 4. This figure shows the expected precursor RNAs of 1800 nt for β3 and β4 and 1500 nt for β1. A larger RNA of 2900-3000 nt is clearly visible with β3 and β4 IVS probes, as well as a faint band of 6500 nt with the β3 probe. The 1300 nt RNAs hybridizing with β3 and β4 are probably processing intermediates. The synthesis in fetal and embryonic erythroid cells of RNAs 3000 to 6000 nt long means that the transcription unit for genes β1, β3 and β4 extends beyond the cap and/or poly A addition sites.

A comparison of Figures 2, 3 and 4 shows that for each gene, the appearance of polypeptide closely follows the appearance of mRNA, which in turn parallels the synthesis of the presumptive precursors. This argues that the primary regulated event in the developmental expression of this gene family is not translation or RNA processing--precursor RNAs do not accumulate in the absence of mRNA, and mRNA does not accumulate in the absence of polypeptide. The simplest interpretation of these data is that the β-like globin genes are under transcription control.

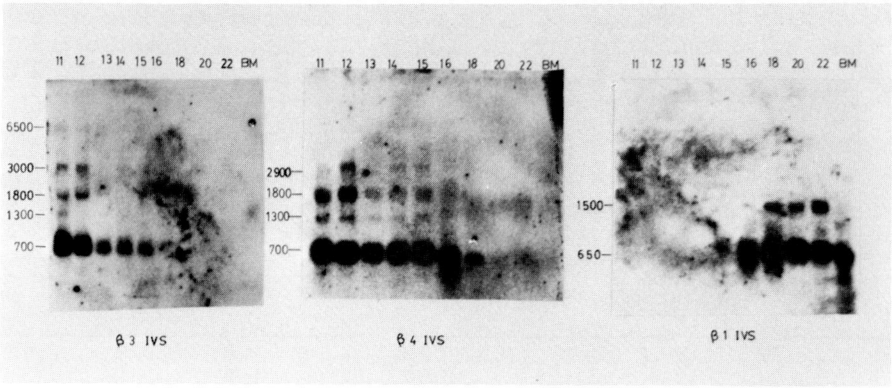

FIGURE 4. Blots of electrophoretically separated RNAs from different days of gestation probed with labeled IVS II from genes β3, β4, or β1.

IV. DISCUSSION

This paper is a preliminary report delineating the temporal pattern of globin gene expression in rabbit embryonic development. We find that production of a set of embryo-specific globins declines half way through gestation, at which time the adult β-globin begins to be produced. This switch occurs before or during the time that the site of erythropoiesis changes from the embryonic yolk sac blood islands to the fetal liver. The adult α-globin gene is actively expressed about four days before the adult β-globin gene turns on. Both precursor and mature mRNAs for genes β3 and β4 are synthesized predominantly in the embryonic erythrocytes, with gene β3 turning off before gene β4. Although this has not been demonstrated yet, we presume that these mature mRNAs are translated into two of the embryo specific globin polypeptides. Similarly, the precursor and messenger RNAs for adult gene β1 are synthesized as the corresponding polypeptide appears in the fetal liver.

The parallel appearance of precursor RNA, messenger RNA, and polypeptide for β3, β4 and β1 during the course of development strongly suggests that regulation is exerted primarily at the transcriptional level. However, these data are limited to the analysis of steady state concentrations of RNAs, and cannot rule out the possibility of very rapid turnover to explain these results. In other words, all three genes could be transcribed at comparable rates at all stages of gestation, but the transcripts are degraded quickly at the times when the polypeptides are not needed. Pulse-chase experiments analyzing the kinetics of globin RNA production in induced mouse Friend cells have favored transcription as the primary level of regulation (Lowenhaupt et al., 1978). These experiments analyze only those events which occur after the time of the shortest pulse, and cannot address possibilities such as processing or degradation of the RNA while it is being transcribed. Although the possibility of very rapid turnover has not been conclusively eliminated, it seems rather contrived to invoke it as a primary regulated event. The simplest interpretation remains that of transcriptional control.

RNA species as large as 3000 or even 6000 nt are transcribed from the rabbit β-like globin genes. These sizes are much larger than the distance from the cap to the poly A addition sites, and we conclude that the transcription units for these genes extend considerably past one or both nucleotides encoding the ends of the mRNA. Bastov and Aviv (1977) reported a β-globin precursor RNA of about 4500 nt in mouse Friend cells, but Haynes et al. (1978) were unable to reproduce this result using purified, cloned cDNA probes. More recently, Reynaud et al. (1980) have observed very large globin RNAs, in the range of 12,000 to 15,000 nt in avian erythroblasts. By analyzing nascent RNA chains synthesized in Friend cell nuclei, Hofer and Darnell (1981) have shown that transcription of the β-globin gene extends at least 1000 nt past the poly-A addition site. Some transcription is also observed 5' to the cap site, but it is not equimolar with the remainder of the transcribed sequences. These data indicate that transcription of the mouse β-globin gene proceeds past the poly-A addition site and that the primary transcript is cleaved at the appropriate position to allow polymerization of adenylic acid on the 3' end.

We have not yet determined the 5' or 3' ends of the large transcripts from the rabbit β-like globin genes. The bulk of the transcription could simply result from read through past the poly A addition site. It is not clear whether the production of these large RNAs is obligatory for formation of mature mRNAs; they could represent occasional, perhaps non-productive, excessive transcription. However, these long transcripts are moderately abundant (easily seen in total RNA), and they may represent normal, early products in the pathway of expression. This would be consistent with the nascent chain data of Hofer and Darnell (1981), and indicates that the globin gene transcription unit exceeds cap to poly A. It is worth noting that we have seen these very large RNAs only in embryonic or fetal erythroid cells; they are not seen in the adult bone marrow by an RNA blot-hybridization analysis. One possible explanation for this result is that the processing of RNA could be noticably slower in prenatal as opposed to adult erythroid tissues.

ACKNOWLEDGMENTS

This work was supported by grant 1 R01 AM 27635-01 from the National Institutes of Health.

REFERENCES

Bastos, R. N. and Aviv, H. (1977) Cell 11, 641-650.
Curtis, P. J. and Weissman, C. (1976) J. Mol. Biol. 106, 1061-1075.
Fritsch, E., Shen, C.-K. J., Lawn, R. and Maniatis, T. (in press) Cold Spring Harbor Symposium on Quantitative Biology, vol. 45.
Goldberg, D. (1980) Proc. Natl. Acad. Sci. U.S.A. 77, 5794-5798.
Hardison, R., Butler, E., Lacy, E., Maniatis, T., Rosenthal, N. and Efstratiadis, A. (1979) Cell 18, 1285-1297.
Hardison, R., Lacy, E., Shen, C.-K. J., Butler, E. and Maniatis, T. (1981) in Organization and Expression of Globin Genes (G. Stamatoyannopoulos and A. Nienhuis, eds.), pp. 89-99. Alan R. Liss, Inc., New York.
Haynes, J. R., Kalb, F. V., Rosteck, P. and Lingrel, J. B. (1978) FEBS Letters 91, 173-177.
Hofer, E. and Darnell, J., (1981) Cell 23, 585-593.
Jelkmann, W. and Bauer, C. (1977) Pflügers Arch. 372, 149-156.
Kinniburgh, A., Mertz, J. and Ross, J. (1978) Cell 14, 681-693.
Kitchen, H. and Brett, I. (1974) Ann. NY Acad. Sci. 241, 653-670.
Lacy, E. and Maniatis, T. (1980) Cell 21, 545-553.
Lacy, E., Hardison, R., Quon, D. and Maniatis, T. (1979) Cell 18, 1273-1283.
Lehrach, H., Diamond, D., Wozney, J. and Boedtker, H. (1977) Biochemistry 16, 4743-4751.
Lowenhaupt, K., Trent, C. and Lingrel, J. (1978) Developmental Biology 63, 441-454.
Maniatis, T., Kee, S. G., Efstratiadis, A. and Kafatos, F. (1976) Cell 8, 163-182.
Maniatis, T., Hardison, R., Lacy, E., Lauer, J., O'Connell, C., Quon, D., Sim, G. K. and Efstratiadis, A. (1978) Cell 15, 687-701.
Reynaud, C., Imaizumi-Scherrer, M. and Scherrer, K. (1980) J. Mol. Biol. 140, 481-504.
Ross, J. (1976) J. Mol. Biol. 106, 403-420.

Rovera, G., Magarian, C. and Borun, T. (1978) Anal. Bioch. 85, 506-518.
Rutherford, T., Clegg, J. B., Higgs, D. R., Jones, R. W., Thompson, J. and Weatherall, D. J. (1981) Proc. Natl. Acad. Sci. U.S.A. 78, 348-352.
Shen, C.-K. J. and Maniatis, T. (1980) Cell 19, 379-391.
Sorenson, G. (1963) Ann. NY Acad. Sci. 111, 44-69.
Southern, E. M. (1975) J. Mol. Biol. 98, 503-517.
Smith, K. and Lingrel, J. (1978) Nucleic Acids Research 5, 3295-3301.
Steinheider, G., Melderis, H. and Ostertag, W. (1972) in International Symposium on the Synthesis, Structure and Function of Hemoglobin (H. Martin and L. Novicke, eds.). pp. 222-235. Lehmans, Munich, Germany.
Thomas, P. (1980) Proc. Natl. Acad. Sci. U.S.A. 77, 5201-5205.
Tilghman, S., Curtis, P., Tiemeier, D., Leder, P. and Weissmann, C. (1978) Proc. Natl. Acad. Sci. U.S.A. 75, 1309-1313.

INTERVENING SEQUENCE MUTATION IN A CLONED HUMAN β^+-THALASSEMIC GLOBIN GENE[1]

Richard A. Spritz[2], Pudur Jagedeeswaran, P. Andrew Biro, James T. Elder, Jon K. deRiel[3], Bernard G. Forget, and Sherman M. Weissman

Departments of Human Genetics and Internal Medicine
Yale University
New Haven, Connecticut

James L. Manley[4], and Malcolm L. Gefter

Department of Biology
Massachusetts Institute of Technology
Cambridge, Massachusetts

ABSTRACT

β globin gene fragments from a patient with homozygous β^+-thalassemia have been cloned and subjected to detailed structural and functional analyses. Restriction endonuclease mapping of the cloned gene framents revealed no deletions or other rearrangements and transcription of the thalassemic gene appeared normal *in vitro*. However, nucleotide sequence analysis of the β^+-thalassemic gene demonstrated a single base change in the body of the small intervening sequence. This nucleotide substitution creates a sequence which is homologous to that of the 3'-splice site of the small intervening se-

[1]Supported by grants from the National Institutes of Health. R.A.S. was supported by fellowships from the Charles E. Culpeper Foundation and the Cooley's Anemia Foundation.
[2]Present address: University of Wisconsin, Madison, WI
[3]Present address: Temple University, Philadelphia, PA
[4]Present address: Columbia University, New York, NY

ISBN 0-12-137420-3

quence. This suggests that either this mutation results in defective post-transcriptional processing of thalassemic β globin mRNA precursors, or that the β^+-thalassemia defect is distant from the β globin structural gene.

INTRODUCTION

The β-thalassemias are a heterogeneous group of genetic anemias characterized by deficient synthesis of β globin chains in the erythroid cells of homozygous individuals. In β^+-thalassemia, the most common type of β-thalassemia, there is a *quantitative* deficiency of *qualitatively* normal β globin chains. Erythroid cells of affected individuals synthesize only 5 to 30% of normal levels of β globin chains (1,2); however, those β chains which are produced appear normal both by peptide mapping and carboxymethylcellulose column chromatography criteria (rev. in 3). Moreover, there is a corresponding deficiency of β globin mRNA in β^+-thalassemic erythroid cells (4-6). This mRNA also appears normal both by cDNA-RNA hybridization criteria and by its ability to direct translation of authentic β globin chains *in vitro* (2). Although the β^+-thalassemias probably result from a wide variety of molecular defects, these features suggest abnormalities of β globin gene transcription or of processing, transport, or stability of β globin mRNA. Indeed, abnormal processing and/or stability of nuclear β globin mRNA precursor molecules has been observed in some cases (7-10).

Family studies have indicated that most β-thalassemia mutations are allelic with, or tightly linked to, the β globin structural locus (1,2). However, restriction endonuclease mapping studies of a number of β^+-thalassemic genes have thus far revealed no abnormalities (11,12), suggesting that detailed nucleotide sequence analysis of β^+-thalassemic genes may be necessary to determine the molecular basis of this group of disorders. Accordingly, we have cloned a β globin gene from a patient with β^+-thalassemia, and have determined the complete nucleotide sequence of this thalassemic gene. Comparison with the DNA sequence of a normal human β globin gene (13) reveals only a single divergent nucleotide in the internal portion of the small intervening sequence. This sequence difference suggests a possible mechanism for abnormal splicing of the β^+-thalassemic nuclear β globin mRNA precursor.

METHODS

Total DNA was prepared from the spleen of a Greek Cypriot patient with typical transfusion dependent β^+-thalassemia. The DNA was digested to completion with *Eco* RI, which cleaves the β globin gene at codon 122, and the fragments were electrophoresed in agarose. The 5.2 kb 5'- and 3.2 kb 3'-β globin gene fragments were identified by standard techniques, eluted from the gel, and cloned in the λgtWES vector system (14). To facilitate subsequent analyses, the 5'-β globin clone, C6, and the 3'-clone, D47, were subcloned in the plasmid pBR322 (15) and the filamentous bacteriophage M13mp2 (16). DNA sequencing was performed by modifications of the methods of Maxam and Gilbert (17), Maat and Smith (18), and Sanger (19). Transcription of the normal (20) and β^+-thalassemic β globin genes by RNA polymerase II was assayed *in vitro* by the method of Manley (21).

RESULTS

Restriction Endonuclease Mapping of the Cloned β^+-Thalassemic β Globin Gene Fragments

Restriction endonuclease mapping of the cloned 5.2 kb 5'- (C6) and 3.6 kb 3'- (D47) β^+-thalassemic gene fragments was performed with a number of enzymes which cleave the β globin locus infrequently. As shown in Fig. 1, the map thus generated agrees completely with that of the normal human β globin locus (20,22-27). The thalassemic and normal 5'-β globin *Eco* RI fragments were also cleaved with *Bgl* II plus *Hin* fI, *Alu* I, or *Hae* III. In each case, the digestion patterns of the normal and thalassemic DNAs were identical (data not shown), indicating that there have been no detectable deletions, insertions, or other rearrangements of the thalassemic β globin locus.

Nucleotide Sequence of the β^+-Thalassemic Globin Gene

The 1904 nucleotide sequence determined (Fig. 2) extends from 216 bases 5'- to the cap site of β globin mRNA to 82 bases 3'- to the polyadenylation site. The human β globin gene is similar in structure to other known mammalian β-like globin genes (28). The protein encoding sequence is interrupted by two intervening sequences. The small intervening

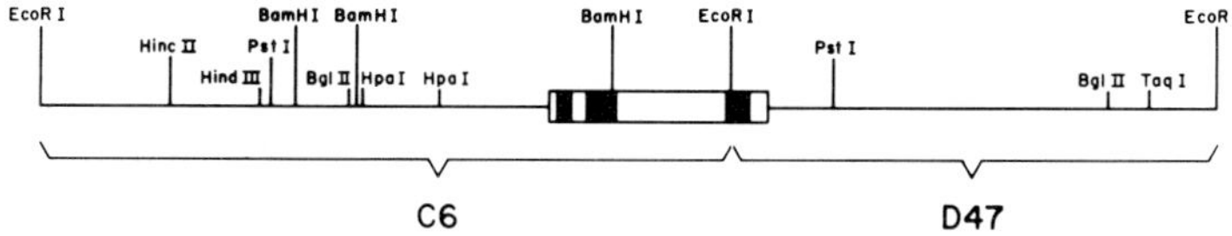

Figure 1. Partial restriction endonuclease cleavage map of cloned β^+-thalassemic β globin gene fragments, C6 and D47. The protein encoding regions are indicated by solid blocks.

```
gctatggggcctagagatatatcttagagggagggctgagggtttgaagtccaactcctaagccagtgccagaagagccaaggacaggtacggctgtcat
             *
cacttagacctcaccctgtggagccacaccctagggttggccaatctactcccaggagcagggagggcaggagccagggctgggcataaaagtcagggca

gagccatctattgcttACATTTGCTTCTGACACAACTGTGTTCACTAGCAACCTCAAACAGACACCATGGTGCACCTGACTCCTGAGGAGAAGTCTGCCG
                                                                  INIValHisLeuThrProGluGluLysSerAlaV

TTACTGCCCTGTGGGGCAAGGTGAACGTGGATGAAGTTGGTGGTGAGGCCCTGGGCAGGttggtatcaaggttacaagacaggtttaaggagaccaatag
alThrAlaLeuTrpGlyLysValAsnValAspGluValGlyGlyGluAlaLeuGlyArg
                                                                   g
aaactgggcatgtggagacagagaagactcttgggtttctgataggcactgactctctctgcctattagtctattttcccacccttaggCTGCTGGTGGT
                                                                                         LeuLeuValVa

CTACCCTTGGACCCAGAGGTTCTTTGAGTCCTTTGGGGATCTGTCCACTCCTGATGCTGTTATGGGCAACCCTAAGGTGAAGGCTCATGGCAAGAAAGTG
lTyrProTrpThrGlnArgPhePheGluSerPheGlyAspLeuSerThrProAspAlaValMetGlyAsnProLysValLysAlaHisGlyLysLysVal

CTCGGTGCCTTTAGTGATGGCCTGGCTCACCTGGACAACCTCAAGGGCACCTTTGCCACACTGAGTGAGCTGCACTGTGACAAGCTGCACGTGGATCCTG
LeuGlyAlaPheSerAspGlyLeuAlaHisLeuAspAsnLeuLysGlyThrPheAlaThrLeuSerGluLeuHisCysAspLysLeuHisValAspProG

AGAACTTCAGGgtgagtctatgggacccttgatgttttctttccccttcttttctatggttaagttcatgtcataggaaggggagaagtaacagggtaca
luAsnPheArg

gtttagaatgggaaacagacgaatgattgcatcagtgtggaagtctcaggatcgttttagtttcttttatttgctgttcataacaattgttttcttttgt

ttaattcttgctttctttttttttcttctccgcaatttttactattatacttaatgccttaacattgtgtataacaaaaggaaatatctctgagatacat

taagtaacttaaaaaaaaactttacacagtctgcctagtacattactatttggaatatatgtgtgcttatttgcatattcataatctccctactttattt

tcttttattttaattgatacataatcattatacatatttatgggttaaagtgtaatgttttaatatgtgtacacatattgaccaaatcagggtaatttt

gcatttgtaattttaaaaaatgctttcttcttttaatatactttttttgtttatcttatttctaatactttccctaatctctttctttcagggcaataatg

atacaatgtatcatgcctctttgcaccattctaaagaataacagtgataatttctgggttaaggcaatagcaatatttctgcatataaatatttctgcat

ataaattgtaactgatgtaagaggtttcatattgctaatagcagctacaatccagctaccattctgcttttattttatggttgggataaggctggattat

tctgagtccaagctaggcccttttgctaatcatgttcatacctcttatcttcctcccacagCTCCTGGGCAACGTGCTGGTCTGTGTGCTGGCCCATCAC
                                                             LeuLeuGlyAsnValLeuValCysValLeuAlaHisHis

TTTGGCAAAGAATTCACCCCACCAGTGCAGGCTGCCTATCAGAAAGTGGTGGCTGGTGTGGCTAATGCCCTGGCCCACAAGTATCACTAAGCTCGCTTTC
PheGlyLysGluPheThrProProValGlnAlaAlaTyrGlnLysValValAlaGlyValAlaAsnAlaLeuAlaHisLysTyrHisTER

TTGCTGTCCAATTTCTATTAAAGGTTCCTTTGTTCCCTAAGTCCAACTACTAAACTGGGGGATATTATGAAGGGCCTTGAGCATCTGGATTCTGCCTAAT

AAAAAACATTTATTTTCATTGCaatgatgtatttaaattatttctgaatattttactaaaagggaatgtgggaggtcagtgcatttaaaacataaagaa

atga
```

Figure 2. Nucleotide sequence of cloned β^+-thalassemic β globin gene. The nucleotide sequence of the "sense" strand is displayed in the 5'→3' orientation. Uppercase letters represent those sequences represented in β globin mRNA; lowercase letters represent the flanking and intervening sequences. The single divergent nucleotide in the normal human β globin gene is displayed above the homologous base in the thalassemic β globin DNA sequence. The 5'-extremity of the known normal β globin DNA sequence (13) is indicated by an asterisk.

sequence occurs between codons 30 and 31, and is 130 nucleotides long. The large intervening sequence is 850 nucleotides in length, and lies between codons 104 and 105.

Comparison of the nucleotide sequence of the β^+-thalassemic gene with that of a normal human β globin gene (13) revealed only a single divergent base. There were no sequence differences in the 5-' or 3'-flanking regions, in the 5'- or 3'—untranslated sequences, or in the large intervening sequence. Moreover, all four of the coding-intervening sequence junctions are normal in the β^+-thalassemic gene. However, in the internal region of the small intervening sequence, 22 bases before the 3'-junction with the adjacent protein encoding region, an adenine replaces a guanine. This substitution creates a sequence that is homologous to that at the authentic 3'-splice junction of the small intervening sequence at six of seven nucleotides (Fig. 3). This sequence might constitute

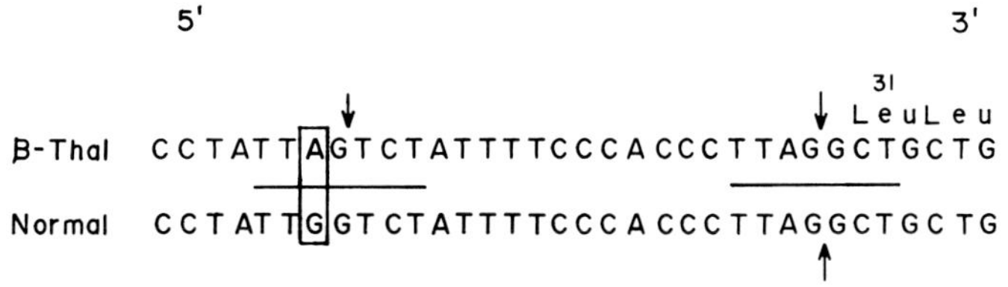

Figure 3. Nucleotide sequence near the 3'-junction of the small intervening sequence and adjacent protein encoding sequence. The partially homologous sequences are indicated by the solid lines, and the single nucleotide change in the β+-thalassemic DNA sequence is indicated by the box. The authentic and putative internal acceptor splice sites are indicated by arrows.

an anomalous acceptor splice signal within the small intervening sequence, resulting in abnormal splicing of some of the thalassemic β globin mRNA precursor molecules at this site.

Transcription of the Thalassemic β Globin Gene In Vitro

The transcriptional capacity of the cloned 5'-*Eco* RI β+-thalassemic fragment was assayed *in vitro* using a cell-free system with RNA polymerase II activity (21). Prior to transcription, the thalassemic and normal β globin templates were cleaved with *Eco* RI. As shown in Fig. 4, both templates

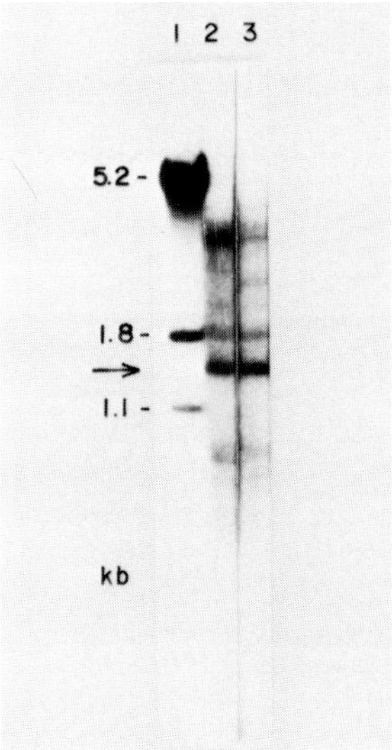

Figure 4. In vitro transcription of normal and β+-thalassemic 5'-Eco RI β globin gene fragments. Lanes: 1, DNA size markers (a mixture of linear SV40 molecules plus 1.1 kb and 1.8 kb Hin d III subfragments of SV40); 2, normal; 3, β+-thalassemic. Arrow indicates β globin-specific transcripts. The labeled 1.8 kb fragment in lanes 2 and 3 consists of end-labeled 18S RNA, and was present in control reactions lacking DNA templates.

were transcribed faithfully in this sytem, and the β^+-thalassemic transcript is quantitatively and qualitatively identical to that of the normal β globin gene. In addition to the major β globin gene transcript that runs between the 1.1 and 1.8 kb size markers, a number of minor transcripts can also be seen. These presumably result from end-to-end transcription of the inserted DNA or of plasmid DNA sequences. Some differences in the size of these large minor transcripts presumably occur due to the different sizes of the respective inserted β globin gene fragments. These results suggest that transcription of this thalassemic gene fragment is normal, at least in this *in vitro* system. Similar results have recently been obtained in other laboratories using other cloned β^+-thalassemic genes (29,30).

DISCUSSION

We have cloned β globin gene fragments from a Greek Cypriot patient with homozyous β^+-thalassemia, whose erythroid cells synthesize reduced amounts of β globin chains. Because the 5'- and 3'-portions of the thalassemic β globin gene were isolated as separate *Eco* RI restriction fragments, it is impossible to be certain that they derive from the same thalassemic allele. However, β^+-thalassemia in the Cypriot population is extremely homogeneous clinically and biochemically (31), suggesting that the patient is probably homozygous for identical β^+-thalassemic alleles.

Detailed restriction endonuclease mapping of the cloned thalassemic β globin gene fragments revealed no detectable deletions, insertions, or other rearrangements.

Nucleotide sequence analysis of the β^+-thalassemic gene revealed only a single nucleotide difference from a normal β globin gene DNA sequence (13). This G to A transition 22 bases upstream from the 3'-junction between the small intervening and adjacent coding sequences creates a sequence which is homologous to the authentic acceptor splice site of the small intervening sequence at six of seven nucleotides. Furthermore, the single nonhomologous nucleotide may not be rigidly specified at the normal 3'-splice junction, as the sequences of the human γ and β globin genes differ from those of the δ and ε genes at this base residue (28). Several authors have noted the existence of considerable sequence complementarity between the regions surrounding RNA splice junctions and the 5'-terminus of the small nuclear RNA, U-1 (32,33), and have suggested that this RNA species may align donor and acceptor splice sites during the RNA splicing process *in vivo*. As shown in Fig. 5, sequence complementarity between U-1 snRNA and the putative abnormal acceptor splice site in the thalas-

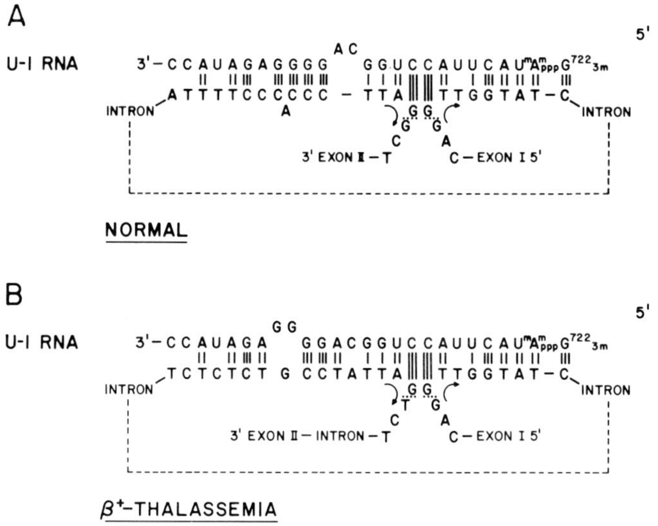

Figure 5. Possible complementary structures formed between normal and thalassemic β globin nuclear mRNA precursors and U-1 snRNA. The nucleotide sequence of U-1 snRNA is that determined by Branlant (34). A, normal β globin mRNA precursor; B, β^+-thalassemic β globin mRNA precursor. Potential base pairing has been maximized by permitting looping out of some unbonded bases.

semic gene is almost as extensive as complementarity of U-1 snRNA with the authentic 3'-splice junction. If this anomalous sequence in the β^+-thalassemic mRNA precursor constitutes a functional acceptor splice sequence *in vivo*, splicing at this site might compete or interfere with RNA splicing at the authentic acceptor site, particularly if recognition of potential splice sites on the precursor RNA molecule occurs by a 5'→3' processive process. A small amount of normally spliced β globin mRNA might thus be accounted for by "leakiness" of the β^+-thalassemic mutation. Any RNAs spliced at the putative abnormal splice site would encode normal β globin amino acid sequences up to codon 29 or 30, depending on the precise site of splicing. These would be followed by a nonsense peptide terminated by an in-phase UAG codon derived from the 3'-border of the small intervening sequence. However, abnormally spliced globin mRNA precursor molecules may well be unstable, or might not be efficiently transported to the cytoplasm, as no such abnormal peptide is detected by carboxymethylcellulose column chromatography of proteins synthesized by reticulocytes of β^+-thalassemic individuals (3). In β^0-thalassemia associated with a nonsense mutation at codon 17 of the β mRNA (34), the predicted abnormal short polypeptide has not been detected and the abnormal β mRNA is presumably unstable since it is present in unexpectedly reduced amounts (34). A mechanism similar to that proposed here may account for atypical splicing of some nuclear mRNA precursor molecules of human growth hormone (35) and chicken ovomucoid (36).

Transcription of the cloned β^+-thalassemic globin gene *in vitro* by RNA polymerase II appears quantitatively and qualitatively normal, suggesting that the single base change in the small intervening sequence does not alter an internal promoter sequence, similar to that defined by Brown and colleagues as directing transcription of the *Xenopus* 5*S* RNA gene *in vitro* by RNA polymerase III (37,38).

In addition to creating a putative anomalous RNA splice site, the base change that we have identified in the β^+-thalassemic gene abolishes an *Eco* PI restriction endonuclease cleavage site (recognition sequence AGACC [39]) which is present in the normal human β globin gene. However, because cleavage of DNA by *Eco* PI is never complete (39), it is unlikely that analysis of amniocyte DNA with this restriction endonuclease will prove useful for the prenatal diagnosis of this β^+-thalassemic allele.

These results suggest that the basic molecular defect in this patient may be an abnormality of RNA processing, due to the mutation in the small intervening sequence of the β^+-thalassemic gene. Alternatively, the molecular defect may be distant from the β globin gene itself. These hypotheses are currently being tested using *in vitro* systems that have the ability to process transcribed mRNA precursor molecules into mature mRNA.

ACKNOWLEDGMENTS

We thank Dr. Tom Maniatis and colleagues for providing subclones of the normal human β globin gene and the normal β globin gene sequence prior to its publication, and Dr. Depak Bastia for supplying T7 exonuclease. We also thank Elaine Coupal, Regina Ezekiel, and Vincent Vellucci for expert technical assistance.

REFERENCES

1. Weatherall, D.J., and Clegg, J.B. (1972). "The Thalassemia Syndromes." Blackwell Scientific Publications, Oxford.
2. Bunn, H. F., Forget, B. G. and Ranney, H. M. (1977). Human Hemoglobins." W. B. Saunders, Philadelphia.

3. Benz, E. J., Jr., and Forget, B. G. (1975). *In* "Progress in Hematology" Vol. 9 (E. Brown, ed.), p. 107. Grune and Stratton, New York.
4. Housman, D., Forget, B. G., Skoultchi, A., and Benz, E.J., Jr. (1973). *Proc. Natl. Acad. Sci. (USA)* 70, 1809.
5. Kacian, D. L., Gambino, R., Dow, L. W., Grossbard, E., Natta, C., Ramirez, F., Spiegelman, S., Marks, P. A., and Bank, A. (1973). *Proc. Natl. Acad. Sci. (USA)* 70, 1886.
6. Benz, E. J., Jr., Forget, B. G., Hillman, D. G., Cohen-Solal, M., Pritchard, J., Cavallesco, C., Prensky, W., and Housman, D. (1978). *Cell* 14, 299.
7. Nienhuis, A. W., Turner, P., and Benz, E. J., Jr. (1977). *Proc. Natl. Acad. Sci. (USA)* 74, 3960.
8. Benz, E. J., Jr., Scarpa, A. L., deRiel, J. K., and Forget, B. G. (1980). *Clin. Res.* 28, 545 A.
9. Maquat, L. E., Kinniburgh, A. J., Beach, L. R., Honig, G. R., Lazerson, J., Ershler, W. B., and Ross, J. (1980). *Proc. Natl. Acad. Sci. (USA)* 77, 4287.
10. Kantor, J. A., Turner, P. H., and Nienhuis, A. W. (1980). *Cell* 21, 149.
11. Flavell, R. A., Bernards, R., Kooter, J. M., deBoer, E., Little, P. F. R., Annison, G., and Williamson, R., (1979). *Nucleic Acids Res.* 6, 2749.
12. Orkin, S. H., Old, J. M., Weatherall, D. J., and Nathan, D. G. (1979). *Proc. Natl. Acad. Sci. (USA)* 76, 2400.
13. Lawn, R. M., Efstratiads, A., O'Connell, C., and Maniatis, T. (1980). *Cell* 21, 647.
14. Tiemeier, D., Enquist, L. W., and Leder, P. (1976). *Nature (London)* 263, 526.
15. Bolivar, F., Rodriguez, R. L., Greene, P. J., Betlach, M. C., Heynecker, H. L., Boyer, H. W., Crosa, J. H., and Falkow, S. (1977). *Gene* 2, 95.
16. Schreier, P. H., and Cortese, R. (1979). *J. Mol. Biol.* 129, 169.
17. Maxam, A. M., and Gilbert, W. (1977). *Proc. Natl. Acad. Sci. (USA).* 74, 560.
18. Maat, J., and Smith, A. J. H. (1978). *Nucleic Acids Res.* 5, 4537.
19. Sanger, F., Nicklen, S., and Coulson, A. R. (1977). *Proc. Natl. Acad. Sci. (USA)* 74, 5463.
20. Lawn, R. M., Fritsch, E. F., Parker, R. C., Blake, G., and Maniatis, T. (1978). *Cell* 15, 1157.
21. Manley, J. L., Fire, A., Cano, A., Sharp, P. A., and Gefter, M. L. (1980). *Proc. Natl. Acad. Sci. (USA)* 77, 3855.
22. Falvell, R. A., Kooter, J. M., de Boer, E., Little, P. F. R. and Williamson, R. (1979). *Cell* 15, 25.
23. Mears, J. G., Ramirez, R., Leibowitz, D., and Bank, A. (1978). *Cell* 15, 15.

24. Fritsch, E. F., Lawn, R. M., and Maniatis, T. (1979). *Nature (London)* 279, 598.
25. Tuan, D., Biro, P. A., deRiel, J. K., Lazarus, H., and Forget, B. G. (1979). *Nucleic Acids Res.* 6, 2519.
26. Fritsch, E. E., Lawn, R. M., and Maniatis, T. (1980). *Cell* 19, 959.
27. Kaufman, R. E., Kretschmer, P. J., Adams, J. W. Coon, H. C., Anderson, W. F., and Nienhuis, A. W. (1980). *Proc. Natl. Acad. Sci. (USA)* 77, 4229.
28. Efstratiadis, A., Posakony, J. W., Maniatis, T., Lawn, R. M., O'Connell, C., Spritz, R. A., deRiel, J. K., Forget, B. G., Weissman, S. M., Slightom, J. L., Blechl, A. E., Smithies, O., Baralle, F. E., Shoulders, C. C., and Proudfoot, N. J. (1980). *Cell* 21, 653.
29. Parker, C. S., Goossens, M., and Kan, Y. W. (1980). *Clin. Res.* 28, 488A.
30. Proudfoot, N. J. Shander, M. H. M., Manley, J. L., Gefter M. L., and Maniatis, T. (1980). *Science* 209. 1329.
31. Modell, C. B. (1980). *In* "Prenatal Approaches to the Diagnosis of Fetal Hemoglobinopathies" (Y. W. Kan and C. D. Reid, eds.), p. 85. DHEW, Washington, D. C.
32. Lerner, M., Boyle, J., Mount, S., Wolin, S., and Steitz, J. A. (1980). *Nature (London)* 283, 220.
33. Rogers, J., and Wall, R. (1980). *Proc. Natl. Acad. Sci. (USA)* 77, 1877.
34. Branlant, C., Krol, A., Ebel, J. P. Lazar, E., Gallinaro, H., Jacob, M., Sri-Widada, J. and Jeanteur, P. (1980). *Nucleic Acids Res.* 8, 4143.
35. Moore, D. DeNoto F., Hallewell, R., Fiddes, J., and Goodman, H. (1981). *J. Supramolec. Struct. and Cell. Biochem.*, Suppl. 5, 402.
36. Stein, J. P., Catterall, J. F., Kristo, P., Means, A. R., and O'Malley, B. (1980). *Cell* 21, 681.
37. Sakonju, S., Bogenhagen, D. F., and Brown, D. D. (1980). *Cell* 19, 13.
38. Bogenhagen, D. F., Sakonju, S., and Brown, D. D. (1980). *Cell* 19, 27.
39. Bachi, B., Reiser, J., and Pirrotta, V. (1979). *J. Mol. Biol.* 128, 143.

STRUCTURES IN INTERGENIC DNA OF NON-ALPHA GLOBIN GENES OF MAN

P. Jagadeeswaran[1]
J. Pan[1]
R. A. Spritz[1]
C. A. Duncan[1]
P. A. Biro[1]
D. Tuan[2]
B. G. Forget[2]
S. M. Weissman[1]

[1]Department of Human Genetics
Yale University School of Medicine
New Haven, CT 06510

[2]Department of Medicine
Yale University School of Medicine
New Haven, CT 06510

ABSTRACT

We have analyzed and compared the nucleotide sequences of the intergenic DNA preceding the human fetal globin and adult δ globin genes. We have at least two categories of reiterated sequence in this globin DNA. One category consists of inverted 300 nucleotide long repetitive "Alu" family DNA sequences. The DNA between the closely spaced paired inverted repeats of Alu sequences may also be repetitive.

The second category of repetitive DNA is a sequence of over five kilobases present between the embryonic and fetal globin genes and pairs are also found downstream from the β chain genes. The sequence across one end of this repeat shows a stretch of approximately 1000 nucleotides bounded by a terminal imperfect inverted repeat of 17 nucleotides beginning with the dinucleotide TG. We suggest that this this element could be an analogue in human genome to

ISBN 0-12-137420-3

movable DNA elements such as a TY1 element of yeast and *copia* of *Drosophila*. A third prominent feature of the new repetitive DNA is the presence of very long purine clusters.

The largest part of the region of human DNA containing the non-α globin genes consists of intergenic DNA. The recognizable signals for transcription initiation *in vitro* and perhaps *in vivo*, for RNA splicing and for polyadenylation, as well as the sequences encoding the 5' and 3' untranslated portions of messenger RNA and the intervening sequences are all adjacent to the coding segments of DNA and in total account for approximately 15% of the DNA within this complex (1). The biological significance of the remaining 85% of the DNA is unknown. The structure of the globin complex is probably representative of that of other gene families in man and higher animals, so the functions and the structure of the bulk of animal cell DNA are largely unknown. Among the intergenic DNA are interspersed repetitive DNA sequences (2-10). While much repetitive intergenic DNA may be "selfish DNA" (11,12) or random accumulations of sequences, there are several hints, from the globin system (13, 14, 15), the histone genes (16) and the yeast cytochrome C genes (17), as well as other eukaryotic genetic systems that suggest that rather remote sequences, sometimes very remote sequences, may have a cis acting influence, either on the levels of gene expression or on the timing of gene expression during differentiation. To provide insight into the structure and organization of intergenic DNA, we have undertaken an extensive analysis of the nucleotide sequence of the intergenic DNA in the human globin gene cluster. We have analyzed contiguous stretches of 7,000 bases of DNA adjacent to and extending upstream from the Gγ fetal globin gene of man and a somewhat longer sequence adjacent to and upstream from the δ globin gene of man. We have also obtained additional partial sequence information on most of the remaining DNA between the fetal and adult and between the fetal and embryonic globin genes. During the course of this analysis, sequence information on the DNA flanking the embryonic ϵ globin gene (18, 19), the fetal γ globin gene (20, O. Smithies, personal communication), and the adult β gene (21) has accumulated and some characterization of certain types of repetitive interspersed DNA in the globin complex (4-10) has been presented. The following note summarizes briefly certain of the salient features of this intergenic DNA.

Materials and Methods

The majority of the sequences were determined by the chemical sequencing methods of Maxam and Gilbert, sometimes supplemented or confirmed by enzymatic sequencing using the Maat-Smith procedure and dideoxy extension procedures of Sanger and Coulson. DNA clones were isolated in λ vectors, both in our own laboratories and in that of Dr. T. Maniatis. The sequence data was analyzed by use of the computer programs of Staden *et al.* (22) and by use of a series of programs developed by one of us (P.J.).

Results

Several notable features of the intergenic DNA can be recognized (Fig. 1). These include:

1. Interspersed repetitive DNA sequences of the Alu family (2-10) in the globin gene cluster.

The "Alu family" consists of repetitive interspersed DNA segments in the human genome that are about 300 nucleotides long. They were recognized independently as a recurring sequence feature in hnRNA, as the moderately rapidly annealing fraction of interspersed DNA in the human genome, and as highly repetitive interspersed RNA polymerase III templates present in the human DNA. The location of these templates within the human non-α globin gene cluster has been described before as have certain general features of this family of reiterated DNA sequences. The sequence of six of the "Alu family" sequences within the globin gene cluster have been determined, four from our laboratory and two by Barralle, Proudfoot and their colleagues (19). Comparative analysis of these Alu family sequences reveals the following salient features (Fig. 2):

(a) Each sequence consists of slightly less than 300 nucleotides terminating at one end by an oligoadenylic acid sequence. These sequences are present in both orientations with respect to globin gene transcription.

(b) Each copy of the Alu family is an imperfect dimer, one-half of which shows strong homology to a major class of interspersed repetitive DNA present in rodent cells.

(c) Prominent sequence features within the Alu family include a strongly conserved 40-base-pair sequence near the 5' terminus or the proximal half of the sequence

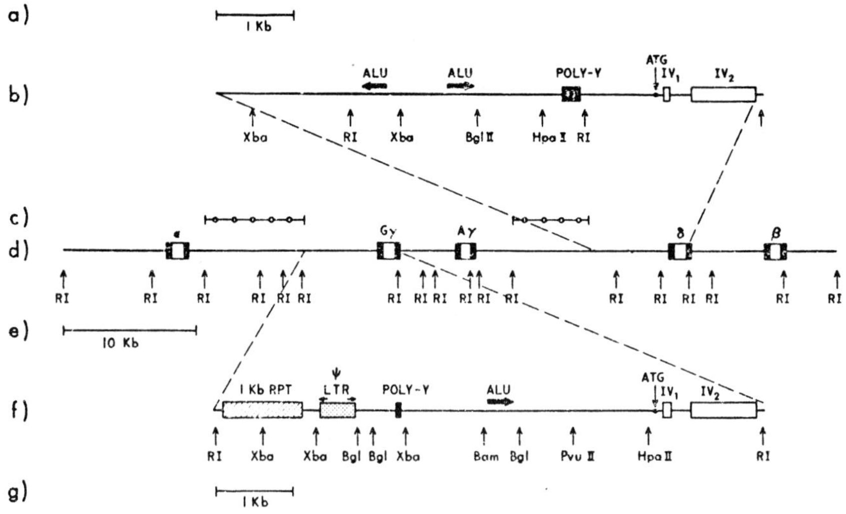

FIGURE 1 Schematic representation of the structural features of the human non-alpha globin gene cluster:

(a) Scale for expanded sequences in (b) and (f). (b) Principal restriction endonuclease sites and structural features in the flanking DNA sequences located upstream from the ζ globin gene. "Alu" refers to 300-nucleotide long reiterated DNA sequence of the Alu I family. "Poly Y" refers to a stretch of polypyrimidines on the sense strand of DNA; "ATG" represents the initiation codon for the ζ gene and IV1 and IV2 represent the two intervening sequences present in the gene. (c) Scale showing regions where incomplete nucleotide sequence information is available for the non-α globin gene cluster. (d) Schematic localization of non-α globin genes along the human chromosome DNA and location of EcoRI restriction endonuclease sites. (e) Scale for (d). (f) Schematic representation of structural features of the DNA located upstream from the human G γ globin gene. Abbreviations as in (b). "ψ LTR" refers to the sequence resembling the long-term repeat of a retrovirus and "1 KB RPT" refers to the repetitive DNA sequence adjoining the ψ LTR.

(in the transcriptional sense).

(d) The Alu family sequences demonstrate an average of greater than 80% homology with one another but there are occasional members with markedly variant or atypical internal sequences, containing multiple tandem repeats of short oligonucleotide sequences.

(e) Many, but not all, Alu family sequences are flanked by direct repeats of single copy DNA sequence. These direct repeats vary in length up to 19 nucleotides. One direct repeat is immediately adjacent to the 5' end of the Alu family sequence. However, the exact distance from the first to the second copy of the direct repeat may vary by a few nucleotides between members of the Alu family.

(f) The majority, but not all of the Alu family sequences are templates for transcription by RNA polymerase III. Transcription begins exactly at the border of the Alu family remote from the oligo(A) sequence and extends across the Alu family sequence into the unique DNA sequence. The 5' portion of the Alu family contains a sequence showing homology to RNA polymerase III promoter sites in other genes. At least one member of the Alu family in the non-α globin gene cluster is not transcribed, presumably due to a base substitution within the proposed promotor site.

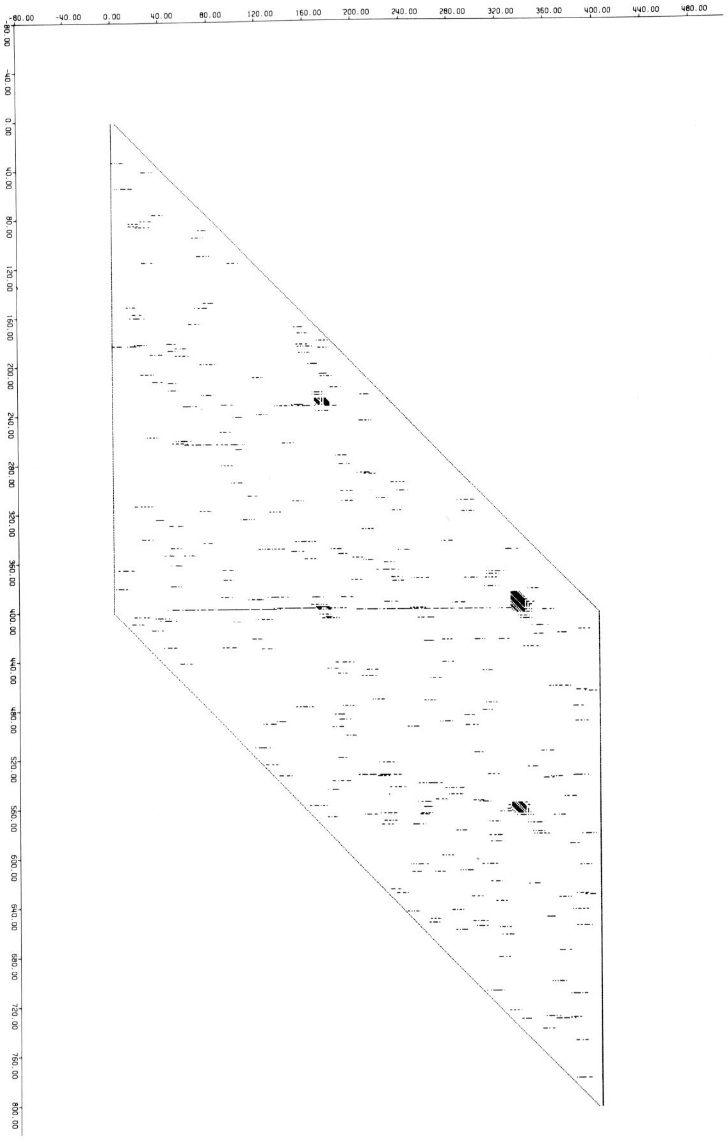

FIGURE 2 Dot matrix showing the similarity between the transcribed (3') Alu family DNA sequence preceding the adult δ globin genes and the Alu family DNA sequence preceding the fetal Gγ globin gene. Vertical coordinates represent the δ globin Alu family sequence and diagonal coordinates correspond to the Gγ globin Alu family sequence.

Dots represent locations of nucleotides that are identical at the corresponding coordinate positions in the two sequences. Dots are printed only when the identical base is part of a run of 10 bases of which seven or more are identical. The most striking feature is the strong homology between the two Alu family sequences shown as the horizontal line for approximately 300 nucleotides. Partial repeats are demonstrated as broken horizontal lines displaced above or below the main homology axis. Deletions are indicated by further vertical displacements of otherwise continuous lines. The graph also demonstrates homology between the 3' end of each Alu family sequence and the 5' end of the other Alu family sequence, as expected from the dimeric structure of Alu family DNA sequences.

Transcripts corresponding exactly to the *in vitro* RNA polymerase III transcripts have not yet been identified *in vivo*. However, there is a representation of Alu family sequences in hnRNA and a lesser representation of Alu family sequences in cytoplasmic polyadenylated RNA. There are also several low molecular weight RNA species partly complementary to Alu sequences. The most prominent of these is the 7S RNA which has been investigated by several groups.

The functions of the Alu family, sequences, if any, are entirely unknown. The presence of direct repeats at either end of the sequence and the marked preservation of sequence between most members of the family suggest that these are transposable DNA elements whose insertion into new segments of DNA may be coupled with a staggered break in the recipient DNA. The exact correlation of the site of transcription initiation by RNA polymerase III *in vitro* with one boundary of the repeat sequence suggests that there is some connection between these events. One can imagine that this repeat sequence has been selected for during evolution so as to preserve the 5'-terminal sequence of the transcript. Perhaps an equally attractive hypothesis, is that, in some manner, the transcription of the Alu sequence is coupled to its ability to be transposed within the genome. It should be noted, however, that direct evidence of transposition of this sequence has not yet been obtained.

2. Non-Alu interspersed repetitive DNA.

We have studied in detail the sequence of over 5,000 nucleotides of repetitive DNA located between the human ε and Gγ globin genes. Segments of DNA sequence complementary to this region are located at many places throughout the genome, including DNA sequences downstream from the 3' end of the ϵ globin gene. Often several subsegments of the repetitive DNA recur together in the same order as that found upstream from the Gγ globin gene. However, there is an inversion in the order of some of the subsegments of the repetitive DNA found downstream from the ϵ globin gene as compared with the order of those subsegments found between the ϵ and Gγ globin genes. In addition, a portion of the repetitive DNA sequence found beyond the 3'end of the ϵ globin gene lacks homology with the reiterated DNA sequence found between the ϵ and Gγ globin genes.

The long repetitive DNA sequence between the ϵ and Gγ globin genes (1 Kb RPT in Fig. 1f) is bounded at its 3'-end by a stretch of approximately 400 nucleotides, demarcated by an imperfect terminal inverted repeat of 17 base pairs. This inverted repeat sequence is markedly GC-rich and is bounded by the dinucleotides TG---CA that have been noted at the ends of the long terminal repeats (LTR's) of proretroviral DNA sequences and transposable DNA elements of *Drosophila* and yeast (23). The sequence of the LTR-like sequence between these two flanking terminal inverted repeats contains some additional direct repeat sequences and shows a very limited, spotty homology to the LTR sequence of Moloney sarcoma virus. However, no clearcut TATA box (24) or polyadenylation signal hexanucleotide can be identified in the LTR-like sequence in the 5'-flanking DNA of the Gγ globin gene.

One of us (PAB) has isolated from a human genomic partial *EcoRI* library a clone that hybridized strongly to the long, repetitive DNA sequence (1 Kb RPT in Fig. 1f) from the γ-globin gene clone. Both of these clones contain LTR-like sequences in similar positions. Comparison of the inverted repeat sequences at the ends of these two different LTR-like sequences reveals that they demonstrate much greater homology at one end of the LTR-like sequence than at the other. The region of the non-globin clone that is complementary to the longer repetitive DNA sequence found between the ϵ and Gγ globin genes (1 Kb RPT in Fig. 1f) has been sequenced and compared to the globin flanking sequence (Fig. 3). The comparison revealed a great deal of homology

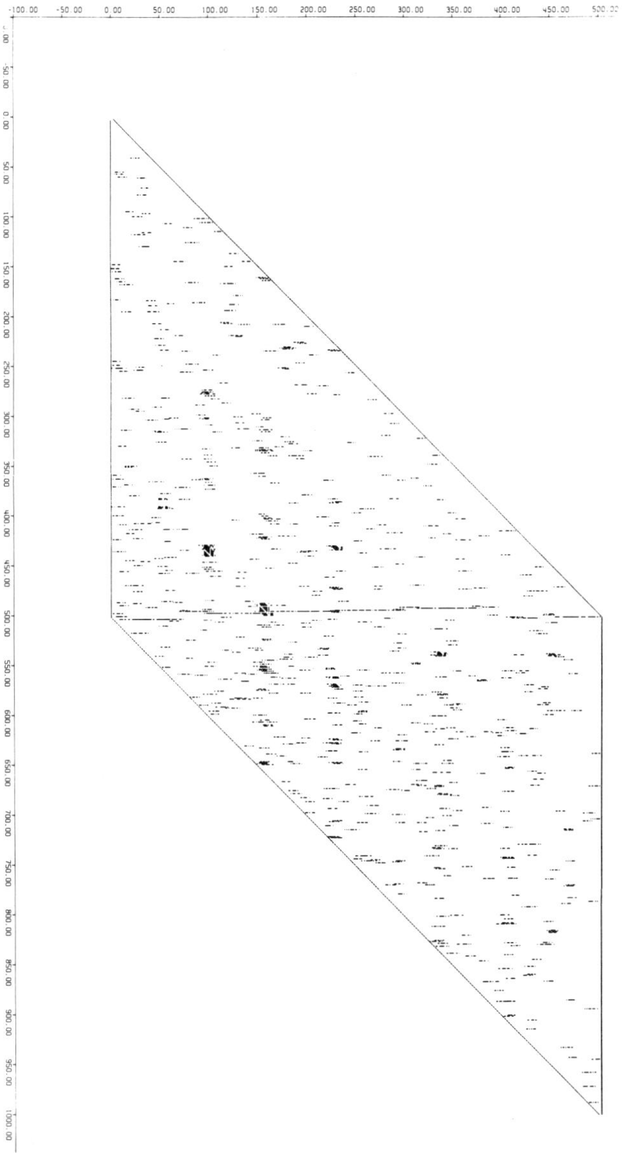

FIGURE 3 Dot homology matrix between 500 nucleotides of the long repetitive DNA sequence (1 Kb RPT in Fig. 1f) found upstream from the Gγ globin gene and 500 nucleotides from the corresponding position of a homologous repetitive DNA sequence isolated from a random genomic clone. The horizon-

tal line of dots indicates areas of homology as in Figure 2. Vertical displacement indicates deletion. Of note is the very strong homology between the main sequences and the lack of evidence of any tandem repeats or polymeric structures in these sequences.

between the two clones in this region (Fig. 3), over a distance of approximately 1Kb. Overall, these structural data are strongly consistent with the possibility that this long, repetitious DNA is the analogue to movable DNA elements such as *copia* in *Drosophila*. However, there is thus far no direct evidence that the DNA sequence has moved around during eukaryotic development, although the distance between the fetal and embryonic genes is larger than that between many other clusters of differentially regulated non-α globin genes in animals, suggesting that the distance between fetal and embryonic genes might have been expanded by insertion of the DNA element. It will be most interesting to see whether one can detect the point in phylogeny at which this element entered into the position between fetal and embryonic globin genes and whether this was correlated with any change in the regulation of globin gene expression.

3. Purine clusters in intergenic DNA.

In addition to the repetitive DNA elements, there appears to be an excess of purine-pyrimidine dinucleotides, particularly short stretches of TG runs scattered in the globin gene cluster and also an excess of short (5 or more) runs of deoxythymidylic acid 5 or more at a stretch in agreement with the observation of Grantham (26). An additional recurrent feature seen in the intergenic DNA upstream from both the Gγ and δglobin genes is a stretch of DNA in which one strand is almost totally purines and the other strand almost totally pyrimidines. This stretch of DNA has a length of approximately 200 nucleotides in the case of the δ globin gene (Fig. 1 and 4), but a length of only 40 or so nucleotides in the region of the δ globin gene.

In addition, the exact position of these purine runs with respect to the 5'-ends of these two genes is not identical. Nevertheless, the marked skewing in distribution of bases on the two strands of DNA suggests that this DNA may have a very specific structural or functional role and perhaps that its physical conformation is different from that of most of the remainder of the chromosomal DNA (27).

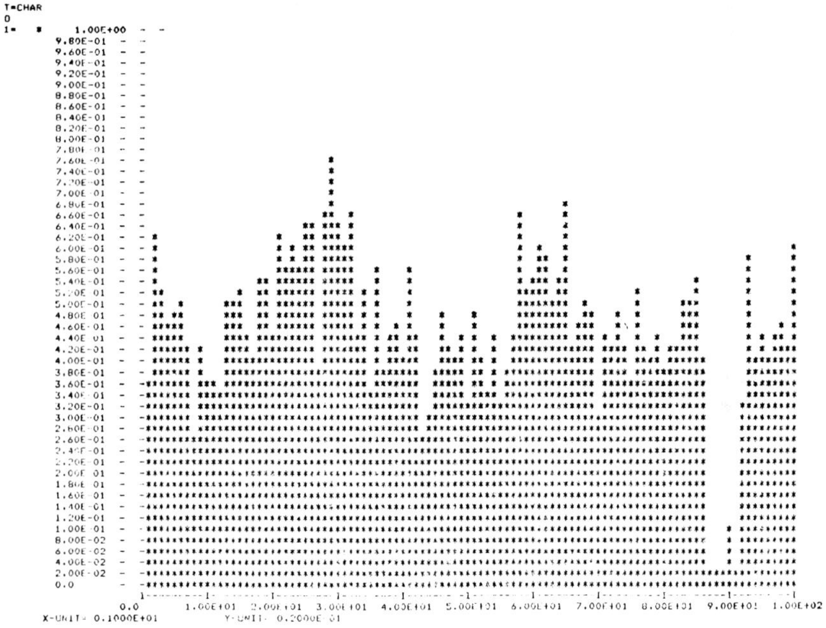

FIGURE 4 Bar graph showing the frequency of purines (A + G) in each successive set of 50 bases upstream from the ζ globin gene. Of note is the region between 1000 and 1200 base pairs upstream from the initiation codon where a stretch of 200 nucleotides is almost totally devoid of

purines. The DNA sequences to either side of this stretch of pyrimidines show a normal abundance of purines.

In conclusion, we have outlined some of the major structural features of long segments of intergenic DNA in the human non- β globin gene cluster. We anticipate that in the not-very-distant future, most or all of the nucleotide sequence of the 50 to 60,000 bases that are part of this cluster will be determined. Our studies demonstrate several kinds of structures at specific sites within intergenic DNA, both in the repetitive and nonrepetitive portions. A major challenge for the near future will be to determine whether any of these sequences correlate with specific higher order structures of chromatin.

ACKNOWLEDGMENTS

We are grateful for the excellent technical help of Julie Yamaguchi and to Richard Wang for supplying restriction endonuclease enzymes. This investigation was supported in part by the following grants of the National Institutes of Health: HL-20922 and P01-GM-20124.

REFERENCES

1. Efstratiadis, A., Posakony, J. W., Maniatis, T. Lum, R. M., O'Connell, C., Spritz, R. A., deRiel, J. K., Forget, G. B., Weissman, S. M., Slighton, J. L., Blechl, A. E., Smithies, O., Barralle, F. E., Shoulders, C., Proudfoot, N. J. Cell 21, 653-668 (1980).

2. Jelinek, W. R., Toomey, T. P., Leinwand, L., Duncan, C. H., Biro, P. A., Choudary, P. V., Weissman, S. M., Rubin, C. M., Houck, C. M., Deininger, P. G., Schmid, C. W. Proc. Natl. Acad. Sci. USA 77, 1318-1402 (1980).

3. Elder, J. T., Pan, J., Duncan, C. H. and Weissman, S. M. Nucl. Acids Res. 9, 1171-1189 (1981).

4. Pan., J., Elder, J. T., Duncan, C. H., and Weissman, S. M. Nucl. Acids Res. 9, 1151-1170 (1981).

5a. Duncan, C. H., Jagadeeswaran, P., Wang, R. R-C., and Weissman, S. M. Submitted to Gene.

5b. Jagadeeswaran, P., Tuan, D., Forget, B. G. and Weissman, S. M. (manuscript in preparation).

6. Weissman, S. M. In "Gene Families of Collagen and other Structural Proteins" (Prockop, D. J. and Champe, P. C., eds), Elsevier-North Holland, New York).

7. Coggins, L. W., Grindlay, G. J., Vass, J. K., Slater, A. A., Montagu, P., Stinson, M. A., Paul, J. Nucl. Acids Res. 8, 3319-3334 (1980).

8a. Fritsch, E. F., Shen, C. K. J., Lawn, R. M., Maniatis, T. Cold Spring Harbor Symposium Quantitative Biology, in press.

8b. Fritsch, E. F., Lawn, R. M., Maniatis, T. Cell 19, 959-972 (1980).

9. Kaufman, R. E., Kretschmer, P. J., Adams, J. W., Coon, H. C., Anderson, N. F., Nienhuis, A. W. Proc. Natl. Acad. Sci. USA 77, 4229-4253 (1980).

10. Adams, J. W., Kaufman, R. E., Kretschmer, P. J., Harrison, M., Nienhuis, A. W. Nucl. Acids Res. 8, 6113-6128 (1980).

11. Doolittle, W. F., Sapienza, C. Nature 284, 601 (1980).

12. Orgel, L. E., Crick, F. H. C. Nature 284, 604 (1980).

13. Tuan, D., Biro, P. A., deRiel, J. K., Lazarus, H., Forget, B. G. Nucl. Acids Res. 6, 2519-2544 (1979).

14. Fritsch, E. M., Lawn, R. M., Maniatis, T. Nature 279, 598-603 (1979).

15. Bernards, R., Flavell, A. Nucl. Acids Res. 8, 1521-1534 (1980).

16. Grosschedl, R., Birnstiel, M. L. Proc. Natl. Acad. Sci. USA 77, 1432-1436 (1980).

17. Smith, M. and Hall, B., personal communication.

18. Baralle, F. E., Shoulders, C. C., Proudfoot, N. J. Cell 21, 621-626 (1980).

19. Baralle, F. E., Shoulders, C. C., Goodbourn, S., Jeffrey, A., Proudfoot, N. J. Nucl. Acids Res. 8, 4393-4403 (1980).

20. Slightom, J. L., Blechl, A. E., Smithies, O. Cell 21, 627-638 (1980).

21. Lawn, R. M., Efstradiatis, A., O'Connell, C., Maniatis, T. Cell 21, 647-652 (1980).

22. Staden, R. Nucl. Acids Res. 5, 1013-1015 (1978).

23. Temin, H. M. Cell 21, 599-600 (1980).

24. Ziff, E., Evans, R. Cell 15, 1463-1475 (1978).

25. Shenk, T. Curr. Topics in Microbiol. and Immunol., in press.

26. Grantham, R. FEBS Lett. 121, 193-199 (1980).

27. Trifonov, E. N. Nucl. Acids Res. 8, 4041-4053 (1980).

SEQUENCES OF A HUMAN FIBROBLAST INTERFERON GENE AND TWO LINKED HUMAN LEUKOCYTE INTERFERON GENES

Shuichiro Maeda
Russell McCandliss
Tsu-Rong Chiang
Lawrence Costello
Warren P. Levy
Nancy T. Chang
Sidney Pestka

Roche Institute of Molecular Biology
Nutley, NJ 07110

ABSTRACT

A human gene library was screened for the presence of interferon chromosomal genes with the use of cloned human fibroblast and leukocyte interferon complementary DNAs (cDNAs) as probes. One fibroblast interferon (IFF) cDNA-related sequence and twelve leukocyte interferon (IFL) cDNA-related sequences were isolated in the screening of 500,000 recombinants. The IFF genomic DNA clone has been sequenced. The sequence of the genomic DNA is virtually identical to the known IFF cDNA sequence and lacks intervening sequences. Among twelve IFL-related DNA clones, eight were shown to be distinct by restriction endonuclease mapping. Sequencing of one IFL DNA revealed that the DNA contains two homologous IFL genes that differ from each other and from the sequence of IFL cDNA used as the probe for screening. This and other data indicate the presence of at least ten chromosomal genes for human IFL.

INTRODUCTION

The human interferons are a family of proteins with potent antiviral activity produced by various cells when exposed to interferon inducers such as viruses. They are classified into three types: fibroblast (IFF), leukocyte (IFL), and immune interferon (IFI) (1). Using recombinant

ISBN 0-12-137420-3

DNA technology, several laboratories have obtained DNA copies complementary to IFF or IFL mRNA (2-9). Several distinct IFL cDNAs sharing more than 80% homology at the nucleotide level in their protein-coding regions have been isolated. This observation is consistent with the existence of different IFL protein species (10-15). In contrast, all IFF cDNA sequences so far reported are almost identical, although the existence of a second type of IFF mRNA which shares only a little sequence homology with the known IFF cDNAs has been suggested (16,17). A comparison of the sequences of an IFF cDNA with one of the IFL cDNAs revealed only 45% homology at the nucleotide level (18). In order to study interferon gene organization and molecular mechanism of interferon induction and activity, we isolated an IFF chromosomal gene and several IFL chromosomal genes from a human gene library. In this report, we describe the isolation and DNA sequence analysis of an IFF gene and two linked IFL genes.

METHODS

Screening of a Human Genome Library

A phage λ Charon 4A recombinant library of the human genome (19), obtained from Dr. T. Maniatis, was screened for IFF and IFL genes by *in situ* plaque hybridization (20) with radiolabeled IFF and IFL cDNAs as probes. Hybridization was carried out essentially as described by Maniatis *et al.* (21) with the exception that 10% dextran sulfate was added to hybridization mixture (22). The IFF cDNA probe used in screening was prepared from the cloned recombinant plasmid 101 containing most of the cDNA sequence for human IFF (8). The IFL cDNA probe was prepared from the plasmid pLeIFA25 which directed the synthesis of human IFL in *Escherichia coli* (23). The cDNA probes were labeled with the use of *E. coli* DNA polymerase I, [α-^{32}P]dATP and [α-^{32}P]dCTP as described by Rigby *et al.* (24). Positive plaques were rescreened and purified. Recombinant phages were grown and purified and phage DNA was extracted according to procedures of Maniatis *et al.* (21). All procedures involving recombinant DNA were performed in accordance with the NIH Guidelines for Recombinant DNA Research.

Analysis of Restriction Digests of Recombinant Phage DNA

Recombinant phage DNA was digested with restriction endonucleases. It was then electrophoresed in agarose gel, transferred to nitrocellulose filter paper as described by Southern (25), and hybridized to radioactive cDNA probes (8) essentially according to procedures of Wahl *et al.* (22).

DNA Sequencing

DNA fragments, cleaved with *Eco*RI restriction endonuclease, were cloned into the *Eco*RI site of the phage M13mp7 (26), and sequenced by the dideoxynucleotide termination technique (27,28) with a synthetic deoxyoligonucleotide primer (Collaborative Research, Inc.) complementary to 12 nucleotides preceding the *Eco*RI site of M13 DNA or appropriate primers obtained from IFF or IFL gene sequences.

RESULTS

Isolation of Human Chromosomal DNA Fragments Hybridizing to Interferon cDNA Probes

The phage λ Charon 4A recombinant library of the human genome (19) was screened by *in situ* plaque hybridization (20) with the mixture of ^{32}P-labeled IFF cDNA (8) and IFL cDNA (23) as a probe. Positive plaques were purified and then hybridized with each of the above radiolabeled cDNA probes separately. One IFF cDNA-related clone (G83) and twelve IFL cDNA-related clones were isolated in the screening of about 500,000 plaques. DNAs were prepared from the positive plaques, and digested with *Eco*RI enzyme. The fragments were separated by 0.8% agarose gel electrophoresis, transferred to nitrocellulose filter paper, and hybridized with the radiolabeled cDNA probe. The size of the *Eco*RI fragments obtained from the cloned genomic DNA and the fragments hybridizing to the cDNA probe are shown in Table I. Clones G57, G73 and G89 give identical restriction and hybridization patterns. Therefore, they appear to be identical. Clone G77B may be derived from a region overlapping with the above three clones because it has three *Eco*RI fragments common in length with the above clones and exhibits the same hybridization pattern. Clones G68 and G77A also yield identical patterns, but are different from all the others. All the other six clones appear to be distinct.

TABLE I. Fragments Obtained by Cleavage of the Genomic DNAs with EcoRI Enzyme[a]

Genomic DNAs	*Class*	*EcoRI Fragments*
G8	*Leukocyte*	*3.3, 3.1, 2.9, 2.2, 1.4, 1.3*
G48	*Leukocyte*	*14, 2.2, 1.4, 0.6*
G55	*Leukocyte*	*7.5, 2.7, 2.3, 2.0, 1.7, 0.9, 0.7, 0.4, <0.1*
G57, 73, 89	*Leukocyte*	*9.0, 3.2, 1.9, 0.8, 0.7, 0.5, 0.4, 0.3*
G68, 77A	*Leukocyte*	*8.5, 4.0, 3.5, 2.5, 1.3, 0.4, 0.2*
G76	*Leukocyte*	*10, 2.2, 1.8, 1.7, 1.2*
G77B	*Leukocyte*	*9.0, 3.2, 2.3, 1.4, 0.7*
G91	*Leukocyte*	*6.5, 5.0, 2.7*
G93	*Leukocyte*	*7.4, 7.2, 2.6, 1.4, 0.1*
G83	*Fibroblast*	*6.0, 4.5, 2.8, 1.9, 1.2, 0.7, 0.6*

[a]*The numbers indicate the size of the fragment in kilobase pairs. The fragments hybridizing with the cDNA probes are underlined.*

Although mutations in propagation of the phage library may change the restriction pattern of the chromosomal DNA, the twelve IFL clones appear to be derived from at least eight distinct regions of the chromosomal DNA because all the differences between the clones cannot be derived from occasional mutations. The 1.9-kilobase pair (kb) *Eco*RI fragment of G83 hybridizing to IFF cDNA and the 2.2, 1.4 and 1.3-kb *Eco*RI fragments of G8 hybridizing to IFL cDNA were purified by electrophoresis on a 5% polyacrylamide gel. The fragments were then subcloned into the *Eco*RI site of the phage M13mp7 for sequencing.

DNA Sequencing

Fig. 1 shows a restriction map of the 1.9-kb fragment in clone G83. Dideoxynucleotide termination sequencing was performed with internal primers obtained from plasmid 101 (8) or from the 1.9-kb genomic DNA fragment itself as presented in Fig. 1. Fig. 2 shows the DNA sequence of the fragment. The sequences of the fragment and the known IFF cDNA or mRNA (4-6,18) are exactly colinear. The 5'-flanking

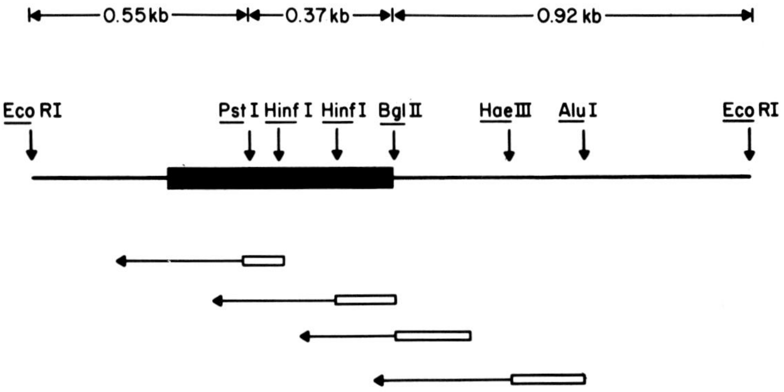

FIGURE 1. Strategy for sequencing a human fibroblast interferon gene. Vertical arrows indicate the restriction sites used for obtaining the primers for dideoxynucleotide termination sequencing. Horizontal arrows represent the direction and length of the DNA sequencing reaction initiating from the primers indicated by open boxes.

sequence contains the sequence ATAAATA corresponding to the "Hogness box" (29). The sequence precedes the putative "cap" site of the gene (5) by 24 nucleotides. The 3'-untranslated region contains the sequence AATAAA 20-22 nucleotides upstream from the site of polyadenylation (4,5,18). The sequence AATAAA, present in most eukaryotic mRNAs, precedes the poly(A) tail by 15-30 nucleotides (30). The sequencing data indicate that the IFF gene does not contain any intervening sequence in the coding region, in the 5'-flanking sequence, or in the 3'-untranslated region.

Fig. 3 shows a map of restriction enzyme-cleavage sites on the 14-kb insert of genomic DNA clone G8. The three fragments hybridizing to the IFL cDNA were sequenced with internal primers obtained from plasmid 104 containing IFL cDNA sequences (8) or a synthetic oligonucleotide primer

```
AAT CGT AAA GAA GGA CAT CTC ATA TAA ATA GGC CAT ACC CAT GGA GAA AGG ACA TTC TAA CTG CAA CCT
                                   -100
                                                                        S1
                                                                        Met Thr Asn Lys
TTC GAA GCC TTT GCT CTG GCA CAA CAG GTA GTA GGC GAC ACT GTT CGT GTT GTC AAC ATG ACC AAC AAG

                                                                    1
Cys Leu Leu Gln Ile Ala Leu Leu Leu Cys Phe Phe Thr Thr Ala Leu Ser Met Ser Tyr Asn Leu Leu
TGT CTC CTC CAA ATT GCT CTC CTG TTG TGC TTC TCC ACT ACA GCT CTT TCC ATG AGC TAC AAC TTG CTT

Gly Phe Leu Gln Arg Ser Ser Asn Phe Gln Cys Gln Lys Leu Leu Trp Gln Leu Asn Gly Arg Leu Glu
GGA TTC CTA CAA AGA AGC AGC AAT TTT CAG TGT CAG AAG CTC CTG TGG CAA TTG AAT GGG AGG CTT GAA
                        100

Tyr Cys Leu Lys Asp Arg Met Asn Phe Asp Ile Pro Glu Glu Ile Lys Gln Leu Gln Gln Phe Gln Lys
TAC TGC CTC AAG GAC AGG ATG AAC TTT GAC ATC CCT GAG GAG ATT AAG CAG CTG CAG CAG TTC CAG AAG
                                                                  200

Glu Asp Ala Ala Leu Thr Ile Tyr Glu Met Leu Gln Asn Ile Phe Ala Ile Phe Arg Gln Asp Ser Ser
GAG GAC GCC GCA TTG ACC ATC TAT GAG ATG CTC CAG AAC ATC TTT GCT ATT TTC AGA CAA GAT TCA TCT

Ser Thr Gly Trp Asn Glu Thr Ile Val Glu Asn Leu Leu Ala Asn Val Tyr His Gln Ile Asn His Leu
AGC ACT GGC TGG AAT GAG ACT ATT GTT GAG AAC CTC CTG GCT AAT GTC TAT CAT CAG ATA AAC CAT CTG
                300

Lys Thr Val Leu Glu Glu Lys Leu Glu Lys Glu Asp Phe Thr Arg Gly Lys Leu Met Ser Ser Leu His
AAG ACA GTC CTG GAA GAA AAA CTG GAG AAA GAA GAT TTC ACC AGG GGA AAA CTC ATG AGC AGT CTG CAC
                                                         400

Leu Lys Arg Tyr Tyr Gly Arg Ile Leu His Tyr Leu Lys Ala Lys Glu Tyr Ser His Cys Ala Trp Thr
CTG AAA AGA TAT TAT GGG AGG ATT CTG CAT TAC CTG AAG GCC AAG GAG TAC AGT CAC TGT GCC TGG ACC

                                                                                166
Ile Val Arg Val Glu Ile Leu Arg Asn Phe Tyr Phe Ile Asn Arg Leu Thr Gly Tyr Leu Arg Asn END
ATA GTC AGA GTG GAA ATC CTA AGG AAC TTT TAC TTC ATT AAC AGA CTT ACA GGT TAC CTC CGA AAC TGA
     500

AGA TCT CCT AGC CTG TGC CTC TGG GAC TGG ACA ATT GCT TCA AGC ATT CTT CAA CCA GCA GAT GCT GTT
                                             600

TAA GTG ACT GAT GGC TAA TGT ACT GCA TAT GAA AGG ACA CTA GAA GAT TTT GAA ATT TTT ATT AAA TTA
                                                                                        700
                                                                            ↓
TGA GTT ATT TTT ATT TAT TTA AAT TTT ATT TTG GAA AAT AAA TTA TTT TTG GTG CAA AAG TCA ACA TGG

CAG TTT TAA TTT CGA TTT GAT TTA TAT AAC CAT CCA TAT TAT AA
                                   800
```

FIGURE 2. Sequence of a human fibroblast interferon gene. The initiation codon for the interferon precursor, the first codon for mature interferon, and the termination codon are enclosed in boxes. The amino acid sequence is presented above the nucleotide sequence. Vertical arrows indicate the putative "cap" site and the polyadenylation site. The possible "Hogness box" and the sequence AATAAA are underlined.

complementary to the M13 DNA region preceding the *Eco*RI site as shown in Fig. 3. Fig. 4 presents the partial DNA sequences of the three fragments and the IFL cDNA used as the probe for obtaining the genomic DNA clones. The sequences are aligned to show sequence homology. The 14-kb insert contains two homologous yet distinct IFL genes. The genomic DNA, G8I-II, is one amino acid longer in the protein-coding sequence than the cDNA and has no intron in the sequence. The 5'-flanking sequence of G8I-II gene contains a sequence TATTTAA, which may correspond to the "Hogness box", 93 nucleotides upstream from the initiation codon for the signal peptide. The sequence of the cDNA also differs from the two genomic DNA sequences.

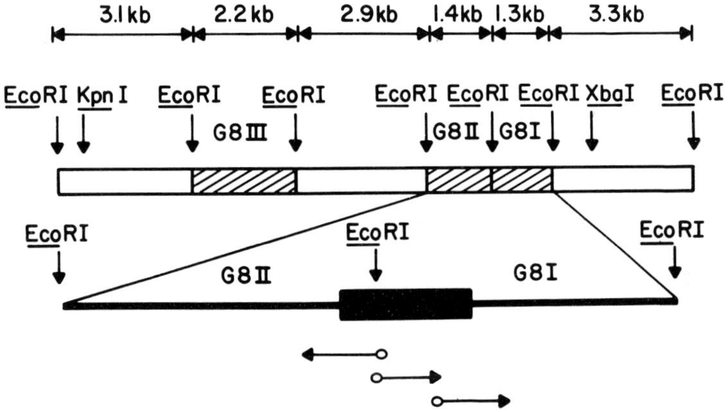

FIGURE 3. Strategy for sequencing human leukocyte interferon genes. Vertical arrows indicate the restriction enzyme-cleavage sites. Hatched boxes represent the EcoRI fragments hybridizing to the cDNA probe. A solid box represents the protein-coding region. Horizontal arrows represent the direction and length of dideoxynucleotide termination sequencing reaction initiating from the primers indicated by open circles.

DISCUSSION

The isolation of eight different human genomic DNA clones containing IFL-related sequences (Table I) indicates the existence of multiple loci for IFL on chromosomal DNA. Partial DNA sequencing data of one of the clones (Fig. 4) showed that the clone contains two homologous, but distinct genes for IFL. The sequence of IFL cDNA used as the probe for isolating the IFL genomic DNA clones also differed from the two IFL genes (Fig. 4). Furthermore, the IFL cDNA sequence is distinguishable from the one reported by Nagata *et al.* (3). The sequence showed that, unlike most of the

```
I-II  AGAAAGCAAAAACAGACATAGAAAGTAAAACTAGGCATTTAGAAAATGGAAATTAGTATGTTCACTATTTAAGACCTATGCACAGAGCAAAGTCTC
                                                                                      S1
                                                                                      Met
I-II  CAGAAAACCTAGAGCCACTGGTTCAAGTTACCCACCTCAGGTAGCCTAGTGATATTTGCAAAATCCCAATGGCCCGGTCCTTTTCTTTACTGATGG
IV              TGAGCCTAAACCTTAGGCTCACCCATTTCAACCAGTCTAGCAGCATCTGCAACATCTACAATGGCCTTGACCTTTGTCTTACTGGTGG
                                                 1
                                                 Cys
I-II  TCGTGCTGGTACTCAGCTACAAATCCATCTGCTCTCTGGGCTGTGATCTGCCTCAGACCCACAGCCTGCGTAATAGGAGGGCCTTGATACTCCTGG
IV    CCCTCCTGGTGCTCAGCTGCAAGTCAAGCTGCTCTGTGGGCTGTGATCTGCCTCAAACCCACAGCCTGGGTAGCAGGAGGACCTTGATGCTCCTGG

I-II  CACAAATGGGAAGAATCTCTCCTTTCTCCTGCTTGAAGGACAGACATGAATTCAGATTCCCAGAGGAGGAGTTTGATGGCCACCAGTTCCAGAAGA
IV    CACAGATGAGGAAAATCTCTCTTTTCTCCTGCTTGAAGGACAGACATGACTTTGGATTTCCCCAGGAGGAGTTT   GGCAACCAGTTCCAAAAGG

I-II  CTCAAGCCATCTCTGTCCTCCATGAGATGATCCAGCAGACCTTCAATCTCTTCAGCACAGAGGACTCATCTGCTGCTTGGGAACAGAGCCTCCTAG
IV    CTGAAACCATCCCTGTCCTCCATGAGATGATCCAGCAGATCTTCAATCTCTTCAGCACAAAGGACTCATCTGCTGCTTGGGATGAGACCCTCCTAG

I-II  AAAAATTTTCCACTGAACTTTACCAGCAACTGAATGACCTGGAAGCATGTGTGATACAGGAGGTTGGGGTGGAAGAGACTCCCCTGATGAATGAGG
IV    ACAAATTCTACACTGAACTCTACCAGCAGCTGAATGACCTGGAAGCCTGTGTGATACAGGGGGTGGGGGTGACAGAGACTCCCCTGATGAAGGAGG

I-II  ACTTCATCCTGGCTGTGAGGAAATACTTCCAAAGAATCACTCTTTATCTAATGGAGAAGAAATACAGCCCTTGTGCCTGGGAGGTTGTCAGAGCAG
III                                                 CTCTTTATCTAATAGAGAGGAAATACAGCCCTTGTGCCTGGGAGGTTGTCAGAGCAG
IV    ACTCCATTCTGGCTGTGAGGAAATACTTCCAAAGAATCACTCTCTATCTGAAAGAGAAGAAATACAGCCCTTGTGCCTGGGAGGTGGTCAGAGCAG
                                                                  END
I-II  AAATCATGAGATCCTTCTCTTTTTCAACAAACTTGCAAAAAAGATTAAGGAGGAAGGATTGAAAACTGGTTCATCATGGAAATGATTCTCATTGAC
III   AAATCATGAGATCCCTCTCGTTTTCAACAAACTTGCAAAAAAGATTAAGGAGGAAGGATTGAAAACTGGTTCAACATGGCAATGATCCTGATTGAC
IV    AAATCATGAGATCTTTTTCTTTGTCAACAAACTTGCAAGAAAGTTTAAGAAGTAAGGAATGAAAACTGGTTCAACATGGAAATGATTTTCATTAAT

I-II  TAATGCATCATCTCACACTTTCATGAGTTCTTCCATTTCAAAGACTCA   CTTC   TATAACCACCACAAGTTGAATCAAAATTTCCAAATGTTT
III   TAATACATTATCTCACACTTTCATGAGTTCTTCCATTTCAAAGACTCA   CTTC   TATAACCA
IV    TCGTATGCCAGCTCACCTTTTTATGA  TCTGCCATTTCAAAGACTCATGTTTCTGCTATGACCATGACACGATTTAAATCTTTTTC AAATGTTT

I-II  TCAGGAGTGTTAAGAAGCATCGTGTTTACCTGTGCAGGCACTAGTCCTTTACAGATGACCA
IV    TTAGGAGTATTAATCAACATTGTATTCAGCTCTTAAGGCACTAGTCCCTTACAGAGGACCA
```

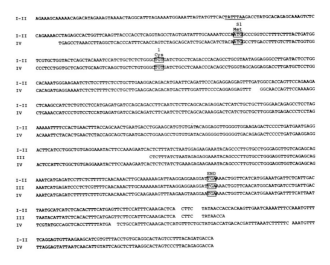

FIGURE 4. Sequences of human leukocyte interferon genes. The gene I-II resides in the EcoRI fragments G8I and G8II and the gene III in the fragment G8III in Fig. 3. The sequence IV is the IFL cDNA used as the probe for screening. The sequences are aligned to give maximal homology with gaps in the coding sequence and in the 3'-untranslated region. The coding sequence of gene III is only partially complete. The initiation codon for the precursor, the first codon for mature interferon, and the termination codon are enclosed in boxes. The possible "Hogness box" of the gene I-II is underlined.

eukaryotic genes so far reported, this IFL gene (Fig. 4, I-II) lacks intervening sequences within the protein-coding region. The homology within the 3'-untranslated regions of the chromosomal genes and the cDNA probe is less than that within the protein-coding regions (Fig. 4). The isolation of many IFL genomic clones and cDNAs indicates the existence of multiple genes for IFL. This is consistent with the isolation of multiple IFL protein species (10-15). The observation that only one of the eight genomic clones was isolated four times (G57, G73, G89, G77B) from the library, whereas only one of the others was isolated more than once, is statistically highly improbable. Therefore, some of these four may be distinct but closely related and it is likely there are, indeed, many additional IFL genes not yet isolated in the human genome. Alternatively, this gene (G57) may be present in the genome in multiple copies.

In contrast to the IFL genes, all the IFF cDNA sequences so far reported (4-6,18) are almost identical, and the sequence of the IFF chromosomal gene reported here (Fig. 2) is also identical to one of the IFF cDNA sequences (4). The sequencing data showed that like the IFL genes reported here and by Nagata *et al.* (31), the IFF chromosomal gene lacks intervening sequences. So far, only a single human IFF protein has been purified (32-34). The available amino acid sequence determined for the IFF protein (32-34) is identical to that predicted from the nucleotide sequence of the IFF chromosomal gene. Therefore, the IFF gene, reported here, appears to be the single gene for human IFF. However, the existence of a second type of IFF mRNA which does not hybridize with the known IFF cDNA has been suggested (16,17). Our IFF cDNA probe would not have detected a chromosomal gene for the second type of IFF.

The above results indicate that human interferon genes represent a multiple gene family and that the diversity of these interferons is a result of distinct genes coding for different interferons. The distinct human interferon species were found to exhibit different biological properties and antigenicities (36). Structural analysis of the individual interferons will provide an understanding of the relationship between their primary structures and their biological activities. Such information may lead to the construction of novel interferons with valuable therapeutic potential. Furthermore, with these genes we will be able to analyze the molecular mechanisms underlying the induction of interferon in human cells.

ACKNOWLEDGMENTS

We thank Dr. T. Maniatis for the human gene library; Dr. Marian Evinger and Mr. Martin Boublik for their superb assistance in screening the human gene library.

REFERENCES

1. Pestka, S., and Baron, S. (1981). *In* "Methods in Enzymology" (S. Pestka, ed.), Vol. 78, in press. Academic Press, New York.
2. Taniguchi, T., Sakai, M., Fujii-Kuriyama, Y., Muramatsu, M., Kobayashi, S., and Sudo, T. (1979) *Proc. Jpn. Acad. Ser. B55*, 464.
3. Nagata, S., Taira, H., Hall, A., Johnsrud, L., Streuli, M., Ecsödi, J., Boll, W., Cantell, K., and Weissmann, C. (1980) *Nature 284*, 316.
4. Derynck, R., Content, J., De Clercq, E., Volckaert, G., Tavernier, J., Devos, R., and Fiers, W. (1980) *Nature 285*, 542.
5. Houghton, M., Eaton, M.A.W., Stewart, A.G., Smith, J.C., Doel, S.M., Catlin, G.H., Lewis, H.M., Patel, T.P., Emtage, J.S., Carey, N.H., and Porter, A.G. (1980) *Nucleic Acids Res. 8*, 2885.
6. Goeddel, D.V., Shepard, H.M., Yelverton, E., Leung, D., Crea, R., Sloma, A., and Pestka, S. (1980) *Nucleic Acids Res. 8*, 4057.
7. Streuli, M., Nagata, S., and Weissmann, C. (1980) *Science 209*, 1343.
8. Maeda, S., McCandliss, R., Gross, M., Sloma, A., Familletti, P.C., Tabor, J.M., Evinger, M., Levy, W.P., and Pestka, S. (1980) *Proc. Natl. Acad. Sci. U.S.A. 77*, 7010.
9. Goeddel, D.V., Leung, D.W., Dull, T.J., Gross, M., Lawn, R.M., McCandliss, R., Seeburg, P.H., Ullrich, A., Yelverton, E., and Gray, P.W. (1980) *Nature 290*, 20.
10. Rubinstein, M., Rubinstein, S., Familletti, P.C., Miller, R.S., Waldman, A.A., and Pestka, S. (1979) *Proc. Natl. Acad. Sci. U.S.A. 76*, 640.
11. Rubinstein, M., Levy, W.P., Moschera, J.A., Lai, C.-Y., Hershberg, R.D., Bartlett, R.T., and Pestka, S. (1981) *Arch. Biochem. Biophys.*, in press.
12. Hobbs, D.S., Moschera, J., Levy, W.P., and Pestka, S. (1981). *In* "Methods in Enzymology" (S. Pestka, ed.), Vol. 78, in press. Academic Press, New York.
13. Zoon, K.C., Smith, M.E., Bridgen, P.J., Anfinsen, C.B., Hunkapiller, M.W., and Hood, L.E. (1980) *Science 207*, 527.
14. Levy, W.P., Shively, J., Rubinstein, M., Del Valle, U., and Pestka, S. (1980) *Proc. Natl. Acad. Sci. U.S.A. 77*, 5102.
15. Allen, G., and Fantes, K.H. (1980) *Nature 287*, 408.
16. Sehgal, P.B., and Sagar, A.D. (1980) *Nature 288*, 95.

17. Weissenbach, J., Chernajovsky, Y., Zeevi, M., Shulman, L., Soreq, H., Nir, U., Wallach, D., Perricaudet, M., Toillais, P., and Revel, M. (1980) *Proc. Natl. Acad. Sci. U.S.A. 77*, 7152.
18. Taniguchi, T., Mantei, N., Schwarzstein, M., Nagata, S., Muramatsu, M., and Weissmann, C. (1980) *Nature 285*, 547.
19. Lawn, R.M., Fritsch, E.F., Parker, R.C., Blake, G., and Maniatis, T. (1978) *Cell 15*, 1157.
20. Benton, W.D., and Davis, R.W. (1977) *Science 196*, 180.
21. Maniatis, T., Hardison, R.C., Lacy, E., Lauer, J., O'Connell, C., Quon, D., Sim, G.K., and Efstratiadis, A. (1978) *Cell 15*, 687.
22. Wahl, G.W., Stern, M., and Stark, G.R. (1979) *Proc. Natl. Acad. Sci. U.S.A. 76*, 3683.
23. Goeddel, D.V., Yelverton, E., Ullrich, A., Heyneker, H.L., Miozzari, G., Holmes, W., Seeburg, P.H., Dull, T., May, L., Stebbing, N., Crea, R., Maeda, S., McCandliss, R., Sloma, A., Tabor, J.M., Gross, M., Familletti, P.C., and Pestka, S. (1980) *Nature 287*, 411.
24. Rigby, P.W.J., Dieckmann, M., Rhodes, C., and Berg, P. (1977) *J. Mol. Biol. 113*, 237.
25. Southern, E.M. (1975) *J. Mol. Biol. 98*, 503.
26. Messing, J., Crea, R., and Seeburg, P.H. (1981) *Nucleic Acids Res. 9*, 309.
27. Sanger, F., Nicklen, S., and Coulson, A.R. (1977) *Proc. Natl. Acad. Sci. U.S.A. 74*, 5463.
28. Smith, A.J.H. (1980). *In* "Methods in Enzymology" (L. Grossman, and K. Moldave, eds.), Vol. 65, p. 560. Academic Press, New York.
29. Goldberg, M. (1979) Thesis, Stanford University.
30. Proudfoot, N.J., and Brownlee, G.G. (1976) *Nature 263*, 211.
31. Nagata, S., Mantei, N., and Weissmann, C. (1980) *Nature 287*, 401.
32. Knight, E., Jr., Hunkapiller, M.W., Korant, B.D., Hardy, R.W.F., and Hood, L.E. (1980) *Science 207*, 525.
33. Okamura, H., Berthold, W., Hood, L., Hunkapiller, M., Inoue, M., Smith-Johannsen, H., and Tan, Y.H. (1980) *Biochemistry 19*, 3831.
34. Stein, S., Kenny, C., Friesen, H.-J., Shively, J., Del Valle, U., and Pestka, S. (1980) *Proc. Natl. Acad. Sci. U.S.A. 77*, 5716.
35. Friesen, H.-J., Stein, S., Evinger, M., Familletti, P.C., Moschera, J., Meienhofer, J., Shively, J., and Pestka, S. (1981) *Arch. Biochem. Biophys. 206*, 432.

36. Pestka, S., Maeda, S., Hobbs, D.S., Levy, W.P., McCandliss, R., Stein, S., Moschera, J.A., and Staehelin, T. (1981). *In* "Cellular Responses to Molecular Modulators" (W.A. Scott, R. Werner, and J. Schultz, eds.), Miami Winter Symposia, Vol. 18, in press. Academic Press, New York.

TRANSCRIPTION AND TRANSLATION OF YOLK PROTEIN mRNA IN THE FAT BODIES OF *DROSOPHILA*

Thomas Barnett
Pieter Wensink

Rosenstiel Basic Medical Sciences Research Center
Brandeis University
Waltham, Massachusetts

ABSTRACT

Immediately after the *Drosophila* female hatches, her fat body begins coordinate and rapid synthesis of three yolk proteins, YP1, YP2 and YP3. The genes coding for these proteins have been cloned and in this paper they have been used to determine the gene repetition frequency and the pattern of transcription in the fat body. Each of the genes is single copy in the fat body tissue. The YP1 and YP3 genes have transcripts ~1.60 and ~1.54 kb in length, respectively, but YP2 has two equimolar transcripts ~1.59 and ~1.67 kb in length. The transcription pattern indicates that the rapid onset of yolk protein synthesis is due to translation of newly synthesized mRNA and not of stored mRNA. For the first 24 hours after the female hatches, both the quantity of fat body RNA complementary to yolk protein genes and the quantity of fat body RNA that can be translated into yolk proteins by a cell-free translation system accumulates in parallel with the fat body's increasing synthesis of the three proteins. From published studies of yolk protein synthesis, we conclude that this transcription is triggered by the hormone ecdysone.

INTRODUCTION

Shortly after a *Drosophila* female emerges from the pupal case, her ovaries undergo a remarkable series of events which transform 16 oogonial cells from a seemingly undifferentiated cell cluster into 15 nurse cells and a primary oocyte that is

ISBN 0-12-137420-3

100,000 fold larger than the initial oogonial cells (King, 1970). During the 75-hour period when this massive growth occurs, the oocyte nucleus remains transcriptionally inactive (Bier, 1963) and the major transcriptional programs of oogensis are carried out by other cells. Ovarian accessory cells, such as nurse cells and follicle cells, synthesize and secrete cytoplasmic factors and chorion membrane proteins, respectively (Hughes and Berry, 1970; Kafatos, 1975). Other major transcription for oogenesis is carried out by a set of extraovarian cells, the fat body cells, which line the abdominal wall. These cells synthesize the abundant egg yolk proteins and secrete them into the female circulatory system.

Synthesis of the three yolk proteins (YP1, YP2 and YP3) occurs rapidly and coordinately after females hatch and continues for several weeks as judged by production of normally developing eggs. The transcriptional and translational control of this rapidly initiated and sustained yolk protein synthesis is not known. We have recently isolated and partially characterized the genes for these yolk proteins (Figure 1) and are now using these genes to describe the pattern of transcription and translation and to examine the mechanism by which this transcription is begun and sustained. This report describes progress we have made in identifying the different mRNAs that code for the three yolk proteins and in describing their transcription and translation during vitellogenesis.

RESULTS AND DISCUSSION

Yolk Protein Genes are Not Amplified during Vitellogenesis

A consideration of the large amount of translatable yolk protein mRNA present in adult flies (unpublished), the negligible lag in YP synthesis following hatching and the rapid appearance of large amounts of yolk protein in the female blood (Kambysellis, 1977) suggests that a special mechanism may be necessary for rapid transcription and/or translation of these genes and their mRNAs. Gene repetition is one possibility. Our previous experiments (Barnett et al., 1980) demonstrated that embryonic and pupal DNA have only single copies of each YP gene. Presumably the germ line also has only single copies. To investigate the possibility that these genes are amplified in the female fat body tissue that is actively synthesizing yolk proteins we extracted DNA from one to twenty-four hour old female abdomens (enriched by dissection for fat body cells) and, for comparison, from similarly staged and dissected male abdomens that do not normally synthesize yolk proteins. Each of the DNAs were digested with endonuclease Hind III (see Figure 1 for a restriction map of each YP gene; also, Barnett *et al.*, 1980) and the resulting fragments were electrophoresed in agarose gels. A southern blot of this gel was hybridized with ^{32}P-labelled YP1, YP2 and YP3 subcloned DNAs (Figure 2). By comparing the intensities of the labelled fragments in female DNA (lane 4) with those derived from either male DNA (lane 5) or from known amounts of each gene fragment electrophoresed in adjacent lanes (1 -3 and 6 -7), it is evident that the three yolk protein genes are represented only once in the DNA of fat body cells. This finding makes unlikely the possibility that transcription from amplified gene copies is responsible for the rapid appearance of large amounts of yolk proteins.

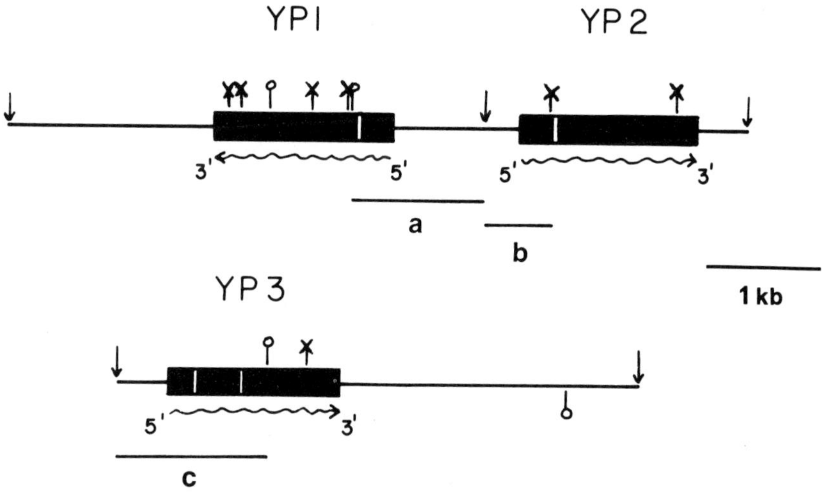

FIGURE 1. A Diagram of the Yolk Protein Gene Regions.
The regions complementary to yolk protein mRNA are shown by the solid black boxes. The open white boxes represent the introns whose presence was determined in this work and whose size and position were determined by M.C. Hung. The direction of transcription (Wensink, unpublished data) is shown by the arrow. Restriction site symbols are Hind III (↓), Pst I (✻), Bam HI (⚲) and Sal I (ȸ). The subclones of each gene are bounded by Hind III sites for pYP1 and pYP2 and by Hind III and Sal I sites for pYP3 (see Barnett *et al.*, 1980). The subclones specifically hybridizing to the 5' end of the different YP gene transcripts are indicated by the lines labelled "a" (subclone pYP1-5'), "b"(pYP2-5'), and "c" (pYP3-5') for the 5' end of the YP1, YP2 and YP3 genes, respectively.

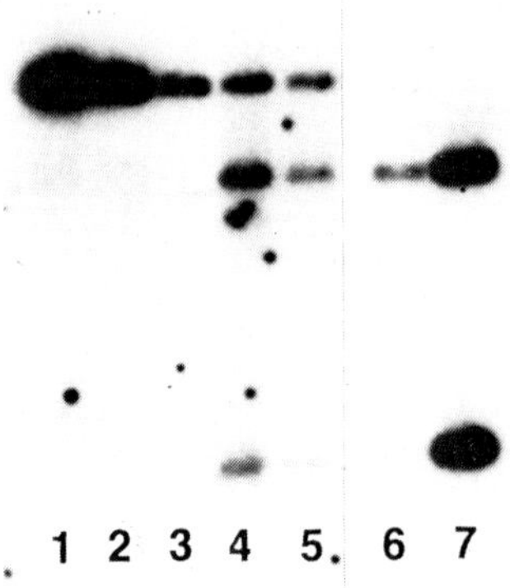

FIGURE 2. YP Gene Copy Number in Female Body Tissue. DNA was extracted from 1- to 24-hour-old female and male abdomens enriched in fat body cells (Barnett _et al._, 1980). Two micrograms of each DNA were digested with _Hind_ III restriction endonuclease and were electrophoresed in a 1% agarose gel (lanes 4, female DNA, and 5, male DNA). Adjacent lanes were loaded with amounts of Hind III digested phage λDmYP5A (a clone containing both YP1 and YP2 genes; lane 6) and λDmYP1C (a clone of the YP3 gene; lane 3) (Barnett _et al._, 1980) to yield the equivalent of a single copy of _the YP_ genes in 2 μg of the genomic DNA. Amounts equivalent to 5 copies of YP1 and 2 gene (lane 7) and 5 and 15 copies of YP3 (lanes 2 and 1) were also included in the gel. The electrophoresed DNA was transferred to nitrocellulose paper and then hybridized with a mixture of ^{32}P-labelled pYP1, pYP2 and pYP3 DNAs. The YP genes are on the X-chromosome (Barnett _et al._, 1980 and Postlethwait and Jowett, 1980) so a two-_fold_ difference between females and males is expected.

Characterization of Yolk Protein Gene Transcripts

The 5' regions of each gene were subcloned (see Figure 1) to provide hybridization probes for specifically assaying each gene's transcript. Homology between the three subcloned segments (pYP1-5', pYP2-5' and pYP3-5') was so slight that southern hybridization signals were not detected and were well below 5% of the signal from self-hybridization (unpublished data).

These hybridization probes were used to detect the yolk protein gene transcripts. Cytoplasmic poly A^+ RNA from adult flies (Barnett _et al._, 1980) was electrophoresed in denaturing formaldehyde-agarose gels (Boedtker, 1971), transfered to nitrocellulose paper (Thomas, 1980) and then hybridized with nick-translated ^{32}P-labelled 5' specific probes. As seen in Figure 3, different-sized RNAs are detected by each probe. Relative to pBR322 DNA size standards, YP1 mRNA is 1.59-1.61 kb in length (lane 2) and the mRNA for YP3 is slightly shorter, 1.53 -1.55 kb (lane 8). Interestingly, there appear to be two YP2 transcripts, 1.59-1.61 and 1.66-1.68 kb in length (lane 4). While these two YP2 transcripts are not well resolved in this lane, longer electrophoresis of a similar gel clearly shows the two bands (lane 6). The lengths of all the YP RNAs are the same whether unfractionated or poly A^+ RNA is used for gel electrophoresis and blotting. The lengths of RNA expected for the sizes of proteins they encode (45,000 to 46,000 daltons) indicates that each of the YP mRNAs contain untranslated regions.

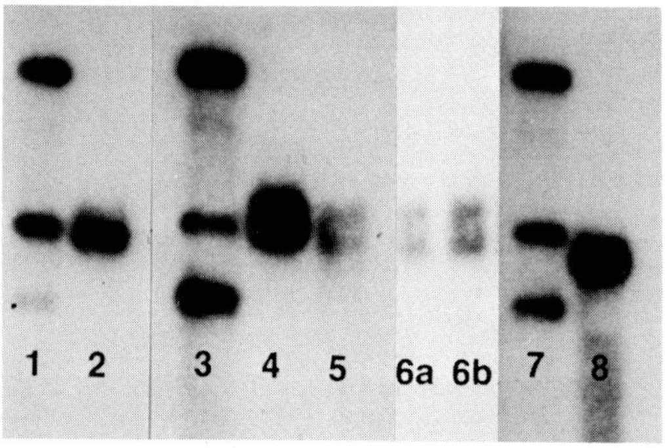

FIGURE 3. Detection of YP RNA Transcripts.

Poly A^+ adult RNA (lanes 2, 4 and 8; 0.1 μg), unfractionated polysomal RNA (Lis _et al._, 1978) from female bodywalls (lane 5, 6a and 6b; 1 μg) and denatured pBR322 DNA fragments of an Ava I/Eco RI/Pst I and a Hinf I digest (Sutcliffe, 1979) (lanes 1, 3 and 7; 0.05 ng) were incubated at 65°C for 5 min in 50% formamide, 2.2 M formaldehyde and 1X gel buffer (20 mM MOPS, 5 mM Na acetate, 1 mM EDTA, pH 7.0), cooled on ice, then electrophoresed in a 2.2 M formaldehyde, 1.5% agarose gel. The nucleic acids were transferred onto nitrocellulose paper in 20 X SSC (Thomas, 1980), rinsed briefly in 4 X SSC, then hybridized with ^{32}P-labelled pYP1-5' DNA (lanes 1 and 2), pYP2-5' (lanes 3, 4, 5 and 6) and pYP3-5' DNA (lanes 7 and 8).

The two YP2 transcripts are of special interest. Our original characterization of the YP genes showed that _in vitro_ translation of hybrid-selected YP2 mRNA yielded _only_ a _single_ polypeptide (Barnett _et al._, 1980). However, when polysomal RNA from fat body _tissue_ is electrophoresed in denaturing gels, two RNA bands are visible after hybridization with a ^{32}P-labelled YP2 5'-specific probe (Figure 3, lane 5). These observations suggest that two YP2 gene transcripts occur *in vivo*, but that all the information needed to synthesize a full-length YP2 polypeptide is present in each of the transcripts.

Our recent experiments have shown that YP mRNAs are also detectable in early and late stage oocytes (data not shown), but the location of the cells responsible for their synthesis (whether follicle or nurse cells) has not yet been determined.

Accumulation of Yolk Protein RNA during Vitellogenesis

Since transcription of the YP genes is not directed by amplified templates, it is likely that the simultaneous appearance of the three yolk proteins in the hemolymph of newly eclosed adult flies proceeds either from mRNA stored during pupation or from high rates of transcription following eclosion. As a first step in understanding some of the transcriptional aspects of YP synthesis, RNAs were prepared from female abdomens at thirteen different times after hatching and were fractionated and detected (Figure 4).

It is clear that at 0 hours no YP RNA is detectable, but that it accumulates rapidly up to 24 hours. This accumulation is parallel to the accumulation of the yolk proteins themselves (Kambysellis, 1977) and rules out any major form of prestored mRNA. Furthermore, the mRNAs for each of the yolk proteins increases similarly up to 24 hours, with a dramatic rise occurring between 12 and 18 hours after hatching. This suggests that the accumulation, and therefore probably the synthesis of the three yolk protein mRNAs, is coordinately and possibly identically, regulated. We are presently quantitating the amounts of each of the three YP RNAs relative to one another and to another known messenger RNA present in the preparations. Our current evidence from a large number of experiments strongly suggests a coordinate accumulation of the three RNA species, at least until 36 hours. At this point the YP RNA population appears to stablilize and then at about 144 to 168 hours it begins a gradual decline. The apparent modulation in levels of YP1 RNA after 120 hours may be real, although it has not been consistently observed. It is observed however, that the amount of YP1 mRNA increases at the time when both YP2 and YP3 mRNAs are declining (Figure 6, 144 and 168 hours).

In order to correlate the accumulation of the yolk protein mRNAs with the appearance of translatable yolk protein mRNA, we determined the ability of mRNAs from staged female

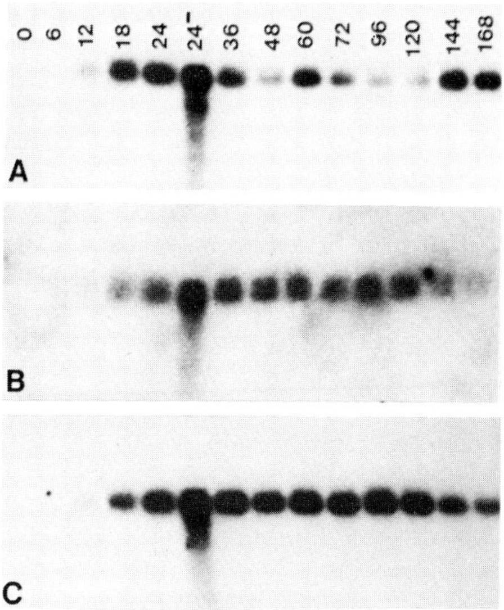

FIGURE 4. Accumulation of YP mRNA after Hatching.
Female adults were staged for the number of hours after hatching shown in Panel A. Body walls containing fat body cells were then dissected and used in the preparation of RNA. Poly A$^+$ RNA was isolated (Barnett *et al.*, 1980) from 1 μg of each RNA preparation. The poly A$^+$ RNA thus obtained from each stage was divided among three 2.2 M formaldehyde-1.5% agarose gels (Panels A, B and C) and electrophoresed. RNAs were then transferred to nitrocellulose paper and hybridized with either ^{32}P-labelled pYP1-5' DNA (Panel A), pYP2-5' DNA (Panel B) and pYP3-5' DNA (Panel C). The lane "24$^-$" is the poly A$^-$ fraction obtained after binding 24 hour poly A$^+$ RNA to oligo dT cellulose.

flies to synthesize yolk proteins in a cell-free translation system (Figure 5). As expected from the RNA accumulation experiment, the ratios between YP1, YP2 and YP3 polypeptides remain constant, further strongly suggesting coordinate increase in translatable transcripts of the three genes.

Synthesis and Secretion of Yolk Proteins

A correlation between the appearance of translatable mRNA and the appearance of newly synthesized yolk proteins was made by comparing the pattern of *in vitro* synthesized YP polypeptides with the set of YPs secreted in an organ culture by fat bodies of staged females (Figure 6). [It should be

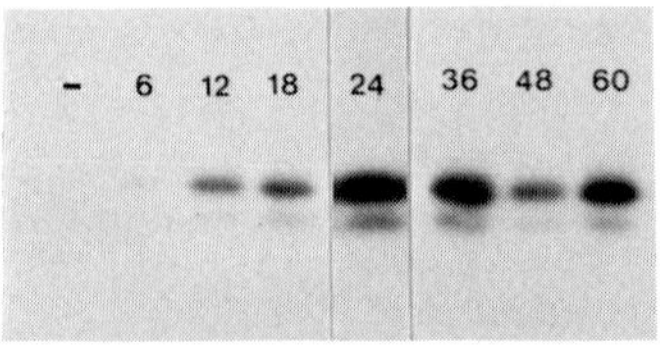

FIGURE 5. Cell-free Translation of poly A^+ RNA from Fat Body Tissue of Staged Female Flies.

20 ng of poly A^+ RNA from staged females prepared as described in the legend to Figure 4 were added to a 25 μl rabbit reticulocyte cell-free system (Paterson *et al*., 1977) containing 50 μCi of ^{35}S-methionine (~1300 Ci/mmole) and incubated for 1.5 hours at 37°C. Yolk proteins were then precipitated by addition of YP antiserum as described by Kessler (1975). Precipitates were electrophoresed in a 10% polyacrylamide-SDS gel (Laemmli, 1970), the gel was fluorographed (Bonner and Laskey, 1974) and exposed to X-ray film. "-" indicates a translation reaction to which no exogenous RNA was added.

noted that although both YP1 and YP2 are the same size after *in vitro* translation, the secreted forms of these two proteins are modified (Warren and Mahowald, 1979) and differ by about 1000 daltons.] For the first 24 hours all three proteins increase in parallel with the accumulation of mRNA (Figures 4 and 5). After this time there is a dramatic decrease in the amount of the yolk proteins secreted into the medium and there is a change in the ratios between the YPs. This occurs despite the continued high levels of translatable yolk protein mRNA and suggests that control is being exerted at the level of yolk protein mRNA translation.

Since large numbers of eggs are just beginning to be made at high rates at 36 hours, it is unlikely that there is an *in vivo* decrease in the amounts of YPs synthesized and secreted. One possible explanation is that there may be a feedback mechanism between the ovaries and fat body tissue and this is disrupted in the organ culture described here. This possibility seems unlikely since Postlethwait has presented convincing data that there is no feedback occurring in agametic female flies (Postlethwait *et al*., 1980). An alternative explanation is that the proteins secreted into the medium are further modified and are no longer recognized by the YP-antibody used in our assay. Relevant to this explanation is a recent report by Warren and Mahowald (personal communication) that yolk proteins in the hemolymph and in the egg are immunologically different.

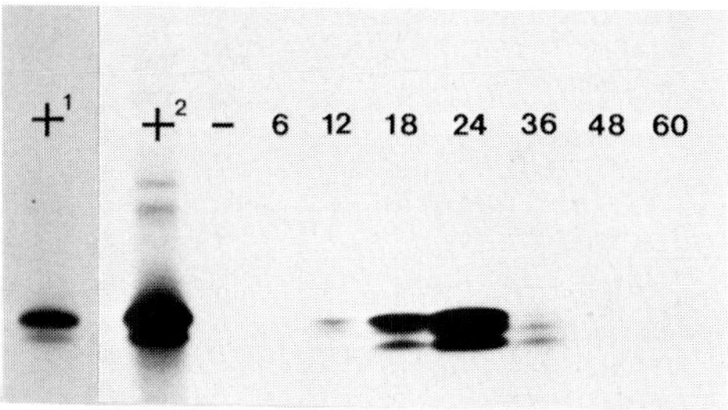

FIGURE 6. Synthesis and Secretion of Yolk Protein by Cultured Body Walls.

Female flies were staged for various numbers of hours as shown above the lanes. During the last hour of staging, body walls from ten females of each stage were prepared and incubated for 1 hour in 25 μl of Robb's Ringers medium (Robb, 1969) containing 1 μCi/ μl of ^{35}S-methionine (~1300 Ci/mmole). After incubation, body walls were removed and the medium was treated as for the supernatants described in Figure 5. Immunoprecipitated yolk proteins were electrophoresed in a 10% polyacrylamide-SDS gel and fluorographed.

The lanes marked with numbers indicate the age in hours of each staged set of animals. The lane marked "$+^1$" shows the immunoprecipitated yolk proteins from a cell-free translation of poly A^+ adult RNA used as a marker; "$+^2$" is a five fold longer exposure of the same lane. The lane marked "-" is the immunoprecipitate from proteins secreted by ten male abdomens into culture medium prepared as described above.

ACKNOWLEDGMENTS

We thank Mark Brennan and Tony Mahowald for a gift of YP specific antiserum, Mien Chie Hung for valuable discussions and communication of unpublished results and Michael Garabedian for restriction site information. We are grateful for support from an American Cancer Society postdoctoral fellowship (TB), a Medical Foundation fellowship (TB) and research grants (PCW and TB) and a Research Career Development Award (PCW) from the National Institutes of Health.

REFERENCES

Barnett, T., Pachl, C., Gergen, J.P. and Wensink, P.C. (1980). Cell 21, 729.

Berk, A.J. and Sharp, P.A. (1977). Cell 12, 721.

Bier, K. (1963). J. Cell Biol. 16, 436.

Boedtker, H. (1971). Biochim. Biophys. Acta 240, 448.

Bonner, W. and Laskey, R.A. (1974). Eur. J. Biochem. 46, 83.

Bownes, M. and Hames, B.D. (1978). FEBS Lett. 96, 327.

Hagenbuchle, O., Bovey, R. and Young, R.A. (1980). Cell 21, 179.

Hughes, M. and Berry, S.J. (1970). Dev. Biol. 23, 651.

Kafatos, F.C. (1975). In Control Mechanisms in Development. Adv. in Exp. Med. and Biol. 62, p. 103.

Kambysellis, M.P. (1977). J. Exptl. Zool. 17, 535.

Kessler, S.W. (1975). J. Immun. 115, 1617.

King, R.C. (1970). Ovarian Development in Drosophila melanogaster. Academic Press, New York and London.

Laemmli, U. (1970). Nature 227, 680.

Lichtler, A.C., Detke, S., Phillips, I.R., Stein, G.S. and Stein, J.L. (1980). Proc. Nat. Acad. Sci. USA 77, 1942.

Lis, J.T., Prestidge, L. and Hogness, D.S. (1978). Cell 14, 901.

McDonnell, M.W., Simon, M.N. and Studier, F.W. (1977). J. Mol. Biol. 110, 119.

Paterson, B.M., Marciani, D. and Papas, T.S. (1977). Proc. Nat. Acad. Sci. USA 74, 4951.

Postlethwait, J.H., Lauge, G. and Handler, A.M. (1980). Gen. Comp. Endocrin. 40, 385.

Robb, J.A. (1969). J. Cell Biol. 41, 876.

Setzer, D.R., McGrogan, M., Nunberg, J. and Schimke, R.T. (1980). Cell 22, 361.

Southern, E. (1975). J. Mol. Biol. 98, 503.

Sutcliffe, J.G. (1978). Nucl. Acids Res. 5, 2321-2322.

Thomas, P.S. (1980). Proc. Nat. Acad. Sci. USA 77, 5201.

ORGANIZATION AND EXPRESSION OF THE ACTIN MULTI-GENE FAMILY IN DICTYOSTELIUM[1]

Michael McKeown[2]
Alan R. Kimmel[3]
Richard A. Firtel[4]

Department of Biology
University of California
La Jolla, California

ABSTRACT

We have shown that at least seven genes of the 17 member actin multi-gene family in Dictyostelium are transcribed. The genes are expressed at different levels and are not coordinately regulated. We have determined the 5' ends of the actin mRNAs and identified sequences common to transcription initiation sites of other Dictyostelium genes. Sequencing of the 3' untranslated regions of the actin genes has revealed subfamilies suggesting certain evolutionary relationships.

INTRODUCTION

Dictyostelium amoebae grow vegetatively on bacteria or in axenic liquid culture. Upon starvation, Dictyostelium initi-

[1]*This work was supported by grants from the National Institutes of Health and the National Science Foundation.*
[2]*National Science Foundation Predoctoral Fellow, National Institutes of Health Predoctoral Trainee.*
[3]*American Cancer Society, California Division Senior Fellow.*
[4]*Recipient of an American Cancer Society Faculty Research Award.*

ISBN 0-12-137420-3

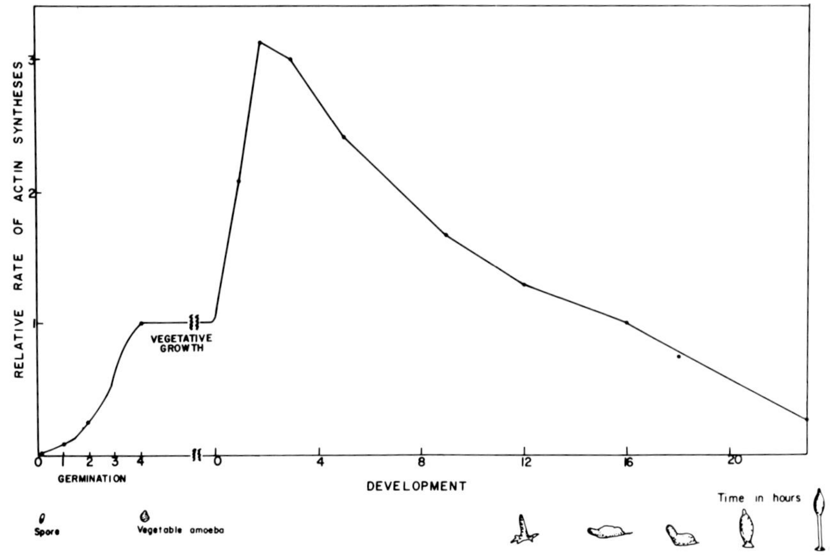

Figure 1. Relative rates of actin protein synthesis during Dictyostelium development. (Summarized from 1-6).

ate a synchronous morphological and biochemical developmental cycle. After approximately six hours, cells become chemotactic to cAMP and migrate toward aggregation centers. By ten hours after starvation they form tight multi-cellular aggregates and then continue to develop into mature fruiting bodies containing spore and stalk cells. Spores can be induced to germinate and release amoebae which re-initiate the cycle.

Actin is a major developmentally regulated protein in Dictyostelium. As shown in Figure 1, there is no detectable actin synthesis in newly activated spores; four hours later, at emergence, actin represents ~1% of the newly synthesized protein (1, 2). This level remains constant throughout vegetative growth. However, when development is initiated by starvation actin synthesis rises three-fold (1-6). The rate of actin synthesis then decreases during the rest of development (1-6). These changes in actin synthesis are accompanied by parallel changes in the relative mass of actin RNA as measured by in vitro translation and RNA excess hybridization (1-6). The rate of synthesis of actin mRNA also varies in an equivalent manner suggesting that the developmental regulation of actin protein synthesis is at the level of transcription.

There are 17 actin genes in Dictyostelium which give rise to two forms of actin mRNA of 1.25 kb and 1.35 kb (7-9).

Table 1. Percentage of Actin RNA Derived from Different Genes In Development.[a]

Actin Gene	0	3	8	13	20
actin 8	18	27	30	19	22
actin 6	23	9	6	10	2
actin 5	5	4	5	3	4
actin M6	1	3	2	2	3
actin 2 sub 2	Ud[b]	Ud[b]	Ud[b]	-	-
actin 4	-	D[c]	-	-	-

Time in development (hr)

[a] average of two experiments; [b] Ud - undetectable, <1% of total actin mRNA; [c] D - detected but not quantitated

These mRNAs encode one major and several minor isoelectric actin forms (1, 2, 7). We have isolated recombinant plasmids containing 10 different actin genes and cDNA plasmids derived from two additional genes. Restriction mapping and DNA blot hybridization have shown that the actin genes comprise a disperse multi-gene family with heterogeneous 3' and 5' flanking sequences (7-12). Sequencing of the 5' ends of different genes has shown that the protein coding sequences are very similar but the untranslated regions of the mRNAs differ substantially in nucleotide sequence.

The non-homologies at the 5' end of actin genes have allowed us to map the sites of initiation of transcription and also to determine the relative levels of expression of different actin genes during development. We have also observed that the actin genes can be classified into sub-families on the basis of their 3' untranslated sequences. Finally, we have identified 5' flanking sequences common to a number of Dictyostelium genes.

RESULTS AND DISCUSSION

A. *Differential Expression of the Actin Genes*

We have developed a modification of the Berk and Sharp S_1 mapping procedure (13) which allows us to examine the expression of individual actin genes throughout development (11).

actin 8	TTCAATAAATAGGATTTTTTTTTATTTTATTTTTTTTTTTCAATACTTAATTCAAATATATATATATAAATCATTTAAAAAUG (Met)
actin 6	TTTCTGAGATTATAAAATGAAATTTTTTTTTTTTTTTTAATTAATTCAAAAAATAATCAAATAAATAAATATAATATAAAAUG
actin 5	GAGATTTGGTATAAATACAAAAAAAAAATTGTTTTTAAATCATTATTAATAAAAACTTAAAATAAATTATATATAAAAAAAUG
actin M6	AAAAAAAGTTTAAAAACTAATTTTTTTTTTTTTTTTTAATCATTCTAATATAAAATAATAAATACAATTAAAAAATAAAAAUG
actin 4	TTTGATTTTTGGATAAATATTAAATTTTTTTTTTTTTCTTATCATTATTCAATCTATAATATATAAAATACAATAAAAAAAUG
actin 7	TTCAGCAATTTAAAATCAAAATTTTTTTTTTTTTTAATCAATTTAAATAAATACTCAATTAAATAATAAAATATAATAAAAUG
actin 2-sub 1	TGTTATTTGGGAAATTATAAATTAAAATTTTCTTATTATAATTTTTAAAAAACTTAAATAAATAATAATAATATAATAAAAUG
actin 2-sub 2	AATTTTTTTTTTTTTTTTTTTTTTTATTTAATTTTTTTTTTTTTTTTATTTTTTTTTTTTTTTAAATATTTTGAAATAATAAAUG
actin 3	AAAAAAAAAATAAAAAACTTTAAAAAATAAACAATAATTTATATATTCAAATAATAATCAAAATAATAAAAAAAAAAAAAAUG

Figure 2. 5' regions of actin genes. Sequences are shown from the AUG initiation codon past the TATA box. TATA box is labeled A and poly(T) region is labeled B. Regions of transcription initiation are shown by stars.

The single-stranded DNA from the 5' end of an actin gene is labeled and hybridized in DNA excess to poly(A)+ RNA. When the hybrids are treated with S_1 nuclease and the fragments separated by electrophoresis, we can identify fragments which result from protection beyond the protein coding sequence common to all actin genes. These fragments arise from protection by the 5' end of the mRNA derived from that gene. Since RNA from other actin genes also hybridizes, some of the labeled DNA will be protected only as far as the initiator AUG codon. The amount of DNA protected by homologous RNA can be compared with the amount protected by heterologous actin mRNA and the ratio of these two values can be used to determine the relative expression of a particular gene at any stage of development. Such results are summarized in Table 1. Clearly, the actin genes are expressed at widely differing levels and all of the genes are not coordinately regulated through development. RNA from the pDd actin 6 gene is present at high levels in vegetative cells but is only a small fraction of actin RNA by late in development. From the results of partial DNA sequencing, we have concluded that all but one of the expressed genes shown in Table 1 (and all of the cDNAs thus far examined) seem to give rise to the major actin form sequenced by Vanderkerckhove and Weber (14). Actin M6 which would give rise to an actin form with two amino acid changes in the first 50 amino acids, is expressed as less than 5% of the actin RNA. RNA has not been detected from actin 2-sub 2, a gene which would yield a protein differing from the standard sequence in 12 out of 105 amino acids now predicted from the DNA sequence.

B. *5'-Actin Sequences*

When S_1 protected fragments are electrophoretically separated adjacent to a DNA sequencing ladder we can predict the site of transcription initiation of actin mRNA. Several 5'-actin termini are detected. One possible explanation is that actin genes have heterogeneous start sites. Figure 2 shows the sequences 5' to the initiation of transcription for five transcribed actin genes. All of these sequences contain two conserved regions. A TATA-like sequence, labeled A, is common to the 5'-end of most eukaryotic genes (15) and a poly (T) sequence, labeled B, is located between the TATA box and the site of transcription initiation for a number of *Dictyostelium* genes. The actin genes have a 5'-untranslated region of 30 to 40 bases.

We have also included other actin 5' sequences. Actins

	Ter1 10 20 30 40
actin 8-1	TAAATTATTTAATAAATAATAAAAAAACAAATTGTTGTAATAATCT(A_{45})
actin 8	TAAATTATTTAATAAATAATAAAAAAACAAATTGTTGTAATAATCTAATAT
actin ITL-1	TAAAC AAATAATTAAAACTAGTGATGAAAGTGCTTCTCACAAAcAATTAT
actin 5	TAAACaAAAaAAAAAAAcCgAGTGATGAAAGcGCTTCTCACAAA ATTAT
actin A1	TAAAtcA TGATGAAAGTGCTTCaCAtAAAAAtAATA
actin 2-sub 2	TAAAtTAAtTAAAAAAAtTTAGTGATGAAAGTGCTTCTCACAcAAAAAtTA
actin 6	TAAACTAAAcAATTAAAATcAGTGATGAAAtgtCTTCTCACActtAAcAat
	50 60 70 80 90
actin 8 cont.	TTTCTTTTTTTTTTAATTTTTTTTTTTTAAATCTTAATAATTATTAAGTTA
actin ITL-1	GtAAAATATaTAATAaaATAc AtTaTTtAATCaTTTTTATTTTTgtTTTA
actin 5	GaAAAATATtUAATAgtATAatAaTtTTaAATCtTTTTTATTTTT TTTA
actin A1	ATAataATaTAaCAATAATAATATTTAAATgTatAATAAAATATAATT A
actin 2-sub 2	tTAtATATgTA CAATAATAAcAATAAAAaCcC AATAAAATATAA A
actin 6	ATA ATATtTA tAtguATAATAATAAAATCtc AATAAAATATAATTctt
	100 110 120 130 140 150
actin 8 cont.	TTTTAATTTTTTTTTTTTTTTTTTTTTTTTTTTTTTTTTTTTTCTATCAAA
actin ITL-1	GTTGTTGaTCTTTATCCGACTaTttAAA AttAATTGT poly(A)
actin 5	GTTGTTG TCTTTATCCGACTtTaaAAAtaaAaaAATTGTA_{11}
actin A1	CTTTTTT TTTaATgGTtGTTGAT CTTTATCCGACCTTA_{20}Tpoly(A)
actin 2-sub 2	CTTTTTTcTTTgATAGTCGTTGAT CTTTATCCGACCTTtA_{14}
actin 6	aTTTTTatTTTttgAaTCGgTtgTtgtCTTTATCCagCCaTcA_{33}

Figure 3. 3' untranslated regions of actin genes. Sequences are shown beginning with the UAA termination codon. Sequences not separated by bars show homology. Small letters indicate bases different from the majority of bases at that position in that group.

7 and 2-sub 1 possess a TATA box and a poly(T) run, but we do not yet have evidence that these genes are transcribed. Actin 2-sub 2 is probably not transcribed and is probably a pseudogene. Clearly it does not possess a TATA box. Another actin gene, actin 3, also does not contain the appropriate TATA and poly(T) run and may also be a pseudogene.

C. 3'-ends of Actin Genes

We have also examined the 3' non-coding regions from a number of actin genes and cDNAs (12). These sequences are compared in Figure 3. The 3' untranslated region of the actin cDNAs are of different lengths. The cDNA plasmids pcDd actin A1 and pcDd actin ITL-1 have 3' untranslated regions of 119nt and 135nt respectively preceding the poly(A) tail while pcDd actin 8-1 has a 3' untranslated region of only 43nt prior to the poly(A) tail. This >75nt difference in length is sufficient to account for the difference between the two size classes of actin mRNA observed (7, 9). As evidenced by pcDd actin A1, the longer actin RNA contains a discontinuous poly (A). Such an organization of Dictyostelium poly(A) has been suggested previously (16-19). A short oligo (A) sequence is encoded in the genome while the poly(A) stretch is added post-transcriptionally. Further support for this idea comes from examination of the genomic clones. The clone pDd actin 8, parent to the cDNA pcDd actin 8-1, has no oligo(A) sequence near the site of poly(A) addition, but all of the other genes shown do have an oligo(A) region beginning 130-140nt beyond the UAA termination codon.

The comparison of the different 3' untranslated regions shown in Figure 3 also reveals that there are at least three sub-families of 3' end sequences. The first sub-family consists of pDd actin 8 which is unrelated in sequence to any of the other actin 3' ends yet sequenced. This gene encodes the shorter mRNA. The other genes, which apparently code for the longer actin mRNA, comprise two sub-families. One is composed of pcDd actin ITL-1 and pDd actin 5; the other is composed of pcDd actin A1, pDd actin 2-sub 1 and pDd actin 6. In addition to the intrafamily homology there is considerable interfamily homology for the first ~45nt of the 3' untrans-

```
               A                  B                *
M4      ATTTATTAATATTTTTATCTTTTTTTTTTTTTTTTAAAAAACTATTTTAGGAGA

               A                 B      **  *
Disc    AACTATAAATAAAAGAAGTTTTTTTTTTTTTTAATCATTAAATTGAAAATCAAA
  I A

               A                 B     **  *
Disc    AAGTATAAAAAGAAAAAAAATTTTTAATTAAAATCATTAAATTGAAAAATTAAA
  I B

               A                 B     **  *
Disc    AAGTATAAAAAGAAAAAAAATTTTTAATTAAAATCATTAAATTGAAAAATTAAA
  I C
               A                   B    **  *
Disc    AACTATAAAAAGAAAATATGATTTTCATTTTAAAATCATTAAACTGAAAAATTA
  I D
               A                         B              *
D 2     AAATATTAAAAAATGAACAAAAGAAAGTTTTTTTTATTTTGAAAACTTATGACC

               A                  B              * *   *    *
actin   TTCAATAAATACGATTTTTTTTTATTTTATTTTTTTTTTTCAATACTTAATTCA
 8
               A               B         **   **
actin   GATTATAAAATGAAATTTTTTTTTTTTTTTTAATTAATTCAAAAATAATCAAAT
 6
               A                B      **   ****  **
actin   TGGTATAAATACAAAAAAAAAATTGTTTTTAAATCATTATTAATAAAAACTTAA
 5
               A                B        *  *
actin   AAGTTTAAAAACTAATTTTTTTTTTTTTTTTTAATCATTCTAATATAAAATAAT
 M6
               A              B          *   * **
actin   TTGGATAAATATTAAATTTTTTTTTTTTTCTTATCATTATTCAATCTATAATAT
 4
```

Figure 4. 5' flanking sequences of transcribed *Dictyostelium* genes. M4 codes for a low abundance mRNA which is associated with a transcribed repeat (21). Disc I A, B, C and D are different genes for the developmentally regulated lectin Discoidin I (22). D2 codes for a Pol II transcribed small nuclear RNA (23). The 5' sequences of the actin genes that are known to be transcribed are also listed. The genes are aligned by the common TATA region, labeled A. Sequences labeled B represent the conserved poly(T) region. Stars show regions of transcription initiation.

lated region of these genes which encode the 1.35 kb RNA. We believe this indicates that the two 3' families were created by a duplication of a common ancestor. The members of two sub-families must have been created by more recent divergence and further duplication.

D. Sequences Common to the 5'-ends of Dictyostelium Genes.

Figure 4 shows all of the known sites of initiation of transcription in Dictyostelium. They all possess a TATA box followed by a poly(T) run which precedes the initiation of transcription. M4 encodes a low abundance class mRNA and is associated with a short interspersed repeat sequence (20). Discoidin I is a lectin important for cell-cell cohesion during Dictyostelium development and is encoded by a four-member multi-gene family (21, 22). D2 is capped small nuclear RNA; recent evidence suggests that such capped species in addition to mRNAs are transcribed by RNA Polymerase II (23). These genes are expressed at widely varying levels and with very different patterns of expression during development; since they all contain a TATA box and poly(T) sequence it is unlikely that these regions are involved in differential gene expression. Rather, such regions may determine the site of transcription initiation (24). We do not observe a conserved CCAAT sequence approximagely 50nt upstream from the TATA box as is seen for the β globin gene family and several other eukaryotic genes.

CONCLUDING REMARKS

We have described the developmental changes in specific actin mRNAs and have determined sequence similarities between the transcription initiation sites for actin and a number of other Dictyostelium genes. We have not been able to identify specific regions involved in the control of actin genes during the life cycle. We feel that the imminent development of transformation and in vitro transcription systems for Dictyostelium will allow us to directly study the relationship between DNA sequence and gene expression.

REFERENCES

1. MacLeod, C., Firtel, R. A. and Papkoff, J. (1980).

Dev. Biol. 76,263-274.
2. MacLeod, C. (1979). Ph.D. thesis. University of California, San Diego.
3. Tuckman, J., Alton, T. A. and Lodish, A. F. (1974). *Devel. Biol.* 40,116-128.
4. Alton, T. A. and Lodish, H. F. (1977). *Devel. Biol.* 60, 180-206.
5. Margolskee, J. P. and Lodish, H. F. (1980a). *Devel. Biol.* 74,37-49.
6. Margolskee, J. P. and Lodish, H. F. (1980b). *Devel. Biol.* 74,50-64.
7. Kindle, K. L. and Firtel, R. A. (1978). *Cell* 15,763-778.
8. Bender, W. Davidson, N., Kindle, K. L., Taylor, W., Silverman, M. and Firtel, R. A. (1978). *Cell* 15,779-788.
9. McKeown, M., Taylor, W. C., Kindle, K. L., Firtel, R. A., Bender, W. and Davidson, N. (1978). *Cell* 15,789-800.
10. Firtel, R. A., Timm R., Kimmel, A. and McKeown, M. (1979). *Proc. Natl. Acad. Sci. USA* 76,6206-6210.
11. McKeown, M. and Firtel, R. A. (1981a). *Cell*, in press.
12. McKeown, M. and Firtel, R. A. (1981b). Manuscript submitted for publication.
13. Berk, A. J. and Sharp, P. A. (1978). *Proc. Natl. Acad. Sci. USA* 75,1274-1278.
14. Vandekerckhove, J. and Weber, K. (1978). *J. Mol. Biol.* 126,783-802.
15. Goldberg, M. (1979). Ph.D. thesis, Stanford University.
16. Jacobson, A., Firtel, R. A. and Lodish, H. F. (1974). *Proc. Natl. Acad. Sci. USA* 71,1607-1611.
17. Firtel, R. A., Jacobson, A., Tuchman, J. and Lodish, H.F. (1974). *Genetics* 78,355-372.
18. Firtel, R. A., Kindle, K. and Huxley, M. P. (1976). *Federation Proc.* 35,13-22.
19. Firtel, R. A. and Jacobson, A. (1977). *International Review of Biochemistry: Biochemistry of Cell Differentiation II, Vol. 15* (J. Paul, ed.), pp. 377-428. University Park Press.
20. Kimmel, A. R. and Firtel, R. A. (1980) *Nucl. Acids Res.*, 8,5599-5610.
21. Rowekamp, W., Poole, S. and Firtel, R. A. (1980). *Cell*, 20,495-505.
22. Poole, S., Firtel, R. A., and Lamar, E. E., and Rowekamp, W. Manuscript in preparation.
23. Wise, J. A. and Weiner, A. M. (1980). *Cell* 22,109-118.
24. Corden, J., Wasylyk, B., Buchwalder, A., Sassome-Corsi, P., Kedinger, C. and Chambon, P. (1980). *Science* 209, 1406-1414.

THE ACTIN GENES OF DROSOPHILA: HOMOLOGOUS PROTEIN CODING REGIONS WITH DISTINCT STRUCTURAL ARRANGEMENTS AND CHROMOSOMAL LOCATIONS[1]

Eric A. Fyrberg[2], Beverley J. Bond*, N. Davis Hershey, Katharine S. Mixter*, and Norman Davidson

Department of Chemistry and Division of Biology*
California Institute of Technology
Pasadena, California 91125

ABSTRACT. We are investigating the structure and expression of the *Drosophila melanogaster* actin genes. Our findings are as follows. There are six actin genes per haploid *Drosophila* genome, and each gene is located in a distinct chromosomal region. The positions of introns within these genes are variable: DmA2 is split in the 5' untranslated region, DmA4 within codon 13, and DmA1 and DmA6 within codon 307. The primary sequence of each *Drosophila* actin is like those of vertebrate cytoplasmic actins, except that in each case a cysteine precedes the three acidic residues of the N-terminus. We relate these findings to recently acquired data from actin genes of other species.

INTRODUCTION

Actin is an important structural component of all eucaryotic cells (for a review see reference 1). In muscle cells actin monomers (m.w. 43,000) polymerize to form precisely arranged thin filaments which are an essential component of the contractile apparatus (2). In nonmuscle cells actin filaments form a cytoskeletal network which apparently integrates the activities of various cytoplasmic elements (3,4).

[1]This work was supported by NIH grants GM 10991-21 and GM 20927.

[2]Present address: Department of Biology, The Johns Hopkins University, Baltimore, Maryland 21218.

ISBN 0-12-137420-3

For several reasons actin genes are particularly interesting subjects for developmental studies. Despite the fact that the amino acid sequence of this protein is highly conserved, multiple actin proteins (isoforms) are known to exist in a variety of metazoans (5,6,7). In several cases, these isoforms are known to be the products of a small family of genes (8,9,10). Synthesis of these actin isoforms is precisely controlled during metazoan ontogeny, indicating that the activities of particular actin genes are regulated by distinct mechanisms.

To better understand the evolution and functioning of this gene family, we have isolated and characterized the actin genes of *Drosophila melanogaster*.

RESULTS

Isolation of All Members of the Actin Gene Family

To isolate the entire *Drosophila* actin gene family, we screened a recombinant bacteriophage library of *Drosophila* genomic DNA with a probe derived from λDmA2, our original actin gene isolate (9). From a screen of 40,000 phages, 30 which gave a strong signal were selected. Restriction maps of these phages indicated that they fell into six distinct classes. To demonstrate that these six classes of phage accounted for all members of the *Drosophila* actin gene family, we digested a representative of each class with EcoRI and separated the digestion fragments on an agarose gel. In an adjacent lane of the same gel an EcoRI digest of genomic *Drosophila* DNA was electrophoresed. DNA fragments were then blotted to nitrocellulose and hybridized with a labeled actin gene probe. As shown in Figure 1, the seven prominent bands of the genomic EcoRI digest can all be accounted for by the six representative phages. We have designated these phages λDmA1-λDmA6 based on the decreasing sizes of the EcoRI fragments which hybridize to the actin probe. Each of the phages contains a single hybridizing EcoRI fragment except λDmA6, which contains two.

Restriction Mapping of Chimeric Phages Containing Genomic Actin Sequences

Restriction maps prepared from λDmA1-λDmA6 (Fig. 2) failed to reveal a particular arrangement of restriction sites which could be used to position the actin structural genes. To better localize the actin coding regions, digested phage DNAs were blotted to nitrocellulose and hybridized to the λDmA2 structural gene probe. These experiments demonstrated that each of the six phages contained a single actin structural gene. The genes were localized within reasonably small (2-6 kb) DNA fragments (see Fig. 2 arrows) which were subcloned in pBR322 and analyzed further.

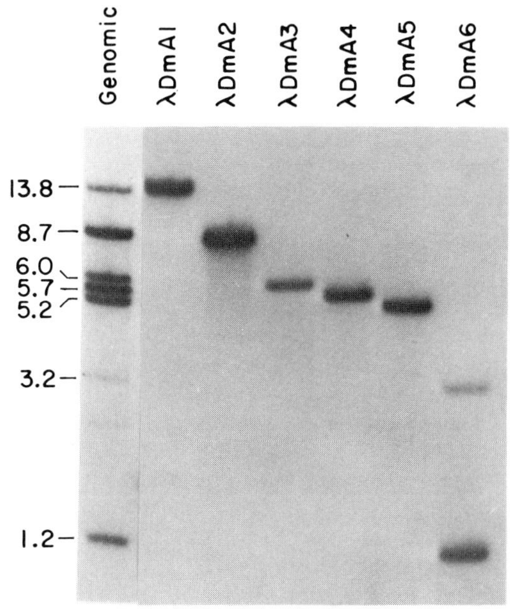

FIGURE 1. Blot-hybridization analysis of λDmA1-λDmA6. An EcoRI digest of genomic Drosophila DNA (left lane) was electrophoresed in parallel with EcoRI digests of λDmA1-λDmA6. Hybridization to the λDmA2 actin gene probe revealed that these six phages contain all of the Drosophila actin genes.

Direct Visualization of Nonconserved Interruptions Within DmA4 and DmA6

Heteroduplexes formed between several pairs of *Drosophila* actin genes are depicted in Figure 3. In each case a duplex segment of 1.1 kb corresponding to the protein coding region terminates in single strand forks. We interpret this to mean that while the protein coding regions of the several genes are homologous, untranslated and flanking sequences are not. In two cases single-stranded loops tentatively identified as intervening sequences were seen: one of 630 nucleotides near the 5' end of DmA4, and one of 360 nucleotides near codon 300 of DmA6. Introns were not seen in other *Drosophila* actin genes, however, these experiments would not detect introns in untranslated regions or very small (<100 nucleotides) introns within coding regions.

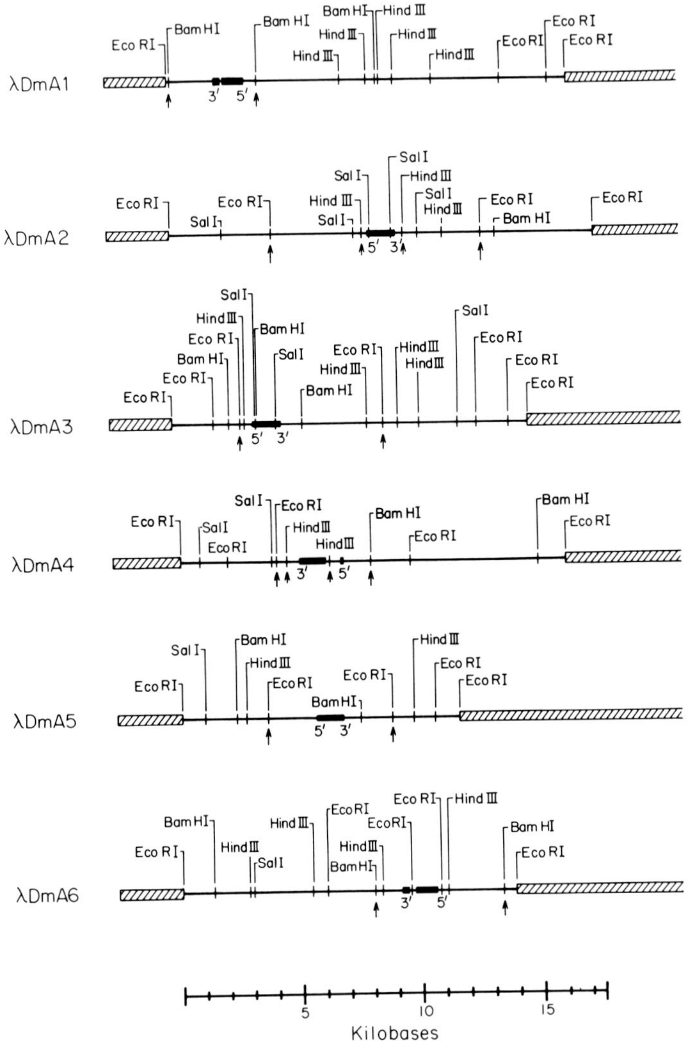

FIGURE 2. Restriction maps of λDmA1-λDmA6. Solid blocks represent the positions of protein coding regions, with interruptions apparent in DmA1, DmA4, and DmA6. Arrows indicate fragments which were subcloned in pBR322. Hatched blocks represent vector sequences.

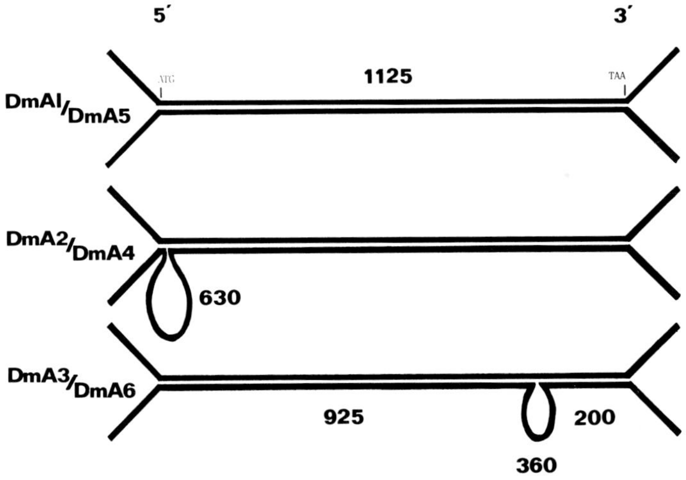

FIGURE 3. A summary of DmA1/DmA5, DmA2/DmA4, and DmA3/DmA6 heteroduplex experiments. These experiments allowed visualization of two intervening sequences, one of 630 nucleotides near the 5' end of DmA4, and one of 360 nucleotides within codon 307 of DmA6. A 60 nucleotide intervening sequence within codon 307 of DmA1 was not seen in these experiments, but was recently discovered by Sánchez, Tobin and McCarthy (personal communication).

Sequence information presented in Figure 4 confirms that the DmA4 and DmA6 introns are not present in other Drosophila actin genes, and additionally localizes an intron within DmA2. In panel A, the sequence near the 5' end of the protein coding region of DmA2 is compared to that of DmA4. As can be seen, DmA4 is split within the glycine codon at position 13, while DmA2 is not. Instead, DmA2 appears to be split eight nucleotides upstream from the start codon, since a eucaryotic intron-exon junction sequence appears at this point. This finding agrees well with our previous R-loop analysis of DmA2 (9).

Nonconservation of the DmA6 intervening sequence is illustrated in panel B. The DmA6 interruption is within the glycine codon at position 307. From the DmA2 and DmA3 sequence data, it is clear that these genes are not interrupted at this position. However, F. Sánchez, S. L. Tobin, and B. J. McCarthy (personal communication) have recently found that DmA1 is split by a 60 nucleotide intron within codon 307, demonstrating that at least one other Drosophila actin gene is split in the analogous position.

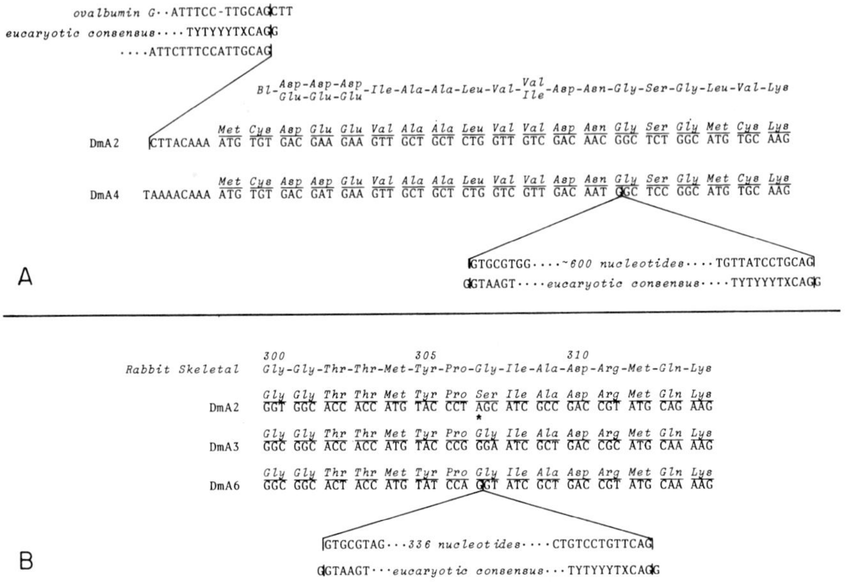

FIGURE 4. Nucleotide sequences surrounding introns within the Drosophila actin genes. Panel A shows that DmA4 is interrupted within the glycine codon at position 13. A comparable intervening sequence is not seen in DmA2, which instead appears to be interrupted 8 nucleotides upstream from the ATG start codon. Panel B demonstrates that DmA6 is interrupted within the glycine codon at position 307, while at least two other genes, DmA2 and DmA3, are not.

All Drosophila Actins are Similar in Sequence to Vertebrate Cytoplasmic Actins

A comparison of the amino terminal tryptic peptides of each Drosophila actin with those of vertebrate skeletal muscle and cytoplasmic actin isoforms is shown in Figure 5. Each of the derived actin amino acid sequences closely resembles the vertebrate cytoplasmic sequences, partial exceptions being DmA1 and DmA6. Therefore Drosophila apparently does not synthesize actins comparable in amino acid sequence to that of vertebrate skeletal muscle.

Figure 5 reveals another surprising result, namely the presence of a cysteine codon following the initiator methionines of each of the Drosophila actin genes. All vertebrate actins begin with an aspartic or glutamic acid residue that is acetylated. It is of course possible that the cysteine residue is cleaved by in vivo processing, and that the mature Drosophila actins begin with the string of acidic residues as do those of vertebrates.

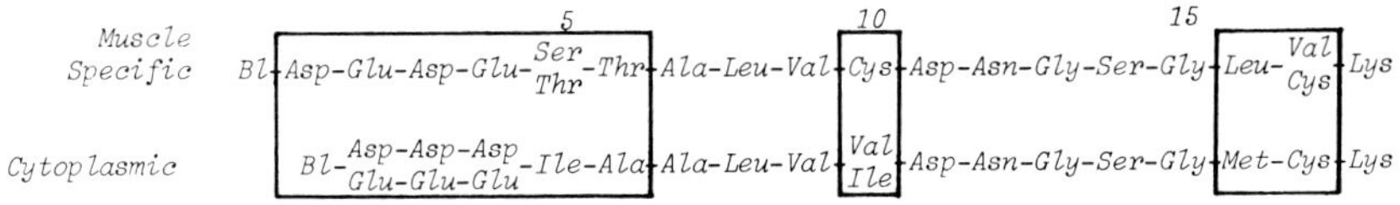

		Met	*Cys*	*Asp*	*Asp*	*Asp*	*Ala*	*Gly*	*Ala*	*Leu*	*Val*	*Ile*	*Asp*	*Asn*	*Gly*	*Ser*	*Gly*	*Met*	*Cys*	*Lys*
DmA1	CCAAG	ATG	TGT	GAC	GAT	GAT	GCG	GGT	GCA	TTA	GTT	ATC	GAC	AAC	GGA	TCG	GGC	ATG	TGC	AAG *
		Met	*Cys*	*Asp*	*Glu*	*Glu*	*Val*	*Ala*	*Ala*	*Leu*	*Val*	*Val*	*Asp*	*Asn*	*Gly*	*Ser*	*Gly*	*Met*	*Cys*	*Lys*
DmA2	ACAAA	ATG	TGT	GAC	GAA	GAA	GTT	GCT	GCT	CTG	GTT	GTC	GAC	AAC	GGC	TCT	GGC	ATG	TGC	AAG
		Met	*Cys*	*Asp*	*Glu*	*Glu*	*Val*	*Ala*	*Ala*	*Leu*	*Val*	*Val*	*Asp*	*Asn*	*Gly*	*Ser*	*Gly*	*Met*	*Cys*	*Lys*
DmA3	ACAAA	ATG	TGT	GAC	GAA	GAG	GTT	GCA	GCT	TTA	GTG	GTC	GAC	AAC	GGA	TCC	GGC	ATG	TGC	AAA
		Met	*Cys*	*Asp*	*Asp*	*Glu*	*Val*	*Ala*	*Ala*	*Leu*	*Val*	*Val*	*Asp*	*Asn*	*Gly*	*Ser*	*Gly*	*Met*	*Cys*	*Lys*
DmA4	ACAAA	ATG	TGT	GAC	GAT	GAA	GTT	GCT	GCT	CTG	GTC	GTT	GAC	AAT	GGC	TCC	GGC	ATG	TGC	AAG
		Met	*Cys*	*Asp*	*Asp*	*Glu*	*Val*	*Ala*	*Ala*	*Leu*	*Val*	*Val*	*Asp*	*Asn*	*Gly*	*Ser*	*Gly*	*Met*	*Cys*	*Lys*
DmA5	CCAAG	ATG	TGT	GAC	GAT	GAG	GTT	GCC	GCA	TTG	GTC	GTG	GAC	AAT	GGT	TCC	GGA	ATG	TGC	AAA
		Met	*Cys*	*Asp*	*Glu*	*Glu*	*Ala*	*Phe*	*Ser*	*Leu*	*Val*	*Val*	*Asp*	*Asn*	*Gly*	*Ser*	*Gly*	*Met*	*Cys*	*Lys*
DmA6	CAAAC	ATG	TGT	GAC	GAA	GAA	GCA	TTC	AGC	CTT	GTC	GTA	GAC	AAC	GGC	TCC	GGC	ATG	TGC	AAG

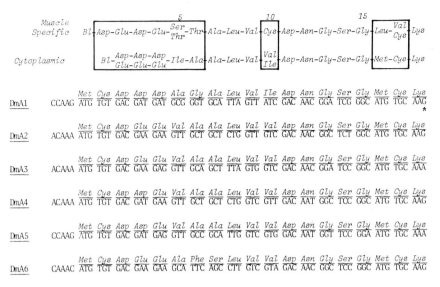

FIGURE 5. Nucleotide sequences of the 5' ends of the Drosophila actin genes. These data reveal that the sequence of each actin amino terminal tryptic peptide resembles those of vertebrate cytoplasmic actins. Thus, Drosophila does not encode an actin whose primary sequence is comparable to that of vertebrate skeletal muscle.

DISCUSSION

Our characterization of the Drosophila actin genes has revealed several unexpected findings. Most striking is the nonconservation of the positions of introns. Of the 10 intron positions within actin genes of yeast (11), Caenorhabditis (12), Drosophila, and sea urchin (13), none is conserved between species. Thus, if introns are deleted from but never inserted into genes, then the primordial actin gene must have had many introns, most of which were subsequently deleted (see 14). Alternatively, many introns may be formed when genomic elements insert into transcribed regions. Further analyses of actin genes will clarify this issue.

The finding that the multiple actin genes of Drosophila are conserved to such a high degree naturally leads one to question the selective value of the six gene copies. We are optimistic that our ongoing developmental analysis of actin gene expression in Drosophila will answer this question.

REFERENCES

1. Pollard, T. and Weihing, R. (1974). In Critical Reviews in Biochemistry, 2, G. Fasman, ed. (Cleveland, Ohio: CRC Press), pp. 1-65.
2. Huxley, H. E. (1969). Science 164, 1356-1366.
3. Wolosewick, J. J. and Porter, K. R. (1979). J. Cell Biol. 82, 114-139.
4. Batten, B. E., Aalberg, J. J., and Anderson, E. (1980). Cell 21, 885-895.
5. Storti, R. V., Coen, D. M., and Rich, A. (1976). Cell 8, 521-527.
6. Garrels, J. I. and Gibson, W. (1976). Cell 9, 793-805.
7. Fyrberg, E. A. and Donady, J. J. (1979). Dev. Biol. 68, 487-502.
8. Fyrberg, E. A., Kindle, K. L., Davidson, N. and Sodja, A. (1980). Cell 19, 365-378.
9. Durica, D. S., Schloss, J. A. and Crain, W. R. Jr. (1980). Proc. Nat. Acad. Sci. USA 77, 5683-5687.
10. Cleveland, D. W., Lopata, M. A., MacDonald, R. J., Cowan, M. J., Rutter, W. J. and Kirschner, M. W. (1980). Cell 20, 95-105.
11. Ng, R. and Abelson, J. (1980). Proc. Nat. Acad. Sci. USA 77, 3912-3916.
12. J. Files, personal communication.
13. Scheller, R. H., McAllister, L. B., Crain, W. R. Jr., Durica, D. S., Posakony, J. W., Britten, R. J., and Davidson, E. H. Submitted for publication.
14. Gilbert, W. (1979). In Eucaryotic Gene Regulation, ICN-UCLA Symposia on Molecular and Cellular Biology, 14, R. Axel, T. Maniatis and C. F. Fox, eds.

A CLUSTER OF DROSOPHILA CUTICLE GENES*

Michael Snyder
Michael Hunkapiller
David Yuen
Don Silvert[1]
James Fristrom[1]
Norman Davidson[2]

Divisions of Biology and Chemistry[2]
California Institute of Technology
Pasadena, California

and

Department of Genetics[1]
University of California
Berkeley, California

I. ABSTRACT

We are studying the sequence organization and expression of the larval cuticle genes of Drosophila. Five major cuticle proteins are synthesized and secreted by late larvae in order to provide a protective coat around the pupa. A 36 kb DNA segment of the Drosophila genome which codes for several larval cuticle genes has been cloned by recombinant DNA techniques. This segment has been localized to chromosomal region 44D and encodes five genes all of which are expressed at the same time of Drosophila development. Four of the genes are clustered within 7.9 kb of DNA and are abundantly expressed in late third instar poly(A)$^+$ RNA. A fifth gene lies 8 kb away from this gene cluster and is expressed at a much

This research was supported by NIH grants S T32 and GM 07616 to M.S.; GM 10991, GM 20927, and BR SG RR07003 to N.D.; and GM 06965 to M.H.

ISBN 0-12-137420-3

lower level. Three of the four abundantly expressed genes have been shown to code for larval cuticle proteins by positive selection and translation of RNA, two-dimensional gel analysis and immunoprecipitations of translated polypeptides. Protein and DNA sequencing studies thus far have confirmed these results for the three genes. Moreover, these sequence comparisons have shown that the three cuticle proteins studied are greater than 35% homologous in amino acid sequence over the portions sequenced. Thus, the cuticle genes encoded at 44D represent members of a family of genes of common ancestry which share the same pattern of developmental expression and reside in a small segment of the Drosophila genome.

II. INTRODUCTION

A fundamental problem in eucaryotic gene expression is how sets of genes are coordinately expressed in a specific cell type and at a specific time of development. In order to understand the underlying processes, we have chosen to study the larval cuticle genes of Drosophila. In this system five major cuticle proteins are coordinately expressed in the epidermal cells of late third instar larvae (Fristrom *et al.*, 1978). These structural proteins are secreted along with another compound, chitin, to form the brown pupal case which surrounds and protects the animal during its metamorphosis. It is probable that the induction of these genes is under the control of the molting hormone, ecdysone, although this has not yet been directly demonstrated. It has been shown that cuticle deposition in many insects is induced by ecdysone, and in Drosophila an enzyme involved in larval cuticle synthesis, dopa decarboxylase, appears to be under ecdysone control (Kraminsky *et al.*, 1980).

In this article, we describe the cloning and characterization of several larval cuticle protein genes. Our results show that several of these coordinately expressed genes share a common ancestry and are clustered in a small segment of the Drosophila genome.

III. RESULTS

A. Selecting Cuticle Clones

A recombinant phage with an insert coding for larval cuticle proteins was isolated by a four step procedure. The details of this procedure will be presented elsewhere. Initially, a random shear Drosophila recombinant DNA library (Maniatis *et al.*, 1978) was screened using a cDNA probe made to poly(A)$^+$ RNA isolated from the integument of late third instar larvae. Cuticle proteins are the most abundant proteins in this tissue. Strongly hybridizing phage were chosen and then tested through two cycles of counterselection. First, we tested the relative intensity with which phage plaques hybridized to cDNA probes made to larval integument poly(A)$^+$ RNA and to embryo poly(A)$^+$ RNA which is not expected to contain cuticle messages. Those phage which hybridized more intensely to the larval integument probe were chosen and subjected to a second cycle of counterselection. Here, phage DNAs were tested for their differential hybridization to embryo, whole larval, pupal, and larval integument cDNA probes. The phage which hybridized strongest to the larval integument probe relative to the other three were put through the final screening.

In this final selection phage DNAs were used to positively select RNA from total late larval poly(A)$^+$ RNA (Ricciardi *et al.*, 1979). This RNA was translated *in vitro* and protein products were analyzed on two-dimensional O'Farrell gels. One clone, denoted λDmLCP1, was found that upon translation yielded four polypeptides with two-dimensional gel migration patterns similar to those of *in vivo* isolated cuticle proteins. The four polypeptides have molecular weights ranging from 11 kilodaltons to 17.5 kilodaltons. Below we present and discuss evidence that 1) there are several genes on this clone; and 2) the proteins they encode are larval cuticle proteins.

More recently, we have cloned sequences flanking λDmLCP1 to test whether they contain additional cuticle genes. These sequences were isolated by the process of chromosomal walking (Yen and Davidson, 1980). In particular one clone, denoted λDmLCP3, which overlaps λDmLCP1 by 3.0 kb and contains 16.0 kb of flanking DNA has been characterized. The restriction maps of the clones used in this study are depicted later in Fig. 2.

B. The Clones Contain a Gene Cluster

Since four polypeptides are translated from RNAs selected by λDmLCP1, we determined how many genes are encoded in the insert using electron microscopic R-looping and Southern blotting. Poly(A)$^+$ RNA from total late third instar larvae was hybridized to λDmLCP1 DNA under conditions favorable for R-loop formation (Kaback *et al.*, 1979). Unhybridized RNA was removed by gel filtration, and the DNA was spread for electron microscopy. As shown in Fig. 1A, multiple genes were observed in λDmLCP1. Three small R-loops each .5 $\pm$ .1 kb in length and thus of a size expected for genes coding for small proteins are clustered within 4.5 kb of DNA. These are named genes II, III and IV in correlation with the cuticle protein they are shown to encode (see below). Genes II (see tracing) and III are separated by .85 $\pm$ .1 kb, while genes III and IV are 1.9 $\pm$ .1 kb apart. Almost all of the λDmLCP1 molecules (n = 268) contain these three

R-loops. No intervening sequences were observed in any of the three genes. In addition, another gene, V, was observed although relatively infrequently; only 5-10% of the molecules contain this R-loop. Since the DNA for this gene is in excess and hybridizations were carried to a Cot at which all the complementary RNA is driven into R-loops (Kaback *et al.*, 1981), the abundance of gene V RNA in late larval

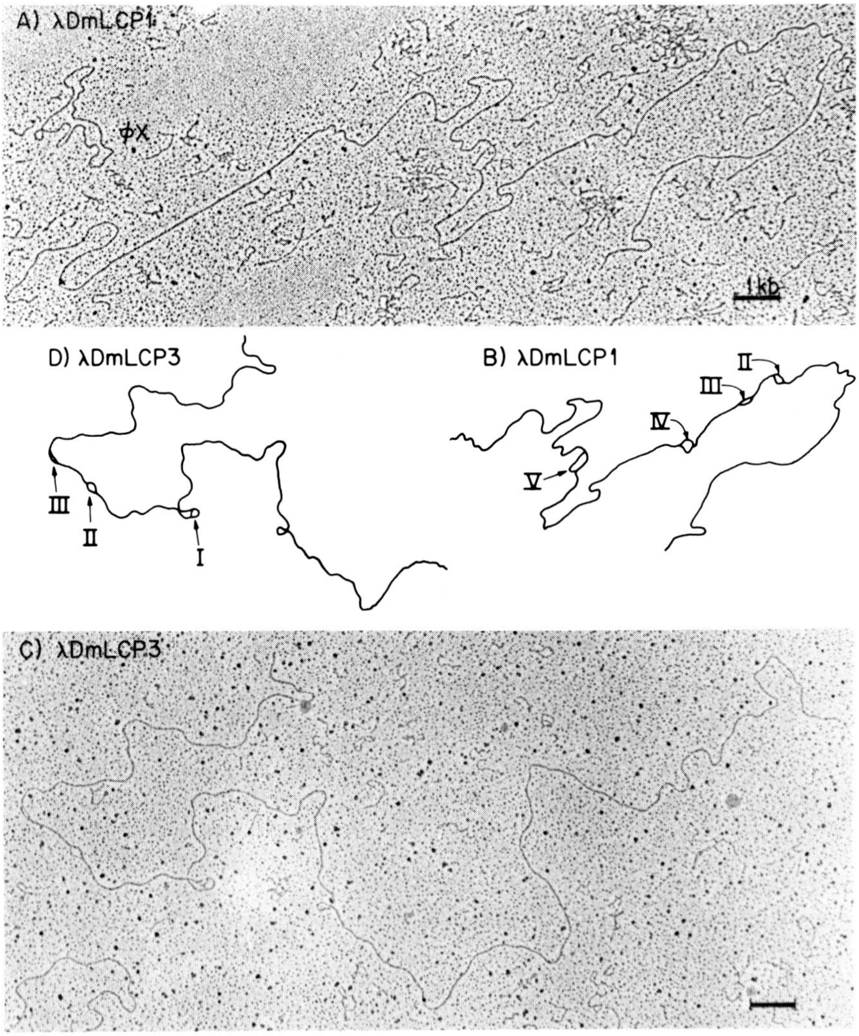

FIGURE 1. R-loops of total late larval poly(A)$^+$ RNA to whole lambda clone DNA. 5 μg of RNA and 100 ng of DNA were used. a) λDmLCP1; b) Tracing of (a); c) λDmLCP3; d) Tracing of (c).

poly(A)$^+$ RNA was estimated to be 2-4 x 10^{-5}. In contrast, the abundance of the RNAs complementary to the other three genes is estimated to be about 10^{-3}, as determined by immunoprecipitations of the translated products for the genes. Gene V is 1.0-1.5 kb in length and is often seen tangled in a fashion consistent with the presence of a small intervening sequence. Because of its low abundance in late third instar larval RNA we have not characterized this gene further.

R-looping of the overlapping clone λDmLCP3 revealed more of the gene cluster (Fig. 1C). Genes II and III are seen on the overlapping sequence shared with λDmLCP1 (see below), and a fifth abundantly expressed gene, gene I, was found. This gene lies 2.85 ± .15 kb from gene II and like the other abundantly expressed genes, it is .5 ± .1 kb in length and has no observable intervening sequences.

The second method with which the presence of multiple genes was established was by restriction endonuclease mapping and Southern blotting experiments (Southern, 1975). Gel blots were performed on restriction endonuclease digests of the phage DNAs and probed with a representative, calf thymus DNA primed cDNA made to total late larval poly(A)$^+$ RNA. The fragments which hybridize as well as the restriction mapping and R-looping data are all summarized in Fig. 2.

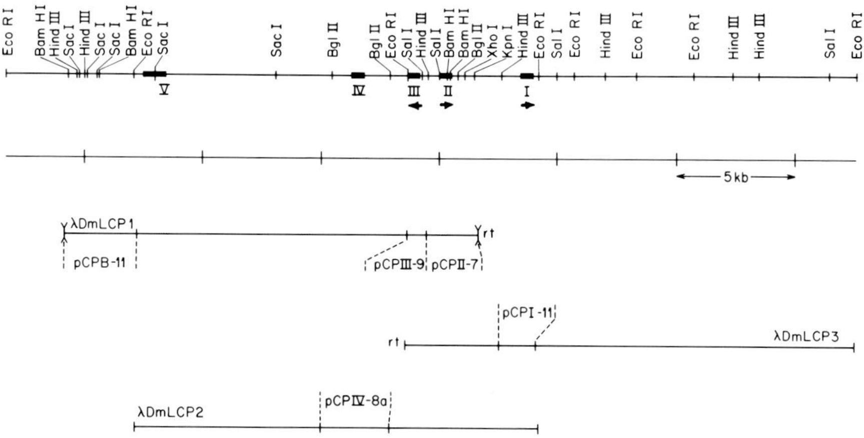

FIGURE 2. The cuticle gene cluster. A restriction map of the region spanned by the lambda clones depicted beneath it. Genes I thru V were deduced from R-looping data and Southern blot analysis (see text). The arrows beneath genes I, II and III indicate the 5′ to 3′ direction of transcription determined from DNA sequencing. Details concerning the derivation of this map will be presented elsewhere.

C. This Gene Cluster Resides At 44D

Four of the five major cuticle proteins have been genetically mapped (Fristrom *et al.*, 1978; and Chihara, Kimbrell and Fristrom, unpublished). Genes for cuticle proteins CP1, CP2 and CP3 reside in the region of polytene bands 44-50 on chromosome 2, the gene for CP5 lies on chromosome 3L, and the gene for CP4 is unmapped. We have cytologically localized this gene cluster on polytene chromosomes by *in situ* hybridization (Gall and Pardue, 1971). As shown in Fig. 3 using λDmLCP1 as a probe, hybridization grains are detected only over the region 44D and at no other sites on the chromosome. Thus, these clones could contain genes for as many as four cuticle proteins. In the next section we show that these genes do in fact code for at least three larval cuticle proteins.

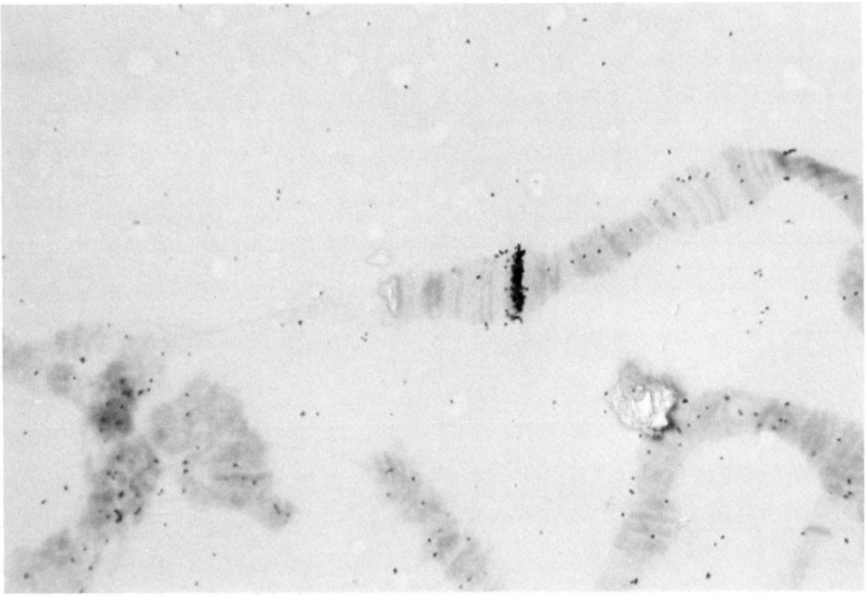

FIGURE 3. In situ hybridization to polytene chromosomes. Grains indicate site of hybridization. The centromere is located in the bottom left corner. 3-Day Exposure.

D. Three of the Genes Encode Larval Cuticle Proteins

There are four abundantly expressed genes on the cloned inserts. Concievably all four could code for larval cuticle proteins. In order to test this, the following experiments were performed. Initially, in data presented elsewhere, we used these clones as well as subclones containing individual genes to select RNA. This RNA was translated *in vitro* and polypeptides were analyzed on two-dimensional gels. Making use of 1) antisera directed against cuticle proteins, 2) RNA isolated from Drosophila strains which make variant cuticle proteins of altered electrophoretic mobilities and 3) the relative position where *in vivo* cuticle proteins and *in vitro* translation products migrate, we established the following: genes I and II encode CP1 and CP2, respectively, and one of the genes, III or IV codes for CP3 while the other probably codes for CP4.

These results have now been verified with protein and DNA sequencing studies. We have sequenced cuticle proteins CP1, CP2 and CP3 at their amino termini and most of genes I, II and III. The details of this work will be presented elsewhere. As shown in Fig. 4, the DNA sequences for genes I, II and III perfectly match the amino acid sequences for CP1, CP2 and CP3, respectively. We have also shown that no other sequences closely related to these genes are present elsewhere in the Drosophila genome. Hence, these genes must encode these proteins.

PROTEIN & DNA SEQUENCES: AMINO TERMINAL ENDS

```
CP1       AsnProProValProHisSerLeuGlyArgSerGluAspValHisAlaAspValLeuSerArgSerAspAspValArgAlaAspGlyPheAspSer
GENE I    AACCCCCCGGTGCCCCATTCCCTAGGCCGTTCGGAGGATGTCCACGCCGATGTCCTTTCCCGATCCGATGATGTTCGTGCCGATGGATTCGATTCC

CP2                           LeuAlaProValSerArgSerAspAspValHisAlaAspValLeuSerArgSerAspAspValArgAlaAspGlyPheAspSer
GENE II                       CTAGCCCCAGTTTCCCGCTCCGATGATGTACACGCTGATGTCCTTTCCCGATCGGACGACGTTCGTGCCGACGGATTCGACTCC

CP3                                                           AsnAlaAsnValGluValLysGluLeuValAsnAspValGlnProAspGlyPheValSer
GENE III                                                      CCAACGCTAATGTGGAGGTCAAGGAGCTGGTCAACGATGTCCAGCCCGATGGCTTTGTCA
```

FIGURE 4. Protein and DNA sequences. Cuticle proteins CP1, CP2 and CP3 were sequenced at their amino termini by the method of Hunkapiller and Hood (1980). The first residue of both CP1 and CP2 was deduced from DNA sequencing. Genes I, II and III were sequenced on both DNA strands by the method of Maxam and Gilbert (1979). A line between the encoded proteins indicates amino acid homology.

E. The Cuticle Genes Are Homologous

Also as noted in Fig. 4, these genes are homologous both in primary amino acid and DNA sequence. Cuticle proteins CP1 and

CP2 are quite closely related; CP2 differs from CP1 only in 6 of the last 8 amino terminal residues. Their homology extends at least 25 residues further into the center of the protein (not shown). In contrast, CP3 is much more distantly related to CP1 and CP2; it shares only 7 of 20 terminal residues with both CP1 and CP2. We note, however, that CP3 also displays more homology toward the central sequence than at the termini.

IV. DISCUSSION

The results presented above show that there is a cluster of four closely spaced genes at chromosomal location 44D which are abundantly expressed in late third instar poly(A)$^+$ RNA. We will present data elsewhere showing that these genes are all expressed at the late larval stage of Drosophila development and not at other embryonic, pupal, and adult developmental stages tested. Moreover, messages for these genes are enriched in larval integument tissue as is expected for cuticle protein genes. Thus these genes are all coordinately expressed with respect to tissue and stage of development. The fifth gene, gene V, which lies adjacent to this cluster is non-abundantly expressed in late larval poly(A)$^+$ RNA.

Our results thus far have indicated that three of the genes, genes I, II and III code for larval cuticle proteins CP1, CP2 and CP3, respectively. It is likely from immunoprecipitation studies that gene IV codes for CP4. This will be determined by further protein and DNA sequencing studies.

The sequencing results thus far have shown that cuticle proteins CP1, CP2 and CP3 show varying degrees of homology. Consistent with this, others (Silvert and Fristrom, in preparation) have shown that all of these proteins as well as CP4, share common antigenic determinants. Moreover, in data to be presented elsewhere we have shown that genes for CP1 and CP2 crosshybridize readily and select the same mRNAs, further confirming their high degree of sequence homology. However, in similar hybridization experiments between genes I or II with gene III, these genes do not crosshybridize. These experiments were performed at criterion where greater than 20% mismatch would not be detected. Our results in Fig. 4 are consistent with this; gene III differs sufficiently in DNA sequence (50%) from both genes I and II at its amino terminus such that it would not be detected by the hybridization experiments. Lastly, we have noted that DNA fragments containing genes III and IV crosshybridize weakly. If gene IV encodes CP4 as suggested by data to be presented elsewhere, this weak crossreactivity would be consistent with the antigenic similarity of these proteins.

An important feature to note is that genes I and II show very striking homology and that gene III is much more distantly related. Two equally plausible mechanisms can account for this: 1) genes I and II are the result of a more recent duplication, or 2) genes I and II are "corrected" relative to each other by a mechanism such as unequal crossing over. In this respect, we note that genes I and II could "correct" each other by unequal crossing over because they lie in a directly repeated orientation. Gene III could not be corrected by such a mechanism with genes I and II, because it lies in the opposite orientation.

The clustering of genes in an organism may play an important role in the coordinate expression of sets of genes. A general mechanism may exist for activating a large chromatin domain for transcription, thereby inducing the coordinate expression of the clustered genes.

ACKNOWLEDGMENTS

M.S. would like to thank P.S., whose help was invaluable in the early stages of this work. Our thanks also to N. Davis Hershey for criticizing the manuscript.

REFERENCES

Fristrom, J. W., Hill, R. J., and Watt, F. (1978). *Biochemistry 19,* 3917.
Gall, J., and Pardue, M. (1971). *In* "Methods in Enzymology" (L. Grossman and K. Moldave, eds.), p. 470. Academic Press, New York.
Hunkapiller, M. and Hood, L. E. (1980). *Science 207,* 523.
Kaback, D. B., Angerer, L. M., and Davidson, N. (1979). *Nucleic Acids Res. 6,* 2499.
Kaback, D. B., Rosbash, M., and Davidson, N. (1981). *Proc. Natl. Acad. Sci. USA,* in press.
Kraminsky, G. P., Clark, W. C., Estelle, M. A., Gietz, R. D., Sage, B. A., O'Connor, J. D., and Hodgetts, R. B. (1980). *Proc. Natl. Acad. Sci. USA 77,* 4175.
Maniatis, T., Hardison, R. C., Lacy, E., Lauer, J., O'Connell, C., Quon, D., Sim, G. K., and Efstratiadis, A. (1978). *Cell 15,* 687.
Ricciardi, R. P., Miller, J. S., and Roberts, B. E. (1979). *Proc. Natl. Acad. Sci. USA 76,* 4927.
Southern, E. M. (1975). *J. Mol. Biol. 98,* 503.
Yen, P., and Davidson, N. (1980). *Cell 22,* 137.

ORGANIZATION AND EVOLUTION OF THE DEVELOPMENTALLY REGULATED SILKMOTH CHORION GENE FAMILIES

Thomas H. Eickbush
C. Weldon Jones[1]
Fotis C. Kafatos

Cellular and Developmental Biology
Harvard University
Cambridge, Massachusetts

I. ABSTRACT

The silkmoth chorion has been used as a model system to study the organization, coordinate expression and evolution of multigene families. A detailed characterization of two genomic clones indicates that each chorion gene is paired with a non-homologous chorion gene which exhibits the same developmental specificity (time of expression). These pairs are divergently oriented and tandemly repeated. Short well-conserved DNA sequences separate the 5'-flanking regions of the gene pairs, while considerable variation in length as a result of frequent insertion/deletion elements characterizes the 3'-flanking regions of the pairs. To better understand the global organization of the silkmoth chorion locus, procedures were developed to clone large continuous regions of chromosomal DNA at this locus. Facilitated by the location of the chorion genes at less than 15 kb intervals, sublibraries of genomic clones were obtained from total chromosomal libraries and repeatedly screened to obtain overlapping clones ("chromosomal walking"). A preliminary account of the organization of the chorion genes encountered in these walks is presented.

[1]Present address: Stanford University, Stanford, California.

ISBN 0-12-137420-3

II. INTRODUCTION

In the last few years it has become recognized that a large number of important gene products in eucaryotes are encoded by families of related genes (multigene families). Well-known examples would now include globin, histone, collagen, immunoglobulin, interferon, actin, tubulin and ovalbumin genes. It is clear, then, that understanding of eukaryotic genomes is dependent upon our understanding the organization, coordinate expression and evolution of multigene systems. We have been addressing these questions using the silkmoth chorion as a model system.

The silkmoth chorion (eggshell) is a complex extracellular structure consisting of over 100 different proteins (1). The proteins are synthesized sequentially according to a strict developmental program by the monolayer of follicle cells, which surround each oocyte throughout its period of maturation in the ovariole. In the american silkmoth, Antheraea polyphemus, and the commercial silkmoth, Bombyx mori, most, if not all, of the chorion proteins are the products of distinct but evolutionarily related genes, as established by (i) direct sequencing of chorion proteins (2, 3, 4); (ii) in vitro translation of chorion mRNA (5, 6); and (iii) hybridization and sequence analysis of cDNA libraries (7, 8, 9).

Members of the chorion multigene families are clustered in the silkmoth genome. In A. polyphemus, two cloned chromosomal segments contain multiple copies of chorion genes (10). In B. mori, genetic analysis by M. Goldsmith and co-workers has established that the majority of the chorion structural genes are clustered within a few map units at one end of chromosome number 2 (11, 12). We have undertaken a detailed structural analysis of the chorion locus, and of its developmental and evolutionary properties, using recombinant DNA procedures. This report summarizes our current knowledge of the fundamental unit of chorion gene organization, as exemplified by the gene 401 and 18 repeats of A. polyphemus, and the approach we have taken towards studying the global organization of the chorion locus.

III. RESULTS AND DISCUSSION

A. Chorion Gene Pairs in A. polyphemus

1. Organization of the Basic Gene Repeat. While as many as 186 A. polyphemus chorion proteins can be resolved on two-dimensional gels, two major molecular weight classes of protein, A (MW 7,000-11,000) and B (MW 11,000-15,000) account for 85 % of the chorion mass (1). To examine in detail the structure and organization of the major chorion genes, a library of A. polyphemus

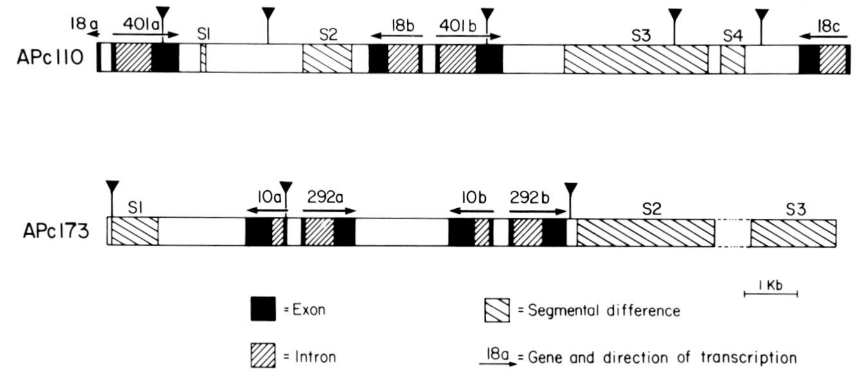

Figure 1 Schematic maps of cloned A. polyphemus chromosomal segments. APc110 (14 kb) contains two complete copies of gene 401, while one complete and two partial copies of gene 18. APc173 (14 kb) contains two complete copies each of genes 10 and 292. The positions of the two exons and single intron of each gene are shown. The genes are paired and transcribed in opposite directions (arrows). Widely stripped regions are inserted segments (S) in the otherwise homologous regions between gene pairs. Inverted triangles represent the location of HindIII restriction sites.

chromosomal DNA was constructed using the charon 4 derivative of phage λ (13, 14). Two cloned 14 kb chromosomal segments, APc110 and APc173, were selected for analysis because they each hybridized to two different cDNA clones. APc110 hybridized to the cDNA clone pc401, which is an abundantly expressed member of the B family, and to clone pc18, which is an abundantly expressed member of the A family; while APc173 hybridized to the major cDNA clones pc10 (B family) and pc292 (A family). Figure 1 shows the location of these genes on the two cloned chromosomal segments: each clone contains multiple copies of an A family member paired with a B family member. Gene pairs 10a/292a, 10b/292b and 18b/401b are entirely represented in these clones, while the 18a/401a and (presumably) 18c/401c pairs are incompletely represented. Genes 18 and 401 encode proteins synthesized during the "late" period of choriogenesis, while genes 292 and 10 encode proteins synthesized during the middle period of choriogenesis (5, 7, 8). Thus A and B genes are paired according to their developmental specificity (time of expression).

The AB gene pairs are divergently transcribed (10). Since

neighboring pairs are similarly oriented (Figure 1), they constitute 2kb direct repeats. The individual repeats are separated by variable lengths of DNA, interposed between the 3' ends of the genes. Homologies and segmental differences in these large 3' regions were established by heteroduplex and restriction endonuclease analysis (13). The 3' flanking regions are clearly homologous but interrupted by insertion/deletion elements (labelled S1 through S4 in Figure 1), which vary in both location and length (0.065 kb to 2.9 kb). Thus the fundamental unit of organization of A and B genes is a long tandem DNA repeat which contains one divergently transcribed AB gene pair, and varies by multiple inserts of non-homologous DNA.

To determine if additional copies of the 401/18 gene pair, other than those found on clone APc110, exist in the genome, a Southern blot analysis of total genomic DNA was conducted (10). By the use of appropriate 5'-specific and 3'-specific probes it was established that 15 $\pm$ 5 copies each of the 401 and 18 genes exist per haploid genome, and that most (probably all) copies of these genes are present as divergent 401/18 pairs. The degree to which these copies of the 401/18 pair are contiguous cannot be evaluated from the Southern blots, but is revealed by chromosomal walking (see below).

2. Sequence Analysis of Chorion Genes. All gene pairs and immediately adjacent flanking sequences in chromosomal clones APc110 and APc173 have been sequenced (13). This analysis revealed that the gene copies found on each segment are not identical. Their DNA sequences differ by both base substitutions and small insertions or deletions (usually reduplications or deletions involving short direct repeats). The intercopy nucleotide sequence divergence for genes 401 and 18 is only 0.9 % while copies of genes 10 and 292 differ by 5.5 %. This finding might suggest that the duplication of genes 10 and 292 is a more ancient evolutionary event than the duplication of genes 401 and 18.

DNA sequence analysis also revealed that each of the eight chorion genes consists of two exons and one intron. In each case the intron interrupts the signal peptide-encoding region between codons -4 and -5 (counting from the mature protein NH_2-terminus). The introns for the different gene copies are quite similar both in sequence and length, although on the average they are clearly more divergent than the copies of exon sequences. In agreement with the consensus sequences, (15), the chorion introns begin with GTPuAG and end with AG.

The positions of the 5' ends of each gene were identified by determining the 5' terminal sequence (cap site) of the corresponding mRNA. For all eight genes the sequence at the cap site is PuTCATT, similar to that determined for sea urchin histone mRNA (16). A remarkable feature of the chorion gene organization are the properties of the 5' flanking DNA which separates the two divergent genes of each pair. This DNA is short and invariant in length for different copies (325 bp for 401/18 and 264 bp for 10/292), as opposed to the long and variable 3' flanking DNA which separates the gene pairs (3.6 or 5.6 kb for 401/18; 1.8 to ≥5.1 kb for 10/292). The 5' flanking DNA is also conservative in sequence: the sequence divergence between copies is intermediate between that of exons and introns (while, by contrast, the 3' flanking DNAs are at least as divergent as the introns). The short 5' flanking DNAs are of special interest because they can be presumed to contain paired, divergent promoters. A canonical Hogness-Goldberg box is found at the expected distance from each cap site. Base compositional features (characteristic % A+T profiles), preceding each cap site, impart to the 5' flanking DNA a weak dyad symmetry, although sequence matches are limited for different genes. Centered approximately at nucleotide -87 beyond the cap site are short sequences which show developmental specificity, i.e. are similar for coordinately expressed genes of different multigene families. Functional tests are needed to determine whether these sequences are in fact temporal specificity elements, analogous to the -35 recognition sites of procaryotic promoters.

B. Chromosomal "Walking" in the Silkmoth Chorion Locus

The demonstrated molecular proximity of at least some copies of chorion gene pairs, and the genetic clustering of many distinct chorion genes (10, 11, 12), indicated that short chromosomal "walks" in the chorion locus would cover many structural genes, and would be invaluable in our attempts to understand the organization and evolution of these multigene families.

"Chromosomal walking" is a technique originally applied to the *Drosophila* genome (W. Bender, P. Spierer and D. Hogness, personal communication) to obtain continuous regions of chromosomal DNA hundreds of kilobases in length. In this technique, a fragment of DNA from one end of a starting genomic clone is isolated, labelled, and used to screen the genomic library. Positive clones are purified, and their DNA isolated and positioned with respect to the starting clone, by means of the respective, overlapping restriction maps. The entire process is then repeated, using as probe a fragment of DNA from the new end.

While several successful attempts have been made to "walk" in the *Drosophila melanogaster* genome, we did not expect to succeed at walking in the silkmoth genome if we followed this straightforward approach. First, the *B. mori* genome is three or four times (17) and the *A. polyphemus* genome seven or eight times (18) larger than that of *Drosophila*. Therefore, a correspondingly larger number of genomic clones would have to be screened at each step. Second, the basic organization of unique and repeated DNA sequences in the silkmoth and fruit fly genomes differ. The *Drosophila* genome contains a long interspersion pattern, with single-copy DNA segments extending an average of 30 kb before being interrupted by repeated DNA elements (17). The two silkmoth genomes, on the other hand, are similar to those of most other organisms, in that they contain a short interspersion pattern ("Xenopus type") of unique and repeated DNA (17, 18). Single-copy DNA sequences in these organisms extend for less than 1.5 kb, on the average, before being interrupted by repeated sequence elements. Third, since the chorion genes are themselves repeated, with homologous DNA at both the 3' and 5' ends of similar genes, most of the DNA probes used to obtain overlapping clones would be repeated elsewhere in the chorion locus. Since any DNA probe containing a repeated element is of limited use in chromosomal walking, finding DNA sequences free of repeated elements would be a time consuming, if not impossible process in certain regions. The following approach was therefore developed to overcome many of the problems associated with walking in the silkmoth chorion locus. This approach should be of general use in the analysis of clustered multigene families in eucaryotes.

C. Chromosomal Walking in *B. mori*

1. General Approach. The following approach was first developed in *B. mori*, where genetic clustering of chorion genes is firmly established. If (as the *A. polyphemus* clones APc110 and APc173 suggest) the chorion structural genes are not more than 15 kb apart on the chromosome, every charon 4 genomic clone which contains a segment of DNA from the chorion locus should contain at least one chorion gene (the size of the average charon 4 clone is 15 kb). Thus, by isolating from a genomic library all clones containing chorion genes, we hoped to establish a sublibrary of overlapping genomic clones which could be repeatedly and easily screened in each step of a genomic walk.

Second, we chose to derive this sublibrary from a partial EcoRI genomic library. Because of the repetitive nature of the chorion locus we expected many of our DNA probes to contain repeated sequence elements. Therefore, a convenient method was needed for separating the positive clones after each screening step into those which are an extension of the area walked, and those that are "jumps" to other areas of the chorion locus. The obvious method is restriction enzyme analysis. This analysis, however, is considerably complicated if the library is generated from randomly restricted chromosomal DNA fragments: the end-fragments of clones from such a library are unpredictable in length, and therefore rather useless in establishing overlaps. By contrast the advantage of an EcoRI partial library is that all clones start and stop at real genomic EcoRI sites. When a step is taken using a partial EcoRI library, the selection, from among the positive clones, of those that are an extension of the chromosomal walk can be quickly and easily made by comparison of EcoRI digests: each overlapping clone must contain the entire EcoRI fragment from which the probe was derived (even if the probe used was only a small segment of the total EcoRI fragment). Most non-overlapping clones should be eliminated by this digest. A further test of whether the remaining clones are truly overlapping is whether they can ultimately be fitted precisely into a continuous chromosomal EcoRI map, encompassing all EcoRI fragments from each putatively overlapping clone. Restriction mapping with additional enzymes can then be undertaken for conformation of the overlaps; such mapping, without the previous screening steps, would constitute a waste of effort.

An additional benefit of a partial EcoRI library became apparent after several rounds of walking had been attempted and nearly 100 positive clones had been characterized. These clones corresponded to both real extensions of the area walked and jumps to other regions of the locus. However, because each clone contained an integral number of "real" EcoRI fragments, the rapidly increasing problem of bookkeeping became a relatively simple matter of recording the size of these fragments. As we progressed in the walk, potentially overlapping EcoRI fragments were easily identified from previous records and retested. Eventually, clones that were originally obtained by "jumps" were also precisely positioned into a continuous EcoRI chromosomal map. In other words, walking sometimes became a matter of fitting together blocks of already isolated overlapping clones.

2. Formation of Genomic Sublibraries. The critical element for success in using a genomic sublibrary to walk in a complex locus, is obviously whether this sublibrary contains overlapping clones spanning the entire locus. We were considerably aided in the formation of our chorion gene sublibrary by the unique developmental features of silkmoth chorion formation. At the peak of choriogenesis, 95 % of the protein synthesized by the follicle cells surrounding the oocyte is chorion (20). Thus total cellular mRNA could be used as template to produce labelled cDNA probes highly enriched for chorion sequences. No mRNA purification steps, which could potentially result in the loss of certain classes of chorion mRNA, were needed. The level of non-chorion mRNA is so low that the use of total mRNA to screen the genomic library would not drastically dilute the chorion sublibrary with clones corresponding to unlinked, extraneous genes which might also be expressed in choriogenic cells.

When labelled cDNA made from total follicular cell mRNA, spanning the entire range from early to late choriogenesis, was used as probe, 246 positive clones were isolated from a total of 60,000 clones of the genomic library (0.41%). This recovery is in good agreement with the genetic distance between the outside chorion markers, relative to the total length of the genetic map in B. mori (12). Assuming a haploid genome size for B. mori of $5x10^8$ bp (17) the recovery of positive clones suggests that the chorion locus may be as much as $2x10^6$ bp in length. This number is obviously quite crude, since it assumes a random fragmentation of the genomic DNA, which almost certainly does not occur in the case of a partial EcoRI digestion. In addition the multiplicity of representation of each DNA region in our sublibrary (see below) suggests that the chorion locus is several-fold shorter than the above estimate.

To further facilitate walking, we formed a second sublibrary, this time using as probe cDNA made from an mRNA preparation specifically enriched for the later stages of choriogenesis. Our reasons for focusing on "late" chorion genes are twofold. First, it is known that late mRNA encodes a somewhat simpler pattern of chorion proteins. Specifically the C proteins, a diverse set of proteins which account for a significant percentage of the complexity but not of the mass of chorion, are only synthesized during the early stages of choriogenesis. Second, independent evidence exists that late chorion genes in B. mori are further clustered within the chorion locus. A spontaneous mutant of B. mori , Gr^B, does not synthesize most of the mRNAs encoding late chorion proteins (6). By Southern analysis using appropriate cloned cDNA probes, it was established that in the Gr^B chromosome a large number of late chorion genes are deleted (9). On the probable assumption that a single deletion event is involved, the Gr^B analysis indicates that many genes expressed late in choriogenesis are tightly clustered within the chorion locus. The screening of 90,000 genomic clones (3 genome equivalents) with cDNA made from late mRNA resulted in the isolation of 281 positive clones (0.39%).

Finally, to ensure that at a minimum we would thoroughly saturate one region of the chorion locus with isolated overlapping clones, a third sublibrary was constructed of chromosomal clones which hybridized at low criterion with a particular cDNA clone. The cDNA clone selected, m1911, corresponds to a very late chorion mRNA sequence, which is not detectable in animals homozygous for the Gr^B mutation. Thus all genes coding for or significantly cross-hybridizing with m1911 are eliminated by the Gr^B DNA deletion. A sublibrary of 81 clones was obtained by screening 150,000 genomic clones with labelled m1911 as probe (0.05%). With these three sublibraries, we hoped to walk by screening at each step only 608 (i.e. 246 + 281 + 81), already plaque purified clones, rather than screening $1.5x10^5$ clones and plaque purifying the positive clones each time.

Figure 2 confirms that these three sublibraries, generated by using as probes total choriogenic mRNA, late choriogenic mRNA and the m1911 cDNA clone, contain different numbers of genomic clones with 1911 genes. The entire set of 608 plaque purified genomic clones were placed in microtiter plates and with the aid of a

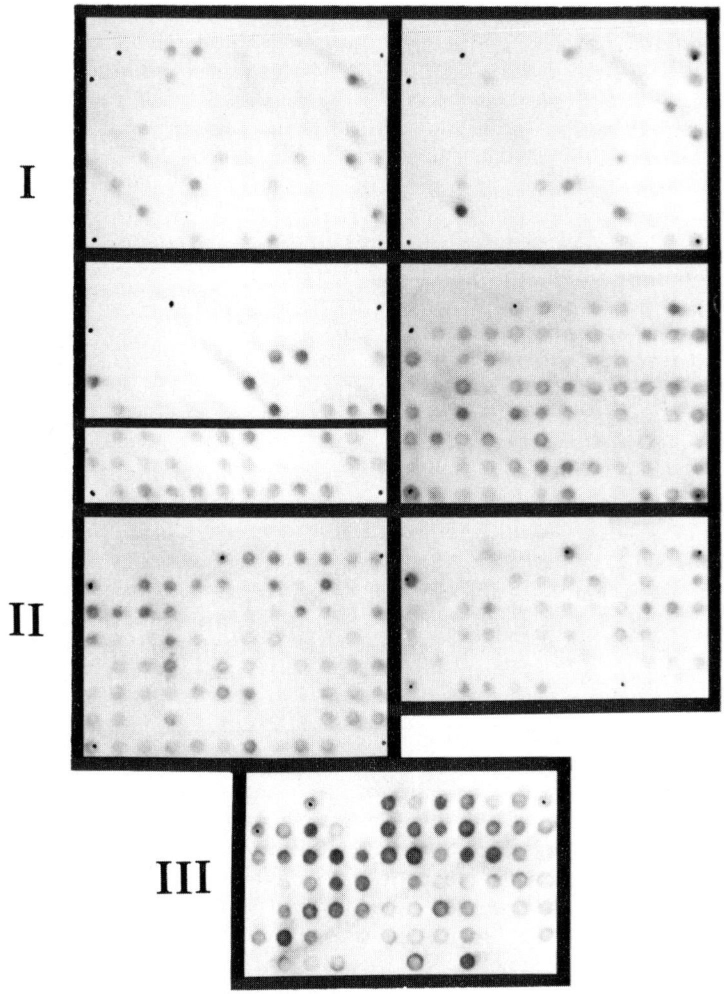

Figure 2 Screen of the B. mori sublibraries for genomic clones bearing 1911-like genes. Roman numberal I refers to the 246 clones obtained with total cDNA, II refers to the 281 clones obtained with "late" cDNA, and III refers to the 81 clones obtained with the cDNA clone on 1911. Hybridization with nick-translated ml911 was conducted at 70°C, 0.3M NaCl.

replicator spotted onto bacterial lawns. DNAs from the resulting plaques were transferred to filters by the standard Benton and Davis procedure (21). Hybridization of the ml911 cDNA clone to these filters at relatively low criterion (70°C, 0.3M salt) indicated which genomic clones contain 1911 or 1911-homologous genes. Of the 246 clones selected with total mRNA, only 45 showed some degree of hybridization (ranging from strong to weak), while 186 out of the 281 clones from the late mRNA sublibrary hybridized to m1911. As expected most of the 1911 sublibrary (71 out of 81) hybridized with m1911. Thus 1911 and its homologous family members represent a significant fraction (approximately 18 %) of the total chorion locus, and account for the major fraction (approximately 66%) of the late chorion region. Again these numbers should be regarded as only rough approximations, since the chromosomal DNA was not cut in a random manner and very large (>20 kb) EcoRI fragments are excluded from this analysis. It should be noted that only 71 clones were obtained from 150,000 clones using m1911 itself as probe, yet 186 clones hybridizing with 1911 could be obtained from the late choriogenic mRNA sublibrary where only 90,000 clones were screened. Presumably this is because some genes are reasonably abundantly expressed and can be recovered with total late mRNA, but are distantly enough related to m1911 so that they are missed in the original m1911 screen. Once these clones are included in the late sublibrary, they can be seen to score as weak positives with m1911.

3. The Chromosomal Walk. We initiated our walk with a clone, Bm623, that hybridized intensely with m1911. By taking small segments of DNA from each end of this initial clone and screening the sublibrary, overlapping clones were readily found. An example of such a sublibrary screen which clearly demonstrates both the usefulness of the chorion sublibraries and some of the difficulties associated with the walk is shown in Figure 3. In this screen a 600 bp fragment of DNA from the end of clone Bm661 was used to screen the sublibraries; this fragment is located between the third and fourth gene pairs from the left in region I, Figure 4 (see below). Fifteen strongly positive clones were obtained along with several weakly positive clones. Of the 15 strongly positive clones, 7 (indicated by the single arrows in Figure 3) were extensions of clone Bm661, while the remaining 8 (indicated by double arrows) were derived from other regions of the chorion locus. As described in the previous section, the separation of clones representing "walks" and "jumps" was easily accomplished by a single EcoRI digestion of DNA from the isolated clones. Examination of the same digests indicated that the 8 clones representing jumps constituted a single set of overlapping clones. The location of this set of clones was eventually established at a distance of approximately 40 kb from the end of clone Bm601 (immediately preceding the first gene pair from the left in region I of Figure 4).

To date we have obtained overlapping genomic clones spanning approximately 350 kb of DNA. This DNA is in four segments which have not yet been physically linked, but which we believe are near

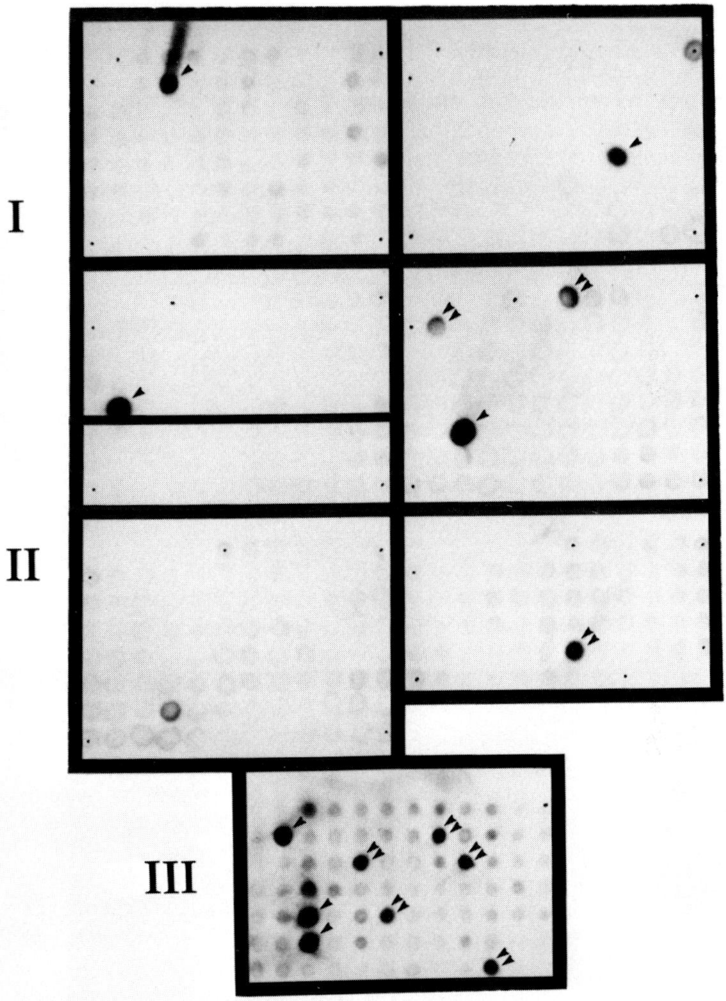

Figure 3 Examples of a screen of the B. mori sublibraries to find overlapping genomic clones ("chromosomal walking"). Roman numerals refer to the three sublibraries described in Figure 2 and the text. The probe in this case was a 600 bp DNA fragment located between the third and fourth (counting from the left) ml911-like gene in Figure 4, segment I. Single arrows indicate real overlapping clones, while double arrows indicate clones which eventually mapped immediately to the left of the left most ml911-like genes of segment I.

each other in the chorion locus. The two largest segments, 140 kb and 125 kb, are diagramed in Figure 4, as regions I and II respectively. The EcoRI fragments from each component clone fit precisely in these maps, and a number of other restriction enzymes have also been tested to ensure that fortuitously similar EcoRI fragments have not led to "jumps" to other regions of the chorion locus. Excluding the ends of the four segments, each EcoRI fragment is represented on the average 10.9 times.

One problem which we were afraid we might encounter while walking in the chorion locus were allelic differences in the location of chorion genes or of endonuclease restriction sites. The DNA used to generate the B. mori library was originally obtained from two sibling adult females (minus ovaries). Consequently, up to four allelic forms of the chorion locus could be represented in the library. The chorion proteins are rapidly evolving among silkmoth species, and significant variations can be seen among wild populations (1), or between inbred strains (11, 12). However, only minor quantitative polymorphism of proteins has been detected within two inbred strains of B. mori (6, 22). Despite the isolation of multiple clones representing each EcoRI fragment, we have not found any evidence for allelic differences, in DNA spacing (i.e. insertion or deletion of DNA), or in the presence and location of EcoRI and HindIII endonuclease restriction sites; this indicates that a high degree of inbreeding has been accomplished in this strain of B. mori.

While approximately 300 kb out of the total 350 kb of DNA were obtained using the partial EcoRI sublibraries described above, occassionally blocks were found, in which no clones extending beyond a certain EcoRI site were present in the sublibraries. We assumed this was an indication that the next EcoRI fragment was very large (>20 kb) hence could not be included in our EcoRI charon 4 library. One such former block is the extreme left end of region I, which extends for 19 kb without an EcoRI site, and a second is the center of region II which is occupied by a 25 kb EcoRI fragment (Figure 4). To clone these two regions a partial EcoRI* library was generated (23). In this procedure, the EcoRI restriction sites in the chromosomal DNA are methylated, and a partial EcoRI digest of the DNA is conducted under conditions where the endonuclease cleaves at the tetranucleotide AATT in addition to the hexanucleotide GAATTC, resulting in a more random fragmentation of the genomic DNA. Prior methylation of the EcoRI sites is necessary, since under all conditions the endonuclease exhibits a marked preference for the full 6 bp recognition sequence. Without methylation a partial EcoRI* library would still be essentially a partial EcoRI library. By screening either this EcoRI* library directly, or additional sublibraries selected using as probes gene sequences nearest the ends of the blocked regions, we were able to walk over these large EcoRI fragments.

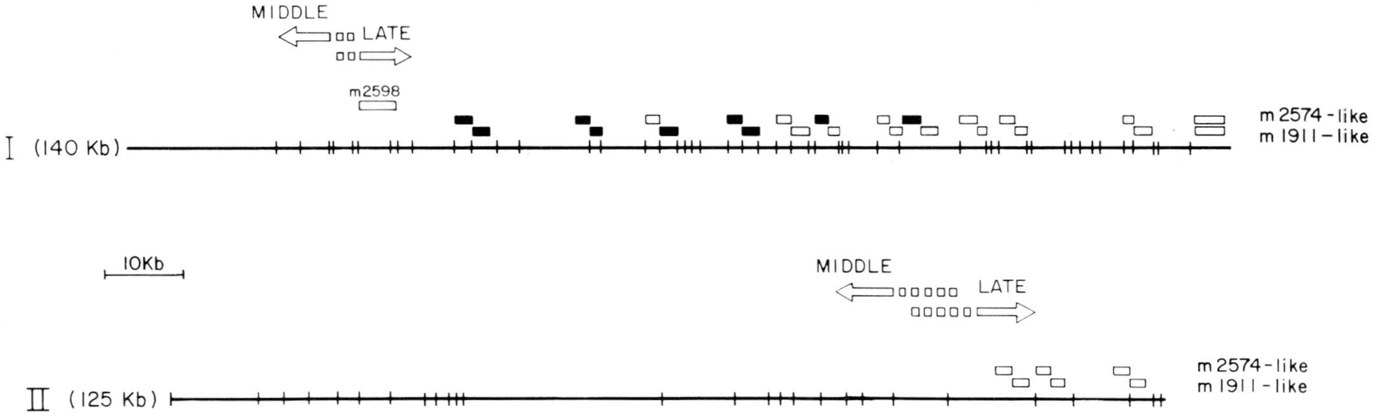

Figure 4 Schematic maps of the two largest segments obtained by the chromosomal walk in *B. mori*. Cross-bars indicate the location of EcoRI restriction sites. Boxes above the EcoRI map represent the location of genes which bear homology to cDNA clone m1911 (lower row) or m2574 (upper row) as determined by restriction mapping and Southern analysis. Solid boxes indicate those genes which exhibit the greatest homology to m1911 or m2574 at high hybridization criteria (85°C, 0.45m NaCl). Arrows refer to chromosome segments bearing genes expressed during the "late" period of choriogenesis or during the "middle" period of choriogenesis. The approximate location of the gene(s) which exhibit(s) homology to the "late" cDNA clone m2598 is indicated by the box labelled m2598.

4. Properties of Regions I and II of the Chorion Locus. Fifteen genes which exhibit hybridization with m1911 are present in regions I and II. Their location, as determined by Southern analysis, is indicated by the lower of the two rows of boxes above the EcoRI map in Figure 4. Twelve of these genes are distributed over a 100 kb region starting from the right end of region I, while the remaining three genes are located on 20 kb of DNA at the right end of region II. Immediately to the left of each m1911-like gene is located another of a different type of gene (upper row of boxes). This second type of gene exhibits homology with the cDNA clone m2574, which represents another mRNA species, like m1911 present very late in choriogenesis and absent in homozygous Gr^B animals (24). The center-to-center spacing of the 1911/2574 pair varies from 3.5 to 14 kb. Since both the 1911 and 2574 gene families are expressed very late in choriogenesis, it is probable that they encode high cysteine proteins; this is in agreement with by the characterization of a related cDNA clone by M. Goldsmith (personal communication). The high cysteine class of proteins characterized by an average of 30 molar percent cysteine, is not found in significant quantities in A. polyphemus. Whatever the nature of the chorion proteins encoded by the 1911 and 2574 gene families, their arrangement in these two regions of the B. mori locus is similar to that found in the A. polyphemus genomic clones described above: two multigene families which exhibit the same developmental specificities are arranged in paired repeats. We do not as yet know the orientation of the genes within each pair (divergent or not); however, within each region the pairs are similarly oriented (as drawn in Figure 4, 2574 is to the left and 1911 to the right in each pair). Thus, as in the A. polphemus clones APc110 and APc173, the gene pairs constitute direct repeats, separated by DNAs of variable length.

The 1911-2574 repeats are not identical. Hybridizations with m1911 and m2574 at high criterion (85°, 0.45M salt) indicate that only four of the 1911 genes and five of the 2574 genes hybridize strongly (indicated by solid boxes in Figure 4). It is interesting to note, that while the four closely homologous 1911 genes are contiguous at the left end of the 1911-2574 portion of region I, the genes showing greatest homology to 2574 are distributed over eight repeats in this region, and tend to be found alternating with more distantly related members of the 2574 family.

All clones containing 1911 and or 2574 genes, which we have isolated from our partial EcoRI sublibraries (a total of 109), can be positioned in the map of Figure 4 and represent one or more of the fifteen 1911-2574 repeats. If any other such repeats exist in the genome, they are not present in these sublibraries, and are thus probably located on >20 kb EcoRI fragments. The right-most pair in region I was in fact recovered from the EcoRI* library, after a block to walking in the EcoRI sublibraries was encountered. Only further walking, especially at the right ends of regions I and II, will indicate if additional 1911-2574 gene repeats are to be found.

Chorion cDNA clone, m2598, represents another chorion mRNA species present late in choriogenesis. It does not significantly cross-hybridize to either m1911 or m2574 (24). When this cDNA clone was used to probe the chorion sublibraries, or used in a Southern analysis of the 350 kb of DNA in our genomic walk, only one homologous gene could be found, at either low or high hybridization criteria. This gene may be extremely interesting, since it is not repeated and is located at the extreme left of the 1911/2574 portion of region I, near what may constitute the border between "middle" and "late" chorion gene domains.

The left 30 kb of region I and 100 kb of region II contain genes which do not exhibit homology with m1911 or m2574, even under low hybridization criterion. These genes, by contrast, hybridize extensively with cDNA clones which represent abundant mRNA species present during the middle period of choriogenesis. This group of cDNA clones encode A and B chorion proteins according to hybrid-selected translation studies (24). Thus, our walk has encountered two regions on the chromosome, each of which contains one block of middle genes and one block of late genes. We are currently attempting to link the two chromosomal regions shown in Figure 4, as well as the two smaller genomic segments not shown. If the right ends of regions I and II are eventually joined, then the entire "late" domain of the chorion locus might be contiguous and embedded between two "middle" domains. If the right ends of regions I and II are not connected, two late domains would exist, perhaps alternating with two middle domains, or perhaps flanking one contiguous middle domain.

In any case, the organization of the chorion locus revealed thus far is intriguing. Contiguous direct repeats of gene pairs presumably represent tandem reduplications, which are undergoing sequence divergence (as revealed by the intensities of Southern hybridizations), and are evolving into distinct although homologous genes. The variable spacing and restriction sites between gene pairs suggest that, as in the case of *A. polyphemus*, the pair may be embedded in a tandem repeat which is undergoing rapid evolution by the acceptance of foreign DNA inserts. As these repeats are evolving into distinct genes, they apparently are maintaining their developmental specificity. However, even more drastic chromosomal events are also affecting the organization of the chorion locus. As discussed in the preceding paragraph, the data of Figure 4 clearly establish that either or both late and middle domains are divided into two parts, containing homologous genes. Therefore, these "subdomains" must have been established by a large chromosomal translocation or inversion.

In *A. polyphemus*, we know that some homologous genes are

expressed at different developmental periods: Genes 401 and 10 are homologous but expressed at different developmental periods, as are genes 18 and 292. How are these developmental specificities dictated, and how do they become altered during evolution? These are among the most important questions to be investigated by further studies on the chorion locus. Provisionally, it is tempting to equate the physically contiguous gene domains with chromatin domains which might become transcribable at a specific stage by some type of global chromatin alteration (25, 26). Might infrequent translocations and inversions move reduplicated gene copies into a different domain, thus initiating a new subfamily, expressed during a changed developmental period?

D. Chromosomal Walking in A. polyphemus

On the basis of Southern blot analysis, 15 ± 5 copies of the 401/18 gene pair exist per haploid A. polyphemus genome (10). In an extension of the B. mori walk summarized above, and in order to discover the degree of clustering of these genes as well as the evolutionary mechanisms of their diversification, a chromosomal walk in the 401/18 region of the chorion locus was initiated. Unlike B. mori, EcoRI restriction sites were known to be rare in the 401/18 gene region (APc110 contains no internal EcoRI sites). A partial EcoRI library would therefore exclude most of the 401/18 genes. Instead restriction sites for the endonucleas HindIII are frequent and indeed represent common landmarks in the 401/18 repeat (see Figure 1). Therefore a partial HindIII library was generated using the λ vector λ788 (27). Since A. polyphemus is not available in inbred strains, the DNA used in the formation of this library was derived from one animal; thus, at most only two allelic forms of the chorion locus should be present in the library. A sublibrary of 349 clones was obtained by screening 750,000 plaques (7.5 haploid genome equivalents) with an equimolar mixture of the cDNA plasmids pc401 and pc18 at high criteria (80°C 0.45M salt). The manner in which this sublibrary was used for walking, and the advantage of a partial HindIII library for easily selecting those clones which are overlapping, are similar to those described above for the partial EcoRI sublibraries of B. mori.

To date 66 genomic clones have been analyzed. Fifty-seven of these clones can be arranged into three regions, totalling 89 kb in length. The HindIII maps of these three regions, containing all or a portion of at least twelve 401/18 pairs, are diagramed in Figure 5. The locations of the 401 and 18 genes are indicated by the upper and lower rows of boxes, respectively. The remaining nine clones are not as yet known to overlap with each other or with the ends of these three regions. If they are indeed non-overlapping,

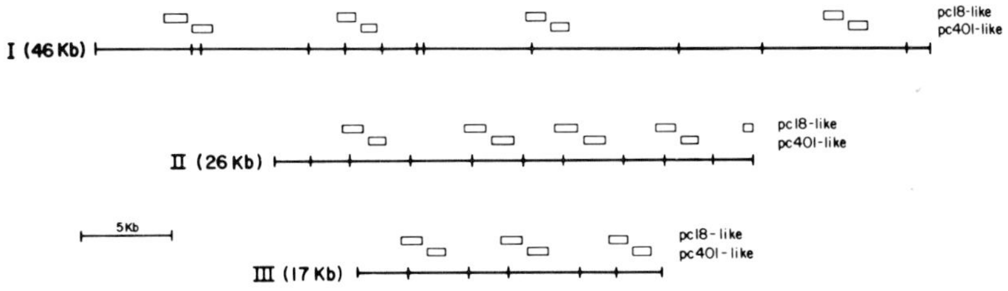

Figure 5 **Schematic map of the three largest segments obtained by the chromosomal walk in *A. polyphemus*. Cross-bars indicate the location of HindIII restriction sites. Boxes above the HindIII map represent the location of genes which bear homology to the cDNA clones pc401 (upper row) or pc18 (lower row) as determined by restriction mapping and Southern analysis**

these nine clones represent another 91 kb of DNA, and 14 additional 401/18 gene pairs (six of which are only partially cloned).

Restriction mapping and Southern analysis indicate that all the 401/18 genes are arranged as divergent gene pairs, and that adjacent gene pairs are invariably oriented in the same direction. Further walking is required to determine whether all 401- and 18-like genes exist in a contiguous tandemly organized domain, or whether major chromosomal translocations or inversions have broken up such a domain. It is already clear that, even if a single 401/18 domain exists, it has undergone extensive diversification. For example, in region I the gene pairs are far apart, while in regions II and III they are closely spaced (average lengths of the 3' flanking DNA 9.9, 3.0 and 3.2 kb, respectively). Furthermore, the distribution of HindIII sites is quite different in the left and right halves of region I. As in the case of *B. mori*, continued analysis of these *A. polyphemus* chorion genes should help us understand better how reduplicated gene copies evolve into developmentally regulated multigene families.

ACKNOWLEDGEMENTS

We are grateful to N.E. Murray and K. Murray for the gift of the λ vector 788, B. Klumpar and M.D. Koehler for help with the figures and S. Foy for secretarial assistance. T. Maniatis and E. Frisch gave us invaluable help in allowing one of us, T.H.E., to visit their laboratory during the construction of the *B. mori* partial EcoRI library. This work was supported by grants from the NSF and NIH to F.C.K. T.H.E. is a Fellow of the Helen Hay Whitney Foundation. C.W.J. was supported by a NSF postdoctoral fellowship and NIH training grant.

REFERENCES

1) Kafatos, F.C., Regier, J.C., Mazur, G.D., Nadel, M.R., Blau, H.M., Petri, W.H., Wyman, A.R., Gelinas, R.E., Moore, P.B., Paul, M., Efstratiadis, A., Vournakis, J.N., Goldsmith, M.R., Hunsley, J.R., Baker, B., Nardi, J. and Koehler, M. (1977). In Results and Problems in Cell Differentiation, 8, W. Beerman, ed. (Berlin: Springer-Verlag), p.45.
2) Regier, J.C., Kafatos, F.C., Goodfleish, R. and Hood, L. (1978). Proc. Nat. Acad. Sci. USA 75, 390.
3) Regier, J.C., Kafatos, F.C., Kramer, K.J., Henrikson, R.L. and Keim, P.S. (1978). J. Biol. Chem. 253, 1305.
4) Rodakis, G.C. (1978). Ph.D. thesis, University of Athens, Athens, Greece.
5) Thireos, G. and Kafatos, F.C. (1980). Dev. Biol. 78, 36.
6) Nadel, M.R., Thireos, G. and Kafatos, F.C. (1980). Cell 20, 649.
7) Sim, G.K., Kafatos, F.C., Jones, C.W., Koehler, M.D., Efstratiadis, A. and Maniatis, T. (1979). Cell 18, 1303.
8) Jones, C.W., Rosenthal, N., Rodakis, G.C. and Kafatos, F.C. (1979). Cell 18, 1317.
9) Iatrou, K., Tsitilou, S.G., Goldsmith, M.R. and Kafatos, F.C. (1980). Cell 20, 659.
10) Jones, C.W. and Kafatos, F.C. (1980). Nature 284, 635.
11) Goldsmith, M. and Basehoar, G. (1978). Genetics 90, 291.
12) Goldsmith, M. and Clermont-Rattner, E. (1979). Genetics 92, 1173.
13) Jones, C.W. and Kafatos, F.C. (1980). Cell 22, 855.
14) Maniatis, T., Hardison, R.C., Lacy, E., Lauer, J., O'Connell, C., Quon, D., Sim, G.K. and Efstratiadis, A. (1978). Cell 15, 687.
15) Breathnach, R., Benoist, C., O'Hare, K., Gannon, F. and Chambon, P. (1978). Proc. Nat. Acad. Sci. USA 75, 4853.
16) Sures, I., Levy, S. and Kedes, L. (1980). Proc. Nat. Acad. Sci. USA 77, 1265.
17) Gage, L.P. (1974). Chromosoma 45, 27.
18) Efstradiatis, A., Crain, W.R., Britten, R.J., Davidson, E.H. and Kafatos, F.C. (1976). Proc. Nat. Acad. Sci. USA 73, 2289.
19) Manning, J.E., Schmid, C.W. and Davidson, N. (1975). Cell 4, 141.

20) Paul, M., Goldsmith, M.R., Hunsley, J.R. and Kafatos, F.C. (1972). J. Cell Biol. 55, 653.
21) Benton, W.D. and Davis, R.W. (1977). Science 196, 180.
22) Nadel, M.R., Goldsmith, M.R., Goplerud, J. and Kafatos, F.C. (1980). Devel. Biol. 75, 41.
23) Kemp, D.J., Cory, S. and Adams, J.M. (1979). Proc. Nat. Acad. Sci. USA 76, 4627.
24) Iatrou, K., Tsitilou, S. and Kafatos, F.C. (in preparation).
25) Garel, A. and Axel, R. (1976). Proc. Nat. Acad. Sci. USA 73, 3966.
26) Weintraub, H. and Groudine, M. (1976). Science 193, 848.
27) Murray, N.E., Brammer, W.J. and Murray, K. (1977). Mol. Gen. Genet. 150, 53.

THE PROTAMINE MULTI-GENE FAMILY IN THE DEVELOPING RAINBOW TROUT TESTES/ANALYSIS OF THE ds-cDNA CLONES

Lashitew Gedamu[1]
Gordon H. Dixon
Michael A. Wosnick

Division of Medical Biochemistry
Faculty of Medicine
The University of Calgary
Calgary, Alberta, Canada T2N 1N4

Kostas Iatrou

Biological Laboratories
Harvard University
Cambridge, Mass. 02138, U.S.A.

I. ABSTRACT

We have synthesized a family of double-stranded cDNAs (ds-cDNAs) using as a template the family of highly purified protamine mRNAs from rainbow trout testis. Individual pure protamine cDNA components were isolated by cloning this family of protamine ds-cDNAs in a plasmid vector (pMB9). Clones containing protamine sequences were characterized by restriction mapping and by a positive hybrid-selected translation assay, which allowed us to correlate particular cDNAs with particular protein components.

To allow more detailed comparisons, complete nucleotide sequences were determined for selected protamine clones. We have detected at least 5 distinctly different coding sequences, which nevertheless show at least 82% homology, and which have probably arisen by repeated gene duplication. These very highly conserved coding sequences do however contain a distinctly variable region near the 5'-end of the mRNA (N-terminus of the protein), corresponding to the major sites of serine

[1]*Present address: University Biochemistry Group, Division of Biochemistry, The University of Calgary, Calgary, Alberta, Canada, T2N 1N4*

ISBN 0-12-137420-3

phosphorylation.

Since the amino acid sequences predicted by our DNA sequences were slightly different from those previously published (1), we have independently determined the amino acid sequences of protamine components C_I, C_{II} and C_{III} from our own source of trout testis. These new peptide sequences are completely consistent with those predicted by our nucleotide sequences.

The 3'-untranslated regions of the protamine mRNAs are, surprisingly almost as highly conserved as the coding regions. Both coding and 3'-noncoding portions appear to be under a similar degree of selective pressure and evolutionary constraint to remain constant.

INTRODUCTION

The developing rainbow trout (Salmo gairdnerii) testis provides an excellent opportunity to study a developmentally regulated multi-gene system, the protamine gene family. Trout protamines comprise at least 3-4 major, extremely arginine-rich polypeptides (1) synthesised at the spermatid stage of testis differentiation which progressively replace the nuclear histones leading to a complete restructuring of sperm chromatin (2). Previous work has suggested that not all protamine components are synthesized at the same rate during testis maturation, and that the pattern of their synthesis is developmentally regulated (3).

In our investigations of the regulation of expression of the protamine multi-gene family, progress has been made in the characterization of the mRNAs coding for protamines (4-10). This family of mRNAs has been isolated and purified in milligram quantities (11) and characterized chemically by partial sequence analysis (10,12,13). Both the poly$(A)^+$ and the poly $(A)^-$ protamine mRNAs (pmRNAs) have been shown to fall into four separable size classes upon electrophoresis in denaturing gels (5,9). Unfortunately cell-free translation studies of the apparently physically separated protamine mRNAs (7,9) as well as hybridization studies of protamine cDNA to poly$(A)^+$ protamine mRNA (8), showed that although enrichment of particular pmRNA components was obtained, complete purification of the individual pmRNA components from one another had clearly not been achieved. This inability to separate cleanly one individual pmRNA component from another and to determine unambiguously the corresponding specific protamine component each encodes is attributed to the very similar size, structure and sequence of the pmRNAs.

In addition to control of expression of the protamine multi-gene family at the transcription level, there is a strong evidence of translational control. This family of pmRNAs have been shown to exist in the form of Ribonucleoprotein particles (14). Free cytoplasmic particles (pmRNP) have been isolated and shown to be totally inactive in *in vitro* translation system and were activated by high salt treatment (15). These pmRNP are found in the postribosomal supernatant fraction of the developing testis cells up to the spermatocyte stage (16). At the spermatid stage of development, these pmRNP's are bound to polysomes and are shown to be fully active in translation (15,16).

More recently, we have purified the individual pmRNA components by cloning the corresponding double-stranded cDNAs (ds cDNAs) prepared against the entire family of purified pmRNAs and have characterised the mRNA components by detailed restriction mapping and nucleotide sequence analysis (17). Clones containing the coding information for protamine components C_{II} and C_{III} have been unequivocally identified using hybrid-selected translation assays (18,19). Our data supports the existence of several different protamine mRNA components with a high degree of homology. However, the predicted amino acid sequences are not entirely consistent with the previously published amino acid sequences determined directly from the polypeptides (1). Our studies further indicated that the genes coding for the protamine components may be grouped into two major families with further subdivision of these two families on the basis of sequence variation at a genetic "hot spot" in the coding regions.

METHODS

Construction and selection of chimeric plasmids containing protamine ds-cDNAs.

Poly(A)$^+$ protamine mRNA (pmRNA) was purified from rainbow trout testis (4,5,11) and ds-cDNA was synthesised under standard conditions (20). The ds-cDNAs were inserted at ECORI site of pMB9 following the poly dA-dT tailing procedure (21) and the hybrid DNA used to transform *E. coli* HBIOI (22). Tetracylcine resistant colonies were picked and screened by colony hybridisation (23), using ^{32}P-labelled protamine cDNA.

Restriction analysis and sequence determination.

DNA containing protamine positive clones were fragmented with restriction endonucleases using conditions suggested by suppliers. Restriction fragments were analysed by electrophoresis on agarose and polyacrylamide gels. DNA fragments were labelled at the 5'-termini with γ-^{32}P-ATP in the presence of T_4 polynucleotide kinase (24,25) after dephosphorylating by treatment with calf intestinal alkaline phosphatase (24). For nucleotide sequence analysis, chemical cleavage of the labelled DNA was performed by the method of Maxam and Gilbert (17).

Hybrid-selected translation

Hybrid-selected translation were performed as described previously (19,26). The mRNA was hybridised to each clone and was translated in a wheat germ cell-free system (7,27) in the presence of ^{3}H-arginine. Labelled products were analysed by chromatography on CM-52 column (4).

Protamine polypeptide sequencing

Purified protamine components containing 50-100 picomoles of material were sequenced following sequential Edman degradation in a Beckman 890C sequencer.

RESULTS AND DISCUSSION

A. *Identification of Clones Containing Protamine cDNA Sequences*

[^{32}P]-cDNA labelled to high specific activity and prepared as described above from the family of highly purified protamine mRNA's (20) was used as a probe for *in situ* colony hybridization (23) to screen for protamine-positive recombinant plasmids. Six positive clones were selected for initial study and were designated pRTP 43, 59, 94, 131, 178 and 242.

B. *Restriction Analysis of the Protamine Positive Clones*

Of the 15 to 20 different restriction enzymes which were tested, very few cleaved in the protamine sequence. This is due not only to the small size of the inserts (250-350 nucleotides each in length) but also to their GC-rich nature (due to the ponderance of arginine residues, whose codons are CGX or AG$^{A}_{G}$ in the protamine coding portion). However, a sufficient number of restriction sites was revealed using enzymes specific for GC-rich sequences to allow a preliminary classification of these clones based on restriction patterns(Figure 1). All plasmids (except the very short pRTP 131) showed three restriction sites in common, reading from the 5'-end, a puGCGCPy site cut by both HhaI and Hae II, a GGCC site cut by Hae III and a CCGG site cut by Hpa II. Later nucleotide sequence analysis showed that the Hha I/Hae II site was in the protamine coding region 41-44 nucleotides from the initiation codon, the Hae III site 84-88 nucleotides from the initiation

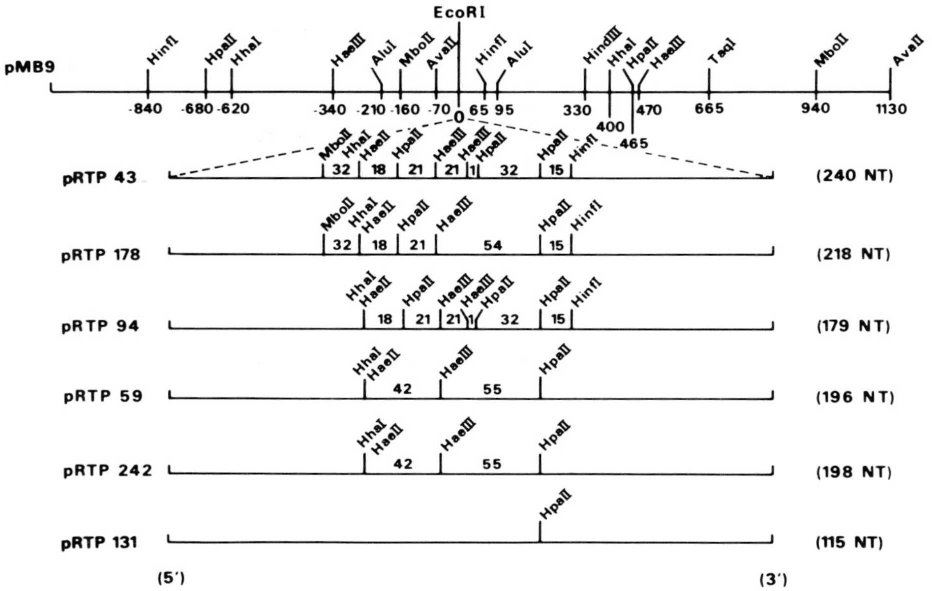

FIGURE 1. Restriction maps of six clones containing protamine ds cDNA sequences. The insertion into plasmid pMB9 was at the Eco R1 site shown at the top. The lengths of the inserts is indicated on the right.

codon and the Hpa II site 43-47 nucleotides downstream from the termination codon in the 3'-non coding region of the mRNA.

One group of plasmids pRTP 43, 178 and 94 showed two additional sites in common, a Hpa II site between the Hha I/Hae II and the Hae III sites and a Hinf I site 15 nucleotides 3' to the rightward Hpa II site. Furthermore, pRTP 43 and 178 the longest inserts have a Mbo II site close to the 5'-end of the protamine coding region and additional Hae III and Hpa II sites very close to the 3'-end of the protamine coding region.

C. *Hybrid-Selected Translation Assays*

In order to determine which cDNA clone encoded which protamine component, individual cloned DNAs were linearized, immobilized on nitrocellulose filters, and hybridized to purified poly$(A)^+$ protamine mRNA, as described in Methods. The hybridizations were performed under increasing degrees of stringency (60°C and 70°C), chosen to minimize cross-hybridization of closely related species. The hybrid-selected mRNAs were then dissociated by heating and translated in the wheat germ cell-free protein synthesizing system (6,31). As shown in Figure 2D, the sequence contained in pRTP 59 hybridizes exclusively at 70°C to an mRNA that encodes protamine component C_{II}. However at the less stringent 60°C (Figure 2A), a C_{III} coding mRNA is also selected. Likewise, in Figure 2E, at high stringency, pRTP 94 encodes only protamine C_{III}, but at lower stringency (Figure 2B) there is cross-hybridizing to C_I and C_{II}. Unfortunately, the high-stringency results were not always so unambiguous. For example, even under the most stringent hybridization conditions, pRTP 242 could not be shown to hybridize selectively to one particular pmRNA component (Figure 2C). The fact that pRTP 242 and pRTP 59 exhibit the same restriction map (Figure 1), and yet respond so differently in the hybridization-translation assay (Figure 2) made it obvious that the only unambiguous method by which these clones could be characterized was through nucleotide sequence analysis.

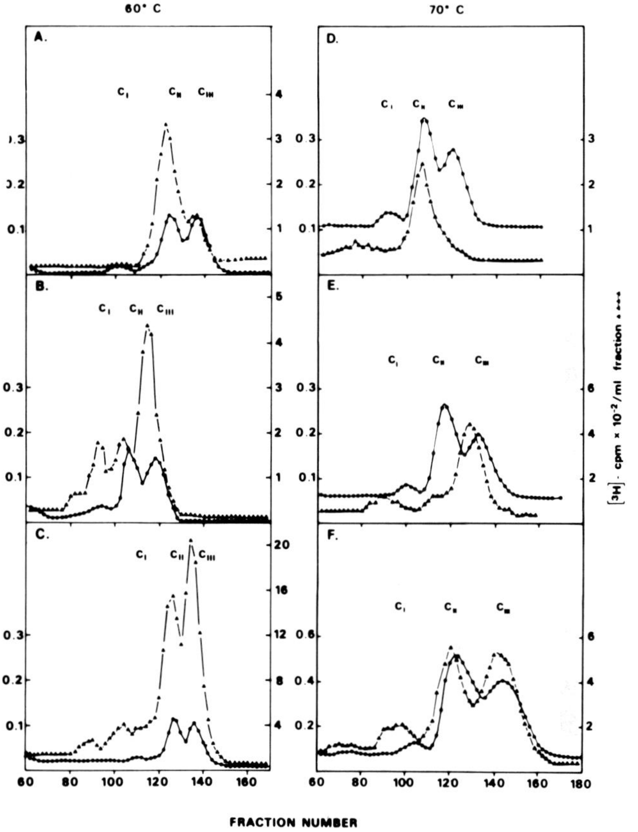

FIGURE 2. Positive hybrid-selected translation analysis of three protamine clones as described in Materials and Methods. Hybrid-selected mRNA translation products (▲-▲-▲) were chromatographed on CM-52 cellulose and compared to authentic trout protamines (●-●-●).

D. Nucleotide Sequence Analysis

1. The Coding Regions. The amino acid coding sequences of five different protamine clones are shown in Figure 3. Only two of the five, pRTP 43 and pRTP 178, actually contain the entire coding sequence, from initiation through termination codons. pRTP 43 in fact extends an additional 27 nucleotides (5'-GATTCATAGTCTTATCTATCAATCACT-3') and pRTP 178 extends an

pRTP 43	MET	PRO	ARG	ARG	ARG	ARG		ALA	SER	ARG	ARG	VAL	ARG	ARG	ARG	ARG	ARG	PRO
	ATG	CCC	AGA	AGA	CGC	AGA	---	GCC	AGC	CGC	CGT	GTC	CGC	AGG	CGC	CGT	CGC	CCC
pRTP 178	MET	PRO	ARG	ARG	ARG	ARG		ALA	SER	ARG	ARG	ILE	ARG	ARG	ARG	ARG	ARG	PRO
	ATG	CCC	AGA	AGA	CGC	AGA	---	GCC	AGC	CGC	CGT	ATC	CGC	AGG	CGC	CGT	CGC	CCC
pRTP 94								ALA	SER	ARG	ARG	ILE	ARG	ARG	ARG	ARG	ARG	PRO
								GCC	AGC	CGC	CGG	ATC	CGC	AGG	CGC	CGT	CGC	CCC
pRTP 59							SER	SER	SER	ARG	PRO	VAL	ARG	ARG	ARG	ARG	ARG	PRO
							TCC	TCC	AGC	CGA	CCT	GTC	CGC	AGG	CGC	CGC	CGC	CCC
pRTP 242							SER	SER	ARG	ARG	PRO	VAL	ARG	ARG	ARG	ARG	ARG	PRO
						GT	TCC	TCC	AGA	CGA	CCT	GTT	CGC	AGG	CGC	CGC	CGC	CCC

pRTP 43	ARG	VAL	SER	ARG		ARG	ARG	ARG	ARG	GLY	GLY	ARG	ARG	ARG	ARG	TERM
	AGG	GTG	TCC	CGG	---	CGT	CGC	AGG	AGA	GGA	GGC	CGC	AGG	AGG	CGT	TAG
pRTP 178	ARG	VAL	SER	ARG		ARG	ARG	ARG	ARG	GLY	GLY	ARG	ARG	ARG	ARG	TERM
	AGG	GTG	TCC	CGG	---	CGT	CGC	AGG	AGA	GGA	GGC	CGC	AGG	AGG	CGT	TAG
pRTP 94	ARG	VAL	SER	ARG		ARG	ARG	ARG	ARG	GLY	GLY	ARG	ARG	ARG	ART	TERM
	AGG	GTG	TCC	CGG	---	CGT	CGC	AGG	AGA	GGA	GGC	CGC	AGG	AGG	CGT	TAG
pRTP 59	ARG	VAL	SER	ARG	ARG	ARG	ARG	ARG	ARG	GLY	GLY	ARG	ARG	ARG	ARG	TERM
	AGG	GTG	TCC	CGA	CGT	CGT	CGC	AGG	AGA	GGA	GGC	CGC	AGG	AGG	CGT	TAG
pRTP 242	ARG	VAL	SER	ARG	ARG	ARG	ARG	ARG	ARG	GLY	GLY	ARG	ARG	ARG	ARG	TERM
	AGG	GTG	TCC	CGA	CGT	CGT	CGC	AGG	AGA	GGA	GGC	CGC	AGG	AGG	CGT	TAG

FIGURE 3. Nucleotide sequences of the amino acid coding portion of five cloned protamine cDNA components. Only the DNA strand with the same sense as the mRNA has been shown, along with the predicted amino acid sequence. The area enclosed in the box is the "variable region" (see text). Since the 5'-boundary of this region is uncertain, a dotted line has been tentatively placed. Dashes are used to indicate gaps which have been introduced for optimal sequence alignment.

additional 5 nucleotides (5'-TCACT-3') beyond the ATG initiation codon. It is significant that the 3 incomplete sequences as well as 2 out of 3 additional protamine coding sequences that have been determined elsewhere (28), all terminate at almost the identical nucleotide position. This probably reflects some tight secondary structure at this point in some of the parent pmRNAs, which cause premature termination of reverse transcription. This further strengthens our view that pmRNAs are very highly structured molecules (8,10,13).

The high degree of homology between these different coding sequences is very striking. In so far as they can be directly compared, these 5 sequences share a minimum overall degree of identity of 82%. The identity between any two selected species can be as high as 99% (as in the case of pRTP 43 and pRTP 178 where only one difference exists in the coding portion). The data further supports our previous report on the sequence diversity of the poly(A)$^+$ pmRNA components which indicated that these components probably possessed a common sequence of substantial length, representing greater than 70% homology, when examined by molecular hybridization (8).

Despite this very high overall sequence conservation, there is one portion of the molecule (enclosed in a box in Figure 3) which is not nearly so well conserved. In this "variable region", many nucleotide substitutions have given rise to several amino acid replacements. In fact,2/3 of the nonconserved nucleotides in these 5 species occur in this region alone. It is unfortunate that we cannot ascertain at this time whether this variable region extends any further toward the 5'-end of the molecule.

Previous work has suggested that not all protamine components are synthesized at the same rate during testis maturation, and that the pattern of their synthesis is developmentally regulated (3). This implies that there might be subtle differences in function for individual components. If such differences exist, they would presumably have to be mediated by sequence differences in the variable region, since the rest of the coding portion seems to well conserved. The fact that protamines undergo phosphorylation at the serine residues found in the variable region (29) coupled with the observation that the number and position of serine residues differ from one sequence component to another (Figure 3), suggests that specific differences in the extent and exact position of serine phosphorylation may be responsible for sublet differences in structure and/or function of the various protamine components.

The codon usage pattern from all protamine sequences determined to date has been investigated. As expected, both the amino acid and codon usages are decidedly non-random. Only

8 of the 20 amino acids are represented at all in the protamines and in every case there seems to be a favored codon assignment for that amino acid. The two most abundant amino acids, arginine and serine, are notable examples. In the case of arginine, the AGG codon is significantly more abundant than the AGA codon (AGG=40; AGA=18). Similarly in the second arginine codon set, CGX, the codon CGC is found twice as often as CGU, and these two together are favored over CGA and CGC by a margin of 6 to 1. In one serine codon series, UCX, only UCC was found (16 times) to the exclusion of the other 3 possibilities. In the second serine codon set, AG_{U}^{C}, only AGC was found in the sequences.

The amino acid sequences of the three major trout protamines have been previously reported by Ando and Watanabe (1). Significantly, not one of the nucleotide sequences we have determined predicts an amino acid sequence entirely consistent with this previous data (Figure 4). Similar discrepancies between amino acid and nucleotide sequence data have been noted in three additional protamine clones (28), making it very likely either that there were some errors in the sequencing of the proteins themselves, or that the source of trout testis from which the polypeptides were purified shows significant genetic variation from our source of material.

In order to find a possible explanation, the protamine polypeptides C_{I}, C_{II} and C_{III} were purified from the same tissue used for the isolation of the pmRNA's and their sequences were determined. The results (Figure 4) clearly show that the amino acid sequences obtained this way are in agreement with those predicted from the nucleotide sequences (see Figure 3) and supports the later possibility.

Clone pRTP 94 hybridizes with a protamine mRNA that codes for C_{III} and its DNA sequence indicates that residue 10 should be Ile: in fact, when the total protamine under peak C_{III} was sequenced, residue 10 showed heterogeneity with Val:Ile:Pro of 1.0:0.3:0.3. Thus C_{III} protamine must be heterogeneous with at least three sub-components present but unresolved under the C_{III} peak. Evidently, pRTP 94 codes for only one of these components possessing a Ile at position 10, however, it is interesting to note in this connection that the C_{III} component selected by pRTP 94 does not chromatograph precisely with the bulk C_{III} peak (Figure 2) indicating that the major component in the C_{III} peak is coded by another mRNA sequence in which residue 10 is a Val. Such a putative mRNA would be represented by the DNA sequence of pRTP 43 which possesses the valine triplet, GTC, at residue 10 instead of the Ile triplet ATC in pRTP 94. Unfortunately pRTP 43 has not yet been tested in the hybridization selection of protamine mRNA's.

PROTAMINE POLYPEPTIDE SEQUENCES

CI*	PRO ARG ARG ARG ARG ARG ALA SER ARG ARG VAL ARG ARG ARG ARG ARG PRO ARG - VAL SER ARG ARG ARG ARG ARG - GLY GLY ARG ARG ARG ARG-COOH
IRIDINE IA (1)	PRO ARG ARG ARG ARG SER SER SER ARG PRO VAL ARG ARG ARG ARG ARG PRO ARG ARG VAL SER ARG ARG ARG ARG ARG ARG GLY GLY ARG ARG ARG ARG-COOH
CII*	PRO ARG ARG ARG ARG SER SER SER ARG PRO VAL ARG ARG ARG ARG ARG PRO ARG - VAL SER ARG ARG ARG ARG ARG ARG GLY GLY ARG ARG ARG ARG-COOH
pRTP 59 (DNA)	SER SER SER ARG PRO VAL ARG ARG ARG ARG ARG PRO ARG - VAL SER ARG ARG ARG ARG ARG ARG GLY GLY ARG ARG ARG ARG-COOH
IRIDINE IB (1)	PRO ARG ARG ARG ARG ARG ARG SER SER SER ARG PRO ILE - ARG ARG ARG ARG PRO ARG ARG VAL SER ARG ARG ARG ARG ARG - GLY GLY ARG ARG ARG ARG-COOH
CIII*	PRO ARG ARG ARG ARG - - ALA SER - ARG PRO VAL (ILE 0.3) (Pro 0.3) ARG ARG ARG ARG ARG PRO ARG - VAL SER ARG ARG ARG ARG ARG - GLY GLY ARG ARG ARG ARG-COOH
pRTP 94 (DNA)	ALA SER - ARG ARG ILE ARG ARG ARG ARG ARG PRO ARG - VAL SER ARG ARG ARG ARG ARG - GLY GLY ARG ARG ARG ARG-COOH
pRTP 43 (DNA)	PRO ARG ARG ARG ARG - - ALA SER - ARG ARG VAL ARG ARG ARG ARG ARG PRO ARG - VAL SER ARG ARG ARG ARG ARG - GLY GLY ARG ARG ARG ARG-COOH

* DETERMINED IN THIS LABORATORY BY AUTOMATIC EDMAN DEGRADATION

(1) ANDO AND WATANABE (1969)

FIGURE 4. Amino acid sequences of protamine components C_I, C_{II} and C_{III} determined by automatic Edman degradation as described in Materials and Methods. A heterogeneity was detected at position 10 in C_{III} where Val, Ile and Pro were found in the ratio 1.0:0.3:0.3.

If, as indicated by the protein sequence analysis of C_{III}, there is a third component possessing Pro at residue 10, it must be at least two nucleotides different to code for Proline whose triplets are CC(X).

The closest correspondence in polypeptide sequence between any of our protamine components and those sequences published by Ando and Watanabe (1) is for C_{II} and Iridine Ia which are identical except that there is a single Arg between Pro_{17} and Val_{19} of C_{II} but two Arg residues in Iridine 1a. This same difference is seen between C_{III} and Iridine 1b but in addition, there are several other differences: the first arginine tract in C_{III} is 4 residues long compared to 6 in Iridine 1b while the Val at residue 10 of C_{III} is replaced by an Ile. However, as noted above, there is microheterogeneity at this position in C_{III} and Ile is represented as a minor component. The exact nature of the discrepancy between our sequences and those previously reported remains unclear; the most likely possibility however, is that the variety of rainbow trout that others have been using (*Salmo irideus*) is genetically distinct from that which we have been utilizing for many years (*Salmo gairdnerii*) despite the fact that taxonomically the two species designations are considered synonymous, *S. irideus* having been renamed *S. gairdnerii*.

2. The Untranslated Region. The nucleotide sequences of the 3'-untranslated region of our protamine cDNA clones are presented in Figure 5. A surprisingly high degree of homology is observed. In fact, comparison of any two specific 3'-noncoding sequences may yield identity relationships as high as 99%. When all sequences are compared, however, after alignment for optimal fit, at least 76% of all nucleotide positions have been conserved. It will be recalled that the coding regions enjoyed a not much higher overall degree of conservation (82%). It seems therefore, that in the trout protamine family, both the coding and the 3'-non-coding regions are under similar degrees of selective pressure and evolutionary constraint to remain constant. The protamines thus provide the clearest example to date in which 3'-untranslated mRNA sequences are apparently not free to drift. This finding is all the more noteworthy when one considers that the protamines themselves have apparently undergone considerable interspecific sequence divergence, even among closely related fish

```
                            20                           40
pRTP  43     ACA-GGCCGGGT---AACCTACCTGAACTA-ACCGCCCCCTACCGG-TTCTCCCTCCAG
pRTP 178     ACA-GGTCGGGT---AACCTACCTGAACTA-ACCGCCCCCTACCGG-TTCTCCCTCCAG
pRTP  94     ACA-GGCCGGGT---AACCTACCTGAACTA-ACCGCCCCCTACCGG-TTCTCCCTCCAG
pRTP 242     ATA-GAATGGGTA-GAACCTACCTGACCTATC-CGCCCCCT-CCGGGTTCACCCTCCC-
pRTP  59     ATA-GAACGGGTA-GAACCTACCTGACCTATC-CGCCCCCT-CCGGGTTCTCCCTCCC-
pRTP 131     ATATGAACGGGTATGAACCTACCTGACCTATCACGCCCCCT-CCGGGTTCTCCCTCCC-

             60                   80                   100                 120
pRTP  43     ACTCGACCACTGGTAGTGCAGAGATGTTAAAAGTCTGCTTAAATAAAAGATGGGCGTTTTAACT (Poly A)
pRTP 178     ACTCGACCACTGGTAGTGCAGAGATGTTAAAAGTCTGCTTAAATAAAAGATGAACGTTTTAACT (Poly A)
pRTP  94     ACTCGACCACTGGTAGTGCAGAGATGTTAAAAGTCTGCTTAAATAAAA
pRTP 242     ----GACCCCTGGTAGTGTAGAGGTGTT-AAAGTCTGCTTAAATAAAAGATGAAC-TTTTAACT (Poly A)
pRTP  59     ----GACCCTTGGTAGTGTAGATGTGTT-AAAGTCTGCTTAAATAAAAGATGGGCGTTTTAACT (Poly A)
pRTP 131     ----GACCCCTGGTGGTGTAGAGGTGTT-AAAGTCTGCTTAAATAAAAGATGGGC-TTTTAACT (Poly A)
```

FIGURE 5. Compilation of nucleotide sequences derived from the 3'-untranslated regions of protamine mRNA's. Only the cloned cDNA strand with the same sense as the mRNA is shown. Dashes (-) have been used to indicate gaps which have been introduced for optimal sequence alignment. Nucleotide position 1 is the one immediately following the G of the TAG termination codon.

species (30).

In the 3'-non-coding region, the non-conserved nucleotides are not distributed totally at random, but seem to be somewhat clustered, tending to separate the molecule into sizeable tracts of conserved sequence. The highest degree of conservation is nearest the 3'-terminus. In fact, the longest tract of conserved sequence, which numbers 20-23 nucleotides (the exact number depends on where a single gap in introduced between positions 91-94 for sequence alignment) is found very near the site of poly(A) addition (Figure 6).

Confirming our earlier report (31), the hexanucleotide sequence 5'-AATAAA-3' which is thought to be a signal involved in polyadenylation, and which is found about 20 nucleotides upstream of the site of poly(A) addition in all eukaryotic mRNAs so far examined (32) is also found in the protamine mRNA's right in the middle of the longest conserved sequence tract. This implies that not only might the AATAAA sequence be important, but that the sequences immediately flanking it may also have an important function.

3. The 5'-Untranslated Region. Since only two of the clones examined even extend into the 5'-untranslated region, very few conclusions can be drawn. However, the 5'-non-coding sequence of pRTP 43 (5'-GATTCATAGTCTTATCTATCAATCACT-3) confirms and extends some previously determined RNA sequences which contained an AUG codon and were believed to orginate from this portion of the molecule (13). From studies on the ribosome protected region of pmRNAs, the 5'-non-coding sequence of pRTP43 is indicated to contain a ribosome recognition site (Gedamu, L. and Dixon, G.H., in preparation).

The cloned protamine genes, in general, show a very high degree of homology and clearly explains our inability to separate the individual pmRNA components from one another using conventional techniques. Careful examination of the nucleotide sequence of the different cDNA clones and data from the restriction analysis and positive hybrid-select translation assay implies that the protamine genes can be grouped into two major families. The two sequence families thus produced seems to be evolving in parallel, since the degree of homology between sequences with in the same family is quite high (>90%), (for example pRTP43, pRTP94 and pRTP178) where as interfamily sequence conservation is significantly lower. Nevertheless, these two families seem not to be diverging very rapidly from one another since at least 70-80% identity still exists between any two sequences compared across family boundaries.

It might be argued that some of the very similar sequences presented in this paper are actually alleles of one another.

Since pooled testes were used as the starting material, this possibility cannot be totally excluded. This explanation might for instance apply to pRTP 178 and pRTP 94 which exhibit only two sequence differences, one each in the coding and non-coding regions. However, in most cases, although great sequence similarity clearly exists, there are enough differences between individual sequences (even those within the same family) to suggest that the allelic explanation by itself is not sufficient.

Now the cDNA clones are well characterised and are more recently used to purify the natural protamine gene to address the problem of organization, number of copies of the genes representing each of the protamine components. In addition, studies on the regulation of protamine gene expression during sperm cell development will be amenable due to the availability of such purified and well characterised cDNA clones.

ACKNOWLEDGMENTS

We thank Dr. F.C. Kafatos for the use of his laboratory facilities during the initial cloning of these protamine cDNAs, and Lydia Villa-Komaroff for advice on the cloning procedures. The generous gift of T_4 polynucleotide kinase and PM2 by Dr. J.H. van de Sande and Mr. B.W. Kalisch is gratefully acknowledged. The technical assistance of Mr. David C. Watson, Mr. Wayne Connor and Ms. Terry Hutcheon is sincerely appreciated. This work was supported by grants from the Medical Research Council of Canada and the National Science Foundation, U.S.A.

Figures 1-5 are taken from a manuscript in press in Nucleic Acids Research.

REFERENCES

1. Ando, T. and Watanabe, S. (1969). *Int. J. Prot. Res. 1,* 221-224.
2. Marushige, K. and Dixon, G.H. (1969). *Devel. Biol. 19,* 397-414.
3. Ling, V., Jergil, B. and Dixon, G.H. (1971). *J. Biol. Chem. 246,* 1168-1176.
4. Gedamu, L. and Dixon, G.H. (1976). *J. Biol. Chem. 251,* 1455-1463.
5. Iatrou, K. and Dixon, G.H. (1977). *Cell 10,* 443-441.
6. Gedamu, L., Iatrou, K. and Dixon, G.H. (1977). *Cell 10,* 443-451.
7. Gedamu, L., Iatrou, K., and Dixon, G.H. (1979). *Biochem. Biophys. Acta. 562,* 481-494.
8. Iatrou, K., Gedamu, L. and Dixon, G.H. (1979). *Can. J. Biochem. 57,* 945-956.

9. Gedamu, L. and Dixon, G.H. (1979). *Nic. Acids. Res. 6,* 3661-3672.
10. Davies, P.L., Dixon, G.H., Ferrier, .LN., Gedamu, L. and Iatrou, K. (1976). *Prog. Nuc. Acids. Res. Mol. Biol. 19,* 135-155.
11. Gedamu, L., Iatrou, K. and Dixon, G.H. (1978). *Biochem. J. 171,* 589-599.
12. Davies, P.L., Ferrier, L.N. and Dixon, G.H. (1977). *J. Biol. Chem. 252,* 1386-1393.
13. Davies, P.L., Dixon, G.H., Simoncsits, A. and Brownlee, G.G. (1979). *Nuc. Acids. Res. 7,* 2323-2345.
14. Gedamu, L., Dixon, G.H. and Davies, P.L. (1977). *Biochemistry 16,* 1383-1391.
15. Sinclair, D.G. (1980). Ph.D. Thesis, University of Calgary.
16. Iatrou, K., Spira, A.W. and Dixon, G.H. (1978). *Dev. Biol. 64,* 82-98.
17. Maxam, A. and Gilbert, W. (1980). *Methods in Enzymol., Vol. 65,* (Grossman, L. and Moldave, K., eds.) Academic Press, N.Y., pg. 499-560.
18. Ricciardi, R.P., Miller, J.S. and Roberts, B.E. (1979). *Proc. Nat. Acad. Sci. USA 76,* 4927-4931.
19. Iatrou,K., Tsitilou, S.G., Goldsmith, M.R. and Kafatos, F.C. (1980). *Cell 20,* 659-669.
20. Efstratiadis, A., Kafatos, F.C.,Maxam, A.M. and Maniatis, T. (1976). *Cell 7,* 279-288.
21. Roychoudhury, R., Jay, E. and Wu, R. (1976). *Nuc. Acids Res. 3,* 101-116.
22. Wensnik, P.C., Finnegan, D.J., Donelson, J.E. and Hogness, D.S. (1974). *Cell 3,* 315-325.
23. Grunstein, M. and Hogness, D.S. (1975). *Proc. Nat. Acad. Sci. USA 72,* 3961-3965.
24. Chaconas, G. and van de Sande, J.H. (1980). In Methods in Enzymol., Vol. 65, (Grossman, L. and Moldave, K., eds) Academic Press, New York, pg 75-85.
25. Gedamu, L. and Dixon, G.H. (1978). *Biochem. Biophys. Res. Comm. 85,* 114-124.
26. Kafatos, F.C., Jones, C.W. and Efstratiadis, A. (1979). *Nuc. Acids Res. 7,* 1541-1552.
27. Roberts, B.E. and Paterson, B.M. (1973). *Proc. Nat. Acad. Sci. USA 70,* 2330-2334.
28. Jenkins, J. (1979). *Nature 279,* 809-811.
29. Sanders, M.M. and Dixon, G.H. (1972). *J. Biol. Chem. 247,* 851-855.
30. Davies, P.L. and Daisley, St. L. (1980). *Proc. Can. Fed. Biol. Soc. 23,* 78.
31. Ferrier, L.N., Davies, P.L. and Dixon, G.H. (1977). *Biochem. Biophys. Acta 479,* 460-470.

32. Proudfoot, N.J. and Brownlee, G.G. (1976). *Nature 263*, 211-214.

32. Proudfoot, N.J. and Brownlee, G.G. (1976). *Nature 263*, 211-214.

ISOLATION AND CHARACTERIZATION OF cDNA CLONES ENCODING MOUSE TRANSPLANTATION ANTIGENS[1]

Michael Steinmetz
John G. Frelinger
Douglas A. Fisher
Kevin W. Moore
Beverly Taylor Sher
Leroy Hood

Division of Biology
California Institute of Technology
Pasadena, California

I. INTRODUCTION

The genes encoding transplantation antigens are contained in a region on mouse chromosome 17 termed the major histocompatibility complex (MHC) or H-2 complex. This complex encodes three families of genes: class I—the transplantation antigens, class II—the Ia antigens, and class III—certain complement components (Fig. 1). Transplantation antigens appear to play a fundamental role in the recognition by T cells of virally altered or neoplastically transformed cells (for a review see ref. 1).

In the BALB/c mouse there are four serologically defined transplantation antigens—K, D, L, and R (2). Transplantation antigens are found on all somatic cells of the mouse. They are highly polymorphic with 56 and 45 alleles having been defined at the K and D loci, respectively (3). The polymorphism of the transplantation antigens is perhaps unprecedented in any other gene system studied to date in mammals. Thus the various inbred strains of mice differ in their MHC genes. The MHC genotype or haplotype of a particular inbred strain of mouse, such as the BALB/c mouse, is indicated by a superscript placed above the antigen designation—e.g., K^d.

[1]*This work was supported by grant GM 06569 from NIH.*

ISBN 0-12-137420-3

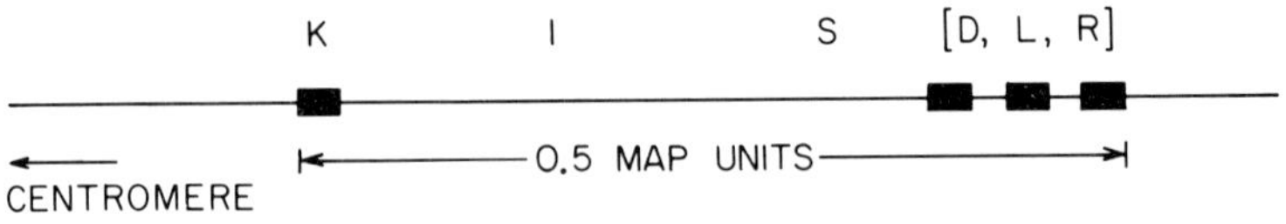

FIGURE 1. Genetic map of the mouse major histocompatibility complex on chromosome 17. K, D, L, and R loci code for transplantation antigens (class I), I loci code for Ia antigens (class II), and S loci code for complement proteins (class III). The relative order of D, L, and R loci is unknown as indicated by brackets. K and D loci are separated by approximately 0.5 centimorgans.

The transplantation antigen is a 45,000 dalton integral membrane protein which is associated with β_2-microglobulin (4). Amino acid sequence studies of transplantation antigens have led to two interesting observations. First, the polypeptide chain seems to be divided into three external domains each of approximately 90 residues and into a transmembrane and a cytoplasmic region. Second, several observations suggest that the transplantation antigens or portions thereof may be homologous to immunoglobulins. For example, the second and the third of the external domains have a centrally placed disulfide bridge spanning about 60 residues just as do all immunoglobulin domains. In addition, amino acid sequence analyses of the transplantation antigens suggest that the third domain may be homologous to the immunoglobulin domains (5). Homology to immunoglobulins has also been found for β_2-microglobulin which has been called a free immunoglobulin domain (6). These observations raise the possibility that immunoglobulins and transplantation antigens share a common evolutionary ancestor.

We have recently reported on the isolation and characterization of three cDNA clones coding for mouse transplantation antigens (7). Other laboratories have isolated cDNA clones coding for human (8,9) as well as mouse transplantation antigens (10). We were interested in isolating genes for mouse transplantation antigens in order to study the organization and expression of the genes in this family.

A. *Three cDNA Clones for Mouse Transplantation Antigens*

Using a human HLA cDNA clone as a probe we isolated three cDNA clones (pH-2I, pH-2II, pH-2III) from two mouse lymphoma cell lines. Determination of the DNA sequence showed that all three cDNAs encode portions of transplantation antigens. Figure 2 shows the extent to which coding (black) and untranslated regions (hatched)

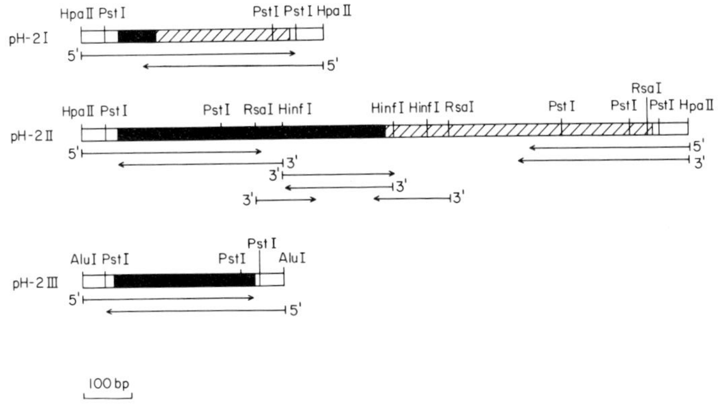

FIGURE 2. Physical map of the three cDNA clones encoding transplantation antigens. The cDNAs were cloned into the Pst I site of pBR322 by dG-dC tailing. Clones pH-2I and pH-2II were obtained from the C14 cell line (d haplotype) and clone pH-2III from the RDM4 cell line (k haplotype). Arrows indicate extent and direction of DNA sequencing with numbers referring to the ends which had been labeled. pBR322 sequences and G-C tails are represented by open boxes, coding sequences by filled boxes, and 3' untranslated regions by hatched boxes. Taken from ref. 7.

are present in the cDNA clones. In Figure 3 a comparison between the complete protein sequence of the K^b transplantation antigen and the amino acid sequences encoded in the isolated clones is given. Clone pH-2I and clone pH-2II, both derived from a d haplotype cell line, contain portions of the coding region as well as 3' untranslated region sequences. These two clones differ from one another in sequence with respect to the coding region and the 3' untranslated region and therefore are derived from two different mRNA molecules. Nevertheless they are highly homologous in the coding as well as in the 3' untranslated region. As shown in Figure 4 the major difference between the two sequences is the presence of a 139 bp insertion in clone pH-2II which gives rise to a longer and therefore different cytoplasmic domain. Hybridization and DNA sequence analyses indicate that this insertion is not a cloning artifact but is also present in genomic genes coding for transplantation antigens. The role of this heterogeneity of the cytoplasmic domain is unclear.

Comparison of the DNA sequences of our clone pH-2I and the H-2 cDNA clone isolated by Kvist et al. (10) from a different mouse cell line also of the d haplotype shows that both clones are identical over the stretch of 96 nucleotides that can be compared and differ in only one position. The restriction maps of the two clones are identical. It is therefore likely that both clones are derived from the same mRNA molecule.

```
K^b       GPHSLRYFVTAVSRPGLGEPRYMEVGYVDDTEFVRFDSDAENPRYEPRARWMEQEGPEYWERETQKAKGNEQSFRVDLRTLLGYYNQSKG   90
PH-2I
PH-2II
PH-2III   R—I————————R———E-
HLA-B7    -S—M——Y-S———R———FIS—————Q—————AS—E———P-I—————D-N—IY-AQA-TD-ES—N-R—————EA
HLA-A2    -S—M——F-S———R———FIA—————Q—————AS?-M———P?I—————D-???-V-AHAHTV———G—R—————EA

K^b       GSHTIQVISGCEVGSDGRLLRGYQQYAYDGCDYIALNEDLKTWTAADMAALITKHKWEQAGEAERLRAYLEGTCVEWLRRYLKNGNATLL  180
PH-2I
PH-2II    —H——————
PH-2III   —————RL——D———W————E————————————————————————A———D———
HLA-B7    ———L-SMY—D—P—————HD—————K——————————RS—————T—Q—QR———A-R———QR—————E——————————E—KDK-E
HLA-A2    ———L-RMY—?————W-F————H—————K—————K———RS——————————Q—————A-RV—QR——————————????????????

K^b       RTDSPKAHVTHHSRPDDKVTLRCWALGFYPADITLTWQLNGEELIQDMELVETRPAGDGTFQKWASVVVPLGKEQYYTCHVYQQGLPQPL  270
PH-2I
PH-2II    —P—————R—EGD————————————————————T————————————————————————N——R—HE——E—
PH-2III
HLA-B7    -A-P—T————PIS-HEA——————————E—————RD—DQT—T——————————R—E——A————S-E—R—————QHE——K—
HLA-A2    ????????????????????????????????????????????????????????????????????????????????????????

K^b       TLRWDEPPSTVSNMATVAVLVVLGAAIVTGAVVAFVMKMRRRATGGKGGDYALAPGSQTSDLSLPDCKVMVHPPHS              346
PH-2I     N——————————————S—M——————
PH-2II    ———EP————D-Y-VI————G————MAII——————— —N———————————————S-EM—R——GDTLGSDWGGAMWT
PH-2III
HLA-B7    -?????????????????????V?-VV-?—————A—C —KSS—————S-SQ-AC-DSAQG-DVSLTA
HLA-A2    ????????????????????????????????????-W —KSSDR———S-SQ-AS-DSAJG-BVSLTA
```

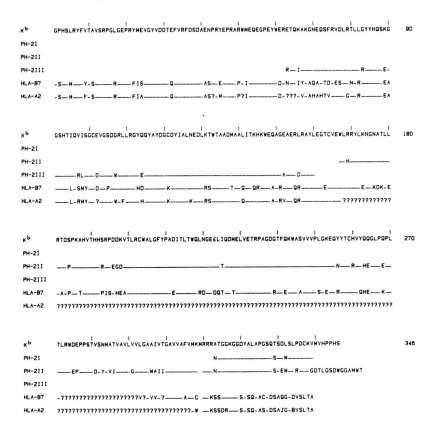

FIGURE 3. Comparison of the amino acid sequence of the mouse K^b molecule, the translated protein sequences of the three H-2 cDNA clones, and the protein sequence of two human transplantation antigens. The mouse H-2K^b sequence (11,12), the human HLA-B7 and HLA-A2 amino acid sequences (8,13), and the translated protein sequences of the H-2 cDNA clones pH-2I, pH-2II, and pH-2III (7) were taken from published data. A gap of one amino acid at position 309 has been inserted into the human sequences as well as into the predicted amino acid sequence of clone pH-2II to achieve maximum homology to the K^b molecule. The single letter code has been used for amino acids: A, Ala; B, Asp or Asn; C, Cys; D, Asp; E, Glu; F, Phe; G, Gly; H, His; I, Ile; J, Glu or Gln; K, Lys; L, Leu; M, Met; N, Asn; P, Pro; Q, Gln; R, Arg; S, Ser; T, Thr; V, Val; W, Trp; Y, Tyr. A dash means identity to the K^b sequence, a question mark indicates that the residue is not known. Numbers refer to the K^b sequence.

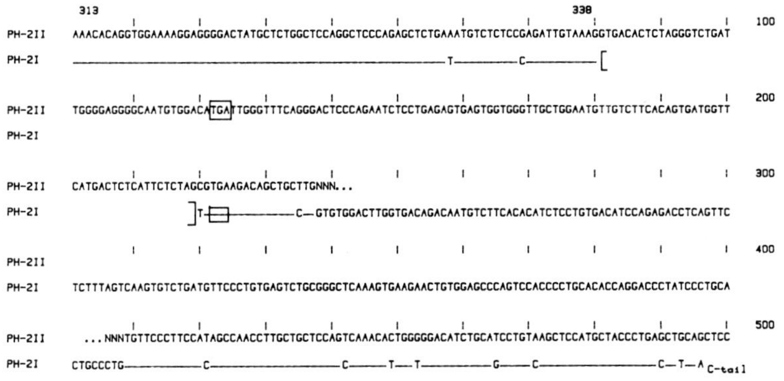

FIGURE 4. A comparison of the DNA sequences of clones pH-2I and pH-2II. The DNA sequences of clone pH-2I and pH-2II are shown in the region where they overlap and are aligned for maximum homology. Only three base substitutions (96% homology) are found in the coding regions of the pH-2I and pH-2II sequences. A gap in the pH-2I sequence, 139 bp in length (indicated by brackets), is required to align the 3′ untranslated regions of these clones. With this alignment the 3′ untranslated regions of the two cDNAs are highly homologous (90%). The sequence of the pH-2II clone from nucleotide position 238 to 408 has not been determined. The termination codons for the predicted amino acid sequences encoded by clones pH-2I and pH-2II are boxed.

Because of the paucity of the amino acid sequence data available for transplantation antigens of the d and k haplotype we are not able to make definite assignments for each of the three cDNA clones. However certain possibilities can be excluded (Table I).

B. *In Mouse Cell Lines One RNA Complementary to the H-2 cDNA Probes Can Be Detected*

We used the H-2 cDNAs to determine the number of cross-reactive messenger species in the mouse lymphoma cell lines by RNA blot analysis. Surprisingly, all three cDNA probes when labeled by nick translation, hybridize to one class of messenger RNA with an approximate length of 1.9 kilobases (kb). Figure 5 shows typical results obtained using a subclone of pH-2II as a probe. The length of this messenger RNA is consistent with the size of the mouse 17S RNA species that has been identified to code for transplantation antigens by in vitro translation and immunoprecipitation (16). The presence of a single band indicates that all cross-hybridizing transplantation

TABLE I. cDNA Clones for Transplantation Antigens[a]

Cell line	Haplotype	Clone	Results from protein and cDNA sequence comparisons
C14	d	pH-2II	not K^d or D^d
		pH-2I	different from clone pH-2II
RDM4	k	pH-2III	not K^k

[a]*The predicted amino acid sequence of clone pH-2II corresponds in 19 out of 19 positions that can be compared (14) to the* L^d *amino acid sequence. It is therefore possible that pH-2II encodes the* L^d *molecule. Taken from ref. 7.*

antigen messenger RNAs (e.g., those for K and D molecules) are approximately of the same size. Alternatively, if messenger RNAs of different sizes do exist, they may not be detected under our hybridization conditions, presumably because of much lower abundance.

C. *The Genes Encoding Mouse Transplantation Antigens are Homologous with Immunoglobulin Genes*

We have developed a computer program to analyze the homology between distantly related genes. Using this program, we were able to establish that the third domain of mouse transplantation antigens shows a 51% sequence homology to the fourth domain of the immunoglobulin C_μ gene (Figure 6). Similar, although somewhat less extensive, homologies can be demonstrated when the third domain of the H-2 clone is compared against the $C_\mu1$, $C_\mu2$, and $C_\mu3$ domains of the immunoglobulin μ gene. Indeed, the third domain of the transplantation antigen appears, if anything, to be more closely related to the C_μ domains than the C_μ domains are with one another. The first part of this region, a 72 nucleotide stretch in clone pH-2II, is strikingly homologous to its counterparts in the domains $C_\mu1$, $C_\mu2$, and $C_\mu4$. Here 21 bases are conserved in all the three immunoglobulin domains analyzed and 20 of the 21 nucleotides are also conserved in the third domain region of the transplantation antigen. This homology has been determined without placing any sequence gaps, thus permitting us to ask how frequently homologies of this extent would be seen if every possible stretch of 72 nucleotides in the pH-2II clone is compared against all possible blocks of 72 nucleotides in the mouse C_μ gene. About 1.1 x 10^6 random comparisons were made and the homology mean for these comparisons was 25% with a standard deviation of 5.4%. The homologies exhibited by the transplantation antigen clone vs. the $C_\mu1$, $C_\mu2$, and $C_\mu4$ domain comparisons fell 5.2, 5.0, and

6.6 standard deviations from the mean and, accordingly, are all highly statistically significant.

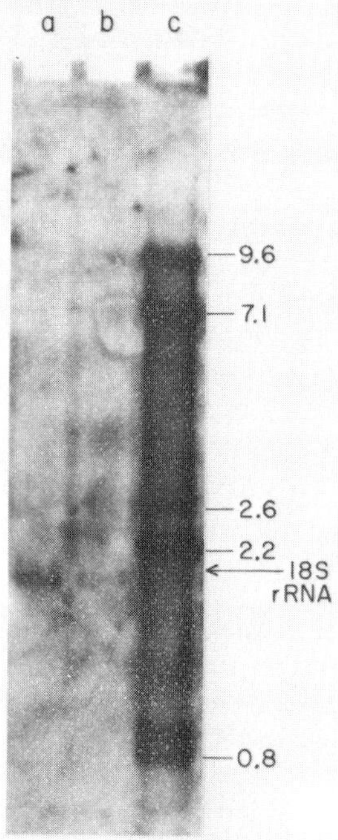

FIGURE 5. Hybridization of a subclone from pH-2II to a Northern blot containing poly(A)$^+$ RNA from C14 and RDM4 cells. Lanes a and b: 1.5 µg of poly(A)$^+$ RNA from the C14 and RDM4 cell lines was subjected to agarose gel electrophoresis in the presence of methylmercury, transferred to a nitrocellulose filter and hybridized with a subclone from pH-2II consisting of a 689 bp Pst I-Pvu II fragment in pBR322. This subclone contained all of the coding region of pH-2II but lacked the repetitive sequences of the 3' untranslated region (see ref. 7). Lane c: Restriction fragments (2.5 x 10^{-17} mol per discrete fragment) of a pBR322 derivative were run in parallel and served as a hybridization marker. Sizes are given in kilobases. In addition, 18S (arrow) and 28S rRNAs were run on the gel as a molecular weight marker. Assuming a size of 2 kb for 18S rRNA (15), we calculate a size of about 1.9 kb for the RNA species that hybridize in both C14 and RDM4 RNA.

```
        199   ValThrLeuArgCysTrpAlaLeuGlyPheTyrProAlaAspIleThrLeuThrTrpGlnLeuAsnGlyGluGluLeuThrGlnAspMetGluLeuValGluThrArgProAla
pH-2II  119   GTCACCCTGAGGTGCTGGGCCCTGGGCTTCTACCCTGCTGACATCACCCTGACCTGGCAGTTGAATGGGGAGGAGCTGACCCAGGACATGGAGCTTGTGGAGACCAGGCCTGCA
Cμ4     1549  -C—AG-C-CC—CT—TGAA——CT——A——GTG—CAG——TTCA—GA—C-ACTCT—C——A—G-A-    TAT-T———TG-CC-GATGCCA
        471   Ala—ValThr—LeuValLys——Ser———SerValGln—LeuGlnArg—GlnLeu—Pro—GluLys    TyrVal—SerAlaProMetPro

        237         GlyAspGlyThrPheGln                LysTrpAlaSerValValValProLeuGlyLysGluGlnAsnTyrThrCysArgValTyrHisGluGlyLeuProGluProLeu
pH-2II  233         GGGGATGGAACCTTCCAG                AAGTGGGCATCTGTGGTGGTGCCTCTTGGGAAGGAGCAGAATTACACATGCCGTGTGTACCATGAGGGGCTGCCTGAGCCCCTC
Cμ4     1663  GAGCCT——CCCC-GG——T-CTTTACCCACAGC-TCCT-A-TGTGACA-A—A-GAATGGAACTCC-GAG—CC—T—C—GT—AGGG—C——CC———AC-C-TAG-G
        509   GluPro—AlaProGly—TyrPheThrHisSerIleLeuThrValThrGluGluGluTrpAsnSerGlyGluThr————Val—Gly———Ala———HisLeuVal

        271   ThrLeuArgTrpGluProProProSerThr
pH-2II  335   ACCCTGAGATGGGAGCCTCCTCCGTCCACT
Cμ4     1783  —GA—GACC-T-    GACAA———
        549   —Glu—ThrVal    AspLys———
```

FIGURE 6. A comparison between the DNA sequences of clone pH-2II and the fourth constant region domain of the mouse immunoglobulin μ heavy chain. A comparison is made between the pH-2II cDNA sequence from nucleotide position 119 to 364 (7) and the fourth constant region exon ($C_\mu 4$) of the μ gene from position 1549 to 1809 (17). This comparison shows that an overall homology of 51% can be achieved by only two gaps in each sequence. The amino acid sequences encoded by the pH-2II and the $C_\mu 4$ DNA sequences are given above and below the nucleotide sequences, respectively. Nucleotides and amino acids that are identical to the pH-2II sequence are indicated by dashes. Taken from ref. 7.

An important point that may be raised at this juncture is whether the homology between immunoglobulins and transplantation antigens is a consequence of divergent or convergent evolution (Figure 7). Our DNA sequence data suggest the two genes arose by divergent evolution. First, 45% of the third bases in the codons compared between the third domain of the transplantation antigen and the $C_\mu 4$ domain are identical to one another. In spite of the uncertainties of codon usage, this conservation of third bases appears statistically significant. Second, of the 20 bases conserved in the 72 nucleotide stretch comparison made above, seven of these positions conserved between the pH-2II and the C_μ sequences represent third-base identities. In both cases it is difficult to explain how a convergent evolution model might explain this striking conservation in the third position of codons. Thus we conclude that immunoglobulins and transplantation antigens have diverged from a common evolutionary ancestor and as such constitute two members of a super-multigene family. An intriguing question is whether other gene families will belong to this super-multigene family. Also, it will be interesting to learn to what extent the evolutionary and regulatory strategies displayed by the immunoglobulin genes are shared by the genes for transplantation antigens.

The location of the homologous stretch of sequence in the transplantation antigens is depicted in Figure 8. Perhaps the reason this sequence is conserved is that a portion of the transplantation antigen must fold into the classic "antibody fold" (19) in order to interact

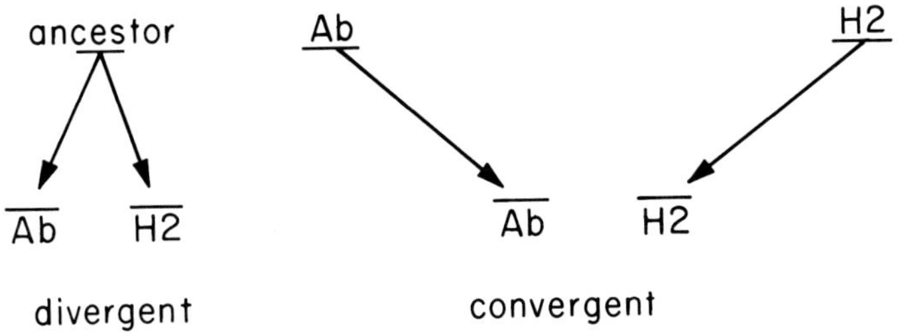

FIGURE 7. Mechanisms for the evolution of homologous sequences. Ab: immunoglobulin genes. H-2: genes encoding transplantation antigens.

effectively in a noncovalent manner with β_2-microglobulin which also is folded into an antibody-like configuration—much as the heavy and light chain domains of immunoglobulins interact with another through a molecular complementarity of their antibody folds. Perhaps the highly conserved residues throughout these domains represent key positions in the folding of the β pleated sheets which constitute the major pattern displayed in the antibody folds.

D. The Genes for Transplantation Antigens Constitute a Multigene Family

Southern blot analyses carried out with a variety of probes and different restriction enzyme digests show that 12-15 bands hybridize to the cDNA probes for mouse transplantation antigens (Figure 9). It is possible that some of the major bands may represent multiple copies. Accordingly, the genes encoding the transplantation antigens belong to a multigenic family.

Immunoglobulin genes undergo rearrangements upon the differentiation of antibody-producing cells. Since transplantation antigens are not expressed endogenously on sperm cells, while they are expressed on all somatic cells of the vertebrate organism, it is appropriate to ask whether the genes encoding transplantation antigens also undergo rearrangements during the differentiation of somatic cells. From Southern blot analyses comparing sperm DNA and liver DNA shown in Figure 9, it is clear that the genes encoding transplantation antigens are not rearranged during differentiation. Since there are at least four different transplantation antigens in the BALB/c mouse one should have a high probability of seeing a DNA rearrangement if this event is central to the expression of the corresponding genes for transplantation antigens. Thus we conclude that the genes encoding

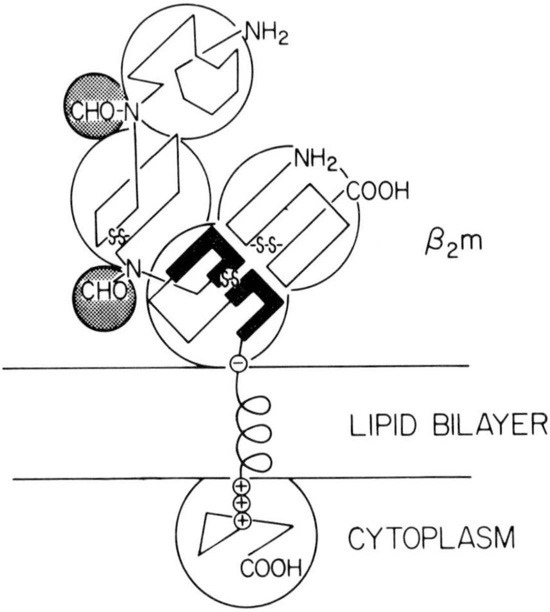

FIGURE 8. A hypothetical model for the association of transplantation antigens with β_2-microglobulin. This model, which was derived from a similar one for human transplantation antigens (18), suggests that the region homologous to immunoglobulins in the transplantation antigen may be associated with β_2-microglobulin. The transplantation antigen is hypothesized to consist of three domains (approximately residues 1-90, 91-180, 181-270) on the outside, a transmembrane portion, and a relatively small domain on the inside of the cell membrane. Two intrachain disulfide bridges (Cys 101-Cys 164 and Cys 203-Cys 259) are present in the second and the third domain of mouse transplantation antigens (11). It is assumed that the two carbohydrate side chains linked to asparagine (N) residues at positions 86 and 176 (11) mark interruption points between the three outer domains. Charged amino acid residues (positions 281 and 308-312) define the transmembrane portion of the molecule.

transplantation antigens do not undergo DNA rearrangements during the differentiation of somatic cells.

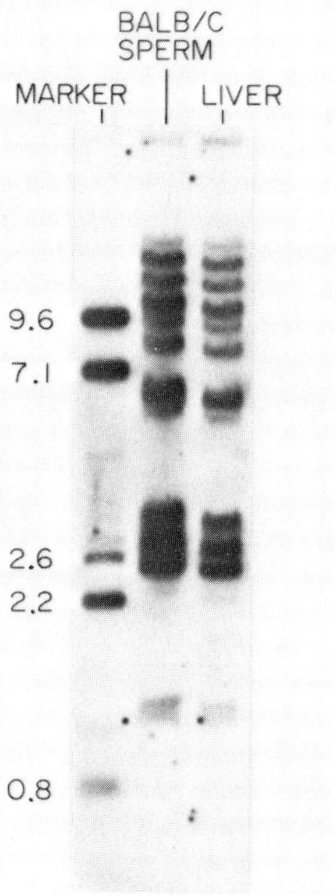

FIGURE 9. Southern blot hybridization of mouse DNA using a cDNA clone encoding a transplantation antigen as a probe. 10 μg of BALB/c sperm or liver DNA was completely cleaved with Bam HI, electrophoresed on a 0.6% agarose gel, and transferred to a nitrocellulose filter. Hybridization was with the subclone from pH-2II containing only the coding sequences (see legend to Fig. 5) which had been labeled by nick translation. Hybridization and washing conditions were as described previously (7). The hybridization marker is the same as in Figure 5. Sizes are in kilobases.

E. *Presence of Species-Associated Sequences Implies that Mouse Transplantation Antigen Genes Diverged from a Common Ancestral Gene After the Separation of Mouse and Human Evolutionary Lines*

Species-associated sequences are those which distinguish most of the genes in a family from one species to another (20). At the level of partial protein sequence analysis of the N-termini of mouse and human transplantation antigens, several species-associated residues were noted (e.g., leucine vs. methionine and serine vs. alanine at positions 4 and 11, respectively). The comparison of the N-terminal 100 residues of two H-2K and two H-2D molecules from the k and b haplotypes with the corresponding regions of the human HLA-A2 and HLA-B7 molecules confirmed the existence of multiple species-associated residues throughout this region (21). Our results permit us to compare the amino acid sequences of three distinct mouse transplantation antigens (K^b, the sequence derived from the pH-2I, and the sequence derived from the pH-2II cDNA clone) with their human counterparts at their C-termini (Figure 3) and once again there is clear evidence of species-associated residues (e.g., positions 314, 320, 322, 323, 325, 328, etc.).

Comparison of the DNA sequences of pH-2I and II with human HLA cDNA sequences again reveals the existence of species-associated sequences (Figure 10). Alignment of part of the pH-2II sequence to the corresponding sequence of the HLA cDNA isolated by Sood et al. (9) shows 82% homology between the two sequences with no nucleotide gaps. However, for the other HLA cDNA sequence which encodes the C-terminal 47 residues, there are nucleotide gaps of three base pairs in the pHLA1 and nine base pairs the pH-2II sequence required to achieve a maximum homology of 74%. It is striking that the same deletion of nine base pairs also is found in clone pH-2I which represents a different transplantation antigen gene than clone pH-2II (compare Figures 4 and 10).

The evolutionary implication of species-associated sequences is that the genes for transplantation antigens of the human and murine evolutionary lines must have undergone extensive gene duplication and deletion or gene correction over the 75 million years of mammalian evolutionary divergence (Figure 11). In order to explain the existence of species-associated residues in multiple genes in a family, the simplest explanation is to assume that all were derived from a single ancestral gene in each evolutionary line after the accumulation of the species-associated differences (Figure 11).

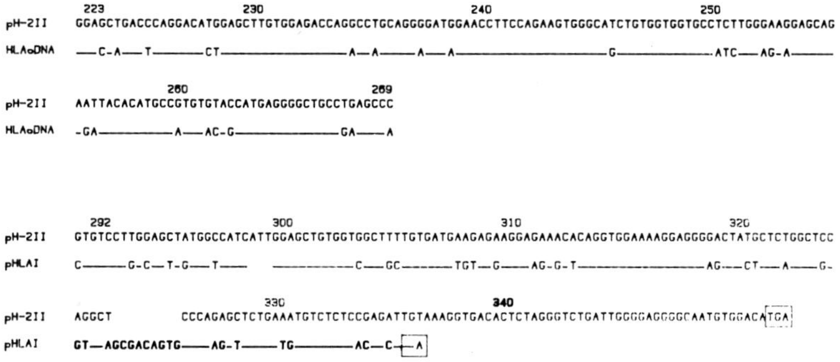

FIGURE 10. A comparison of the DNA sequences of mouse and human cDNA clones encoding transplantation antigens. The HLA cDNA sequence has been published by Sood et al. (9) and the pHLA1 sequence by Ploegh et al. (8). Dashes indicate identity to the DNA sequence of the pH-2II cDNA clone. To achieve maximum homology a gap of three nucleotides was inserted into the pHLA1 sequence at codon 299 and a gap of nine nucleotides into the pH-2II sequence at codon 226. Codons that terminate the predicted amino acid sequences of the clones are boxed. Codon numbers refer to the K^b amino acid sequence.

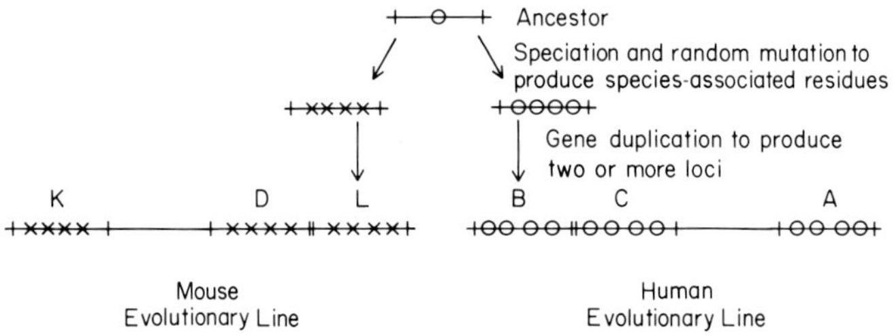

FIGURE 11. A hypothetical model for the evolution of genes with species-associated sequences for mammalian transplantation antigens. The X's and O's indicate species-associated sequences in the mouse and human evolutionary lines, respectively. The symbols K, D, L, B, C, and A represent the mouse and human transplantation antigens.

II. SUMMARY

We have shown by Southern blot analysis that the genes encoding transplantation antigens constitute a multigene family. Two explanations for the multiplicity of these genes are possible. First, there may be many more genes encoding transplantation antigens than previously suspected, as well as pseudogenes. Perhaps some of the functional genes are expressed at such low levels that they cannot be detected by the serologic and cytotoxic assay methods currently available to the immunogeneticists. In addition the lack of suitable recombinants may make identification of these genes difficult. An alternative explanation is that this family of genes may include genes coding for various T-cell differentiation antigens. They are homologous to transplantation antigens in that they, too, are constituted of 45,000 dalton integral membrane proteins associated with β_2-microglobulin. Indeed the T-cell differentiation antigens TL, Qa1, Qa2, Qa3, and Qa4 all fall in this particular category and correspondingly could be encoded by genes homologous to the genes of transplantation antigens. There is no protein sequence data available on any of these T-cell differentiation antigens; therefore, their sequence homology to the transplantation antigens is unknown at this time. It is thus extremely important to have phenotypic data to compare with the genotypic analysis of the DNA sequences. As mentioned earlier in this paper, it is virtually impossible to make any direct assignments of our three cDNA clones because there are insufficient protein sequence data available on the transplantation antigens to permit an unambiguous assignment. In the future, the protein sequence analysis of these transplantation antigen will be critically important in the definition of this multigene family. Alternatively, a correlation between genotype and phenotype may be approached by transformation experiments in which the gene being analyzed is expressed in a cell of a different phenotype and correspondingly can be analyzed by serologic or cytotoxic methods. Preliminary reports of the success of this approach have come from several laboratories including our own (R. Goodenow, unpublished data). Transformation experiments will allow us not only to identify unequivocally the gene products of the major histocompatibility complex, but also to carry out functional studies to determine in detail the nature of their respective functions. Clearly the genes encoding the transplantation antigens offer an exciting opportunity to analyze a complex, multigenic family at the level of organization, gene expression, evolution, and function.

ACKNOWLEDGMENTS

This work was supported by National Institutes of Health Grant GM 06965. MS is the recipient of a fellowship from the Deutsche

Forschungsgemeinschaft and JGF and KWM are supported by NIH postdoctoral fellowships. DAF and BTS are NIH trainees.

REFERENCES

1. Zinkernagel, R. M., and Doherty, P. C., Adv. Immunol. 27, 51 (1980).
2. Hansen, T. H., Ozato, K., Melino, M. R., Coligan, J. E., Kindt, T. J., Jandinski, J. J., and Sachs, D. H., J. Immunol., in press.
3. Klein, J., Science 203, 516 (1979).
4. Vitetta, E. S., and Capra, J. O., Adv. Immunol. 26, 147 (1978).
5. Strominger, J. L., Orr, H. T., Parham, P., Ploegh, H. L., Mann, D. L., Bilofsky, H., Saroff, H. A., Wu, T. T., and Kabat, E. A., Scand. J. Immunol. 11, 573 (1980).
6. Peterson, P. A., Cunningham, B. A., Berggard, I., and Edelman, G. M., Proc. Nat. Acad. Sci. USA 69, 1697 (1972).
7. Steinmetz, M., Frelinger, J. G., Fisher, D., Hunkapiller, T., Pereira, D., Weissman, S. M., Uehara, H., Nathenson, S., and Hood, L., Cell 24, 125 (1981).
8. Ploegh, H. L., Orr, H. T., and Strominger, J. L., Proc. Nat. Acad. Sci. USA 77, 6081 (1980).
9. Sood, A. K., Pereira, D., and Weissman, S. M., Proc. Nat. Acad. Sci. USA 78, 616 (1981).
10. Kvist, S., Bregegere, F., Rask, L., Cami, B., Garoff, H., Daniel, F., Wiman, K., Larhammar, D., Abastado, J. P., Gachelin, G., Peterson, P. A., Dobberstein, B., and Kourilsky, P., Proc. Nat. Acad. Sci. USA, in press.
11. Martinko, J. M., Uehara, H., Ewenstein, B. M., Kindt, T. J., Coligan, J. E., and Nathenson, S. G. Biochemistry 19, 6188 (1980).
12. Uehara, H., Coligan, J. E., and Nathenson, S. G., Biochemistry, in press.
13. Orr, H. T., Lopez de Castro, J. A., Parham, P., Ploegh, H. L., and Strominger, J. L., Proc. Nat. Acad. Sci. USA 76, 4395 (1979).
14. Nairn, R., Nathenson, S. G., and Coligan, J. E., Eur. J. Immunol. 10, 495 (1980).
15. Wellauer, P. K., Dawid, I. B., Kelley, D. E., and Perry, R. P., J. Mol. Biol. 89, 397 (1974).
16. Dobberstein, B., Garoff, H., Warren, G., and Robinson, P. J., Cell 17, 759 (1979).
17. Kawakami, T., Takahashi, N., and Honjo, T., Nucleic Acids Res. 8, 3933 (1980).
18. Krangel, M. S., Orr, H. T., and Strominger, J. L., Scand. J. Immunol. 11, 561 (1980).
19. Poljak, R. J., Amzel, L. M., Avey, H. P., Chen, B. L., Phizackerley, R. P., and Saul, F., Proc. Nat. Acad. Sci. USA 70, 3305 (1973).

20. Silver, J., and Hood, L., in "Contemporary Topics in Molecular Immunology" (H. N. Eisen and R. A. Reisfeld, eds.), v. 5, p. 35. Plenum Publishing Corporation, New York (1976).
21. Nathenson, S. G., Uehara, H., Ewenstein, B. M., Kindt, T. J., and Coligan, J. E., Ann. Rev. Biochem., in press.

DEVELOPMENTAL GENETICS OF THE BITHORAX COMPLEX IN DROSOPHILA[1]

E. B. Lewis

Division of Biology
California Institute of Technology
Pasadena, California

ABSTRACT

The bithorax gene complex (BX-C) is a gene cluster of at least eight loss-of-function loci and at least four cis-regulatory regions that control the level of development of the majority of the body segments. Many BX-C mutants exhibit strong polar position effects which extend over several genes of the complex and are subject to further modification or "transvection" by rearrangements which disturb chromosome pairing in the vicinity of BX-C. The pattern of cuticular and tracheal structures in a given embryonic segment is controlled by the number and specificity of action of the particular BX-C genes which become active in that segment. Regulation of the BX-C genes, themselves, appears to be negative and mediated by a repressor thought to be coded for by a trans-regulatory gene, Polycomb. Affinity of cis-regulatory regions of BX-C for such a repressor is assumed to determine the level of activity of each gene of the complex and can account for the wild-type pattern of segmentation if there is a gradient in repressor (or inducer) concentration in the early embryo.

INTRODUCTION

The body segmentation pattern of Drosophila provides a model system for studying how genes regulate development. In this and other segmented organisms, whether invertebrate or vertebrate, a linear array of redundant segments is laid down early in embryogenesis. Much of the later development of the organism consists of

[1]*This work was supported by USPHS grant HD 06331.*

ISBN 0-12-137420-3

little more than the gradual diversification of those segments to produce the head, thoracic and abdominal regions of the body. In *Drosophila* the bithorax gene complex (BX-C) plays a major role in bringing about that diversification in the thoracic and especially the abdominal regions of the fly (1).

This paper deals with the current status of our knowledge of the bithorax complex. The principal categories of mutant types and their map locations are reviewed in order to illustrate the genetic basis for a molecular mapping of the complex now underway in the laboratories of D. Hogness and W. Bender. How the BX-C genes function in development and how the complex itself is regulated can only be treated at the present time in terms of abstract models, one of which is elaborated briefly.

I. EARLY HISTORY OF THE COMPLEX

Interest in the developmental aspects of the bithorax mutants developed only gradually. When the current studies were begun in 1946, there already existed three types of mutants, all spontaneous in origin (2). The first type was represented by three alleles: bx, found by Bridges in 1915; bx^3 of Stern (1925); and bx^{34e} of Schultz (1934).

All are involved in a transformation of the anterior portion of the third thoracic segment, T3, toward the corresponding portion of the second, T2. In order of increasingly extreme transformations of that type they seriate as follows: $bx < bx^{34e} < bx^3$. In addition, bx is a highly variable allele, while bx^{34e} and bx^3 are remarkably uniform in their expression. A second type of mutant was found in 1919 by Bridges, who named it bithoraxoid (bxd) on the basis that it has wing-like halteres, qualitatively different from those of bx, and it complements bx; that is, bx/bxd is wild type. A third type of mutant was then discovered by Hollander in 1934. It is slightly dominant (has slightly enlarged halteres), acts as a recessive lethal, and fails to complement fully any of the bx mutants; hence, it was given the symbol bx^D. Since it fails to complement bxd as well, bx^D seemed to pose a paradox. Either bx^D was a double mutant or bxd was to be regarded as an allele of bx. By 1946, the concept of multiple allelism had clearly outlived its usefulness for such cases, and, as a result, the three types of bithorax mutants promised to provide intriguing materials for testing the hypothesis that they constituted not a multiple allelic, but a pseudoallelic, series; or, in modern parlance, that they represented a gene complex, or gene cluster, or multi-gene family -- three more or less synonymous terms signifying a group of closely linked genes with similar effects.

It soon became evident from crossing-over studies (3) that bx^D is allelic to neither bx nor bxd but instead occupies a separate locus between them (Fig. 1). Apparent allelism had resulted from cis-trans position effects and bx^D was renamed Ultrabithorax, Ubx. A remarkable X-ray induced dominant mutant, Contrabithorax, was then discovered. It maps between bx and Ubx and transforms T2 towards T3. A new recessive type of mutant, postbithorax, pbx, which must have been induced simultaneously with Cbx, could be separated by crossing over from Cbx and was found to occupy a fifth locus distal to that of bxd (3). The pbx mutant effects an extreme transformation of posterior T3 toward posterior T2. Thus, a striking specialization was evident: bx mutants transform only anterior, and pbx, only posterior, regions of T3. The double mutant, bx^3 pbx, was constructed and found, not unexpectedly, to produce for the first time a four-winged fly having a virtually complete conversion of T3 to T2, although the

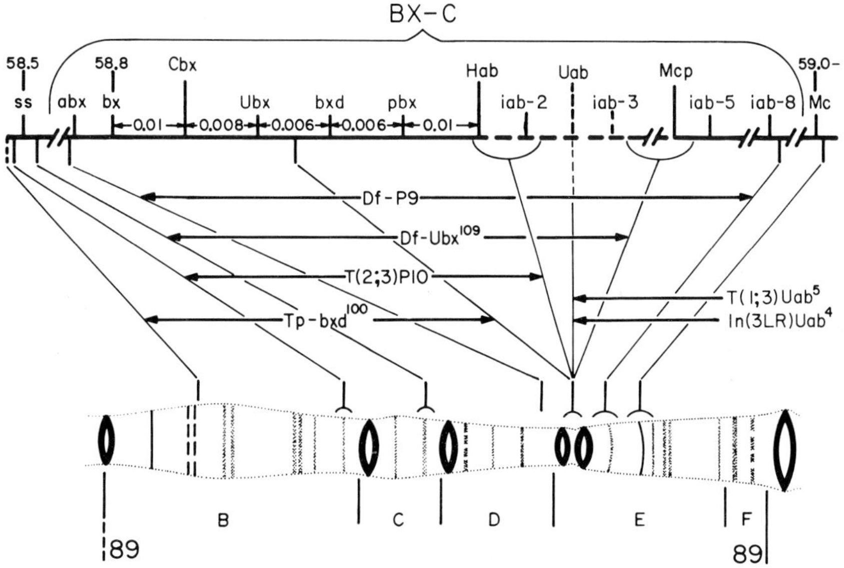

FIGURE 1. Correlation of genetic and cytological maps of the bithorax complex (BX-C).

abx = anterobithorax; bx = bithorax; Cbx = Contrabithorax; Ubx = Ultrabithorax; bxd = bithoraxoid; pbx = postbithorax; Hab = Hyperabdominal (formerly Cbxd); Uab = Ultra-abdominal; iab = infra-abdominal; Mcp = Miscadastral pigmentation; ss = spineless, and Mc = Microcephalus, are flanking markers except that Mc acts as if it is a small duplication for iab-8⁺ and possibly additional genes at the right end of the complex.

wings and thorax proper of T3 were somewhat smaller than those of T2 (4). A careful study of bxd had already shown that it is also associated with a transformation of posterior T3 towards posterior T2 (5). In addition, however, bxd flies have a thoracic-like transformation of the first abdominal segment; whereas, that segment is wild type in pbx flies. The puzzling feature that a single mutant, bxd, can effect two quite distinct types of transformations was finally cleared up by means of a simple cis-trans test with pbx. Thus, bxd/pbx has a wild-type first abdominal segment but a moderate pbx phenotype; whereas, bxd pbx/+ + is completely wild type. Hence, of the two types of mutant transformations associated with bxd the thoracic-like modification of the first-abdominal segment is attributable to the bxd mutant gene itself, while the pbx transformation is attributable to a weakening of the dominance of the pbx^+ gene located cis to bxd.

II. CURRENT STATUS OF THE GENETIC MAP OF BX-C

The chief additions to the genetic map of BX-C (Fig. 1) since its last publication (1) are three loci: anterobithorax (abx); infra-abdominal-2, iab-2; and iab-5, evidence for which will be briefly summarized below. There now exist 8 loci that are characterized by having mutant alleles that, for the most part, are all of a recessive loss-of-function type: the mutant resulting in a segmental transformation from a more posterior to a more anterior level of development. The Ubx mutant, although a slight dominant, is also of this type since Df/+ and Ubx/+ have the same phenotype -- a slightly enlarged haltere -- where Df stands for any deficiency that includes the Ubx^+ gene, such as that derived from $Tp\text{-}bxd^{100}$ or T(2;3)P10 (Fig. 1).

A curious rule is evident thus far; namely, that the more distal a locus is in the chromosome the more posterior in the organism is the type of segmental development controlled by that locus; however, bxd is a clear exception to this rule, being proximal rather than distal to pbx. Whether such a rule, if indeed it is a significant one, reflects the evolutionary origin of the complex through a process of repeated gene duplications starting at one end and/or whether it has some functional significance is not known. Another rule that is also only suggestive at this time is that a set of four cis-dominant regulatory mutants, Contrabithorax, Cbx, Hyperabdominal, Hab, Ultra-abdominal, Uab, and Miscadastral pigmentation, Mcp, appear to be located very close, if not immediately adjacent, to the structural-like genes which they regulate.

A. *The anterobithorax Locus*

The discovery of a separate locus to the left of bx was prompted by an observation that the bx^3 mutant is associated with two types of more or less independently varying transformations. Although the principal effect in a bx^3 homozygote is on T3, as already noted, T2 is also affected secondarily and in an opposite manner. The latter effect consists of a slight but consistent underdevelopment of a region in anterior T2, designated here as the "presutural area." That area is also consistently underdeveloped in the otherwise strongly transformed anterior T3 of the bx^3 homozygote. That the effect on T2 is a different type of function from that seen in T3 is evident when a duplication of bx^+ (and Ubx^+) from Tp-bxd^{100} (Fig. 1), or a duplication for the full complement of wild-type genes of the complex, is added to the homozygous bx^3 genotype. Such a duplication genotype ($bx^3/bx^3/+$) has a wild-type T3 but retains the reduced prescutal effect on T2. Hence, the T3 towards T2 transformation seen in bx^3 homozygotes acts as a recessive loss of function while the effect on T2 acts as a dominant gain of function. The dual transformation effected by bx^3 could be accounted for if there exists an anterobithorax gene, abx^+, which becomes weakly activated or derepressed in some way by bx^3. A likely candidate for an abx type of mutant was a previously existing bx-like mutant that had been designated bx^7 (6). The homozygote for this X-ray induced mutant has a highly variable and often asymmetric transformation of anterior T3 toward anterior T2, ranging from virtually no transformation to one that is more extreme than found in the bx^3 homozygote. Moreover, in homozygous bx^7, a presutural area frequently arises in T3, even when T3 is not otherwise well transformed toward T2.

Another indication that bx^7 might represent the sought after abx mutant was the finding that bx^7 and bx^3 show a strong transvection effect (7). Thus, the phenotype of bx^7/bx^3, which is weak and variable and somewhat like that of the bx^1 homozygote, is made much more extreme in the presence of structural heterozygosity for any rearrangement which strongly disrupts pairing proximally to BX-C. Since in previously known examples of BX-C transvection (7) nonallelic pairs of mutants have been involved (e.g., Ubx in trans with bx, bx^{34e} or bx^3; bx^3 in trans with pbx; Cbx in cis with Ubx; or in a recently discovered and unpublished case, bxd in trans with pbx^2), a recombinational analysis of bx^7 and bx^3 was undertaken. On the basis of one wild-type and one double mutant crossover, bx^7 has recently been found to lie roughly 0.01 map unit proximal to the bx locus. The symbol for bx^7 has been changed to abx not only because of this mapping evidence but also because of the finding that, although residual variability of abx still remains, the abx bx^3 double mutant homozygote often has a well-formed presutural area on T3 and on T2, as would be expected when an abx^+ gene cis to bx^3 has been replaced by an abx mutant gene. Of particular interest is the phenotype of the triple mutant, abx bx^3 pbx, which when homozygous as well as hemizygous results in a four-winged fly in which the thorax and wings of T3 approach very closely in size as well as structure those of T2.

B. The infra-abdominal-2 Locus

In a previous study (1) two types of evidence were described which pointed to the existence of an iab-2$^+$ gene: (a) the behavior of Hab which seems to activate or derepress in cis an iab-2$^+$ gene with resultant transformation of T3 toward A2, the second abdominal segment, and to a weaker extent, of the first abdominal segment, A1, towards A2; and (b) the T(2;3)P10 rearrangement which damages or has a position effect of iab-2$^+$ resulting in reduction in the width of the ventral setal belts of A2 (and segments more posterior to A2) in embryos hemizygous for this rearrangement, namely, T(2;3)P10/Df-P9 embryos. Recently Kuhn *et al.* (8) have described a recessive iab-2 mutant which produces in adults a weak transformation of the expected type: transformation of A2 toward A1. This mutant, designated, iab-2, has been kindly made available to us by Dr. Kuhn, and preliminary mapping evidence has shown that it maps between the locus of bxd and that of Mcp; since its order with respect to Mcp and a hypothetical iab-3 gene has not been determined, the genetic map is shown as dashed in this region (Fig. 1).

C. The infra-abdominal-5 Locus

Studies of M. Crosby of this laboratory have provided evidence for a recessive iab-5 mutant (9). The existence of an iab-5$^+$ gene had been inferred from the behavior of the Mcp mutant which has the phenotype expected if iab-5$^+$ is activated or derepressed in A4 as well as A5 (1). An ethyl methane sulfonate (EMS)-induced revertant of Mcp obtained by Crosby was shown by her to contain a recessive allele, designated iab-5^{C7}, which maps immediately distal to Mcp (Fig. 1). The iab-5^{C7}, after separation from Mcp by crossing over, produces, when hemizygous, a moderately strong transformation of A5 towards A4 (and to a lesser extent of A6 and A7 toward A4 as well). Since iab-5^{C7} is associated with reduction in crossing over in the Mcp-Mc map interval, it is possible that the mutant is associated with a small rearrangement which may also damage other iab$^+$ genes to the right of Mcp.

D. Other infra-abdominal Loci

Evidence for the existence of an iab-3$^+$ and an iab-8$^+$gene has been presented elsewhere (1). No point mutants are known, however, for iab-3, nor for Uab, a presumptive cis-regulatory region for iab-3$^+$; therefore, the loci of iab-3 and Uab are designated with dashed vertical lines in Fig. 1. The order of iab-2, iab-3, and Uab with respect to Hab and each other is also unknown, which is indicated in Fig. 1 by the dashed portion of the genetic map. Additional evidence for an iab-8 locus is the finding by Kiger (10) that the original Uab^1 mutant, which is nearly lethal when homozygous has a recessive phenotype of loss of genitalia. This dominant mutant is of EMS origin and appears cytologically normal; nevertheless, in extensive tests it greatly reduced recombination over a region extending from the locus of a bxd-like mutant, induced simultaneously with Uab, and the locus of Mc; hence, Uab is evidently a very minute rearrangement, accompanied by either mutations or position effects at the bxd and iab-8 loci. Recently, Kuhn (11) has found that strains of the third chromosome tumorous-head mutant, tuh-3, contain another possible recessive iab-8 mutant associated with loss of genitalia and testicular defects that he has shown (11) localizes to the iab-8 region on the basis of being uncovered by Df-P9 and not by Df-Ubx^{109}. We find that this presumptive iab-8 point mutation maps genetically between Mcp and Mc.

III. SUPPRESSIBILITY OF BX-C MUTANTS

One of the first occurrences of a suppressor mutation was the suppressor of Hairy-wing, su-Hw, found by Bridges in 1923 and located by him at 54.8 in chromosome 3 (2). Although this mutant was lost, another allele, su^2-Hw, was isolated from a balanced bx^3 strain, where it was located cis to bx^3 (2). Specific alleles, all of spontaneous origin, at a variety of loci throughout the genome have been found to be suppressible by homozygous su^2-Hw; such as, y^2, sc, dm, ct^6, lz, Bx^2 and ci (12). Within BX-C among the spontaneous mutants tested, bx^{34e}, bx^3, bxd, bxd^{51j} and bxd^{55i} (a weak allele, unpublished) are suppressed while bx, Ubx and Ubx^{61d} (Gloor, see ref. 4) are not. Among EMS-induced alleles tested; namely, bx^8 (5), Ubx^{51} (a weak allele, unpublished, which maps between bx^{34e} and bxd), 11 other Ubx-like and two weak Ubx-like mutants, all cytologically normal, none is suppressed. Finally, among X-ray induced mutants tested; namely, abx, Cbx and pbx, none is suppressed.

In one instance the action of su^2-Hw is dominant instead of recessive; namely, its suppression of a specific cut allele, ct^K (13). We have found (unpublished) that a heterozygote for a deficiency for the su^2-Hw gene, Df(3)red^{P52}, in the form Df/+, is as effective in suppressing ct^K as is su^2-Hw/+. Thus, su^2-Hw is unlikely to be coding for an altered tRNA in the manner expected for a typical translational suppressor. Either su^2-Hw affects translation indirectly or another mechanism of suppression is involved. Molecular analysis of BX-C mutants suppressible by su^2-Hw may be expected to provide an explanation for the allele specificity of this unusual type of suppressor.

IV. THE ROLE OF BX-C IN DEVELOPMENT

Early in the analysis of BX-C it became evident that a given mutant gene was interfering with the state or level of development (L) which one or more segments would otherwise normally achieve rather than interfering with the development of a specific segment as such. Thus, comparison of a bx with a bx bxd homozygote revealed that the bx mutant partly prevents an LT2 → LT3 type of transformation not only in the anterior part of T3, as already described, but also in the anterior part of A1, whenever that part is prevented by the bxd mutant from achieving LA1 (5). In modelling the functional role of the BX-C genes in development, we have therefore assumed that a given wild-type gene elaborates a substance which in some way brings about, or contributes to, a specific transformation such as LT2 → LT3. The current status of the model will be illustrated by reviewing the major transformations that have thus far been characterized.

Analysis of the role of individual BX-C genes in bringing about intersegmental transformation is facilitated by the circumstance that even in genotypes in which the entire complex is missing the organisms survive until the end of embryonic development. Furthermore, at that point the cuticular and tracheal systems are already sufficiently well differentiated that the level of thoracic or abdominal development in most of the segments of the body can be determined. As a result it is possible to use two grossly abnormal genotypes to recognize extreme boundary conditions for the range of levels of development that a given body segment can achieve. Thus, when all of the genes of the complex are absent, as they are in homozygotes for certain deletions such as Df-P9 (Fig. 1), segments T3 and all of the first seven abdominal segments transform toward LT2, as judged by tracheal and cuticular patterns recognized in late embryonic stages at about the time of death of such genotypes. [The status of A8 is uncertain in that it has possible head-like transformations in the form of rudimentary plates corresponding perhaps to portions of the mandibular apparatus (1).]. At the other extreme all of the thoracic segments, possibly some of the head segments, and all of the abdominal segments, approach LA8 (or achieve it in the case of A8) in homozygotes for Pc^3, an extreme allele of Polycomb (2;14), a trans-regulatory gene believed to code for a repressor of BX-C (1;15).

In addition to the two extremes, LT2 and LA8, one might expect a segment to be capable of approaching or achieving any of eight additional levels corresponding to the levels of development found in the wild-type organism in the eight segments from T3 to A7 inclusive. The way in which six of such levels are believed to be derived from LT2, the presumptive ground state, is diagrammed in Fig. 2. Evidence in support of the interpretation of wild-type gene functions shown in that figure comes from several types of genotypes. The most valuable approach has been one which involves constructing genotypes in which only certain wild-type genes of the complex are present, the remaining genes of the complex being completely absent. The most useful rearrangements available for such a purpose are those involving insertions of portions of BX-C elsewhere in the genome; namely, $Tp(3)bxd^{100}$ and T(2;3)P10 (Fig. 1). The former rearrangement, which will hereafter in this paper be abbreviated as Tp-100, consists of a transposition of a region, shown in Fig. 1, from the right arm (3R) of the third chromosome into section 66C of the left arm (3L). The region transposed to 3L is abbreviated as Dp-100 and contains only the proximal, or leftmost in Fig. 1, portion of BX-C extending up to, and including, Ubx$^+$; while the distal portion of the complex, abbreviated as Df-100, remains in 3R.

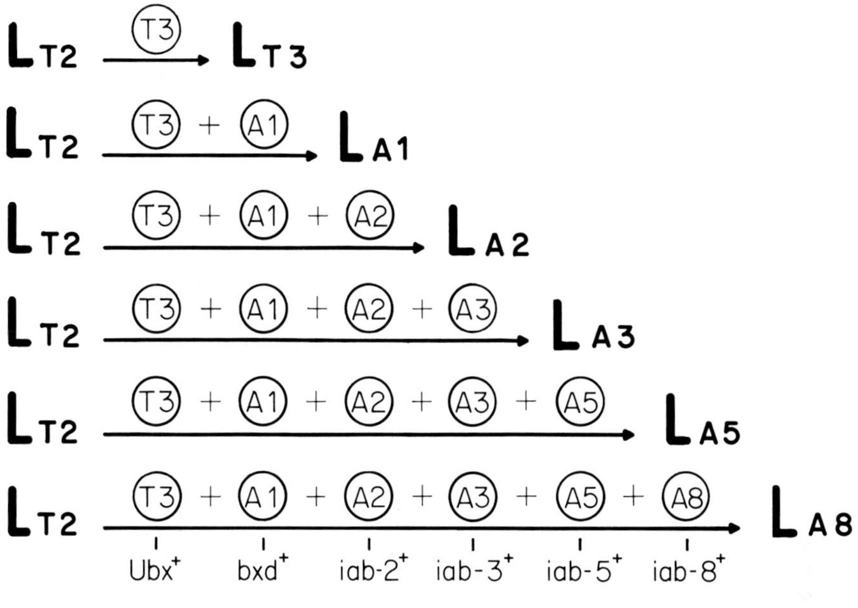

FIGURE 2. The postulated role of BX-C substances in controlling levels of development.

LT2 and LT3 designate 2nd and 3rd thoracic levels of development, respectively; LA1 to LA8 designate 1st to 8th abdominal levels, respectively. Symbols of the wild-type BX-C genes are shown at the bottom of the figure, directly under symbols of substances which are postulated to be elaborated by those genes; e.g., the Ubx$^+$ substance is symbolized by enclosing "T3" in a circle; omitted from the diagram, in this case, is the contribution of abx$^+$, bx$^+$ and pbx$^+$ genes, all of which are involved in mediating the transformation to LT3, presumably by elaborating additional substances. For simplicity, all substances are assumed to act additively in bringing about the designated transformations.

A. The LT2 → LT3 Transformation

Embryos which are homozygous for Df-P9 and therefore lack all of the BX-C genes have been compared with embryos which are of the same genotype except for the addition of Dp-100. As described elsewhere (1), embryos of the former genotype have a primitive tracheal system with discontinuous dorsal tracheal trunks in all segments from T2 to A8; while embryos of the latter genotype exhibit the typical wild-type pattern; namely, continuous dorsal tracheal trunks terminating in T2 and in A8.

Conversely, the effect of loss of the proximal portion of BX-C has been studied by comparing embryos which are homozygous for Df-P9 with those of genotype, Df-100/Df-P9. Embryos of the latter type are deficient for the proximal portion of BX-C, from abx to Ubx, inclusive, and have been found (1) to have a discontinuous dorsal tracheal trunk only in segments T2, T3 and A1; indeed, such embryos closely resemble those of homozygotes for the Ubx point mutant, except that the latter survive and reach later stages of development when distinct anterior spiracles (not developed to a significant extent in embryonic stages) are well developed not only in T2, as in wild type, but in T3 (and A1), as well. Explicit note needs to be taken at this point that Tp-100/Df-P9 animals have a tracheal trunk pattern like that seen in wild type; indeed, such animals survive to the adult stage (and exhibit an extreme bxd and extreme pbx phenotype). Thus the defects in the dorsal tracheal trunk in Df-100/Df-P9 are attributable to the deficiency and not to any mutation or position effect that might have accompanied Tp-100 at its time of origin (from an X-rayed third chromosome, wild-type except for a marker mutant, ri). Hence, it has been inferred (1) that, as far as the dorsal tracheal trunk system is concerned, the presence of Ubx^+ and of wild-type alleles of other, more proximal, BX-C genes suffices to establish continuity of trunks in T3 and to suppress potential formation of a pair of anterior spiracles in that segment. Furthermore, such an effect may be thought of as a transformation of T3 from LT2 towards LT3. In other words, just as T3 in the wild-type adult has, for example, halteres in place of wings, so T3 in the wild-type embryo and larva has continuous dorsal tracheal trunks in place of sections of that trunk terminating in a pair of spiracles. In addition, on the basis of finding that Dp-100 alone suffices to restore continuity of that trunk in all segments between T2 and A8, it has been inferred (1) that once the gene or genes which effect that function become active in T3 they do so also in all segments posterior to T3.

That Dp-100 is able to contribute another component of the LT2 → LT3 transformation is known from studies of the effect on adult flies of adding that duplication to either a homozygous bx^3, or a homozygous Ubx, genotype. In the Dp-100 bx^3/bx^3 case, T3 is wild type, instead of having the extreme transformation of anterior T3 to LT2 that is typical of homozygous bx^3. In the Dp-100 Ubx/Ubx case (5), anterior T3 is wild type (except for the very slight enlargement of the anterior haltere typical of Ubx/+). In addition, in this latter case, strong bxd and pbx mutant phenotypes remain, as expected on two grounds; namely, (a) Ubx is known (3) to effect, in cis, strong inactivations of the functions of bxd^+ and pbx^+ genes, and (b) the available evidence indicates that those genes are absent in Dp-100, being present instead in an inactivated state in the distal (or Df-100) portion of Tp-100. It follows that at least three genes, abx^+, bx^+ and Ubx^+, are involved in controlling an LT2 to LT3 transformation in the anterior compartment of T3; while at least one additional gene, pbx^+, is involved in controlling the corresponding transformation in the posterior compartment. Whether each gene is making a specific substance remains to be determined; for the sake of simplicity, Ubx^+ is arbitrarily shown (Fig. 2) as the only gene producing a substance that controls LT2 → LT3.

B. *The LT2 → LA1 Transformation*

The bxd mutants of the complex tend to cause A1 to remain at a thoracic level of development not only in adult states (4) but in embryonic and larval stages as well (1). In these latter stages, three types of cuticular effects are seen, especially in bxd (or bxd^{51j}) hemizygotes; namely, 1) the abdominal type of ventral setal band (VSB) normally present in A1 of wild type larvae is converted to a thoracic-type VSB, closely resembling that found in T3; 2) a pair of ventral pits (VP) which normally occurs in each of the three thoracic segments of wild type appears not only in A1, but in A2 to A7, inclusive, as well; and 3) a pair of Keilin organs (KO) which also are found only on thoracic segments of wild type are partially developed in A1 (1) (often two hairs being present instead of the three that make up each of these sensory organs).

Evidence, recently obtained, suggests that the constellation of phenotypic effects on A1 of bxd mutants involves at least one more locus than that of bxd, itself. The evidence is based on an analysis of bxd^{111}, a bxd mutant obtained by irradiation of a wild-type (Canton-S) strain. This mutant is like all other bxd mutants of X-ray origin thus far obtained in having 1) a more extreme bxd (and pbx) phenotype in the adult stage when tested in the hemizygous condition, or opposite a bxd or Ubx mutant, than the spontaneous bxd mutants have under the same conditions; and 2) an associated chromosomal rearrangement with a breakage point within BX-C that separates the two heavy double bands of 89E (Fig. 1). Hemizygotes for each of seven such rearrangements have been examined in first-instar larvae and in all but the bxd^{111} case the typical triad of cuticular defects described for the bxd hemizygote is present. The tested rearrangements, in addition to Tp-100, already described above, are as follows, where an abbreviated symbol of the type adopted in the Tp-100 case is used and the other breakage point(s) is indicated in parentheses after the symbol of the rearrangement: T(3;4)-101 (101); In-106 (72D-73A); Tp-110 (91D to 92A inserted into 89E); T(1;3)-111

(89E3-4 to 90B inserted into 4D); In-113 (69C); and In-114 (94A-C). Among a total of 23 hemizygous T(1;3)-111 first-instar larvae examined, VSB in A1 was not T3-like; but instead, in 10 cases, was intermediate between T3 and A1 in a somewhat variegated pattern, and in the remaining 13 cases was similar to that found in A1 of hemizygous wild-type (DfP9/+) larvae (whose VSB on A1 is slightly sparser than that on A1 of homozygous wild-type larvae); VP were present in all animals usually on all segments from A1 to A7, inclusive; and KO were partially developed in all animals on A1.

The basis for the unusual phenotype of T(1;3)-111 hemizygotes has been explored by the method of overlapping deletion analysis with the results shown in Table 1. In an earlier study (1), the function of bxd^+ in embryonic cuticular structures was inferred from the phenotypes of two genotypes; namely, a genotype designated A in Table 1, Dp-100 Df-P9/Df-P9, in which bxd^+ function is either impaired or abolished; and a genotype designated C in that table, Dp-P10/+; Df-P9/Df-P9, in which bxd^+ (and pbx^+) activity is unimpaired as far as can be detected. In both of these genotypes, the only other BX-C genes present in the animal are those in the overlapping portions of the duplications; that is, the wild-type alleles of Ubx and loci proximal thereto. From the phenotypes of genotypes A and C, it was inferred that bxd^+ has at least three functions in the embryonic cuticle: 1) suppression of ventral pits on all abdominal segments; 2) promotion of A1-like VSB on A1, and of an A1 component to development of VSB in the remaining abdominal segments; and 3) partial suppression of KO in all abdominal segments.

The insertional nature of T(1;3)-111 permits a third type of deletion hemizygote to be constructed that contains all loci proximal to the breakage point in 89E of that translocation. The phenotypic properties of this Df-111/Df-P9, or "B" genotype are shown in Table 1. It is evident that genotype B resembles genotype A in failing to suppress VP on abdominal segments, but retains the ability possessed by genotype C to promote A1-type VSB development and partial suppression of KO on those segments. (Among 20 embryos of genotype B scored, 15 showed VSB of an A1-type on all abdominal segments while five showed VSB on A1 and A2 that were only partially transformed from T3 towards A1 in a variegated fashion; however, on the remaining abdominal segments of those five animals, the VSB was like that on A1 of the wild-type hemizygote.)

The simplest interpretation of these results is that genotype C contains two loci whose wild-type alleles bring about the triad of embryonic cuticular effects previously attributed solely to bxd^+. On such an interpretation, the function of the more distally located wild-type allele would be to promote the first effect of that triad (on VP); while the more proximally located wild-type allele would promote the remaining two effects (on VSB and KO). Since the spontaneous bxd

Table 1. Comparison of embryonic phenotypes of three overlapping deletion genotypes with respect to pattern of cuticular structures on abdominal segments, A1 to A8, inclusive.

Genotype	Ventral pits	Type of ventral setal bands	Number of hairs on Keilin organ
A. $\frac{\text{Dp-100 Df-P9}}{+ \quad \text{Df-P9}}$	Present	T2-like	3
B. $\frac{\text{Df-111}}{\text{Df-P9}}$	Present	A1-like	1 (or 2)
C. $\frac{\text{Dp-P10}}{+}$; $\frac{\text{Df-P9}}{\text{Df-P9}}$	Absent	A1-like	1 (or 2)

mutants behave phenotypically as if they lack wild-type function with respect to all three effects such mutants can most easily be interpreted as alleles of the proximal, or bxd, locus that weaken the dominance of the wild-type allele of the distal locus. For the purposes of the present discussion, the distal locus will be given the tentative symbol iab-1, and the proximal locus will be considered to be the original bxd locus. On such an interpretation the adult bxd mutant phenotype (thoracic-like A1) would be attributable, paradoxically, to the iab-1 locus since T(1;3)-111, although nearly, but not quite, wild-type with respect to VSB on A1, has a typical extreme adult bxd (and pbx) phenotype when homozygous or hemizygous. Since hemizygotes for the rearrangement-associated bxd mutants, with the exception of T(1;3)-111, also are associated with the entire triad of bxd mutant effects on the embryonic cuticle as well as an extreme adult bxd (and pbx) transformation, they evidently also have breakage points proximal to iab-1.

Another indication of separable genetic units as the basis of the bxd phenotype comes from the observation that the Uab^5 translocation is associated with a recessive inability to suppress ventral pits on the abdominal segments. Although such an inability is typical of all known bxd mutants, Uab^5 fails to show any other aspects of the bxd mutant phenotype. The possibility that the dominant effect of Uab^5 -- an adult transformation, figured elsewhere (1) of A1 and A2 towards A3 -- might mask any recessive bxd effect has been ruled out by recombining Uab^5 with other appropriate rearrangements to generate one derivative having only the proximal portion of Uab^5 with respect to the breakage point in BX-C (Fig. 1), and another derivative having only the corresponding distal portion. The former derivative acts as if it contains a fully wild-type bxd function; while the latter derivative carries the dominant Uab^5 effect. (It should be noted that the results of the analysis of the two halves of the Uab^5 translocation indicates that the breakage point of that rearrangement occurs between bxd and iab-2; uncertainty in regard to the map location of the postulated Uab locus and of the Uab^5 and Uab^4 breakage points is indicated in Fig. 1 by means of. dotted lines extending from the genetic map.)

Definitive evidence for the existence of additional genes controlling the various components of the bxd transformation will require identifying mutants of such genes which, unlike bxd^{111} and Uab^5, are not associated with gross chromosomal rearrangements. Therefore, the achievement of LA1 is shown in Fig. 2 as requiring the presence of only a single bxd^+ substance, symbolized (A1). Inasmuch as the Ubx^+ substance, (T3), appears to act in A1 to suppress anterior spiracle formation in that segment, it is assumed that this latter substance is also required in achieving LA1 in the first abdominal segment.

An alternative interpretation would be that LT3 is an intermediate state between LT2 and LA1. Such an interpretation has been discussed elsewhere (3) and considered to be unlikely not only on evolutionary grounds but also on the basis that in the case of the bx^3 pbx (or the more recently derived abx bx^3 pbx) homozygote an extreme adult transformation of T3 towards T2 occurs while A1 remains wild type; that is, A1 still achieves LA1 when transformation of that segment to T3 presumably has been at least partially blocked. A caveat, however, is that in such a homozygote Ubx^+ is still partially, if not fully, functional. Since all existing Ubx mutants are associated with a weakened dominance of bxd^+ in cis, it has not been possible to obtain animals in which bxd^+ is functional,and Ubx^+ is not; it cannot be excluded that LT3 is an intermediate state between LT2 and LA1.

C. The Remaining Transformations

Fig. 2 also shows the present view of the role played by BX-C substances in producing the remaining body segment transformations. Although alternative interpretations for many of these transformations have been assumed (1); for example, LT2 → LA5 vs. LA4 → LA5 for the function of iab-5^+, it now seems more reasonable to postulate that the attainment of a given level of segmental development involves the summation of the contribution of a number of BX-C gene functions.

A specific example of the role played by the various BX-C substances is depicted in Fig. 3. The bxd^+ substance, (A1), is assumed to accomplish a number of effects: a highly specific suppression of ventral pits (VP); development of possibly a specific portion of the ventral setal band (VSB) and maintenance of a continuous dorsal

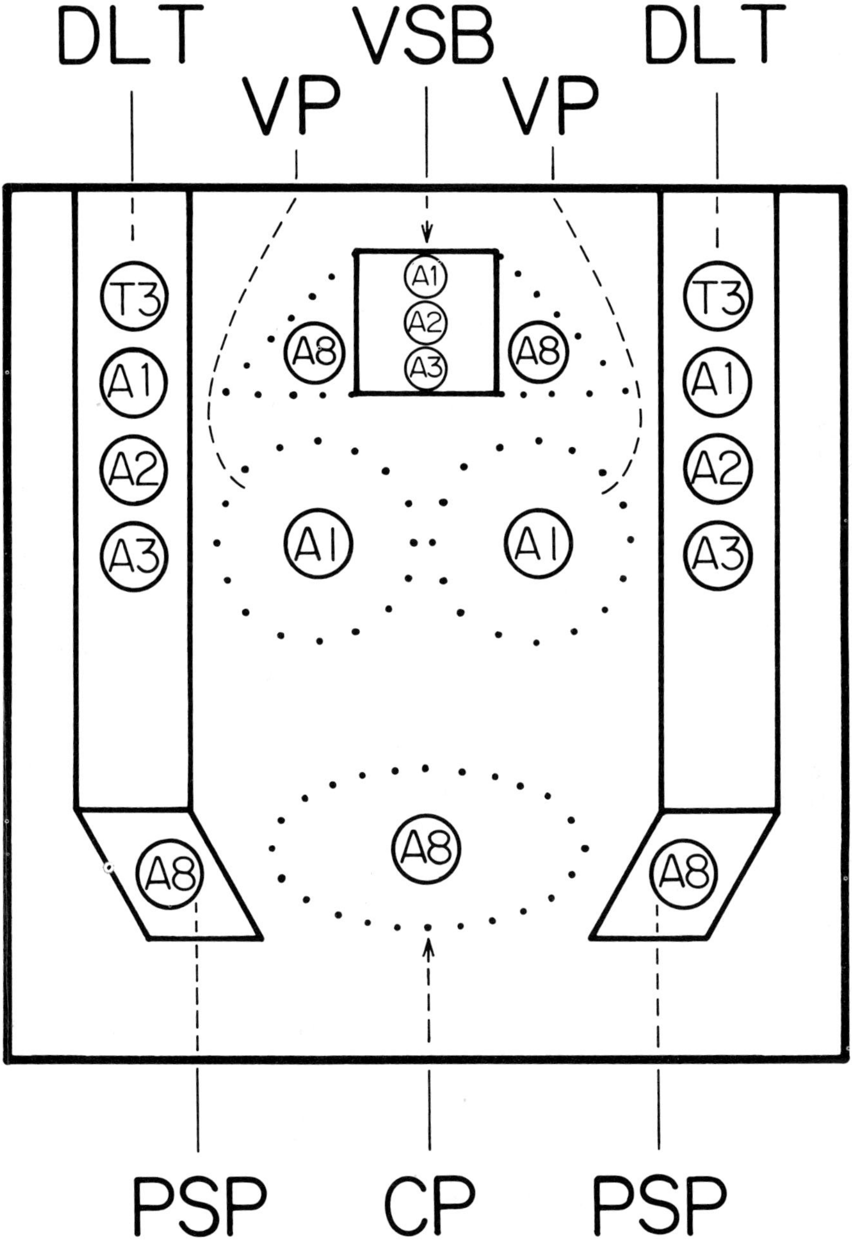

FIGURE 3. Interpretation of state of BX-C gene activity within wild-type embryonic segment A8.

Presence or absence of specific structures is shown as being under the control of specific BX-C substances (see legend of Fig. 2 for description of symbols). DLT = dorsal longitudinal tracheal trunk; VSB = ventral setal band; VP = ventral pit; PSP = posterior spiracle; CP = chitnized plates. Absence of structures (VP, CP and lateral edges of VSB) is indicated by dotted outlines.

tracheal trunk (DLT), a property which it is assumed to share with many other genes of the complex (1), including not only the wild-type Ubx, iab-2, and iab-3 genes, as depicted, but also others not shown; such as, iab-7^+ which appears to be active in restoring the trunk between A6 and A8 in the Ubx^{109} hemizygote (1). On the basis of deletion analyses and the properties of the homozygous Pc mutant (1), the wild-type iab-8 substance, (A8), is assumed to be highly specific in effecting a posterior spiracle (PSP); in truncating the VSB of A8 (shown by the dotted triangular area surrounding the square that depicts VSB in A8 in FIg. 3); and finally in suppressing the potential formation of chitinized plates (CP), which arise in the absence of the iab-8 region of the complex (1). At least two additional loci, iab-2 and iab-3, play some role in elaborating the VSB on A8 although in what order and in what specific ways is not known.

The picture that emerges is one of a mosaic of structures (and presumably functions as well) each of which is under the control of a specific BX-C gene, or in some cases, several genes working in concert. The precise relationship between compartments (19) and groupings of structures affected by specific genes is not clear. It has, of course, long been recognized that the bx and pbx mutants seem to restrict their effects in T3 to anterior and posterior compartments, respectively, both at the morphological and cellular level (20). The actual state of activity of wild-type alleles of bx and pbx *in situ* and in regenerated compartments (21) is expected to become accessible to analysis at the molecular level. Whether a given gene is actually switched on or off in a binary manner (22), seems unlikely in the case of the BX-C genes which in at least some cases show a gradation of activity between being off or fully on, depending on the tissue or segment involved. Finally, it will be of interest to learn how BX-C and the Antennapedia gene complex (23, 24) interact in laying down the fundamental body segmentation plan and whether the ground state for a segment is mesothoracic (T2) or prothoracic (T1) (25).

V. REGULATION OF BX-C IN THE WILD-TYPE ORGANISM

A diagrammatic representation of the postulated state of activity of some of the principal BX-C genes in the wild-type embryo is shown in Fig. 4. In the vertical column at the left the relevant thoracic and abdominal segments are depicted using the same highly stylized representation of structures shown in Fig. 3 for segment A8.

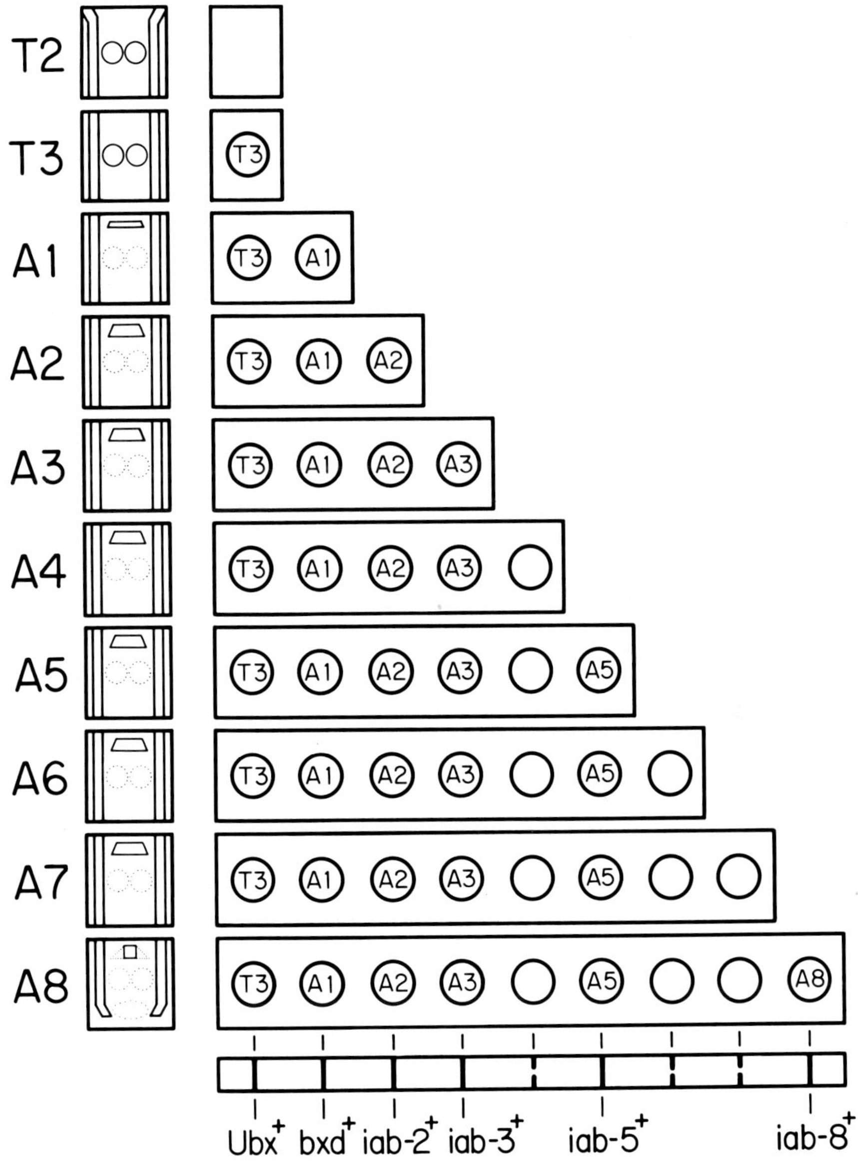

FIGURE 4. Postulated state of BX-C activity in thoracic (T) and abdominal (A) segments of the wild-type embryo.

Squares along left margin signify segments with: paired dorsal tracheal trunks (vertical bars, terminating in T2 and A8); paired ventral pits (presence shown as solidly, absence as dottedly, outlined circles); and, ventral setal bands (trapezoids in A1 to A7, or small square in A8, signify regions with coarse, hooked setae; not shown are corresponding regions in T2 and T3 with tinier setae). Dotted ellipse in A8 signifies absence of chitinized plates. BX-C gene activity is shown by presence or absence in rectangles to the right of segments of symbols for postulated BX-C substances (see Fig. 2 legend).

In each segment commencing with T3 and proceeding posteriorly, successively more gene products are shown as being elaborated in the rectangle to the right of that segment. Whether more than one substance is elaborated in a given cell and whether each substance is cell specific are not yet determined.

The wild-type pattern has been postulated to come about as the result first of a gradient along the body axis in the concentration of a repressor (1) elaborated by the wild-type allele of the Pc gene, or alternatively, the concentration of an inducer (4). At present these alternatives cannot be decisively distinguished, nor can a gradient in a positive trans-regulatory substance produced by a gene such as Rg-bx (16-18) be ruled out completely. Therefore, somewhat arbitrarily, negative trans-regulation is assumed to operate with a gradient in the repressor somehow being laid down very early in development with the highest concentration occurring in the anterior, and the lowest in the posterior, portion of the animal. Secondly, a given BX-C gene is assumed to be regulated in cis by a region whose affinity for repressor is adjusted in such a way that the more distal the gene is in the chromosome, the more intense is the affinity of the cis-regulatory region of that gene for repressor. Thus, in T3 of wild-type embryos, the affinity of such a region for Ubx^+ is postulated to be so weak that this gene is derepressed in all segments commencing with T3 and posteriorly. On the other hand, the affinity in the case of the cis-regulatory region for $iab\text{-}8^+$ is assumed to be so strong that in wild-type embryos that gene only becomes derepressed in A8, where the level of repressor is postulated to be effectively minimal. In the case of hemizygous Pc embryos, or, as recently found, in the case of embryos homozygous for a deletion for the Pc region (78E), $iab\text{-}8^+$ becomes active in all of the thoracic and abdominal segments, and probably some head segments as well, presumably owing to total lack of Pc repressor except for that derived maternally.

Thus far cis-regulatory regions have been identified genetically by recombinational analysis only in the case of Cbx and Mcp, with Hab and Uab being other examples yet to be analyzed. The Cbx mutant has been interpreted as the cis-regulatory region for Ubx^+ (1;3). Alternatively, since Cbx arose simultaneously with pbx as a single X-ray induced event, Cbx has been interpreted as an insertion of the wild-type allele of pbx between the bx and Ubx loci, "with accompanying escape of that gene from repression" (16). The existence of these and still other dominant cis-regulatory mutants of the BX-C complex provides a unique opportunity to examine the underlying mechanism of gene activation and repression in higher organisms.

ACKNOWLEDGMENTS

I thank Susan Ou, Josephine Macenka and Tony Ramey for technical assistance and Drs. Loring Craymer and Ian Duncan for helpful discussions.

REFERENCES

1. Lewis, E. B. (1978). *Nature* **276**, 565-570.
2. Lindsley, D. L. and Grell, E. H. (1968). "Genetic variations of Drosophila melanogaster." Carnegie Inst. Wash. Publ. No. 627.
3. Lewis, E. B. (1955). *Amer. Nat.* **89**, 73-89.
4. Lewis, E. B. (1963). *Amer. Zool.* **3**, 33-56.
5. Lewis, E. B. (1951). *Cold Spring Harbor Symp. Quant. Biol.* **16**, 159-174.
6. Lewis, E. B. (1980). *Drosophila Information Service* **55**, 207-208.
7. Lewis, E. B. (1954). *Amer. Nat.* **88**, 225-239.
8. Kuhn, D. T., Woods, D. F. and Cook, J. L. (1981). *Mol. Gen. Genet.* **181**, 82.
9. Crosby, M. (personal communication).
10. Kiger, J. (personal communication.
11. Kuhn, D. (personal communication).
12. Lewis, E. B. (1967). *In* "Heritage from Mendel" (R. A. Brink, ed.), pp. 17-47. University of Wisconsin Press, Madison, Wisconsin.
13. Johnson, T. K. (1976). Ph.D. Dissertation, University of Texas at Austin.
14. Lewis, P. H. (1947). *Drosophila Information Service* **21**, 69.
15. Puro, J. and Nygren, T. (1975). *Hereditas* **81**, 237-248.
16. Lewis, E. B. (1968). *Proc. XII Int. Cong. Genetics* **2**, 96-97, Science Council of Japan, Tokyo.
17. Garcia-Bellido, A. and Capdevila, M. P. (1978). *Symp. Soc. Dev. Biol.* **36**, 3-21.
18. Ingham, P. and Whittle, R. (1980). *Molec. Gen. Genet.* **179**, 607-614.
19. Garcia-Bellido, A., Rippoll, P. and Morata, G. (1973). *Nature New Biol.* **245**, 251-253.
20. Garcia-Bellido, A. and Lewis, E. B.(1976). *Develop. Biol.* **48**, 400-410.

21. Adler, P. N. (1981). *Develop. Genetics* **2** 49-73.
22. Kauffman, S. A. (1973). *Science* **181**, 310-318.
23. Kaufmann, T. C., Lewis, R. and Wakimoto, B. T. (1980). *Genetics* **94**, 115-133.
24. Denell, R E., Hummels, K. R., Wakimoto, B. T. and Kaufmann, T. C. (1981). *Develop. Biol.* **81**, 43-50.
25. Morata, G. and Kerridge, S. (1981). *Nature* **290**, 778-781.

GENETIC DISSECTION OF EMBRYOGENESIS IN *CAENORHABDITIS ELEGANS*

Randall Cassada, Edoardo Isnenghi, Kenneth Denich, Khosro Radnia, Einhard Schierenberg, Kenneth Smith and Günter von Ehrenstein+

Department of Molecular Biology,
Max-Planck-Institute for Experimental Medicine,
Göttingen, Federal Republic of Germany

ABSTRACT

The complex process of embryogenesis in the simple nematode *Caenorhabditis elegans* is invariant from animal to animal. Cell lineages have been studied by direct observation of individual cells in living embryos using Nomarski differential-interference -contrast microscopy (1). To dissect events genetically involved in embryogenesis, we have isolated a set of 36 recessive temperature-sensitive (ts) mutants in 30 separate *emb*-genes, which cause arrest of embryonic development (2). The fraction of *emb* mutants among total ts lethals and the recurrence frequency (second alleles) allowed two independent estimates of 200-500 genes essential for embryogenesis (of a total of about 2000 essential genes). So far 54 *emb* genes have been detected (2,3,4), still far from genetic saturation. We have mapped 25 new *emb* genes (resolution of 1 recombination unit). Surprisingly, 10 *emb* genes are clustered near gene *unc-32* on linkage group III. Higher resolution mapping here, using deletions, is under way. We have tested the mode of expression (the necessity and/or sufficiency for normal embryogenesis) of the wild-type alleles of these 30 genes in the pa-

+ Deceased December 26, 1980.

ISBN 0-12-137420-3

rents and zygote by performing genetic crosses in which a wild-type allele appears in various configurations, and then determining at the restrictive temperature (25° C) the effect on the viability of the resulting progeny genotypes (3,4). A majority of the emb genes are of the maternal-expression-necessary class (18 of 30 genes studied), in agreement with the results from other similar mutants (3,4). For 3 genes, neither maternal nor zygotic expression is sufficient (both necessary?). We have also found 2 zygotic-necessary-and-sufficient genes. For 1 gene paternal expression is partially sufficient. The remaining 7 are of the parental-or-zygotic-expression-sufficient class. We have ordered the ts mutants sequentially in development by temperature shift experiments and according to their arrest stage (terminal phenotypes). Their cellular and subcellular properties are being studied to identify the cellular processes defective in the mutants, and ultimately the mechanisms for genetic control of cell behavior in embryogenesis. We are finding a variety of defects in early cell lineages, including the timing of embryonic cell divisions similar to those already described in another set of mutants (5).

INTRODUCTION - GOALS AND PROPERTIES OF THE SYSTEM

We are investigating the embryogenesis of a very simple animal, the nematode *C.elegans*. We look at intact, living embryos, to try to understand the complexity of phenomena *in situ*, to sort the various phenomena into individual developmental steps. Using a high resolution light microscopy it should be possible to observe and delineate "unitary" cellular processes. Ultimately we seek to understand the underlying molecular processes and genetic controls behind the cell processes. To most efficiently sort genes into pathways, using the T4 bacterophage paradigm, we first look at the phenotype and later try to define what the individual steps within a developmental pathway are. *Caenorhabditis elegans* is favorable for such an approach, because it has a small number of genes and a small number of cells. Nonetheless it is complex enough to be an interesting model for higher organisms. The principal experimental tool has been the use of mutants. Our goal has been to isolate as many mutants and to define as many genes essential for embryogenesis as we conveniently could and study their phenotypes.

One of the main take-home lessons of the study so far is that there is a large maternal component to embryogenesis, probably even more than in other organisms. Nematodes have been known from classical studies (6,7) to have a determinate (mosaic) development, which may have to do with the strong maternal component. Embryogenesis is not only how one cell becomes a multicellular multi-tissue animal but also what is the mechanism for producing this one cell? Most mutationally detected genes in C.elegans are maternally expressed and act early in embryonic development or pre-embryonically in oogenesis (2,3,4).

C.elegans has been pioneered by Sydney Brenner at the Cambridge MRC Laboratory (8,9). The system has been recently reviewed (10,11). The simplicity of C.elegans is a large part of its appeal, as illustrated first by its life cycle (Table 1).

TABLE 1: Quantitative aspects of the life cycle in C.elegans.

	fertilization	embryogenesis	hatch	4 molts	adult
time:		12 hours		2 days	
	1 cell		550 cells		950 cells (plus germ cells)
weight:	1 μg	1 μg	1 μg		200 μg
length:	50 μm	50 μm	200x10 μm		1500x70 μm

It is a free-living, non-parasitic nematode, and grows on agar plates with E.coli as food source. A complete generation time is 3 days. Embryogenesis itself takes 12 hours at 25° from fertilization to hatching and involves production of 550 cells. Post-embryonically four cuticle molts without metamorphosis produce an adult similar to the just-hatched worm. Most cells are post-mitotic at hatching. But a small fraction (about 10 %) divide post-embryonically to produce additional adult structures, mostly nerves and sexual structures including the gonad and a final total of some 950 cells (12), approximately double that of the embryonic worm, plus another 2500 germ line cells arising in the gonad (13). Post-embryonic events have been described (12) including gonad development (14). The adult produces some 300 progeny over a couple of days, so growth of a population is quite rapid. A single petri dish can contain up to 10^5 individuals. In liquid cultures, for biochemical studies, one readily obtains yields of 10 g/l. The egg cleaves into smaller and smaller cells, with no increase in the egg mass (1 μg) during embryogenesis. The hatched worm grows some 7-fold in length and 200-fold in mass to adulthood, so the cells of the young adult are on the average 100-fold larger than the same cells in the hatched larva. The developing embryo can be followed at the cellular level through embryogenesis under a coverslip thanks to its small size, transparency, and minimal requirements from the environment (1,15). The entire post-embryonic cellular events have also been followed similarly (12). Development as seen at the cellular level is extremely reproducible from individual to individual, a common feature in mosaic nematode development. This allows us to describe developmental events after looking at only a few individuals, for example for comparing mutant and wild type events, or even allows us to follow one group of cells in one individual, another group in another and make a composite picture of the two which is valid and reliable. As a particularly useful application, it is possible to take a single embryo, fix and embed it, take complete serial sections for EM and from the electron micrographs reconstruct the cells, knowing one embryo will be essentially like another at that same stage (16). The embryo can be observed in ultra-structural detail in various ways to relate directly such events in any given stage to events visible in other, living embryos with the same number of cells. The reproducibility and non-regulative features suggest that the absence of a single gene product in a mutant should lead to a distinct defect, which may be visible at the cellular level. In a more regulative system one might expect compensation for a missing product and masking of the defect. Many nematodes have similarly convenient cellular properties, but *Caenorhabditis elegans* is, in addition, particularly convenient for genetic experiments. It is quite easy to get mutants with visible (morphological or behavioral) phenotypes and map these. The current map (17,18) includes over 300 genes, distributed over 6 linkage groups. Besides the short generation time and the small size of the organism, the sexual system (Figure 1) has distinct advantages for genetics.

Normal cultures consist nearly exclusively of self-fertilizing hermaphrodites, otherwise normal "females" that transiently produce a supply of some 300 sperm as late juveniles. Once mature, the adults produce oocytes that increase in size as they ripen, passing assembly-line fashion down each of the two gonads and through the spermatheca for fertilization. Development continues *in utero* and after laying. This self-fertilization produces inbred homozygous stocks without the need for mating, especially favorable for obtaining homozygous mutants after mutagenesis. But for genetics, i.e. to do crosses for

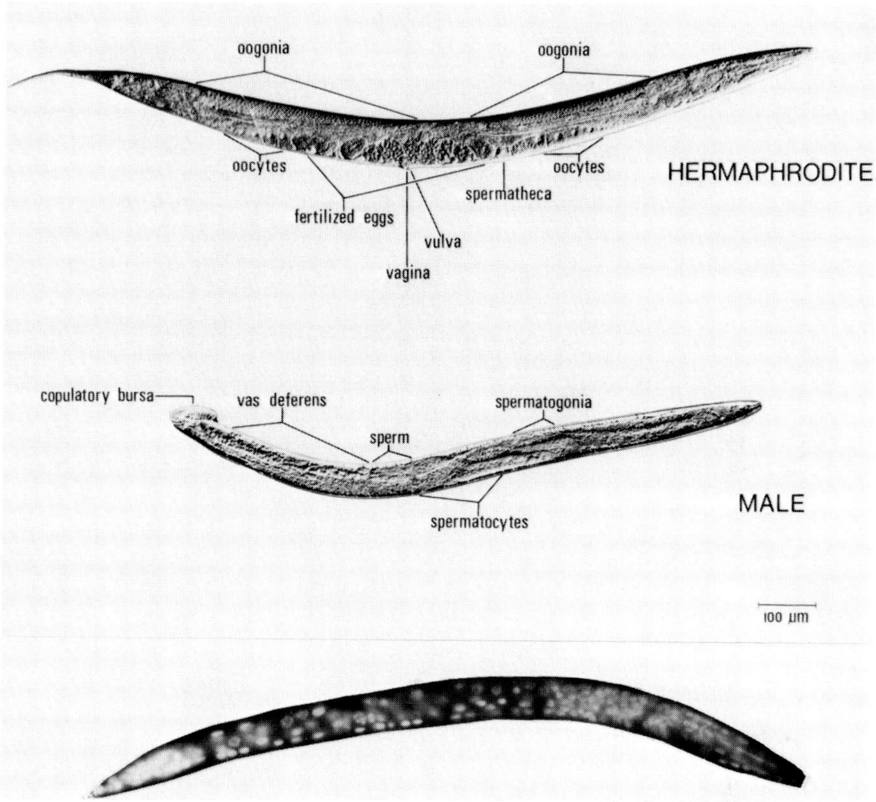

FIGURE 1: The reproductive system and cellular anatomy of C.elegans from top to bottom, adult hermaphrodite and male by light microscopy (after 15) and hermaphrodite juvenile by UV-induced fluorescence after Hoechst 33258 staining (26).

complementation and mapping, sex is necessary. The diploid hermaphrodites have two sex chromosomes, XX, but spontaneously (0.3 % frequency (19)) lose an X meiotically to produce XO diploid males, which can mate efficiently, competing out the hermaphroditic sperm (20) and producing 50 % males among the cross progeny. An additional practical advantage is that the nematodes, which have adapted evolutionary to a soil habitat, survive freezing quite well. Mutant stocks are routinely frozen (9) and stored indefinitely over liquid nitrogen, avoiding the risks of transfer error and genetic drift. It has been estimated (9) from mutational frequencies that there are some 2000 total essential genes in C.elegans. The DNA complexity is 80,000 kbp (21).

CELLULAR ASPECTS OF NORMAL WILD-TYPE EMBRYOGENESIS

C.elegans embryogenesis is nearly completely described. Extending classical nematode studies (6,7,22), nearly all embryonic cell divisions have been followed (1,15,23,24), noting time and cleavage direction, as well as cell movements through the first half of embryogenesis, where essentially all cell divisions occur (Figure 2). This was possible using Nomarski

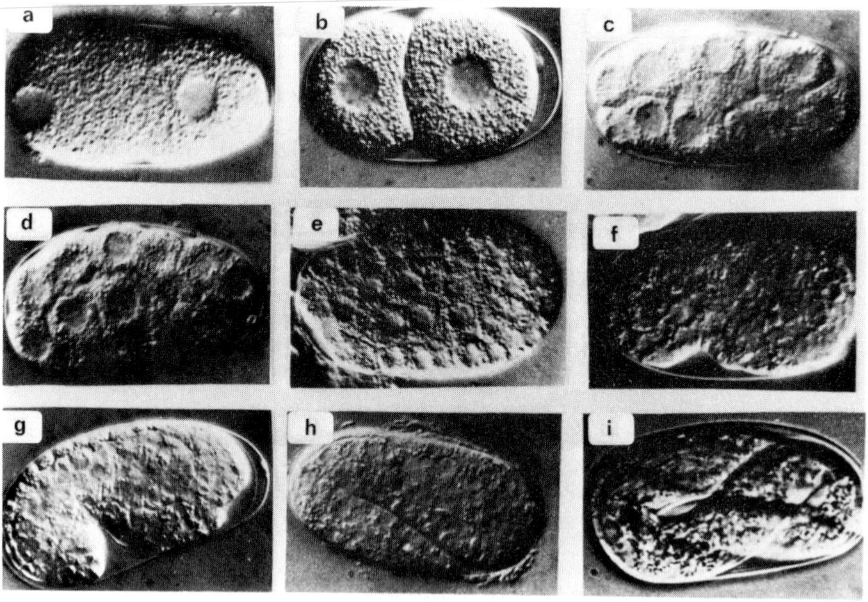

FIGURE 2: Stages in normal (wild type) embryogenesis (15,24). anterior, right; dorsal, top. a) pronuclei during zygote formation b) 2-cell c) 26-cell d) 28-cell e) ca. 300-cell f) ca. 540-cell lima bean g) tadpole h) plum i) 550-cell pretzel stage

differential interference optics, in which only a thin section of the embryo is visible, and continuously focusing up and down through the embryo, with recording on video tape for thorough analysis. The tracing of the individual cells through the second half of embryogenesis (Figure 2f-i), i.e. through morphogenesis, cell elongation and tissue formation to the finished worm is now being completed (Figure 3) (23). Despite some difficulties following individual nerve cells in the densely packed head ganglia around the nerve ring (Figure 1, bottom panel), a complete embryonic description will be soon available. Together with the already finished post-embryonic lineages (12,14), the whole cellular lineage and cell origins and fates will be known from fertilization to adulthood. Following fertilization and zygote formation, a sequence of asymmetric cleavages ensues (Figure 2b), each of which segregates a somatic precursor cell and a smaller stem-like cell, which generates the remaining somatic precursor cells and inherits the potential to become a germ cell precursor. This pattern and the general fate of these various somatic precursor cells were studied classically in Ascaris (6,7,22). The somatic precursor cells AB, MSt, E, C, and D (Figure 3) shown in the lineage tree were thought to produce exclusively primary ecto-

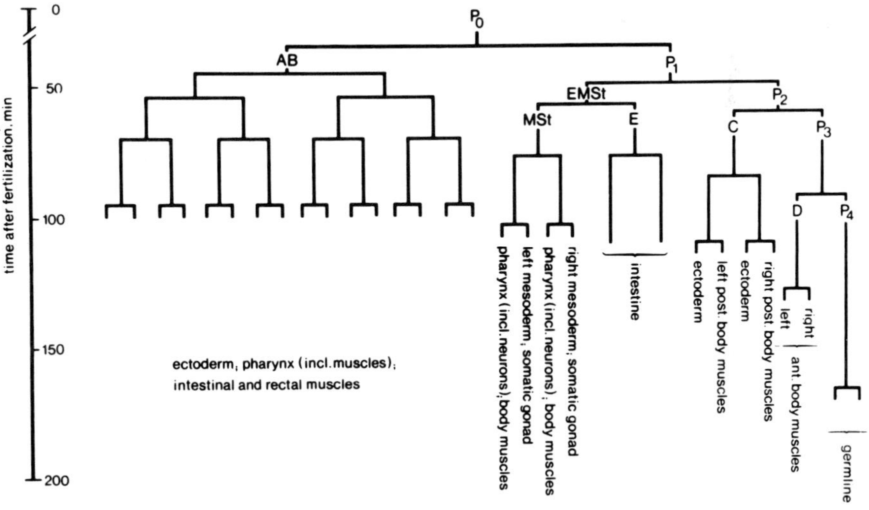

FIGURE 3: Cleavage pattern generating the somatic precursor cells AB, MSt, E, C, D and the germ line precursor P_4, and the eventual fates of the cells in these lineages (23).

derm, mesoderm, endoderm and secondary ectoderm, mesoderm and germ line. (The primary-secondary repeat may correspond to two primitive segments, anterior and posterior (15)). The recent studies (23) show that the classical view was somewhat oversimplified, although basically correct. Specifically, AB cells produce not only ectoderm but also some mesoderm; MSt, conversely, produces some ectoderm, as well as mesoderm. These minority cells may be reprogrammed. Or it may be they are derived from ooplasm near the inital division plane separating AB from MSt cytoplasm and get partitioned to the "wrong" side of the first division membrane. This implies that the division plane is skewed relative to a hypothesized gradient of anterior-posterior determinants determining, in this instance, ectoderm vs. mesoderm. Such a gradient is hypothesized because all the divisions establishing the somatic precursor cells partition the ooplasm roughly orthogonal to the anterior-posterior axis (15), i.e. into anterior-posterior compartments, the posterior cell in each case being the germ line precursor. Once established, each somatic cell lineage, i.e. the precursor and its descendants divide rapidly and rhythmically. The daughters within a lineage maintain division synchrony, independent of their position, as if each lineage inherits a cell cycle clock (1,15). Moreover, the rate of cell division is slower for each lineage, proceeding more or less linearly from anterior to posterior. This could be interpreted as if there were waves of division from anterior to posterior, or more likely as if the clocks in more posterior lineages are progressively slower. As one consequence, the anterior divisions produce many more (also smaller) cells than the posterior lineages. In the initial divisions within a lineage there is apparently further segregation into smaller anterior-posterior compartments, evident from the ultimate anatomical fates of the various sublineages (23,25). This suggests further segregation of anterior-posterior determinants or information. The anatomical decisions involved include other relationships besides anterior-posterior, namely left-right, dorsal-ventral, inside-outside (15,24). Most division axes have some anterior-posterior component. There are two interesting aspects to these decision-making divisions. First there seems to be a polarity relationship so that the anterior daughter in an anterior-posterior division is always determined to be left or dorsal and not right or ventral. Second, there is a sequence in which various types of anatomical decisions are made which is different in different lineages, but similar in the two putative segments of the embryo. For example, the two ectoderm lineages AB and C follow a similar pattern (24). In analyzing the embryonic cellular phenomena, the goal is to reduce the events to individual cell processes, assumed to correspond to the action of individual gene pathways. This then serves as a reference system for observing mutants to try to specify in which pathway a defect has been produced and look for specific changes in the rules or patterns of divisions. Moreover, mutants have been identified in which the patterns of the post-embryonic lineages are altered (26,27). The post-embryonic defects give us clues as to what sorts of embryonic changes may be possible. For example, these post-embryonic alterations occur at the very end of the division program, i.e. in the final divisions.

ISOLATION AND GENETIC ANALYSIS OF TEMPERATURE-SENSITIVE EMBRYONIC ARREST MUTANTS

By analogy with T4 (28) and Drosophila (29), it appeared that temperature sensitive (ts) lethal mutants would be the most useful. Others (3,4) had already successfully isolated such embryonic mutants in C. elegans and we wanted to extend these methods to obtain larger numbers of mutants and thus get more of an overview. Following ethyl methane sulfonate mutagenesis, 10,000 segregated F_2-generation clones were grown up at a permissive temperature of 16°C and replicated to 25°C to look for ts non-growth. From some 300 ts lethal mutants so obtained we were able to identify 36 which accumulate embryos at 25°C and behave genetically as single Mendelian recessives. By complementation analysis we found that the 36 mutants define 30 emb genes, i.e. genes essential for embryogenesis. We also obtained 2 sets of emb mutants from others (3,4). We have complemented all 3 sets of emb mutants against each other and find that they comprise 54 total emb genes, including 9 with multiple alleles (2). The frequency of multiple allele or gene recurrence can be analyzed by Poisson statistics to give a number of 200 total genes essential for embryogenesis isolable by these methods. Although we have tried to correct for the higher mutability of some genes in making this calculation, it is probably still a slight underestimate. Nonetheless it is surprising that such a small number of genes are required for such a complex process as the production of a 550-cell multi-tissue animal. The small number encourages us that genetic saturation for emb genes will occur in the foreseeable future, allowing a better overview of the roles of the individual genes. Classes of genes which occur in multiple, redundant copies or have less than essential functions can perhaps be identified later.

Further genetic mapping has been done (2,17) relative to

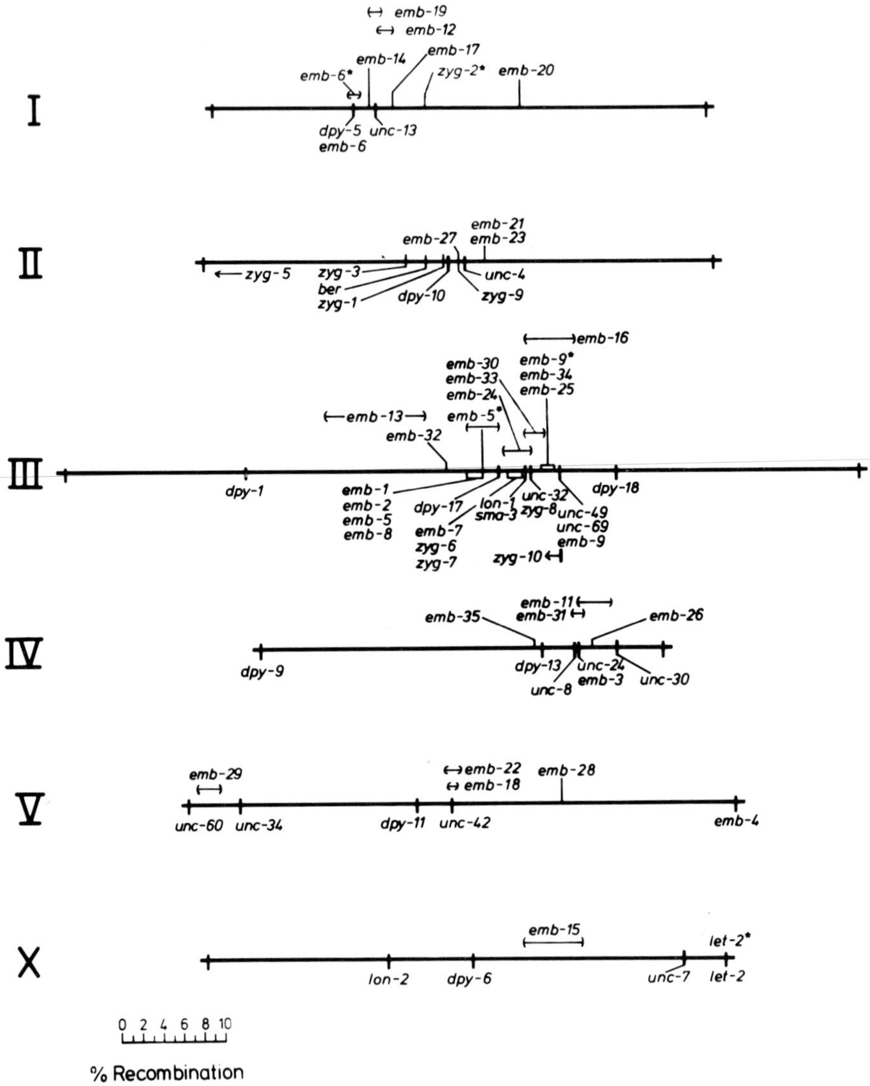

FIGURE 4: Simplified genetic map of C.elegans with all emb genes so far identified, (some called zyg). emb-1 through emb-10, Miwa et al. (4), emb-11 through emb-35, Cassada et al. (2), zyg-1 through zyg-10, Hirsh and Vanderslice (30) and Wood et al. (3), let-2, Meneely and Herman (31). Complementation (2) indicates emb-7 is identical with zyg-4 and emb-9 with zyg-6.

at least 2 different known marker genes with 2-factor tests for distances and 3-factor tests for ordering nearby genes, to produce the map shown in Figure 4. The mapping resolution is less than 1 map unit but still contains many uncertainties indicated by the bars (instead of giving the best current data) where gene order is uncertain. Besides the 25 new emb genes, 20 emb genes mapped by others (3,4) are shown. The emb genes are generally distributed like other genes, dispersed over all 6 linkage groups with considerable central clustering. There is one exception. Within 1 map unit of unc-32 on linkage group III there are 10 emb genes. This pronounced clustering is consistent with some of these 10 genes being adjacent, perhaps due to the presence of a complex locus. To investigate this possibility further, overlapping deletions are being isolated to map this region further. These should help in ordering the known genes and eventually in saturating the region for emb genes.

CELLULAR PROPERTIES OF ts EMB MUTANTS

The properties of the mutants in the 30 genes described above have been studied (32,33) particularly at the cellular level, to try to order them into developmental pathways, which could then be analyzed at the molecular level. Figure 5 shows examples of various classes of arrest stages or terminal phenotypes seen when mutant embryos arrest development. These are comparable to various wild type stages in Figure 2. One mutant stops at the one-cell stage where the pronuclei disappear. Others stop during proliferation, a majority as a rather undifferentiated ball of several hundred cells; either at the end of proliferation or early in morphogenesis and some at the end of morphogenesis. Most mutants undergo visible alterations and appear abnormal, even monstrous, due to the expression of the remaining functional gene products before the absence of one gene function causes arrest. This is especially pronounced for the early morphogenesis arrest mutants. Some of the proliferation-arrest mutants arrest early, others later, and a given mutant may show considerable variability in its arrest stage. One mutant arrests at more or less the 16-cell stage after a normal division pattern. This is interesting because at this stage the divisions by the stem-like posterior cell to produce the precusor founder cells are completed. This may mean different specific gene products (still present in the mutant) are required for these early divisions. Another mutant produces no cells, i.e. has no cytokinesis, but can

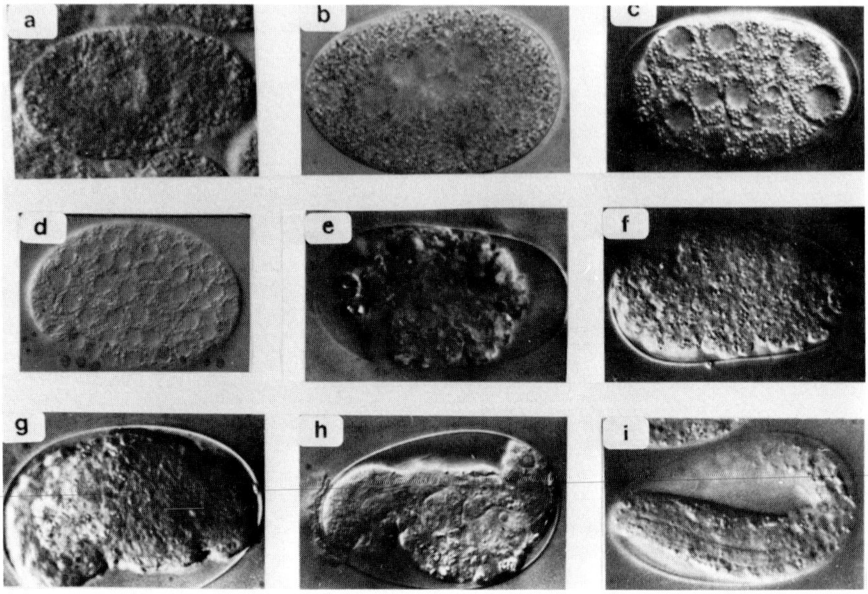

FIGURE 5: Examples of various classes of arrest stages of emb mutants a) one cell with pronuclei (emb-27); b) one-cell, multinuclei (emb-30); c) 16-cell (emb-6); d) variable proliferation with osmotic (eggshell?) defect (emb-14); e) mid-to-late proliferation (emb-23); f) 500-cell late proliferation (emb-13); g) lima bean, early morphogenesis (emb-18); h) morphogenesis, twitching (emb-29); i) pretzel (emb-24).

produce up to 16 nuclei in a more or less normal division sequence. In a frequent arrest class, a flat, soft, osmotically sensitive egg is produced with a visibly defective eggshell, rather than a defect in embryogenesis per se. From the biochemistry of the eggshell from other nematodes we expect a complex structure (with protein, chitin, and lipid-like layers), thus presumably requiring a large number of genes. The arrest stages relative to normal stages for the 30 genes studied are presented in Figure 6. Not shown are the arrest stages for 3 pretzel-arresting mutants isolated, which behaved anomalously genetically, suggesting they might be chromosome rearrangements (not further characterized) (2).

These terminal phenotypes yield considerable information in terms of assigning mutant genes to gene function pathways. Mutants in the same pathway should accumulate embryos at the same stage; those in different pathways, at different stages. Of course, we cannot distinguish between genes in different path-

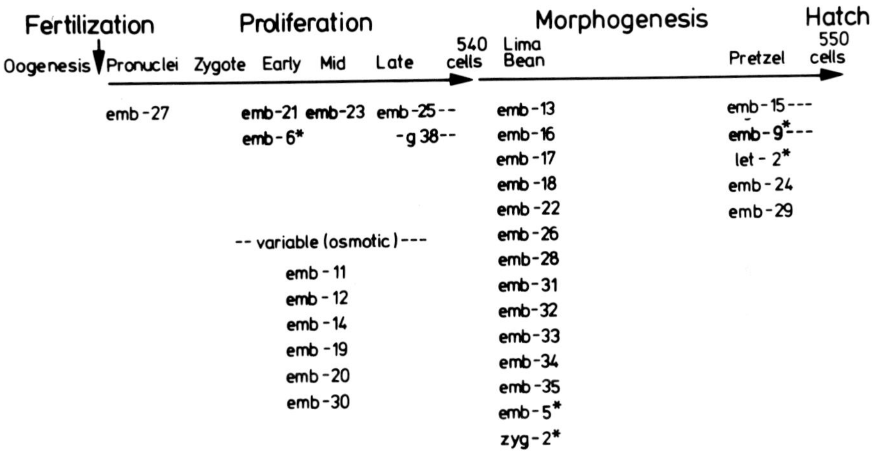

FIGURE 6: Arrest stages of a representative set of emb mutants, compared to normal stages of embryogenesis.

ways for which the arrest stage is similar without more careful examination. The arrest stage suggests possible roles for the genes, i.e. which pathways could possibly be involved. For example, the mutants arresting in proliferation are presumably deficient in pathways needed for cell division. Those arresting in morphogenesis may lack determinants needed for differentiation and cell-cell interactions. Those arresting late may be defective in processes needed for cuticle formation or hatching. Further observation is required to sub-classify each class into different pathways and cell processes and to order genes within one pathway sequentially.

We are following cellular events in detail in the mutants from fertilization onward, so far at least to the 100-cell stage for 27 mutants. All mutants display some visible, fairly specific defect. The maternal mutants (see below) show very early defects, specifically in cell or nuclear membrane formation, rate of mitosis, or the position of the first cleavage plane. Defects in polar body formation and extra pronuclei also have been observed. So far no striking pattern change has been detected, but we expect such defects to arise in later divisions, as discussed. As described above, there seem to be cell cycle clocks in the wild type embryos. Certain mutants show defects in these timing processes (33). In wild-type embryos, the most anterior lineages divide 3 times an hour and the most posterior less then once per hour in earlier divisions, with slowing down in later divisions in all lineages. Most of the mutants are rather similar to wild type in that they have an anterior to posterior gradient of division rates, but most are overall slower than wild type. One mutant is unusual in that it completes many if not all divisions in a fairly normal sequence but with a defective clock, 5-fold slower than wild type. It ends up as a ball of cells, which dies. One mutant divides faster than wild type anteriorly and slower than wild type posteriorly. Still another mutant, conversely, divides slower anteriorly and faster posteriorly relative to wild type and, probably related to this, produces an extra posterior division, namely in the germ line from 2 to 4 cells (not seen in wild-type embryogenesis). In these latter two mutants the plane of the first division is shifted so there is an incorrect proportion of cytoplasm in the daughter cells. The mutant with faster anterior divisions has more anterior cytoplasm, and the mutant with another posterior division has more posterior cytoplasm. This suggests models where the ratio of cytoplasm to nucleus may be determining for cell division rates, perhaps involving histones or histone mRNA (34).

In several mutants, there are rather specific timing defects in the E cells, i.e. endoderm or gut precursors (5,33). The number of cells in the assembled gut at hatching is 20. At the 24-stage, when there are 2 E cells, there is a gastrulation (1,15) in which the E cells (the largest cells present at this stage) migrate ventrally and posteriorly into the interior of the embryo (Figure 2c). After gastrulation (Figure 2d) these cells begin to divide again, eventually giving rise to 20 cells and forming a tube, the gut primordium. This gut primordium may be important for organizing some aspects of the morphogenesis of the embryos. Most notably the pharynx appears to form on the anterior end of the gut. Mutants in 6 genes have been identified in which the timing and/or direction of the early E-cell divisions or the timing of E-cell migration is altered (5,33). Subsequently, the gut forms abnormally in many cases. On the other hand, from the cases where some defects are compatible with normal gut formation, one may infer something about the limits of the normal process. Visible events specific to the E-cells seem to be particularly sensitive to disturbance in mutants. If the 2 E cells divide before they migrate in, they have difficulty ever getting inside. Even, if the axes of the divisions are incorrectly oriented, rotation or migration can occur so that a normal gut forms in some gastrulation mutants.

The gut cells, after the 8 E-cell stage, display tissue specific autofluorescence due to the accumulation of tryptophan catabolites in small granules also visible as birefringent granules under polarized light (35). One might expect to find defects in tissue-specific gene product expression in the emb mutants. Therefore mutant embryos developing at the restrictive high temperature have been examined for the presence of this specific fluorescence, looking for mutants in which either the E cells fail to produce the characteristic fluorescence (the requisite enzyme or substrate is absent) or in which other cell types spuriously fluoresce. Neither of these alterations has been discovered so far; in all mutants examined the E cells and only the E cells fluoresce. Using for example fluorescent antibodies to tissue-specific molecules, one can look for the timing and patterns of tissue-specific gene expression, and potentially detect pattern mutants. Recessive non-lethal mutants have been isolated (35) in which the gut fluorescence is altered in color and intensity. Using these mutants it has been possible to demonstrate the induction of genetic mosaics in the gut, following X-irradiation (36). In these mosaics, part of the gut is phenotypically normal and part is mutant. Using such mosaic techniques we can look at the emb mutants to try to identify the tissue focus of the defect.

MODE OF GENE EXPRESSION AND TEMPERATURE SENSITIVE PERIODS IN EMB MUTANTS

Finally, several tests have been done to ask about the mechanism of gene expression in the 30 emb genes described here. First a series of 3 progeny tests (37,4) has been used to determine the mode of gene expression. By determining the fates of progeny starting with various configurations of wild type and mutant cytoplasm and nuclei, one can determine the sufficiency (and necessity) of maternal, zygotic and paternal wild type expression for embryo survival. These tests sort the 30 genes studied into 4 classes (32) as a) strict maternal, i.e. maternal necessary and sufficient, b) facultative i.e. either maternal or zygotic sufficient, c) both required or d) strictly zygotic. No strict paternal rescue mutant was found but one facultative maternal mutant showed partial paternal rescue. Most genes, as mentioned above, are strictly maternal. Among the 18 genes of this class are the eggshell genes, as might be expected. For 7 genes, either maternal or zygotic wild type expression is sufficient. It seems reasonable that for such genes in the normal wild-type homozygote the mother supplies most of the product, at least during embryogenesis. As already discussed, it is not surprising that most of the genes essential for embryogenesis are maternal since a) embryogenesis is very rapid, b) there is no increase in mass during embryogenesis, and c) everything appears very determinate and strictly reproducible.

Shifts up and down have also been done (32) to determine the normal and defective execution stages (4), respectively for these 30 genes. These can also be thought of as the end and the beginning of the usually given temperature-sensitive period. These periods have been related to the maternal tests and arrest stages of phenotypes (2,32). Examples of 9 genes from the various classes are shown in Figure 7. Some of the

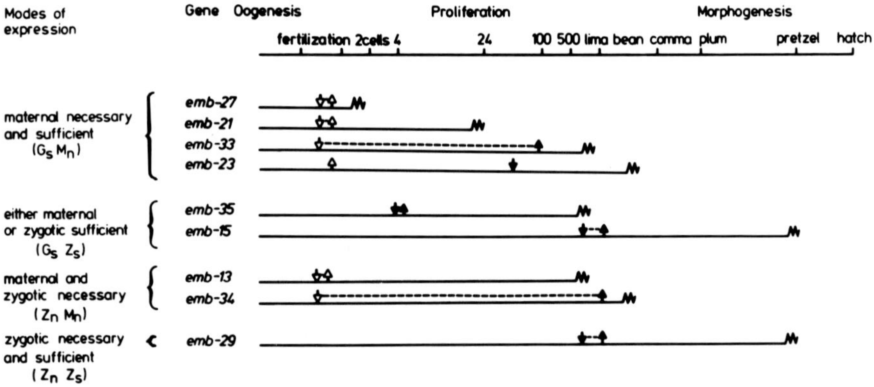

FIGURE 7: Examples of various classifications of emb mutants as to mode of expression, terminal phenotype (W) and temperature-sensitive periods (TSP) (↑ end of TSP from shift up, ↓ beginning of TSP from shift down; open arrow indicates a possible, earlier time, not yet determined).

maternally acting genes have their temperature-sensitive period wholy before (or at) fertilization, although they arrest much later. Other maternal mutants have temperature-sensitive periods in embryogenesis, and these periods can be surprisingly long in duration, suggesting the maternal gene product is thermolabile and/or has not acted at times very long after fertilization. Most of the maternal mutants arrest during or just at the end of the proliferation phase (2,32), in agreement with the results of others (4,3). The zygotic mutants stop late and have later, shorter temperature-sensitive periods. One maternal mutant (emb-23) has the end of the temperature-sensitive period apparently before the beginning. This can be interpreted (30) to mean that a short time at the permissive temperature at any time during the long period between the arrows is enough to allow sufficient wild-type gene function for successful embryogenesis. Another mutant (emb-13) has the temperature-sensitive period at or before fertilization even though zygotic gene function is required. To explain this result, one could speculate that the oocyte (prezygotic) genome may already be functioning before fertilization, i.e. during oogenesis. Those mutants arresting at the beginning of morphogenesis have been examined for detectable muscle twitching. As indicated, about half the mutants arresting here show twitching, indicating that muscle differentiation has occurred, a characteristic of wild-type development at this early morphogenesis phase. Continuing the lineage studies beyond the 100-cell stage may reveal visible cellular defects which can be correlated to the temperature-sensitive period and the eventual fate of the mutant embryo. Molecular studies on the mutants should help to determine the extent of normal differentiation, for example the role of histone gene expression (34,38) in embryogenesis.

In summary, we have investigated a simple embryonic system involving a small number of cells (550) and a small number of genes (200). Many of the events in wild-type embryogenesis have already been observed. We have isolated a representative collection of ts embryonic arrest mutants in 30 genes. Characterization of these mutants genetically and phenotypically helps to elucidate the role of maternal and zygotic gene expression in each of the developmental steps visible at the cellular level. Future studies of differentiation and cell-cell interaction using, for example, monoclonal antibodies should bring more understanding of the molecular bases of these cellular processes and their genetic control.

ACKNOWLEDGMENTS

For capable and conscientious technical assistance we thank Ingrid Ostermeyer. We especially thank Johji Miwa for valuable advice, experimental help, sharing results before publication, and providing mutants prior to publication. We thank David Hirsh and Bill Wood for helpful discussions, sharing results before publication, and providing mutants. We thank Bob Edgar and John Sulston for helpful discussion. We also thank Mechthild Ziemer for photography and Brigitte Knoke for typing. E.I. was supported by the Deutsche Forschungsgemeinschaft. K.D. and K.R. are the Max-Planck-Society pre-doctoral and post-doctoral research fellows, respectively.

REFERENCES

1. Deppe, U., Schierenberg, E., Cole, T., Krieg, C., Schmitt, D., Yoder, B., von Ehrenstein, G. (1978). Proc.Nat.Acad. Sci. USA, 75, 376.
2. Cassada, R., Isnenghi, E., Culotti, M., von Ehrenstein, G. (1981). Develop.Biol., 84, 000.
3. Wood, W.B., Hecht, R., Carr, S., Vanderslice, R., Wolf, N., Hirsh, D. (1980). Develop.Biol. 74, 446.
4. Miwa, J., Schierenberg, E., Miwa, S., von Ehrenstein, G. (1980). Develop.Biol. 76. 160.
5. Schierenberg, E., Miwa, J., von Ehrenstein, G. (1980). Develop.Biol. 76, 141.
6. Boveri, T. (1899). Die Entwickelung von Ascaris megalocephala mit besonderer Rücksicht auf die Kernverhältnisse. in: Festschrift Kupffer, p. 383. G. Fischer, Jena.
7. zur Strassen, O. (1896). Arch.Entw.Mech. 3, 133.
8. Brenner, S. (1973). Brit.Med.Bull. 29. 269.
9. Brenner, S. (1974). Genetics 77, 71.
10. Edgar, R.S., and Wood, W.B. (1977). Science 198, 1285.
11. Riddle, D. (1978). J.Nematol. 10, 1.
12. Sulston, J.E., and Horvitz, M.R. (1977). Develop.Biol. 56, 110.
13. Hirsh, D., Oppenheim, D., and Klass, M. (1976). Develop. Biol. 49, 200.
14. Kimble, J., and Hirsh, D. (1979). Develop.Biol. 70, 396.
15. Schierenberg, E. (1978). Die Embryonalentwicklung des Nematoden Caenorhabditis elegans als Modell. Ph.D. Thesis, Göttingen.
16. Krieg, C., Cole, T., Deppe, U., Schierenberg, E., Schmitt, D., Yoder, B., and von Ehrenstein, G. (1978). Develop. Biol. 65, 193.

17. Herman, R.K., and Horvitz, H.R. (1980). in: "Nematodes as Biological Models" (B.M. Zuckerman, ed.) Vol. 1, p. 227. Academic Press, New York.
18. Herman, R.K., Horvitz, H.R., and Riddle, D.L. (1980). in "Genetic Maps" (S.J. O'Brien, ed.) Vol. 1, p. 183. Laboratory of Viral Carcinogenesis National Cancer Institute NIH, Bethesda.
19. Hodgkin, J. (1974). Genetic and anatomical aspects of the C.elegans male. Ph.D. Thesis, Darwin College, University of Cambridge, Cambridge, England.
20. Ward, S., and Carrel, J.S. (1979). Develop.Biol. 73, 304.
21. Sulston, J., and Brenner, S. (1974) Genetics 77, 95.
22. zur Strassen, O. (1959). Zoologica 107, 1.
23. von Ehrenstein, G., Sulston, J.E., Schierenberg, E., Laufer, J.S., and Cole, T. (1981). in: "International Cell Biology 1980-81" (H.G. Schweiger, ed.), p. 522. Springer Verlag, Berlin.
24. von Ehrenstein, G., and Schierenberg, E. (1980). in "Nematodes as Biological Models" (B.M. Zuckerman, ed.), Vol. 1, p. 1. Academic Press, New York.
25. Sulston, J., and Schierenberg, E., unpublished.
26. Albertson, D., Sulston, J., and White, J. (1978). Develop. Biol. 63, 165.
27. Horvitz, H.R., and Sulston, J.E. (1980). Genetics 96, 435.
28. Edgar, R.S., and Lielausis, I. (1964). Genetics 49, 649.
29. Suzuki, D.T., Kaufman, T., Falk, D., and the U.B.C., Drosophila Research Group (1976). in "The Genetics and Biology of Drosophila" (M. Ashburner and E. Novitski, eds.), Vol. 1a, p. 207. Academic Press, New York.
30. Hirsh, D., and Vanderslice, R. (1976). Develop.Biol. 49, 220.
31. Meneely, P.M., and Herman, R.K. (1979). Genetics 92, 99.
32. Isnenghi, E., Cassada, R., Smith, K., Rhadnia, K. and von Ehrenstein, G., unpublished.
33. Denich, K., Schierenberg, E., Isnenghi, E., Cassada, R., unpublished.
34. von Ehrenstein, G., Schierenberg, E., and Miwa, J. (1979). in "Cell Lineage, Stem Cells and Cell Determination" (N. Le Douarin, ed.) INSERM Symposium No. 10, p. 49. Elsevier/North Holland, Amsterdam.
35. Siddiqui, S.S., and Babu, P. (1980). Mol.Gen.Genet. 139, 21.
36. Siddiqui, S.S., and Babu, P. (1980). Science 210, 330.
37. Hirsh, D., Wood, W.B., Hecht, R., Carr, S., and Vanderslice, R. (1977). in "Molecular Approaches to Eucaryotic Genetic Systems" (J.N. Abelson, G. Wilcox, and C.F. Fox, eds.), ICN-UCLA Symposia on Molecular and Cellular Biology, Vol. 8, p. 347. Academic Press, New York.
38. Certa, U., Cassada, R., and von Ehrenstein, G., this Symposium.

Genes Coding for Metal-Induced Synthesis of RNA Sequences Are Differentially Amplified and Regulated in Mammalian Cells

Ronald A. Walters[1]
M. Duane Enger
Carl E. Hildebrand
Jeffrey K. Griffith

Genetics Group
Los Alamos National Laboratory
Los Alamos, NM 87545

I. ABSTRACT

We have isolated three variant cell lines which survive cadmium (Cd^{++}) concentrations 10-200 fold greater than that which kills parental Chinese hamster cells (line CHO). Cd^{++} treatment of the variants induces the synthesis of a highly abundant poly A^+ RNA class which directs the synthesis of metallothionein in a cell free translation system. Hybridization of cDNA complementary to these inducible, highly abundant RNA sequences ($cDNA_a$) with RNA from variant cells showed that: (i) the induced abundant class has a total complexity of ∽2000 NT; (ii) Cd^{++} induction increases the cellular concentration of these sequences ∽2000-fold above preinduction levels in each of the variants; (iii) most, if not all, of these sequences are expressed constitutively in uninduced cells. Cd^{++} induction of sensitive CHO cells increases the cellular concentration of only a subset of the sequences inducible in resistant cells and then only to a level 100 fold higher than in uninduced cells; the remainder of the sequences could not be induced to a measureable extent. In addition, only ∽50% of the sequences are constitutively expressed at measureable levels in uninduced CHO cells. Hybridization of $cDNA_a$ with genomic DNA from the three resistant variants showed that genes coding for the induction specific RNA sequences are amplified ∽10-fold in Cd^r20F4 cells, ∽4 fold in Cd^r30F9 cells, and unamplified in Cd^r2C10 cells relative to CHO. While sensitive CHO cells can tolerate only 0.2μM Cd^{++}, Cd^r30F9, Cd^r20F4, and Cd^r2C10 cells are resistant to 40μM, 26μM, and 2μM Cd^{++} respectively. Thus, gene amplification alone cannot be responsible for the observed resistance of the variant cell lines.

1 Supported by the U.S. Department of Energy

ISBN 0-12-137420-3

II. INTRODUCTION

Cadmium is a toxic metal that persists in the environment, and the level of Cd^{++} to which individuals in industrialized nations have been exposed has already resulted in in vivo body burdens less than an order of magnitude lower than that known to produce overt toxicity (1). Although the mechanism(s) by which a cell or organism ameliorates the cytotoxic effect of cadmium is not known, it is thought that the synthesis of metallothioneins (MT), small metal binding proteins, may play an important role by sequestering Cd^{++} in a non-toxic form (2-3).

In order to define the role(s) played by cellular processes in cadmium detoxification, the molecular events associated with cadmium exposure have been examined in Cd^{++} resistant and Cd^{++} sensitive Chinese hamster cells. Three cadmium resistant variants have been selected by stepwise culture in increasing Cd^{++} concentrations. Compared to the parental cell (line CHO), each of the resistant variants (i) has a higher cadmium toxic threshold, (ii) can synthesize more inducible MT and (iii) can accumulate more inducible translatable MT mRNA (4-5). While these results are consistent with a role for MT in the acquisition of a cadmium resistant phenotype, they reveal very little of the underlying molecular mechanisms responsible for Cd^{++} resistance. We have initiated studies designed to probe cellular responses to Cd^{++} treatment and report here that Cd^{++} treatment of the resistant variants induces the synthesis of a highly abundant poly A^{+} RNA class, a major portion of which is metallothionein mRNA. Not only is this RNA class differentially regulated in resistant variant cells but the structural genes encoding these inducible RNAs are differentially amplified in the resistant variants. However, there is not a direct correlation between the degree of gene amplification and resistance. Although the acquisition of Cd^{++} resistance may be a considerably more complicated phenomenon than originally expected, these resistant variants provide models useful for the study of both regulation of inducible gene function and the factors responsible for Cd^{++} detoxification.

III. METHODS

The conditions of cell culture, derivation of resistant variants, and properties of the three resistant variant Chinese hamster cell lines Cd^{r}30F9, Cd^{r}20F4, and Cd^{r}2C10 have been described elsewhere (3,5). In none of the variants derived from the parental CHO cell was cadmium resistance due to a failure of cells to transport cadmium from the extracellular medium. In each of the experiments reported here, Cd^{++} induction conditions were those previously shown to result in maximum production of translatable metallothionein mRNA: Cd^r30F9 cells - 40μM $CdCl_2$ for 4h; Cd^r20F4 cells - 20μM $CdCl_2$ for 4h; Cd^r2C10 cells - 2μM $CdCl_2$ for 8 h; CHO cells - 2μM $CdCl_2$ for 11 h (5).

We have previously described the conditions for poly A^+ RNA extraction, synthesis of ^{3}H-labeled cDNA (~7×10^7 cpm/μg), cDNA-poly A^+ RNA hybridization, and the S1 nuclease assay for cDNA-RNA hybrid formation (6). All values of Rot and Cot were corrected to the standard salt concentration (7).

A tracer ($cDNA_a$) complementary to Cd^{++} inducible, highly abundant poly A^+ RNA was prepared by reacting cDNA copied from induced Cd^r20F4 poly A^+ RNA to an ERot of 210 with a 1000-fold mass excess of poly A^+ RNA from uninduced CHO cells. The unreacted fraction was isolated from hydroxylapatite (HAP) and then reacted with a 10,000-fold mass excess of poly A^+RNA from induced Cd^r20F4 cells to ERot of 0.07. The double strand fraction was isolated from HAP, hydrolyzed with 0.3N NaOH (18h, 37°), dialyzed, and precipitated with ethanol and tRNA carrier.

Genomic DNA was isolated from nuclei of the same cells from which the cytoplasmic poly A^+ RNA was obtained. Nuclei were prepared as previously described (6), and DNA was isolated by the method of Kedes <u>et al</u>. (8) winding the DNA out of ethanol at the final step. The DNA was sheared by homogenization to a 300 nucleotide (NT) double strand length. For hybridization, aliquots were prepared which contained 600 μg genomic DNA, 7 μg ^{14}C-labeled genomic DNA (cells grown in 0.0075 μCi/Ml ^{14}C-thymidine for 3 cell doublings), and 13 pg of ^{3}H-$cDNA_a$. Samples were melted and annealed at 68° in 10 mM Tris (pH 7.5) - 1.5M NaCl - 0.2% sodium dodecyl sulfate - 2mM EDTA, and hybrid formation was assayed by HAP chromatography.

IV. RESULTS

We recently showed that the poly A^+ RNA from Cd^{++} induced Cd^r20F4 cells contains a highly abundant RNA class undetectable in uninduced CHO cells (9). We have exploited this difference in abundance to isolate from total induced Cd^r20F4 cDNA a tracer ($cDNA_a$) complementary to induction specific RNA sequences (see "Methods" for additional detail). A comparison of the hybridization to induced Cd^r20F4 RNA of $cDNA_a$ and total Cd^r20F4 cDNA is shown in Fig. 1A. The $cDNA_a$ hybridized, with an $ERot_{1/2}$ of 0.04, to the most abundant RNA class which comprises ~5% of the total induced Cd^r20F4 poly A^+ RNA. Comparison of the data in Fig. 1A with that

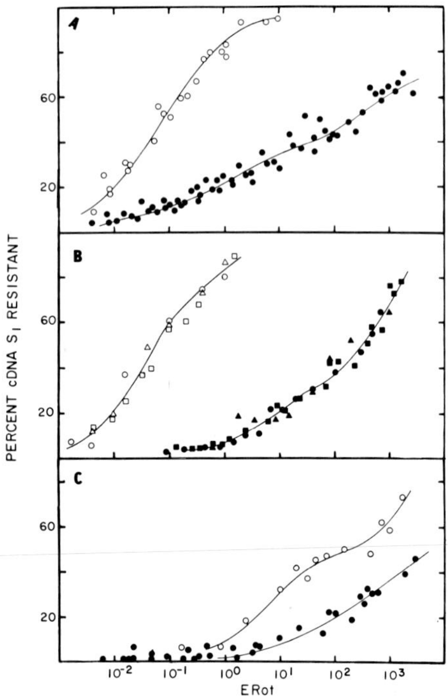

Figure 1. Hybridization of cDNA to poly A^+ RNA. (A) Hybridization to Cd^{++} induced Cd^r20F4 RNA of total Cd^r20F4 cDNA (closed circles) and $cDNA_a$ (open circles). (B) Hybridization of $cDNA_a$ to RNA from Cd^r30F9 (circles), Cd^r20F4 (triangles), and Cd^r2C10 (squares) which had been either Cd^{++} induced (open symbols) or not induced (closed symbols) prior to RNA extraction. (C) Hybridization of cDNAa to RNA from Cd^{++} induced (open circles) or uninduced (closed circles) CHO cells.

measured earlier (6) for the hybridization of the kinetic standard chicken globin cDNA with its template ($ERot_{1/2}$ of 0.0018, 1820NT complexity) indicates that the total complexity of the induced class is ∿2000NT. It will be noted that, at an ERot of 1, more than 80% of $cDNA_a$ is hybridized to induced Cd^r20F4 RNA while less than 4% is hybridized to uninduced CHO RNA (Fig. 1C) and that the concentration of RNA sequences complementary to $cDNA_a$ is at least 10,000 fold higher in induced Cd^r20F4 than uninduced CHO RNA. Furthermore, the shape of the $cDNA_a$-induced Cd^r20F4 RNA hybridization curve and its span of ERot indicate that the RNA sequences complementary to $cDNA_a$ are present at nearly equal concentrations. We have cloned and determined the nucleotide sequence of DNA molecules complementary to $cDNA_a$. As will be reported elsewhere, the nucleotide sequence of one of the cloned DNAs is consistent with that predicted from the published amino acid sequence of mammalian metallothionein from other species (2). The extent to which the synthesis of RNA sequences complementary to $cDNA_a$ could be induced by cadmium was measured in each of the three cadmium-resistant variants. Each cell type was exposed to cadmium under conditions known to induce the maximal synthesis of translatable metallothionein mRNA. As illustrated in Fig. 1B, there is little difference among the three resistant variants in either the constitutive level or the maximal level to which the RNA sequences complementary to $cDNA_a$ could be induced by cadmium treatment. Comparison of the $ERot_{1/2}$ of the $cDNA_a$ reactions with RNA from induced and uninduced cells indicates that cadmium treatment induces an ∼2000-fold increase in the concentration of these sequences. The curve defining the reaction of $cDNA_a$ with RNA from uninduced cells (Fig. 1B) appears to show some biphasic character. If the biphasic character is real, the $ERot_{1/2}$ of the total reaction would slightly overestimate the relative constitutive concentration of the majority of the RNA sequences.

The reaction of $cDNA_a$ with RNA from cadmium induced and uninduced sensitive CHO cells is shown in Fig. 1C. In contrast to the cadmium resistant variant cells (Fig. 1B), the curve defining the reaction of $cDNA_a$ with induced CHO RNA is clearly biphasic. Only ∼50% of the sequences complementary to $cDNA_a$ could be induced to an appreciable extent and then only to a maximum of 100 fold higher than the constitutive concentration. The remainder of the sequences hybridized with kinetics similar to that of the least abundant RNA (see Fig. 1A). It should be noted that the concentration of the subset of sequences induced in CHO is still $<1\%$ that in the induced variant cells. Compared to the resistant variants, the constitutive concentration of these sequences in CHO is at least 10 fold lower at the point where hybridizations were terminated due to thermal instability of the RNA.

One way by which the resistant variant cells could attain an increased capacity for the synthesis of induction specific RNA sequences is by amplification of their respective structural genes. We tested for this possibility by annealing $cDNA_a$ to genomic DNA isolated from nuclei of each of the cell types. To be absolutely certain that the results could not be compromised by different rates of annealing of driver DNA, each hybridization mixture contained, as an internal standard, ^{14}C-labeled genomic DNA; thus, the rates of annealing of both driver genomic DNA and $cDNA_a$ were assayed simultaneously. As shown in Fig. 2A, the rates of hybridization of genomic DNA from each of the cell types were experimentally

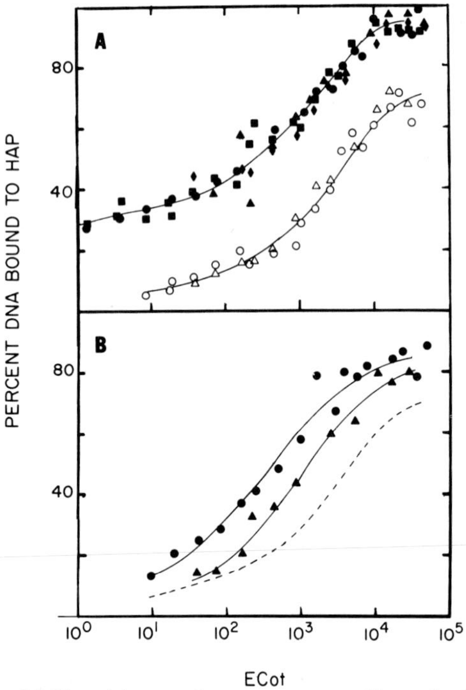

Figure 2. Amplification of genes encoding induction specific RNA sequences. (A) Closed symbols represent annealing of ^{14}C-labeled genomic DNA to driver DNA isolated from Cd^{r}30F9 (circles), Cd^{r}20F4 (squares), Cd^{r}2C10 (triangles), and CHO cells (diamonds). Open symbols represent annealing of cDNA$_a$ to genomic DNA from Cd^{r}2C10 (circles) and CHO cells (triangles). (B) cDNA$_a$ was annealed to genomic DNA from Cd^{r}20F4 (circles) and Cd^{r}30F9 cells (triangles). For comparison, the dashed line shows the annealing of cDNA$_a$ to DNA from Cd^{r}2C10 and CHO cells and is reproduced from panel A.

indistinguishable. cDNA$_a$ was hybridized to genomic DNA from CHO, Cd^{r}2C10 (Fig. 2A), Cd^{r}20F4, and Cd^{r}30F9 cells (Fig. 2B). The kinetics of reaction of cDNA$_a$ to CHO DNA and Cd^{r}2C10 DNA were experimentally indistinguishable and were similar to that of the single copy component of genomic DNA (Fig. 2A). However, cDNA$_a$ reacted with Cd^{r}20F4 DNA and Cd^{r}30F9 DNA 10 fold and 4 fold faster, respectively, relative to CHO DNA (Fig. 2B). These data show that, relative to CHO, the structural genes coding for induction specific RNA sequences are amplified 10-fold and 4-fold, respectively, in the genomes of Cd^{r}20F4 and Cd^{r}30F9 and are unamplified in the genome of Cd^{r}2C10. Since cDNA$_a$ annealed to CHO DNA with kinetics similar to that of single copy DNA, the respective structural genes must be present at no more than a few copies per genome in cadmium sensitive CHO cells from which the cadmium resistant variants were derived. Gene amplification is likely not the result of formation of high heteroploid cells during variant selection since the total DNA content of resistant variant cells is the same as that of parental CHO cells based on flow microfluorometric DNA analysis (data not shown).

V. DISCUSSION

Chronic exposure of sensitive cells to increasing concentrations of particular toxic agents can result in the acquisition of specific resistant phenotypes. Alt et al. (10) and Wahl et al. (11) have shown that mammalian cells resistant to methotrexate and N-(phosphonacetyl)-2-aspartate could be derived by continuous culture in progressively increasing concentrations of the drugs. In each case, resistance was accompanied by (i) an increased production of the enzyme which ameliorated the cytotoxic effects of the drug, (ii) an increased production of the mRNA coding for the enzyme, and (iii) amplification of the structural gene encoding the enzyme to an extent sufficient to completely account for the increased production of the respective mRNA. In some respects, the scenario developed for the acquisition of a cadmium resistant phenotype displayed by the Chinese hamster variants studied here is similar to that outlined above for acquisition of resistance to methotrexate and N-(phosphonacetyl)-2-aspartate. In each case, the cadmium resistant variant cell can synthesize more of the protein, metallothionein, putatively associated with amelioration of cadmium cytotoxicity (5) and can accumulate a minimum of 100 fold higher concentrations of poly A^+ RNA sequences (which include the metallothionein mRNAs) induced by Cd^{++} treatment (Fig. 1B) when compared to the sensitive parental CHO cell (Fig. 1C); in two of the three resistant variants, the structural genes encoding the induction specific RNAs have been amplified.

However, for each of the properties the cadmium resistant variants share with methotrexate and N-(phosphonacetyl)-2-asparatate resistant variants, there are other important properties that are quite different. These include the observations that: (i) each of the cadmium resistant variants synthesize approximately the same amount of induction specific sequences when maximally induced (Fig. 1B); (ii) while the induced concentration of these sequences in each of the resistant variants is $\sim$2000-fold higher than the constitutive level, the structural genes encoding these RNA sequences are not amplified in Cd^r2C10 (Fig. 2A); (iii) differences in the extent of gene amplification, 10 fold and 4 fold, in Cd^r20F4 and Cd^r30F9 cells, respsectively, are not accompanied by similar differences in the ability to synthesize Cd^{++} induced RNA sequences or in resistances to Cd^{++}; (iiii) although sensitive CHO and resistant Cd^r2C10 cells each have the same number of structural genes, CHO cells can synthesize only a subset of the induction specific RNAs (Fig. 1C) and then to a level <1% of that in Cd^r2C10 cells.

Since sensitive parental CHO cells can tolerate <0.2μM Cd^{++} while the independently derived Cd^r30F9, Cd^r20F4, and Cd^r2C10 cells are resistant to 40μM, 26μM, and 2μM Cd^{++}, respectively, it would appear that neither gene amplification nor the ability to synthesize induction specific RNAs is directly responsible for the acquisition of a Cd^{++} resistant phenotype. Nonetheless, this should not be interpreted to suggest that either of these parameters are unimportant and play no role in resistance. Certainly each of the resistant variants is more proficient than CHO in the synthesis of both metallothionein and induction specific RNA sequences after cadmium treatment (Fig. 1B). Further, the induction specific RNA sequences are differentially regulated in the resistant variants and sensitive CHO (Fig. 1C). It may be that other factors also contribute to heavy metal detoxification and resistance (5, 12-13). Each of these factors may be necessary but not alone sufficient for acquisition of resistance. Resolution of the role played by each of the factors will be an important step in understanding heavy metal detoxification.

ACKNOWLEDGMENTS

We are pleased to acknowledge the expert technical assistance of J. L. Hanners, J. G. Tesmer, and B. B. Griffith.

REFERENCES

1. Piscator, M., in "Trace Elements in Human Health and Disease, Essential and Trace Elements" (A.S. Prasad, ed.), p. 431. Academic Press, New York (1976).

2. Kägi, J., in "Separatum Aus: Metallothionein" (J. Kägi and M. Nordberg, eds.) p. 42. Experientia Supplementum 34 (1979).

3. Hildebrand, C.E., Tobey, R.A., Campbell, E.W., and Enger, M.D. (1979) <u>Exptl. Cell Res.</u> 124, 237.

4. Enger, M.D., Griffith, B.B., and Hildebrand, C.E. (1980), <u>J. Cell Biol.</u> 87, 270a.

5. Enger, M.D., Ferzoco, L.T., Tobey, R.A., and Hildebrand, C.E., J. Toxicol. Environ. Health (in press).

6. Walters, R.A., Yandell, P.M., and Enger, M.D. (1979) Biochemistry 18, 4254.

7. Britten, R.J., Graham, D.E., and Neufeld, B.R. (1974) Methods Enzymol. 29(Part E) 363.

8. Kedes, L.H., Cohn, R.H., Lowry, J.C., Chang, A.C.Y., and Cohn, S.N. (1975) Cell 6, 359.

9. Walters, R.A., Hildebrand, C.E., Enger, M.D., and Griffith, J.K. (1980) J. Cell Biol. 87, 271a.

10. Alt, F.W., Kellems, R.E., Bertino, J.R., and Schimke, R.T. (1978) J. Biol. Chem. 253, 1357.

11. Wahl, G.M., Padgett, R.A., and Stark, G.R. (1979) J. Biol. Chem. 254, 8679.

12. Chen, R.W. and Ganther, H.E. (1975) Environ. Physiol Biochem. 5, 378.

13. Levinson, W., Opperman, H., and Jackson, J. (1980) Biochim. Biophys. Acta 606, 170.

METALLOTHIONEIN-I GENE AMPLIFICATION IN CADMIUM-RESISTANT MOUSE CELL LINES

Larry R. Beach
Kelly E. Mayo
Diane M. Durnam
Richard D. Palmiter

Howard Hughes Medical Institute Laboratory
Department of Biochemistry
University of Washington
Seattle, Washington

ABSTRACT

Four murine cell lines were selected in a stepwise manner for resistance to the toxic metal Cd. Because the heavy-metal binding protein metallothionein (MT) has been implicated in metal detoxification, we measured the amount of MT mRNA, the rate of MT gene transcription and the number of MT genes in these Cd^R cell lines. Three of the cell lines have expressible MT genes; selection for Cd-resistance in each of these results in increased MT mRNA production as well as MT gene amplification. The fourth cell line does not express or amplify its MT genes in response to Cd.

INTRODUCTION

Resistance to a wide variety of cytotoxic compounds has been obtained in cell culture by gradually increasing the concentration of the toxin over a period of many cell generations. This selection scheme frequently results in increased production of the target enzyme(s) affected by the toxin. In those cases that have been analysed with specific gene probes, there is a corresponding amplification of both

ISBN 0-12-137420-3

the mRNA and the genes coding for the affected protein. The first examples of gene amplification arising from this type of gradual selection were obtained with specific inhibitors of dihydrofolate reductase (DHFR)(1) and aspartate transcarbamylase (2).

Cells resistant to Cd have been obtained using a similar selection scheme (3,4). These cells have elevated levels of metallothionein (MT). Unlike the examples mentioned above, MT is not an enzyme but a small cysteine-rich protein that sequesters Cd, thereby protecting other proteins from inactivation by Cd (5). To study the regulation of MT genes, we isolated clones containing the mouse metallothionein-I (MT-I) gene (6) and used specific probes derived from these clones to show that the MT-I gene is regulated at the transcriptional level by heavy metals and by glucocorticoids (7-9). We also used these probes to show that Cd resistant (Cd^R) Friend erythroleukemia cells produce elevated amounts of MT-I mRNA and have amplified MT-I genes (10). More recently, we have analyzed three additional Cd^R cell lines. In this report, we compare and contrast these lines and the parental lines from which they were derived (Cd^S), focusing upon the amplification and expression of the MT-I genes in these cells.

Table 1. Comparison of Cd^S and Cd^R cells.

Cd-sensitive (S) or Cd-resistant cells that were withdrawn from Cd for 1 to 2 weeks (R) were used for each assay. For the ED_{50} assay, cells were incubated with varying concentrations of Cd for 14 hr and then labeled for 1 hr with ^3H-thymidine. The concentration of Cd that inhibited incorporation by 50% is indicated (10). Relative MT-I gene number was determined as indicated in Figure 2 and reference 10. The number of MT-I genes in parental cell lines (S) was assumed to be 2 for each cell line. The MT-I gene number in Cd^R FE and hepa 1A cells takes into account the increased DNA content of these cells relative to the parental cell lines. Measurements of MT-I gene transcription and mRNA accumulation were made from cells induced for 1 hr and 8 hr, respectively (10). The dexamethasone (Dex) concentration used was 100 nM in all cell lines; the Cd concentration used varied as follows: FE^S, 30 μM; FE^R, 60 μM; hepa $1A^S$, 30 μM; hepa $1A^R$, 150 μM; $S180^S$, 20 μM; $S180^R$, 120 μM; $W7^S$, 10μM and $W7^R$, 10μM.

Table 1. Comparison of Cd^S and Cd^R Cells

			FE	hepa 1A	S180	W7
ED_{50} *(μM Cd)*	S		12	25	8	11
	R		40	150	30	22
MT-I gene *(number per cell)*	S		2	2	2	2
	R		12	60	20	2
Chromosome number per cell *Average (range)*	S		40 (38-43)	58 (56-64)	70	40
	R		70 (65-217)	99 (82-117)	70	40
Extrachromo-somal elements *Average (range)*	S		0	0	0	0
	R		3 (2-9)	85 (41-150)	0	0
MT-I gene transcrip-tion *(ppm)*	S	Control	40	25	10	0
		Dex	60	30	80	0
		Cd	170	130	135	0
	R	Control	80	95	110	0
		Dex	170	95	270	0
		Cd	915	575	1370	0
MT-I mRNA *(molecules per cell)*	S	Control	250	30	50	0
		Dex	850	60	450	0
		Cd	1500	1500	1300	0
	R	Control	730	1740	450	0
		Dex	1530	2580	1000	0
		Cd	7220	33400	6800	0

RESULTS AND DISCUSSION

Cd Resistance

Four murine cell types were selected for Cd resistance. Hepa 1A (hepatoma), S180 (sarcoma), W7 (thymoma) cells were adapted for growth in 70 μM, 30 μM, and 10 μM $CdSO_4$, respectively. Friend erythroleukemia cells (FE, proerythroid) were selected for resistance to 145 μM $CdSO_4$ plus 80 μM $ZnSO_4$. To compare the level of Cd resistance in these cell lines we determined the rate of DNA synthesis in cells cultured with increasing concentrations of $CdSO_4$. The "effective dose" of $CdSO_4$ at which incorporation of ^{3}H-thymidine into DNA was 50% of control cells (ED_{50}) is listed in Table 1. For this assay and others described later, Cd^R cells were withdrawn from Cd for one to two weeks before the assay. Since withdrawn cells contain little or no Cd, each of the Cd^R cell lines could be evaluated independently of the Cd concentration to which they had been adapted.

Each of the Cd^R cell types has a higher ED_{50} than the parental Cd^S line. In Cd^R cells selected with Cd alone there is a positive correlation between ED_{50} values and the Cd concentration to which the cells were adapted. Although Cd^R FE cells were adapted to medium containing 145 μM Cd, they have a lower ED_{50} value than Cd^R hepa 1A cells which were adapted to only 70 μM Cd. This is a consequence of supplementing the medium with Zn which decreases the toxicity of Cd, probably by protecting Zn-metalloenzymes from Cd substitution. Indeed, when Cd^R FE cells are transferred to medium containing only Cd, equivalent growth is obtained using 60 μM Cd.

MT-I Gene Amplification

To measure the relative number of MT-I genes, DNA was cleaved with *Eco* R1, electrophoresed on agarose gels, blotted onto nitrocellulose paper and hybridized to a MT-I specific probe. The relative number of MT-I genes was determined from the ratio of Cd^S to Cd^R DNA which resulted in autoradiographic bands of equal intensity.

Cd^R FE cells were shown previously to contain 6-fold more MT-I genes than Cd^S FE cells (10). The S180 and hepa 1A Cd^R cells have amplified their MT-I genes 10 and 30-fold, respectively (Table 1). Although none of these cell lines are truly diploid, we assumed for comparative purposes that each Cd^S cell type contains two MT-I genes per cell. The numbers of MT-I genes per Cd^R cell listed in Table 1 take into account the observation that resistant FE and hepa 1A cells have approximately twice as much DNA as the respective Cd^S cells. There is a positive correlation between Cd resistance (ED_{50}) and the number of MT-I genes per cell.

W7 cells are unique because the MT-I gene in these cells is hypermethylated and not inducible (11). As a consequence, Cd is very toxic for growth of W7 cells and selection for resistance was started at 1 μM instead of 10μM Cd as was used for the three other cell lines. When W7 cells were selected for resistance to 10 μM Cd, the MT-I gene remained transcriptionally inactive and unamplified. This indicates that amplification of the MT-I gene does not occur in the absence of MT gene expression and suggests that these cells have attained Cd resistance by another mechanism. Uptake of ^{109}Cd by Cd^R W7 cells is the same as the parental cell line. Perhaps Cd^R W7 cells have amplified genes for a different Cd-binding protein or a rate-limiting enzyme inhibited by Cd.

Chromosomal Analysis

To determine if karyotypic changes occurred in the process of becoming Cd resistant, we analyzed the chromosomes in the Cd^S and Cd^R cell lines. Giemsa stained metaphase chromosomes from Cd^S and Cd^R FE, S180 and hepa 1A cells are shown in Figure 1. The chromosome counts from these and many other preparations are summarized in Table 1. Cd^R FE cells contain an average of 3 very small chromosomes (Figure 1B) that persist in withdrawn cells. Cd^R hepa 1A cells contain many double minute chromosomes (Figure 1D). The small chromosomes in Cd^R FE cells and the double minute chromosomes in Cd^R hepa 1A cells differ in appearance. The double minutes look like paired dots in Giemsa stained metaphase spreads, whereas the small chromosomes in Cd^R FE cells resemble very small telocentric chromosomes. The chromosome analysis shown in Fig. 1 also shows that the number of chromosomes nearly doubled in Cd^R hepa 1A and FE cells. This change is not prerequisite for Cd resistance, since the Cd^R S180 cell chromosomes are indistinguishable from Cd^S cells (Fig. 1, E + F). Moreover, we have characterized another independently derived Cd^R FE cell line which has amplified MT-I genes but no detectable chromosomal differences from Cd^S FE cells. The next step in this type of analysis is to determine the chromosomal location of the MT-I genes in these cells.

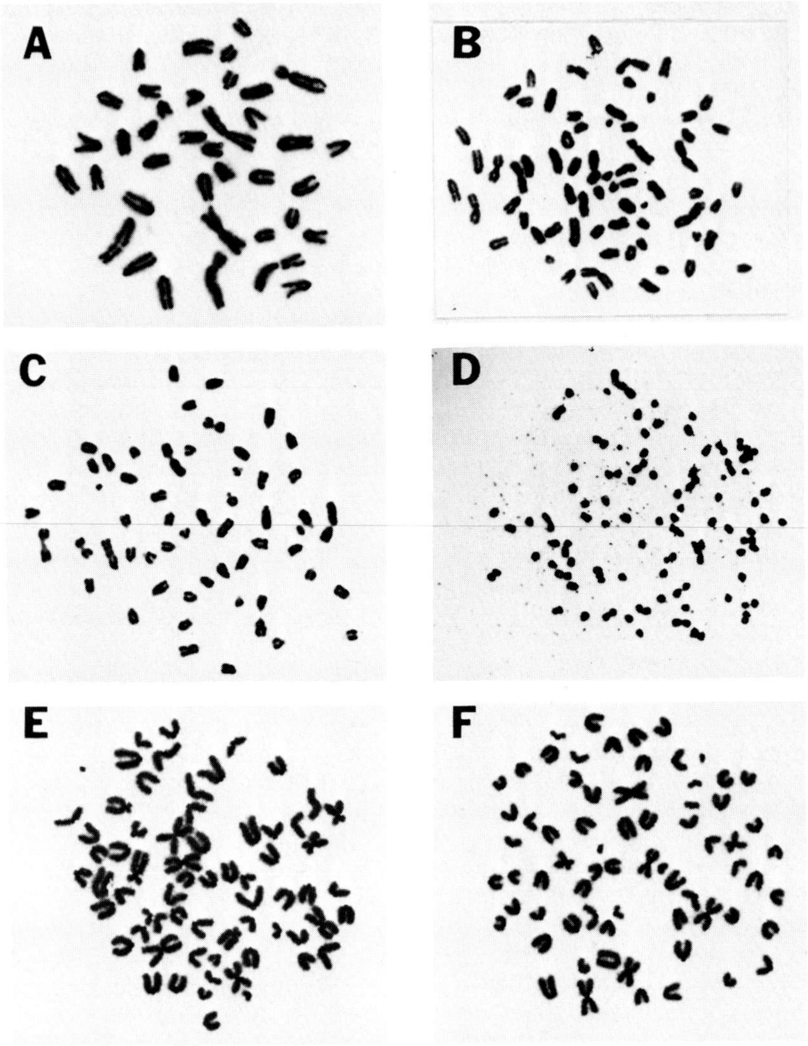

Figure 1. Metaphase chromosomes. Metaphase spreads from vinblastine treated cells were prepared and Geimsa stained as described (21). A. Cd^S FE. B. Cd^R FE. C. Cd^S hepa 1A. D. Cd^R hepa 1A. E. Cd^S S180. F. Cd^R S180.

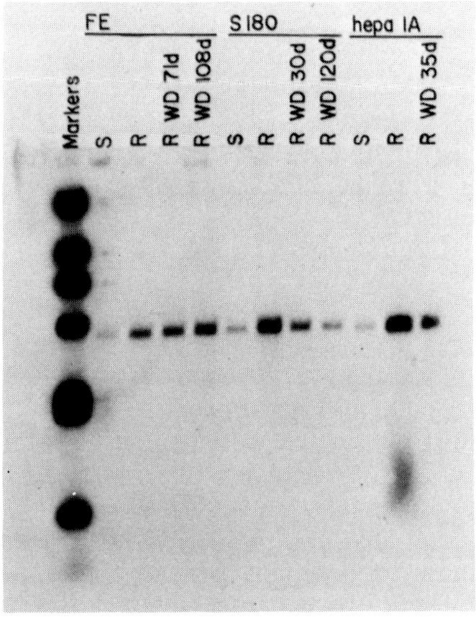

Figure 2. Relative gene copy number and stability of MT-I genes.

Three micrograms of DNA from each Cd sensitive (S) and resistant (R) cell line was digested with Eco RI, electrophoresed on a 0.7% agarose gel, blotted and hybridized to nick translated MT-I probe as described (10). Hind III digested and end-labeled lambda phage DNA was included as molecular weight markers (lane 1), the sizes in kilobase-pairs are 23.7, 9.5, 6.6, 4.3, 2.2, 1.9 and 0.6. WD, withdrawn from Cd for the days indicated.

Stability of the Amplified Genes

The amplified MT-I genes in Cd^R hepa 1A and S180 cells are unstable. Cd^R S180 cells lost approximately one half of their amplified genes after 30 days of withdrawal and contained no amplified genes by 120 days of withdrawal (Figure 2, lanes 7-9). Approximately two-thirds of the amplified genes in Cd^R hepa 1A cells were lost after 35 days of withdrawal (Figure 2, lanes 11 and 12). In contrast, the amplified MT-I genes in withdrawn Cd^R FE cells were stable; there was no loss of MT-I genes after 108 days of withdrawal from Cd (Figure 2, lanes 3-5).

Regulation of Amplified MT-I Genes

To characterize the functional capacity of MT-I genes in both Cd^S and Cd^R cells, we measured the relative rate of MT-I gene transcription and the number of MT-I mRNA molecules per cell after maximal induction with glucocorticoids (dexamethasone) or $CdSO_4$. No MT-I gene transcription or mRNA accumulation was detectable in either Cd^S or Cd^R W7 cells (Table 1). However, both parameters were higher in each of the Cd^R cell lines with amplified MT-I genes. Cd-induced MT-I gene transcription was 5-10 fold higher and MT-I mRNA levels were 5-22 fold higher in Cd^R cells than in Cd^S cells (Table 1). The increases in MT-I transcription and MT-I mRNA generally correspond to the extent of MT-I gene amplification.

In contrast, dexamethasone-induced MT-I transcription and MT-I mRNA levels do not increase in proportion to the extent of MT-I gene amplification. In S180 cells, where this phenomenon has been studied in most detail (12), our results are consistant with amplified MT-I genes being essentially unresponsive to glucocorticoids. The explanation for this surprising result is currently unknown.

Concluding Remarks

Three of the four cell lines studied induce MT-I mRNA in response to Cd or dexamethasone (Table 1). In each of these inducible cell lines, selection for Cd resistance leads to MT-I gene amplification. These observations provide the best evidence that MT plays an important role in the detoxification of Cd.

The amplification of MT-I genes is similar in several respects to the amplification of DHFR and aspartate transcarbamylase genes. In each case, gradually increasing the selective pressure results in progressive amplification of the target gene; it has not been possible to achieve high levels of gene amplification in a single step. Furthermore, in all cases, the amplified DNA sequences are considerably larger than the target gene. In the case of the MT-I gene, the amplified region is at least 50 times larger than the gene (10, 12). These similarities are indicative of a common mechanism of amplification.

Schimke et al. (13) suggested that DNA sequences are randomly duplicated at a low frequency in cultured cells. They proposed that at any one time a small number of cells will contain duplicate genes for a target protein and these cells will have a selective advantage in the presence of the cytotoxic agent. The initial duplication and further amplification could occur by such mechanisms as unequal crossover (14), uptake of DNA from dead cells (15) or disproportionate replication (16). The initial processes involved in either stable or unstable gene amplification may be the same. Recently it has been proposed that unstably amplified genes evolve into a stable form from the integration of unstable double minutes into large circularized forms which fragment and subsequently integrate into chromosomes to form stably amplified genes within homogeneously staining regions (17). It is unlikely that these mechanisms would result in rapid gene amplification and are therefore consistent with the requirement for gradual selection.

Gene families are thought to arise by duplication and divergence of a single primordial gene. A number of gene families have been characterized, including α-amylase genes (18), globin genes (19) and the related genes for growth hormone, prolactin and placental lactogen (20). The amplification of MT-I genes that are regulated by Cd but not by glucocorticoids may be an example of such evolutionary divergence. Analysis of the process and consequences of gene amplification should provide valuable insights into the evolution of gene families.

ACKNOWLEDGMENTS

We thank Abby Dudley for creative secretarial assistance and Doug Ross for the preparation and photography of chromosome spreads. This work was supported by the National Institutes of Health grants HD-09172, GM-07270 and ES-05174.

REFERENCES

1. Alt, F.W., Kellems, R.E., Bertino, J.R., and Schimke, R.T., J. Biol. Chem. 253, 1357 (1978).
2. Wahl, G.M., Padgett, R.A., and Stark, G.R., J. Biol. Chem. 254, 8679 (1979).
3. Rugstad, H.E., and Norseth, T., Biochem. Pharmacol 27, 647 (1978).
4. Hildebrand, C.E., Tobey, R.A., Campbell, E.W., and Enger, M.D., Exp. Cell Res. 124, 237 (1979).
5. Kägi, J.H.R., and Nordberg, M., "Metallothionein" Birkhaüser Verlag, Basel (1979).
6. Durnam, D.M., Perrin, F., Gannon, F., and Palmiter, R.D., Proc. Natl. Acad. Sci., USA 77, 6511 (1980).
7. Durnam, D.M. and Palmiter, R.D., J. Biol. Chem., in press (1981).
8. Hager, L.J. and Palmiter, R.D., Nature, in press (1981).
9. Mayo, K.E. and Palmiter, R.D., J. Biol. Chem. 256, 2621 (1981).
10. Beach, L.R., and Palmiter, R.D., Proc. Natl. Acad. Sci., USA 78, 2110 (1981).
11. Compere, S., and Palmiter, R.D., Cell, in press (1981).
12. Mayo, K.E., and Palmiter, R.D., manuscript submitted.
13. Schimke, R.T., Kaufman, R.J., Alt, F.W., and Kellems, R.F., Science 202, 1051 (1978).

14. Smith, G.P., Cold Spring Harbor Symp. Quant. Biol. 38, 507 (1973).
15. McBride, W.O., and Ozer, H.L., Proc. Natl. Acad. Sci. USA, 70, 1258 (1973).
16. Tartoff, K.D., Ann. Rev. Genet. 9, 370 (1975).
17. Bostock, C., and Tyler-Smith, C., Biochem. Soc. Trans. 9, 93P (1981).
18. MacDonald, R.J., Crerar, M.M., Swain, W.F., Pictet, R.L., Thomas, G., and Rutter, W.J., Nature 287, 117 (1980).
19. Weatherall, D.J. and Clegg, J.B., Ann. Rev. Genet. 10, 157 (1976).
20. Shine, J., Seeburg, P.H., Martial, J.A., Baxter, J.D. and Goodman, H.M., Nature 270, 494 (1977).
21. Deaven, L.L. and Petersen, D.F., Methods Cell Biol. 8, 179 (1974).

THE ROLE OF DOUBLE MINUTE CHROMOSOMES IN UNSTABLE METHOTREXATE RESISTANCE

Peter C. Brown and Robert T. Schimke

Department of Biological Sciences
Stanford University
Stanford, California

ABSTRACT Murine 3T6 cells resistant to 50 μM MTX and having approximately 50-100 DHFR gene copies per cell lose these amplified sequences and cellular resistance to MTX with prolonged growth in MTX-free medium. Coincident with these losses are the losses of DMs and DHFR sequences in fraction enriched in DMs by filtration. Additionally, the correlation between DHFR enzyme content, gene copy number and DM content was established by analysis of populations of cells prelabeled with a fluorescein derivative of MTX and sorted for fluorescence by fluorescence activated cell sorting.

INTRODUCTION

Double minutes (DMs) are small, paired extrachromosomal elements which have been described in a variety of metaphase preparations of tumor tissue (1,2) and in cell lines of mouse and human origin (3,4,5,6,7). While the function of DMs in these cells is unknown, their persistence suggests that they provide a selective advantage to cells that contain multiple DMs.

Permanent cell lines which have been selected in increasing concentrations of methotrexate (MTX) become resistant to high concentrations of the drug and have been

Supported by research grants from the National Cancer Institute (CA16318), the National Institute of General Medical Sciences (GM14931), and (PCB) supported by a post-doctoral fellowship from Damon Runyon-Walter Winchell Cancer Fund.

ISBN 0-12-137420-3

characterized as having amplified the structural gene coding for dihydrofolate reductase (DHFR), the target enzyme inactivated by MTX (8). Because of selective gene amplification of the DHFR genes, cells overproduce DHFR and survive in normally toxic concentrations of MTX. Resistance to MTX is often unstable and approximately 50 percent of the amplified DHFR sequences are lost within 20 generations in MTX-free medium (9). Characteristically, these cells contain multiple DMs and we have presented preliminary evidence which suggests that the amplified DHFR genes are localized to these DMs (10). In this report, we substantiate these findings for a MTX resistant mouse fibroblast line (3T6) and describe the relocation of amplified DHFR genes to chromosomes in a murine S-180 cell line coincident with stabilization of MTX resistance.

RESULTS

Unstably Amplified DHFR Genes are Associated with DMs. In Figure 1 is shown a metaphase preparation of 3T6 cells which have emerged after having been selected in increasing concentrations of MTX. The cell line (3T6 R50) is resistant to 50 μM MTX and has been shown to have 80-100 copies of the DHFR structural gene (11). In addition to necessarily overexposed chromosomes, one may see numerous DMs, none of which are seen in sensitive parental cells. By cytologic comparison with *Drosophila* chromosome IV, DMs appear to contain on the order of 200-1000 kb of DNA (P. Brown, unpublished).

In Figure 2, we have employed a new technique to demonstrate the association of amplified DHFR genes with DMs in 3T6 cells. Resistant cells were incubated with a fluorescent derivative of MTX (MTX-F) under conditions in which resulting cellular fluorescence is proportional to DHFR overproduction (14). Cells were analyzed and sorted by fluorescence activated cell sorting (FACS II) into four subpopulations. For each group, relative DHFR gene amplification and DMs per cell were determined along with fluorescence values for each subpopulation. From the distribution of these parameters in the subpopulations encompassing the entire population of 3T6 R50 cells, extensive heterogeneity is immediately apparent. That this heterogeneity is reflected in variable resistance to MTX has been demonstrated in unstably resistant murine S-180 cells (15). More importantly, the comparison of DHFR overproduction, DHFR gene amplification and DMs per cell shows good correlation throughout the fractionated populations and suggests that most if not all amplified DHFR genes are associated with DMs.

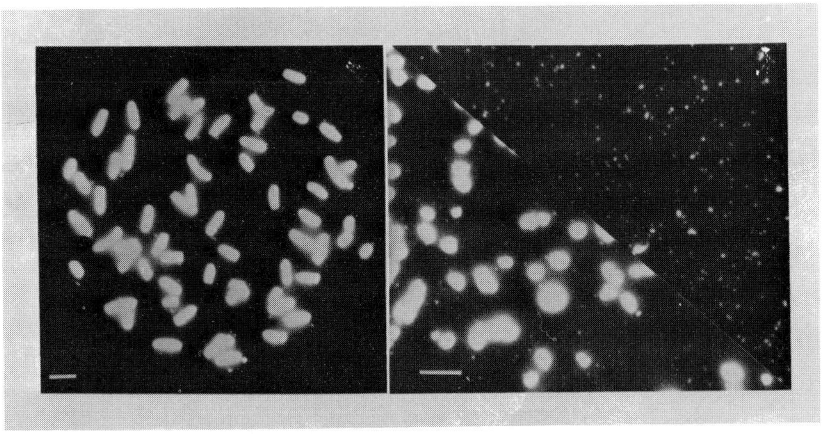

FIGURE 1. MTX resistant 3T6 cells have DMs which may be enriched by filtration. Panel A. Metaphase spread of 3T6 R50 cells stained with Hoechst 33258 (12). Panel B. Composite showing total metaphase chromosome preparation (lower left) prepared by modification (11) of published procedures (13) and DM-enriched fraction (upper right) obtained in the filtrate after passage of the total chromosome preparation through a 1 micron pore diameter Nuclepore filter. Scale = 5 μM.

Enrichment of DMs by Filtration and the Loss of DHFR Genes with Growth in MTX Free Medium. We sought improvements in the purification of DMs beyond our previous efforts (10). *Drosophila* chromosome IV had been purified by filtration through Nuclepore membrane filters (16) and this technique, when coupled with improved methods of chromosome isolation (13), has resulted in the purification of a DM fraction from 3T6 R50 cells which is enriched approximately 20-30 fold in amplified DHFR sequences relative to total chromosomal DNA (Table I).

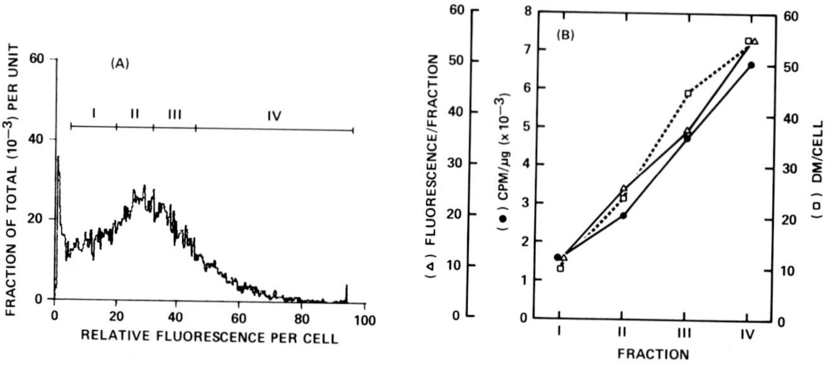

FIGURE 2. DHFR gene amplification and DMs in subpopulations of 3T6 R50 cells sorted according to DHFR overproduction. Panel A. Cells were labeled with MTX-F and sorted according to fluorescence into 4 subpopulations, I-IV, each comprising about 25% of the total. From each group, cells were prepared for DNA extraction and karyotype analysis. MTX sensitive cells display <2 fluorescence units per cell. Panel B. DHFR sequence abundance in each fraction was determined as in Table 1 (●). Also plotted are average DMs per cell (minimum of 25 randomly selected metaphases examined, (□), and average fluorescence of each subpopulation sorted (△).

Table 1. Loss of DM associated DHFR sequences with growth in MTX-free medium

generations minus MTX	total DNA cpm/ug[a]	total DNA percent	DM DNA cpm/ug[a]	DM DNA percent	DM/cell[b]	percent
none	490	100	9500	100	28(±17)	100
17	300	61	4250	44	13(±12)	46
34	120	24	3250	34	8(±13)	28
47	45	9	1000	10	5(±11)	18

a. Determined by hybridization of nick translated cloned DHFR cDNA to filter bound DNA from total chromosomal DNA and DNA enriched in DMs by filtration(11). Data is expressed relative to the amount of DNA applied to each filter corrected for hybridization to equivalent amounts of sensitive cell DNA.
b. DMs were determined in at least 75 randomly selected metaphase spreads. Standard deviation is in parentheses

Table 1 further shows that with growth in the absence of MTX cells lose amplified DHFR sequences in total DNA and in DM-enriched DNA and in DMs per cell with approximately equal facility. This correlation becomes apparent upon comparison of normalized hybridization values and DM counts (percent of starting) at each point in the experiment. Taken together, these results including the co-purification of DMs and DHFR sequences by filtration further establish the localization of unstably amplified DHFR genes to DMs.

Stabilization of MTX Resistance with Continued Growth in MTX. An analysis of a cloned subline (R_1A) of MTX-resistant murine S-180 cells soon after isolation and after approximately 600 generations in the presence of MTX (R_1C) revealed that stabilization of resistance had occurred and that stabilization was coincident with the disappearance of DMs (10). Since the DHFR gene copy number in the two populations is essentially identical (120-140 genes/cell) we performed in situ hybridizations in order to determine whether the amplified DHFR genes in R_1C cells were clustered at presumptive chromosomal locations.

While accurate calibrations have not been completed, we believe that the in situ hybridization method used has sufficient sensitivity to detect a localized cluster of at least 10 copies of the DHFR gene. Even greater sensitivity is claimed by others using similar protocols (17, G. Wahl, pers. communication). Therefore, DMs which probably have no more than 4 copies of the DHFR gene (11) would be undetectable even if the underlying DM could be distinguished from an exposed grain of the photographic emulsion. A comparison of in situ hybridization patterns of R_1A and R_1C cells reveals a predominance of grains clustered on multiple chromosomes in R_1C cells and a paucity of comparable clusters in the unstable R_1A cells. We interpret these data to indicate that stabilization of resistance is contingent upon relocation of amplified DHFR genes from extrachromosomal elements (i.e., DMs) to predominantly clustered chromosomal loci.

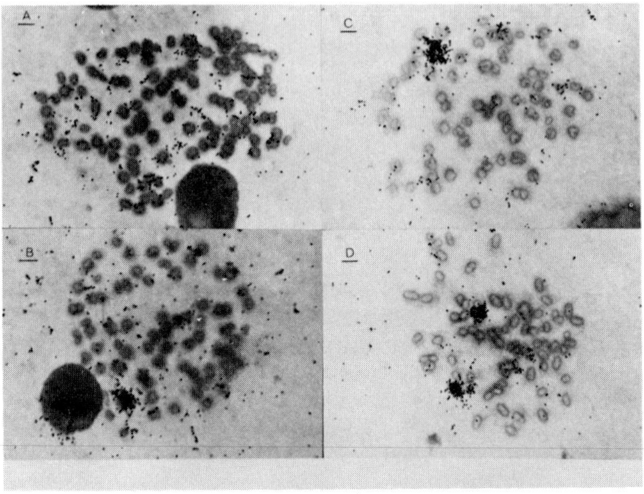

Cells	DMs per cell (percent)						Distribution of grain clusters (percent)			
	0	1-10	11-20	21-30	31-40	40	0	1	2	3
R_1A	29	33	11	8	3	16	62	30	8	0
R_1C	77	17	3	3	0	0	7	35	51	5

FIGURE 3. Distribution of DMs and DHFR gene clusters in S-180 cells. Murine S-180 cells were hybridized *in situ* with ^{125}I-labeled cloned DHFR cDNA (30) and clusters of grains determined in 138 randomly selected metaphase spreads of R_1A (Panel A, no clusters; Panel B, one cluster) and of R_1C (Panel C, one cluster; Panel D, two clusters). DM counts were made on slides prepared in parallel.

DISCUSSION

Association of Unstably Amplified DHFR Genes with DMs. We have shown through a series of correlations that unstably amplified DHFR genes are localized to DMs. This approach was necessitated because of our inability to directly demonstrate this fact by *in situ* hybridization. The evidence which substantiates the association of DHFR genes to DMs is as follows: 1) Copurification of DMs and amplified DHFR sequences. 2) Correlation of DHFR gene amplification and DMs in cells sorted for DHFR overproduction. 3) Similar kinetics of loss of amplified DHFR genes in total DNA and in DM-enriched DNA and in DMs in cells grown out of MTX selection.

DMs are self replicating elements which lack centromeric regions and therefore are not segregated proportionately at mitosis (3,18,19). Undoubtedly some DMs and associated DHFR genes are lost at this stage of the cell cycle and this loss contributes to the instability of the amplified state. In addition, we have observed that unstably resistant cells with high DHFR gene copy number grow more slowly as a population than do cells with intermediate or low DHFR gene amplification (15). This observation implies that daughter cells which acquire fewer DMs inherit a selective growth disadvantage in the absence of MTX and eventually predominate. Conversely, in the presence of MTX the selective advantage is reversed and cells with greater DHFR gene amplification prevail.

Stabilization of Amplified DHFR Genes. From an initially unstable clonal population (R_1A) a population of cells emerged under continuous MTX selection which exhibited stable MTX resistance and DHFR gene amplification (R_1C). Coincident with stabilization was the loss of DMs and a reassignment of DHFR genes to chromosomal loci. An attractive hypothesis explaining this phenomenon is that DMs in the R_1A cells coalesced and integrated at distinct chromosomal loci. A driving force favoring this association is the efficiency and stability acquired with physical linkage to the centromere. In this context the behavior of DMs is not unlike the stabilization of selectable genes which have been introduced into cells by DNA and chromosome mediated gene transfer (17,20,21). Coalescence of DMs may have been observed in murine cell lines (6,22), and it remains to be determined whether chromosomal sites of DHFR gene amplification described here correspond to the locus of the unamplified DHFR gene as is probably the case in CHO cells (23) or whether the sites of putative integration are more random as with transfected DNA (17).

It is not clear if the chromosomal clustering that we have observed in stabilization of R_1A cells is homologous to homogeneously staining regions (HSR) which have been described in MTX resistant cell lines (24,25,26,27) as well as in cell lines with no known drug resistance (5,28). Detailed karyotypic studies are complicated due to the highly aneuploid nature of R_1C and R_1A cells. The possibility remains, however, that the DHFR gene clusters seen in the R_1C cells are indeed HSRs which arose independently during the period of DM-mediated DHFR gene amplification and eventually supplanted the DM mode of amplification.

In our experience, instability of MTX resistance is the rule in cells which have recently been selected in MTX (11,29). In mouse 3T3 and 3T6 cells, DMs appear during the initial selection (30, P. Brown, unpublished data) and increase in frequency and number throughout subsequent MTX selections. In contrast, DM-like elements are observed extremely infrequently in newly selected CHO cells (29) and never in PALA resistant Syrian hamster cells (G. Wahl, pers. communication) although both of these lines contain HSRs encompassing amplified DHFR and CAD genes, respectively. We cannot say whether our failure to unequivocally show DMs in the CHO cells is a result of their extremely transient nature or whether gene amplification proceeds exclusively within the chromosomal structure in these cells. In mouse, however, it seems clear that DMs mediate the initial stages of DHFR gene amplification resulting in MTX resistance. Elsewhere we have presented models which attempt to rationalize the formation of DMs and HSR as they pertain to the development of drug resistance (30).

REFERENCES

1. Marinello, M.J., Bloom, M.L., Doeblin, T.D., Sandberg, A.A. (1980) N. Engl. J. Med. 303:704.
2. Reichman, A., Riddel, R.H., Martin, P., Levin, B. (1980) Gastroenterology 79:334-339.
3. Levan, G., Mandahl, N., Bregula, V., Klein, G., Levan,A. (1976) Hereditas 83:83-90.
4. Barker, P.E., Hsu, T.C. (1978) Exp. Cell Res. 113: 457-458.
5. Balaban-Malenbaum, G., Gilbert, F. (1977) Science 198:739-742.
6. Cowell, J.K. (1980) Cytogenet. Cell Genet. 27:2-7.

7. Quinn, L.A., Moore, G.E., Morgan, R.T., and Woods, L.K. (1979) Cancer Research 39:4914-4924.
8. Alt, F.W., Kellems, R.E., Bertino, J.R., Schimke, R.T. (1978) J. Biol. Chem. 253:1357-1370.
9. Schimke, R.T., Kaufman, R.J., Alt, F.W., and Kellems, R.F. (1978) Science 202:1051-1055.
10. Kaufman, R.J., Brown, P.C., Schimke, R.T. (1979) Proc. Natl. Acad. Sci., U.S.A. 76:5669-5673.
11. Brown, P.C., Beverley, S.M., and Schimke, R.T. submitted for publication.
12. Latt, S.A. (1975) Chromosomes Today Vol.5, pp 367-394.
13. Blumenthal, A.B., Dieden, J.D., Kapp, L.N., and Sedat, J.W. (1979) J. Cell Biol. 81:255-259.
14. Kaufman, R.J., Bertino, J.R., Schimke, R.T. (1978) J. Biol. Chem. 253:5852-5868.
15. Kaufman, R.J., Brown, P.C., and Schimke, R.T. submitted for publication.
16. Hanson, C.V. (1975) in New Techniques in Biophysics & Cell Biology, Eds. Pain, R.H., and Smith, B.J. J. Wiley and son, London, pp 43-83.
17. Robins, D.M., Ripley, S., Henderson, A.S. and Axel, R. (1981) Cell 23:29-39.
18. Barker, P.E., Drwinga, H.L., Hittelman, W.N., and Maddox, A.M. (1980) Exp. Cell Res. 130:353-360.
19. Levan, A. and Levan, G. (1978) Hereditas 88:81-92.
20. Perucho, M., Hanahan, D. and Wigler, M. (1980) Cell 22:309-317.
21. Klobutcher, L.A. and Ruddle, F.H. (1979) Nature 280: 657-660.
22. Levan, A., Levan, G., and Mandahl, N. (1978) Cytogenetics and Cell Genetics 20:1-6.
23. Roberts, M., Huttner, K.M., Schimke, R.T., and Ruddle, F.H. (1980) J. Cell Biol., in press.
24. Nunberg, J.N., Kaufman, R.J., Schimke, R.T., Urlaub, G., Chasin, L.A. (1978) Proc. Natl. Acad. Sci., U.S.A. 75:5553-5556.
25. Dolnick, B.J., Berenson, R.J., Bertino, J.R., Kaufman, R.J., Nunberg, J.H., Schimke, R.T. (1979) J. Cell Biol. 83:394-402.
26. Biedler, J.L., Spengler, B.A. (1976) Science 191: 185-187.
27. Bostock, C.J., and Clark, E.M. (1980) Cell 19:709-715.
28. George, D.L. & Francke, U. (1980) CytoGenet. Cell Genet. 28:217-226.
29. Kaufman, R.J. & Schimke, R.T. submitted for publication.
30. Schimke, R.T., Brown, P.C., Kaufman, R.J., McGrogan, M., & Slate, D.L. (1981) Cold Sprg. Harbor Symp.Quant.Biol. Vol. 45, in press.

MODULATION OF THE STRUCTURE OF THE NUCLEOSOME CORE PARTICLE DURING DEVELOPMENT

Robert T. Simpson
Lawrence W. Bergman

Developmental Biochemistry Section
NIAMDD
National Institutes of Health
Bethesda, Maryland

ABSTRACT: Conservation of sequence is a hallmark feature of the inner nucleosomal histones in adult metazoans, as is conservation of structure of the nucleosomal core particle. In contrast, histone variants arise from post-synthetic modification or occur in sperm and early embryos of sea urchins. Ef fects of such altered histones on several static and dynamic features of core particle structure have been determined, using acetylated histones, *S. purpuratus* sperm, blastula and pluteus. For each case, significant modulation of core particle structure in terms of DNase I cutting site susceptibility, low ionic strength unfolding, and/or thermal denaturation, was observed. Histone variants have demonstrated effects on chromatin structure, even at the most basic level, the core particle; we suggest that these structural alterations likely have functional consequences *in vivo*.

I. INTRODUCTION

Chromatin is organized as a tandem repeating array of subunits called nucleosomes (for review, see 1). While certain features of chromatin structure, for example, the length of DNA associated with each nucleosome, are variable, it has been commonly accepted that the structure of the nucleosome core particle is quite constant when it is isolated from a variety of tissues and organisms. This canonical core part-

ISBN 0-12-137420-3

icle is modeled as a squat cylinder, 11 nm in diameter and 5.5 nm in height, consisting of an inner octomer of histones (2 each of H2A, H2B, H3 and H4) wrapped by 146 base pairs (bp) of DNA in a shallow helical path with a circumference of about 80 bp. Characteristic sites of accessibility or lack of same to nucleases reflect both the periodicity of the DNA helix and interactions of histones with the nucleic acid. Wrapping of DNA in the core particle is determined only by interactions with the four smaller histones; the particle may easily be reassociated in a form identical with the original from isolated DNA and histones.

The sequences of the inner histones from adult metazoans show evidence of remarkable conservation (for review see 2); apparently even minor changes must lead to alterations in function which are unacceptable to the cell. Certain situations do exist, however, where variant forms of these histones occur. Best known are the post-synthetic modifications of these proteins, including acetylation, phosphorylation, ADP-ribosylation, methylation, and covalent attachment of another protein (1,2). In view of the marked conservation of sequence of the histones, one might anticipate that each of these several modifications should have significant effects on histone function.

Different subtypes of H1, H2A, and H2B are found in the sperm of animals whose histones are not replaced during spermiogenesis by protamine-like proteins (3). A distinct set of subtypes of these proteins is found in early embryos of sea urchins and other species (4-7). Further, during sea urchin embryogenesis, synthesis of these early histones ceases at about blastula and a second embryonic set is synthesized (7-11); both sets of histones persist through pluteus as constituents of chromatin.

The four smaller histones are proteins with two structural domains (2). The carboxyl-terminal two-thirds is a more globular, less basic region and is thought to be involved in the proper protein-protein interactions to form the nucleus of the core particle, although other experiments demonstrate that this region also provides a significant portion of the stabilization of DNA binding (12,13). The amino-terminal third of these proteins is likely less organized in structure and is highly enriched in basic residues. It is this latter portion of the histones which is affected by chemical modifications and is the site of (most of) the primary sequence alterations which occur in histone subtypes (2).

It is our contention that these alterations in structure of the four inner histones are likely to lead to altered functional properties for these proteins. For the inner histones, it is likely that their function *is* structure, the proper dynamic packaging of DNA in the nucleus. Here, we summarize effects of several types of histone changes on structure of the core particle. While all particles discussed share with the canonical core particle the presence of a histone octamer and 146 bp of DNA, other static and dynamic features of structure differ from the canonical nucleosome for each. Thus, variations in inner histones modulate the structure of the chromatin core particle. Plausible relationships of structural differences to *in vivo* functional differences are discussed.

II. PARAMETERS MEASURED AND THEIR SIGNIFICANCE

A. Mapping DNase I Cutting Site Susceptibility

DNase I creates single strand nicks in chromatin DNA, leading to a population of molecules which migrate on gel electrophoresis as a series of bands of size 10.4 x N, where N is an integer (14). The sites at which the nuclease cuts DNA within the core particle can be determined by labeling the 5' ends of the DNA (in the intact particle) with radioactive ATP and polynucleotide kinase, digesting with the nuclease, electrophoresing the denatured DNA and localizing the isotope by autoradiography (15-17). The lengths of the labeled fragments are the distances from the 5' end at which nicking occurred; relative susceptibilities of different cutting sites can be determined by study of the rate of appearance of a given length fragment during the time course of a digestion. Sites with low cutting susceptibilities are probably sites which interact strongly with histone amino-terminal regions (12).

B. Reversible Conformational Changes of Core Particles

1. Unfolding of Core Particles at Low Ionic Strength. At ionic strength 0.1, the core particle is a compact structure, sedimenting at about 11S. As the ionic strength is lowered, the particle undergoes a shape change, becoming a more expanded, slower sedimenting entity (18). In contrast to early studies which suggested that this transition might occur in two steps, more recent investigations demonstrate a two-state process (19), shown schematically in the upper left of Figure 1. This transition is fully reversible on restoration of ionic strength to higher values. Wu et al. modeled the structure of the expanded core particle at very low ionic strength as a flat disk, 15 nm in diameter and 5 nm in height, with nearly a full turn of DNA wound on an expanded histone core, as shown in the lower left of Figure 1 (19). This transition involves disruption of some histone-histone interactions, since covalent crosslinking of the histone into a stable octamer precludes the conformational change (18,19).

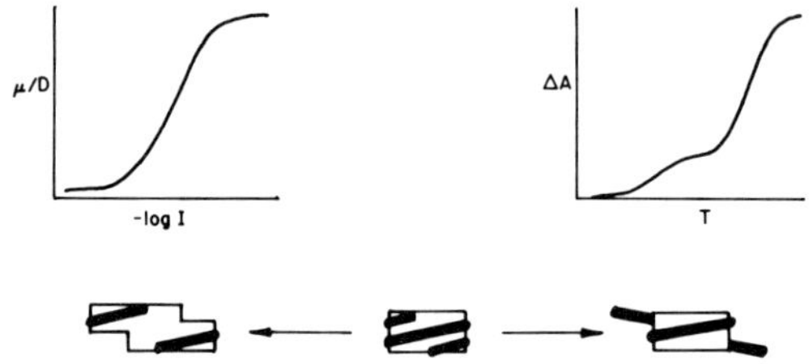

Figure 1. Diagrammatic representation of two reversible conformational changes for the nucleosome core particle.

2. Unspooling and melting of the ends of core particle DNA. Melting of core particles occurs in (at least) two distinct phases; reversible melting of about 25-30% of the total at 60-65° is followed by irreversible melting of the remainder of the DNA at about 76-77° (20), shown schematically in the upper right of Figure 1. A variety of types of hydrodynamic and spectroscopic data suggest that the DNA which melts in the reversible phase is unspooled from its normal position, tightly opposed to the histone octomer, at temperatures about 10° below the actual melting (20-22). No detectable changes in protein structure occur during the first phase of melting and the temperature of the melt is not markedly affected by covalently crosslinking the histone octomer (23). Using S1 nuclease digestion of core particles labeled at the 5' ends, we confirmed the suggestion of Weischet et al. (20) regarding the mechanism of the reversible melting phase; it involves unspooling and melting of 20-25 bp DNA at each end of the core particle segment (21). Figure 2 diagrams the core particle with unspooled DNA ends in the lower right. Since the remainder of core particle structure is not disrupted, the conformational change is freely reversible on lowering temperature.

III. RESULTS

A. Acetylated Histones

Hyperacetylated H3 and H4 can be produced *in vivo* by inhibiting the activity of histone deacetylase with sodium n-butyrate (24). Treatment with butyrate leads to an increased susceptibility of nuclear DNA to digestion with DNase I (25). Further, core particles with acetylated histones are preferentially excised from chromatin by micrococcal nuclease versus particles with less acetylated histones (25). To study the effects of histone acetylation on core particle structure in more detail, we have used a simple sequence DNA to construct chromatin which allows higher resolution data than that obtained with random sequence DNA (26). Histones were associated with poly(dA-dT) by salt gradient dialysis and core particles prepared from the semi-synthetic chromatin by conventional means. Absence of sequence heterogeneity in these particles allows more facile detection of effects of the modified proteins on core particle structure.

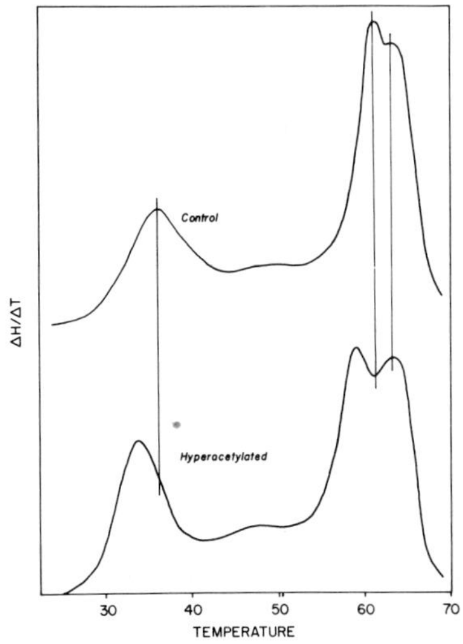

Figure 2. Derivative melting profiles for core particles containing poly(dA-dT) and control or hyperacetylated histones.

Figure 2 compares the thermal denaturation profiles (as derivatives) of core particles containing hyperacetylated histones and those of controls. Three well resolved transitions are present. The lowest corresponds to melting of about 40 bp of DNA, the two upper result from melting of about 80 bp. As indicated by the index lines in the figure, the highest temperature transition is unaffected by the histone modification. Both the first portion of the irreversible transition and the reversible transition are destabilized by the presence of hyperacetylated histones. By crosslinking studies, H3 and H4 are thought to have their main sites of interaction with DNA near the ends and at the central segment, 60-80 bp from the ends (27). Acetylation of histones decreases positive charge, leading to weakening of electrostatic interactions between protein and DNA. Acetylation of H3 likely facilitates unspooling of the ends and the reversible melting; acetylation of H4 and/or H3 also destablizes interactions which lead to the first half of the irreversible transition.

Correlated with these dynamic changes in core particle structure resulting from hyperacetylated histones, we find static changes in histone-DNA interactions measured by mapping DNase I susceptibility. In line with the reasoning presented above, we would expect histone acetylation to increase cutting rates in those regions where the modified histone amino-terminal regions interacted with the DNA. Experimentally, we find that cutting rates are increased at the ends of core particle DNA and at the sites 20, 60, and 80 bases from the 5' end, when particles containing hyperacetylated histones are compared with those having largely unmodified histones (28, L.W. Bergman and R.T. Simpson, unpublished).

B. *Strongylocentrotus purpuratus Sperm Histones*

The inner histones of sea urchin sperm differ from those of adult tissues in higher organisms (3); H2A has a faster mobility on SDS electrophoresis while H2B, present as two subtypes in a 2:1 ratio, migrates much more slowly, behind H3. The sequence of sperm H2B in another sea urchin reveals that the variant histone is much more highly basic and longer; these variations occur almost exclusively in the amino-terminal region (2).

We have isolated core particles from sperm of *S. purpuratus* using minor modifications of techniques employed for other tissues. These particles contain an octamer of inner histones and 146 bp of DNA (29), in common with the canonical core particle. The sperm particle also undergoes the low ionic strength unfolding in a fashion exactly equivalent to the core particle from calf thymus (19,29); the transition is centered at ionic strength 1.2 mM.

In contrast, two other features of the sperm particle differ from those of calf thymus or chicken erythrocyte core particles. As shown in Figure 3, the thermal denaturation

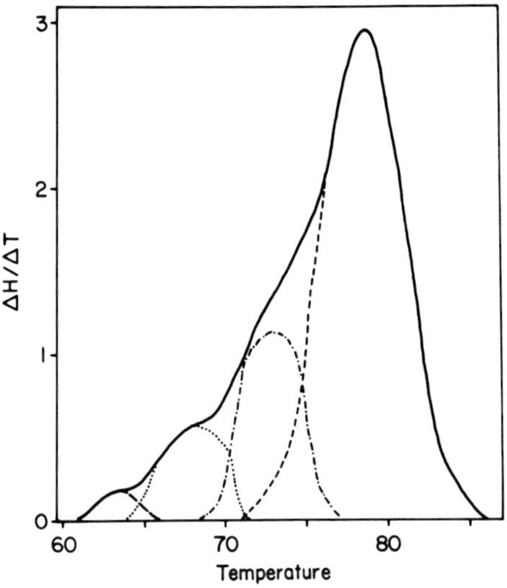

Figure 3. Derivative plot of thermal denaturation of the sperm core particle. Four transitions were resolved by eye.

profile for the sperm core particle differs by the main, irreversible transition being at higher temperatures, 79° as opposed to 76-77° for typical core particles. More markedly different is the stabilization of the reversible transition afforded to the sperm particle by the variant forms of H2A and H2B (29). Resolution of the derivative melting profile shows that three lower temperature transitions are present. The first is small and likely due to a fraction of DNA tails longer than core particle length. The second and third, occurring at temperatures of 68.5° and 73°, respectively, have a total magnitude equal to the reversible 60-65° transition of the chicken erythrocyte core particle (20). We suppose that these arise from the same unspooling and melting of 20-25 bp segments at the ends of core particle DNA. However, in the sperm particle, variant histones stabilize this melting such that it requires temperatures nearly equal to those required for total disruption of the core particle. The two transitions may result from the two forms of H2B present in sperm histones.

We have compared the susceptibilities of DNase I cutting sites for sperm and erythrocyte core particles using both native nucleosomes (29) and core particles formed with poly(dA-dT) (26). Figure 4 shows cutting site autoradiograms for these two particles when poly(dA-dT) is used as the nucleic acid. Samples were digested in identical fashion for increasing lengths of time (from left to right) with DNase I. The

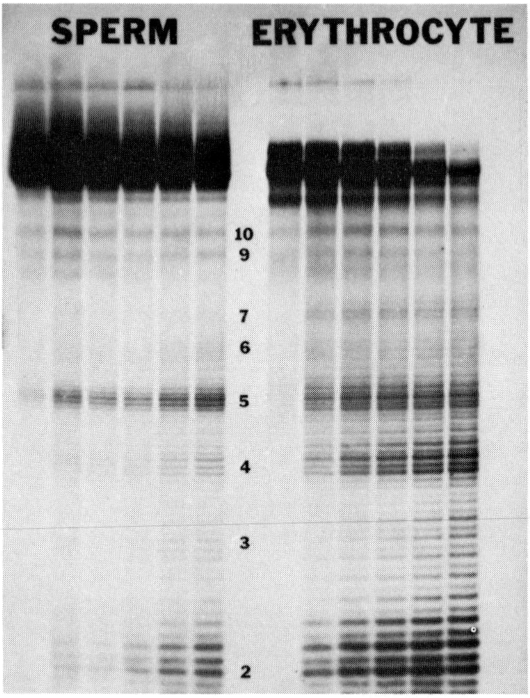

Figure 4. DNase I cutting site maps for core particles.

band numbers are the distance in bases from the 5' end divided by 10. General features of cutting site susceptibilities described above are evident in the map for the erythrocyte particle, their visualization is enhanced by the single base resolution afforded through use of the synthetic DNA. Cutting sites are evident throughout the particle but preferential cutting is observed at sites spaced at multiples of 10 from the 5' end of the DNA segment. Sites 30, 60, 80 and 110 bases from the end are relatively infrequently cut when compared to the other, favored cutting sites (15-17).

The sperm core particle exhibits significantly different cutting site susceptibilities from the erythrocyte particle. Major changes are noted for the sites 20 and 40 bases from the 5' end. These are favored cutting loci in the erythrocyte core particle; in the sperm particle they are cut with frequencies nearly as low as the disfavored site at 30 bases from the end. Additionally, the site 10 bases from the ends is cut with lower frequency in the sperm particle (29).

Inspection of the helical path of DNA in the core particle (30) and the sites where different histones are thought to interact with DNA (27) reveals that the loci where cutting is decreased in the sperm particle are clustered in space and are sites where H2B interacts with the nucleic acid. We suggest that the longer, basic, amino-terminal region of this histone in sperm binds DNA more strongly than the H2B of the canonical core particle, leading to the observed changes in cutting site susceptibility for DNase I and the stabilization of the core particle in the reversible melting transition.

C. *Early and Late Histones of S. purpuratus Embryos*

During embryogenesis in sea urchins, at least two distinct sets of histones are synthesized and associate with chromatin (4-7) [we ignore the cleavage stage histones identified by Newrock et al. (7)]. Prior to blastula an early set of H1, H2A and H2B are synthesized, partially from stored maternal mRNA (8-11). Then, there occurs a switch; early histone synthesis ceases (although these histones remain associated with DNA) and synthesis of a late set of these three histones begins, these being made using newly synthesized embryonic RNA (11). Several subtypes of these latter histones are present (6); since we can not isolate core particles containing homogeneous histone subtypes, we will consider them as a single, "late" class. Sequence information for the late histones is not available. Protein sequences for the early histones (deduced from DNA sequences) demonstrate major differences between H2B of embryo and calf thymus (31). Most, but not all, differences occur in the amino-terminal third of the protein.

We have isolated core particles from *S. purpuratus* at two stages of development, hatching blastula and pluteus. The former have inner histones which are over 95% the early forms of H2A and H2B, judged by electrophoresis on Triton-acid-urea gels (32). The latter have 25% early and 75% late histones; this ratio is that expected on the basis of cell numbers at the two stages, assuming dilution of the early histones during development. For the three parameters measured, these core particles differ from one another and/or from the canonical core particle.

Thermal denaturation profiles for core particles containing early and late histones are similar in shape to that of the chicken erythrocyte core particle (20); for all cases about 30% of the melting occurs in an early phase. However, for both sets of sea urchin core particles, the temperature required for both the reversible and irreversible transitions is decreased from that required for the erythrocyte particle (Table I). The variant histones present in sea urchin embryos thus alter the stability of the core particle; the sea urchin particle is both more easily unspooled and more easily totally disrupted than the particle of typical adult tissues.

Table I. Melting Temperatures of Core Particles

Source	T_{m1}	T_{m2}
Chicken erythrocyte	*62.0°*	*77.0°*
S. purpuratus blastula	*57.0°*	*73.5°*
S. purpuratus pluteus	*55.0°*	*73.5°*

An effect of the embryonic histones is also observed when the unfolding of the core particle at low ionic strength is examined. Figure 5 shows that the sedimentation coefficient of pluteus core particles decreases as ionic strength is lowered, exactly as that for the sperm particle (29), which, in turn, unfolds nearly identically to the calf thymus particle (19). In contrast, the core particle from blastula stage sea urchin embryos undergoes this low ionic strength conformational change more readily than the other core particles. The midpoint for the transition for blastula core particles is at an ionic strength about 3-fold higher than that for pluteus or sperm core particles. As discussed previously, this conformational change likely involves disruption of the normal histone-histone interactions in the core particle. The different histones in blastula and pluteus core particles thus

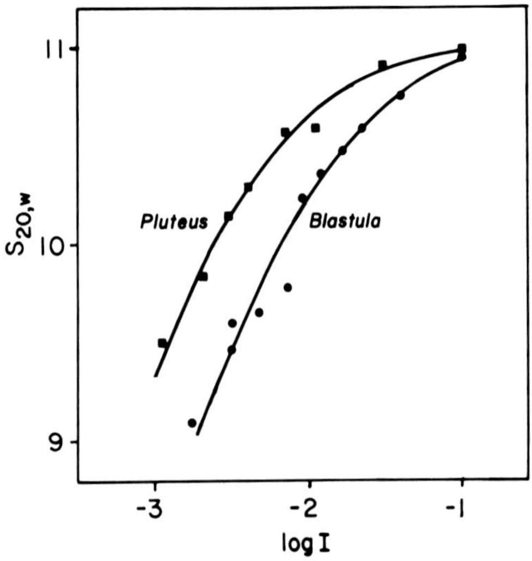

Figure 5. Unfolding of Blastula and Pluteus Core Particles.

lead to a lesser stability of histone-histone interactions in the former, relative to the latter. We will return to plausible biological implications of this difference between blastula and pluteus core particles later.

Mapping of DNase I cutting site susceptibilities has been carried out with core particles from blastula and pluteus (we have not yet extended such studies to the use of semi-synthetic core particles to rule out the effect of DNA modifications on the results obtained).

The rate of digestion of blastula core particle DNA by DNAase I is about twice that of the pluteus particle, as judged by disappearance of 146 base length 5' end labeled DNA. While both particles share common features of DNase I susceptibility with the canonical core particle, cutting at 10 base intervals and relative infrequency of cutting at 30, 60 80, and 110 bases from the 5' end, certain differences do exist between the pair of embryonic core particles. Figure 6 shows scans of autoradiograms of blastula and pluteus core particles labeled at the 5' end and digested for 4 or 8 min, respectively, with DNase I. The blastula core particle has a cutting site map similar to that of core particles from HeLa or chicken erythrocyte (15-17), although the 30 and 70 base sites are more frequently cut in the embryo particle. The pluteus core particle evidences less frequent cutting by the nuclease at the sites 40, 70, and 100 bases from the 5' end of the DNA and an increased, but poorly defined, frequency of cutting around 60 bases from the end. While the differences between these two populations of core particles are

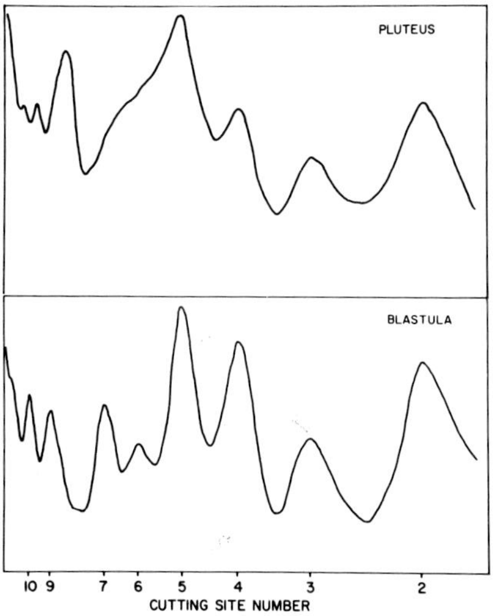

Figure 6. Cutting sites in blastula and pluteus particles.

less than those between sperm and erythrocyte particles (29), they are, nevertheless, likely significant, for they far exceed any differences in cutting site susceptibility that we have observed between erythrocyte and HeLa core particles.

IV. DISCUSSION

A. Modulation of Core Particle Structure

Evolutionary conservation of histone sequences is a hallmark of these proteins. In adult metazoan organisms, sequences of H3 and H4 are essentially constant and those of H2A and H2B vary only slightly (2). Since H3 and H4 are thought to "frame" the nucleosome (1), it comes as no surprise that the length of DNA and its basic wrapping in the core particle are rigidly conserved structural features of chromatin. H2A and H2B complete the structure of the core particle and, as we show here and elsewhere (29), are capable of modulating core particle structure.

In contrast to a generally assumed constancy of structure of the core particle, we now know of a number of compositional features which alter demonstrably the structure of chromatin, even at this most basic level. To date, core particle structure has been shown to vary from the canonical model due to 1) variant forms of H2A and H2B (29), 2) presence of H1 (33), 3) presence of nonhistone proteins (34,35), 4) postsynthetic modification of histones (28), and 5) differing DNA sequence (26). Thus, variations in all components of chromatin have demonstrated structural effects at the core particle level.

We realize that for no case do we know that the detected structural variations have any significance in terms of chromatin function during replication or transcription. However, given the generally conserved nature of the histones and the correlated conserved structure of the core particle, it is our bias that the presence of histones which differ from the usual (in sequence or by modification) alters chromatin structure in a biologically significant fashion. We also note that changes in the inner histones may, in addition, lead to alterations in chromatin structure at other levels which we have not yet examined. Note particularly the suggestion of McGhee et al. (36), based on several lines of evidence, that the amino-terminal regions of the inner histones might be involved in particle-particle interactions leading to stabilization of the next level of DNA folding in chromatin. Overall, it is our impression that many types of nucleosomes can and do exist in chromatin; some will differ in a static fashion while other will have dynamic structural differences.

B. Biological Relevance of Structural Modulation

Comparison of the biology of tissues with different core particle structures offers a clue as to plausible mechanistic roles for the variations we have observed. For this purpose, let us consider structural changes which affect low ionic strength unfolding, on the one hand, and melting behavior and DNase I susceptibility, on the other.

The only particle thus far studied which differs from the others in low ionic strength unfolding is that from sea urchin blastula (Figure 5). This structural change is thought to involve partial disruption of the histone octomer with expansion of the core particle (19). Biologically, the early sea urchin embryo is distinguished from the other tissues studied by its rapid rate of division; the first seven divisions occur in about 10-12 hours and thereafter the division rate slows markedly with about two or three divisions between blastula and pluteus over a 3-4 day period (37). While histone octomers are thought to remain conserved through replication (38), it is not unreasonable to assume that some degree of opening of the protein core of the nucleosome is required for events at the replication fork in eukaryotic cells. Perhaps the early histone variants in sea urchin are specialized to allow unfolding of the core particle in a fashion suitable for replication at a fast rate.

Melting transitions and DNase I susceptibility are altered in different directions from the properties of the canonical core particle by different histone variants in sea urchins and other tissues. Histone variants present in embryos or histone acetylation lead to more facile melting and increased accessibility of some nuclease sensitive sites. In contrast, chromatosomes, containing H1, and sperm core particles have higher melting temperatures (particularly for the reversible transition) and decreased susceptibility to the nuclease.

Based on the size of RNA polymerase and the close side by side approximation of the DNA strands in a core particle, it seems to us unlikely that transcription will proceed through a nucleosome. More likely, we feel, is a transient disruption of the structure followed by its reconstruction after the transcriptional event. A reasonable candidate for the initial phase of this disruption is unspooling of the end segment of DNA from its normal position bound tightly to the core histones (21,29,33). Temporal correlations of histone acetylation with gene activation and the high synthetic rates of early embryos, on the one hand, and association of H1 with inactive chromatin and the transcriptional inactivity of sperm nuclei, on the other, make possible a nice correlation of biological function with modulation of particle structure (as reflected by the melting profiles).

Similarly, nucleases are probes for accessibility of DNA in segments of chromatin. As such, they are valid (although naive) models for the availability of chromosomal DNA to interaction with other proteins of more biological significance, polymerases, regulatory molecules, etc. In general, we find that core particles from more active environments appear to have higher nuclease susceptibilities; the converse is true of core particles from transcriptionally inert sources.

The implications we make from the current studies are drawn from study of complex systems, all with large genomes in varying states of transcriptive and/or replicatative activity. It seems obvious that direct tests of the possible role of core particle structure modulation in function of chromatin requires study in detail of defined, small, functional segments of chromatin.

ACKNOWLEDGMENT

We thank Mrs. Bonnie Richards for preparation of the manuscript.

REFERENCES

1. McGhee, J.D., and Felsenfeld, G. (1980) *Annu. Rev. Biochem. 49,* 1115-1156.
2. Isenberg, I. (1979) *Annu. Rev. Biochem. 48,* 159-191.
3. Easton, D., and Chalkley, R. (1972) *Exptl. Cell Res. 72,* 502-506.
4. Seale, R., and Aronson, A.I. (1973) *J. Mol. Biol. 75,* 633-645.
5. Ruderman, J.V., and Gross, P.R. (1974) *Develop. Biol. 36,* 286-298.
6. Cohen, L.H., Newrock, K.M., and Zweidler, A. (1975) *Science 190,* 994-997.
7. Newrock, K.M., Alfageme, C.R., Nardi, R.V., and Cohen, L. H. (1977) *Cold Spring Harbor Symp. Quant. Biol. 42,* 421-431.
8. Skoultchi, A., and Gross, P.R. (1973) *Proc. Natl. Acad. Sci. USA 70,* 2840-2844.
9. Gross, K.W., Jacobs-Lorena, M., Baglioni, C., and Gross, P.R. (1973) *Proc. Natl. Acad. Sci. USA 70,* 2614-2618.

10. Newrock, K.M., Cohen, L.M., Hendricks, M.B., Donnelly, R. J., and Weinberg, E.S. (1978) *Cell 14*, 327-336.
11. Childs, G., Maxson, R., and Kedes, L.H. (1979) *Develop. Biol. 73*, 153-173.
12. Whitlock, J.P., Jr., and Simpson, R.T. (1977) *J. Biol. Chem. 252*, 6516-6520.
13. Whitlock, J.P., Jr., and Stein, A. (1978) *J. Biol. Chem. 253*, 3857-3861.
14. Noll, M. (1974) *Nucleic Acids Res. 1*, 1573-1578.
15. Simpson, R.T., and Whitlock, J.P., Jr. (1976) *Cell 9*, 347-353.
16. Noll, M. (1977) *J. Mol. Biol. 116*, 49-71.
17. Lutter, L.C. (1978) *J. Mol. Biol. 124*, 391-420.
18. Gordon, V.C., Knobler, C.M., Olins, D.E., and Schumaker, V.N. (1978) *Proc. Natl. Acad. Sci. USA 75*, 660-663.
19. Wu, H-M, Dattagupta, N., Hogan, M., and Crothers, D.M. (1979) *Biochemistry 18*, 3960-3965.
20. Weischet, W.O., Tatchell, K., van Holde, K.E., and Klump, H. (1978) *Nucleic Acids Res. 5*, 139-160.
21. Simpson, R.T. (1979) *J. Biol. Chem. 254*, 10123-10127.
22. Simpson, R.T., and Shindo, H. (1979) *Nucleic Acids Res. 7*, 481-492.
23. Stein, A., Bina-Stein, M., and Simpson, R.T. (1977) *Proc. Natl. Acad. Sci. USA 74*, 2780-2784.
24. Riggs, M.G., Whittaker, R.G., Neumann, J.R., and Ingram, V.M. (1977) *Nature 268*, 463-464.
25. Simpson, R.T. (1978) *Cell 13*, 691-699.
26. Simpson, R.T., and Kunzler, P. (1979) *Nucleic Acids Res. 6*, 1387-1415.
27. Mirzabekov, A.D., Schick, V.V., Belyavsky, A.V., and Bavykin, S.G. (1978) *Proc. Natl. Acad. Sci. USA 75*, 4184-4188.
28. Simpson, R.T. (1978) *Cell 13*, 691-699.
29. Simpson, R.T., and Bergman, L.W. (1980) *J. Biol. Chem. 255*, 10702-10709.
30. Finch, J.T., Lutter, L.C., Rhodes, D., Brown, R.S., Rushton, B., Levitt, M., and Klug, A. (1977) *Nature 269*, 29-36.
31. Sures, I., Lowry, J., Kedes, L.H. (1978) *Cell 15*, 1033-1044.
32. Alfageme, C.R., Zweidler, A., Mahowald, A., and Cohen, L.H. (1974) *J. Biol. Chem. 249*, 3729-3736.
33. Simpson, R.T. (1978) *J. Biol. Chem. 254*, 5524-5531.
34. Mardian, J.K.W., Paton, A.E., Bunick, G.H., and Olins, D.E. (1980) *Science 209*, 1534-1536.
35. Sandeen, G., Wood, W.I., and Felsenfeld, G. (1980) *Nucleic Acids Res. 8*, 3757-3778.

36. McGhee, J.D., Rau, D.C., Charney, E., and Felsenfeld, G. (1980) *Cell 22*, 87-96.
37. Hinegardner, R.T. (1967) in "Methods in Developmental Biology", Wilt, F.H., and Wessells, N.K., eds., Crowell, New York.
38. Leffak, I.M., Grainger, R., and Weintraub, H. (1977) *Cell 12*, 837-846.

EVIDENCE FOR A STRUCTURAL ROLE OF COPPER IN HISTONE-DEPLETED CHROMOSOMES AND NUCLEI

Ulrich K. Laemmli, Jane S. Lebkowski
and Catherine D. Lewis

Departments of Molecular Biology and Biochemistry
University of Geneva

ABSTRACT. Recent evidence suggests that metalloprotein interactions are important in maintaining one level of organization of the DNA in histone-depleted chromosomes and nuclei. When treated with the metal chelators, 1,10-phenanthroline, neocuproine, or thiols, but not with EDTA, histone-depleted chromosomes and nuclei unfold their compacted DNA as evidenced by a reduction in their sedimentation coefficient. For nuclei, the s value is reduced by such chelation from a fast form I (18000s) to a slow form II (8500s). With chromosomes, a similar shift from a fast form I (5500s) to a slow form II (1500s) is observed. It is unlikely that this relaxation of the DNA is due to the possible degradative processes of 1,10-phenanthroline known to occur under some conditions.

Metal-depleted chromosomes and nuclei prepared in thiol buffers provided a means to identify the metal involved. Such treatments generate histone-depleted structures of the slow form II. Addition of as little as 5×10^{-7} M Cu to metal-depleted chromosomes restores the fast sedimenting form I. Other metals tested (Mn, Co, Hg, Zn) clearly do not have this effect. This selective effect of Cu and the removal of the bound metal by the chelators 1,10-phenanthroline and neocuproine, but not by EDTA, suggests

ISBN 0-12-137420-3

that this metal-protein interaction is specific and of importance for chromosome and nuclear structure.

I. INTRODUCTION

In metaphase chromosomes, histone and nonhistone proteins package the entire eukaryotic genome for ready distribution to daughter cells. The basic nucleoprotein fiber of both metaphase chromosomes and interphase nuclei has a dimension of about 250 Å and is composed of the repeating structural unit called the nucleosome (1). The sole structural determinants of this basic fiber appear to be the histone proteins. Elucidation of the long range organization of this fiber is formidable task. The relatively large size of chromosomes, the complexity of their proteins, and their enormous length of DNA, all contribute to the difficulty of studying these structures. Electron micrographs of intact chromosomes reveal a hopeless array of fibers, and there is no simple approach to the understanding of their packaging arrangement. Thus, analysis of chromosome structure must be carried out in a stepwise manner, by dissociating chromosomes into stable and homogeneous intermediates which can be characterized both structurally and biochemically. One striking feature of chromosomes is their "spongelike" behaviour. Chromosomes swell up to 50 times if they are exposed to conditions which solubilize chromatin (low ionic strength and chelation of divalent cations). Despite this increase in volume, the swollen chromosomes retain their metaphase morphology and regain their compact form if original ionic conditions are restored (2). Therefore, structural components must exist in chromosomes which crosslink the chromatin fiber into the classical metaphase shape.

Several reports indicate that the chromatin fiber of interphase nuclei is subjected to similar topological constraints. "Nucleoids" (protein-deficient cells) released from cells by lysis in NaCl and detergent exhibit a biphasic alteration in sedimentation rate with exposure to increasing concentrations of ethidium bromide (3,4,5). This evidence indicates the presence of chromatin attachment sites in interphase nuclei.

The presence of these DNA constraints has also been confirmed by Igo-Kemenes and Zachau (6), noting that the maximum length of soluble chromatin obtained by mild nuclease digestion is about 75 kb. They concluded that chromatin is attached to a framework, and that the 75 kb soluble fragment results from digestion between adjacent points of DNA attachment. The length of the constrained DNA (proposed to be loops) was also estimated to be about 85 kb in *Drosophila* tissue culture cells (3) and about 220 kb in HeLa cells (4).

Evidence for the existence of such a framework in metaphase chromosomes, composed of scaffolding proteins, comes from the finding that it is possible to remove all histone and a large number of nonhistone proteins by a variety of methods without the destruction of the overall metaphase morphology. Removal of the histones and many nonhistone proteins is achieved by competition with the polyanion dextransulfate/heparin or with 2M NaCl. The histone-depleted chromosomes can be isolated by sucrose gradient sedimentation (4000 - 7000s), and are insensitive to RNaseA. Mild trypsin treatment destroys these structures (7).

When viewed in the electron microscope, the histone-depleted chromosomes show a central scaffold surrounded by a halo of DNA. The scaffolding consists of two fibrous backbones, one for each chromatid, joined at the centromere to which the DNA is attached in a radial loop arrangement. The average loop length is 23 μm (8).

The scaffold can be isolated as a stable structure, containing less than 0.1% of the total DNA, by nuclease digestion of chromosomes prior to removal of the histones. These scaffolds have the same stability properties and contain the same nonhistone proteins as the DNA containing histone-depleted chromosomes (9).

Based on these findings, a simple chromosome model has been proposed in which the long-range order of the nucleoprotein fiber is determined by the scaffolding proteins (10). These proteins form a central network extending from one end of the chromatid to the other and hold the DNA in radially arranged loops. The role of the histones in this model is to condense the DNA loops into shorter, thicker, loops surrounding the fibrous scaffold.

We have recently obtained additional evidence for a radial loop model by studying thin sections of chromosomes (11). Under conditions which chelate divalent cations, chromosomes swell by a factor of about 4 in width and the basic 250 Å fiber relaxes to the thin 100 Å fiber. Cross-sections show a central area from which the chromatin fibers emerge in a radial fashion, often forming loops which are 3-4 μm long. Chromosomes fixed in the presence of 1 mM $MgCl_2$ are more compact, and consist of the 250 Å chromatin fiber. A radial loop arrangement can also be deduced in these chromosomes, although the loops are more difficult to visualize due to the compactness of the structure. This paper briefly reviews recent studies showing the involvement of metalloprotein interactions in stabilizing the scaffold of both chromosomes and nuclei.

II. RESULTS

A. Sedimentation studies with histone-depleted chromosomes and nuclei isolated by various methods.

The best evidence for the existence of special scaffolding proteins which maintain the histone-depleted chromosomes in a compact form comes from sedimentation studies (7). These studies provide an assessment of the heterogeneity of a population of chromosomes and illustrate possible changes induced by various means in the degree of folding of the DNA in the histone-depleted chromosomes. Electron microscopy alone is not sufficient to analyze the structure of dissociated chromosomes due to the inherent difficulty in the interpretation of the micrographs.

For sedimentation studies (^{3}H)-thymidine labeled chromosomes or nuclei are diluted into a low ionic solution containing dextran-sulfate and heparin and layered immediately on top of a sucrose gradient (7). During this treatment virtually all the histones and many nonhistone proteins are dissociated from the chromosomes by competition with polyanions. Despite this extraction, the DNA remains compacted by the so-called scaffolding proteins in a highly folded fast sedimenting form.

Initial studies with histone-depleted chromosomes were carried out with chromosomes isolated in the presence of hexylene glycol (Hex chromosomes).

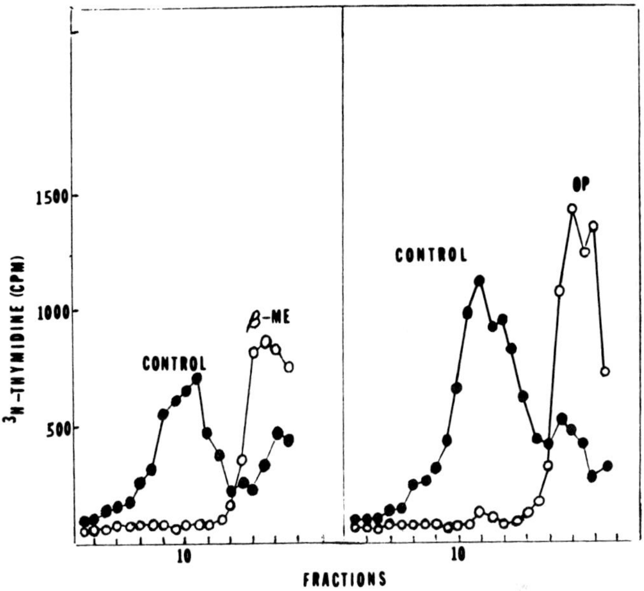

FIGURE 1. Partial unfolding of histone-depleted chromosomes with β-mercaptoethanol and 1,10-phenanthroline. Metaphase chromosomes were isolated from colcemid-treated Hela cells by a modification of the aqueous method of Marsden and Laemmli (11). The chromosomes are stabilized in this procedure with a modified reticulocyte standard buffer containing 5mM $MgCl_2$. The chromosomes were purified by a combination of sucrose and "Percoll" gradients. A detailed description of this procedure will be published elsewhere (13). We would like to point out: (a) all operations were carried out at 4^oC, (b) the nonionic detergent digitonin was used instead of NP40, and (c) the antioxidant thiodiglycol (Pierce) was included in all solutions at a final concentration of 1%. For a typical gradient we mixed about 5μl of isolated (^{3}H)-thymidine-labeled chromosomes (OD_{260}

about 1) with 200μl of lysis mixture: (10mM Tris/HCl, pH9.0, 10mM EDTA, 0.1% digitonin, 2mg/ml of dextran sulfate and 0.2mg/ml heparin). The sample was layered on a linear 5 to 30% sucrose gradient. Other details were as previously described (7). Centrifugation was at 5000 rpm in an SW50.1 rotor for 120 min. at 4°C. Panel (a): (●) gradient profile of the control sample (form I) not treated with β-mercaptoethanol; (O) gradient profile following addition of 14mM β-mercaptoethanol to the lysis mixture (form II). Panel (b): (●) gradient profile of the control; (O) gradient profile following addition of 3mM 1,10-phenanthroline to the lysis mixture.

Due to possible impurities in hexylene glycol (e.g. aldehydes) which could have stabilized the histone-depleted chromosome, it was important to repeat these experiments with chromosomes isolated by other methods. Two different chromosome preparations were used. The first was the aqueous method of Marsden and Laemmli (11) in which chromosomes are stabilized with a modified reticulocyte standard buffer (RSB) containing 5 mM $MgCl_2$ and 0.5 mM $CaCl_2$ (RSB chromosomes). The second was a modification of the method of Blumenthal et al., (12) in which the chromosomes are stabilized by the polyamines, spermine and spermidine, rather than by divalent cations.

The sedimentation profile of the histone-depleted chromosomes isolated by the RSB nethod is similar to that derived from chromosomes stabilized with hexylene glycol. The histone-depleted chromosomes sediment in both cases as a broad peak between 4000 and 7000s (compare Figure 1 in this paper to Figure 1 of ref. 7). Examination of these structures by fluorescence and electron microscopy confirms that the histone-depleted chromosomes isolated without hexylene glycol retain their metaphase morphology and contain a network of proteins along the chromatid axis to which the DNA is bound (13).

There is, however, a major difference between these two types of chromosomes. While histone-depleted chromosomes isolated from Hex-chromosomes are stable if β-mercaptoethanol is added to the lysis solution (7), addition of thiols to histone-depleted RSB chromosomes leads to dissociation and unfolding of these structures as shown by sedimentation and fluorescence microscopy (13). Figure (1) shows that addition of β-mercaptoethanol leads to an upward shift of the gradient profile due to the unfolding of the mitotic DNA. A similar observation is made with extracted nuclei (14). The DNA of histone-depleted nuclei sediments as a peak of about 18000s, while thiol treatment converts this structure to a slower form (form II) of about 8500s. Figure (2) shows the gradient profile of both these structures

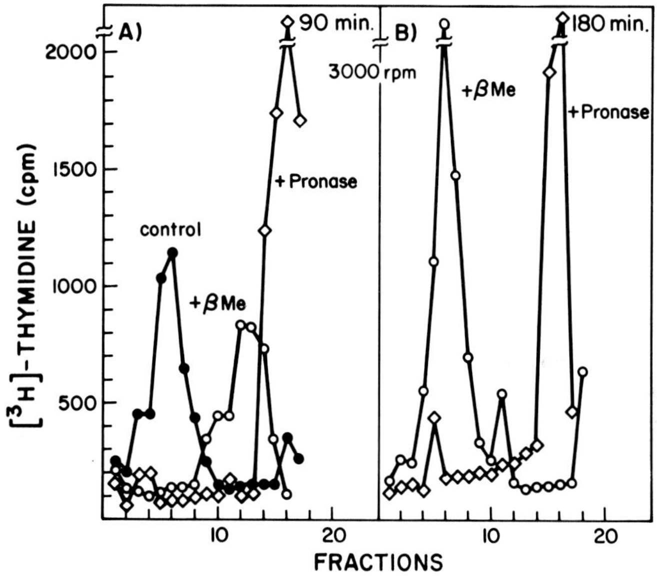

FIGURE 2. Sedimentation of histone-depleted nuclei. (^3H)-thymidine-labeled nuclei were isolated in a reticulocyte standard buffer containing 5mM $MgCl_2$ using 0.1% digitonin to disperse the cells. Details will be published elsewhere (14). For gradient studies, 3μl of nuclei (OD_{260}=5) were mixed with 200μl of a high salt lysis mixture containing 2M NaCl, 10mM Tris, pH9.0, 10mM EDTA, 0.1% digitonin and gently layered on a 5 to 20% linear sucrose gradient prepared in the same buffer. The antioxidant thiodiglycol at a final concentration of 1% was present in all isolation and extraction

solutions.

Identical results were obtained if the low salt extraction procedure described in Figure 1 was used. Panel (a): (●) gradient profile of the histone-depleted nuclei (form I) not treated with β-mercaptoethanol; (O) gradient profile (form II) following addition of 14mM β-mercaptoethanol to the lysis mixture; (◊) gradient profile following addition of pronase (5μg/ml) to the lysis mixture. Centrifugation was for 90 min at 3000 rpm in panel (a) and for 180 min. in panel (b). The untreated control (●) sediments to the bottom of the gradient under the conditions of panel (b) and was omitted from the graph.

after two different sedimentation times. Addition of β-mercaptoethanol unfolds only partially the DNA of histone-depleted nuclei, producing a shift from about 18000 to 8500s. A proteolytic enzyme is required for complete dissociation (14 and Figure 2).

At present, it is not known whether the β-mercaptoethanol treated histone-depleted chromosomes are stable structural intermediates or whether the unfolding is complete. Despite this reservation, the fast sedimentating structure will be referred to as form I and the slower sedimenting structure as form II for both chromosomes and nuclei.

The discrepancy between Hex and RSB chromosomes with respect to their sensitivity was investigated further. It was found that the elimination of the sensitivity to thiols is not caused by hexylene glycol but rather by an incubation step at 37°C in the presence of $CaCl_2$. Both histone-depleted nuclei and RSB chromosomes sedimented in their fast form I despite addition of β-mercaptoethanol (13,14) if they had been treated previously with $CaCl_2$ at 37°C. An example of such an experiment is shown in Figure 3. The histone-depleted nuclei run as the fast form I despite addition of β-mercaptoethanol if they are briefly incubated at 37°C in the presence of 0.5mM $CaCl_2$. We will show elsewhere that the conversion of form I to II by thiols is accompanied by the loss of a subset of proteins and that this loss is prevented following exposure to $CaCl_2$ (14). It should be pointed out that addition of $CaCl_2$ at 4°C is not sufficient to abolish sensitivity to β-mercaptoethanol, providing a rationale for the difference observed between Hex and RSB chromosomes. The original procedure for the isolation of Hex chromosomes requires an incubation of the cells at 37°C (15). Elimination of this step and the use of $MgCl_2$ instead of $CaCl_2$ results in the isolation of chromosomes sensitive to thiols (13).

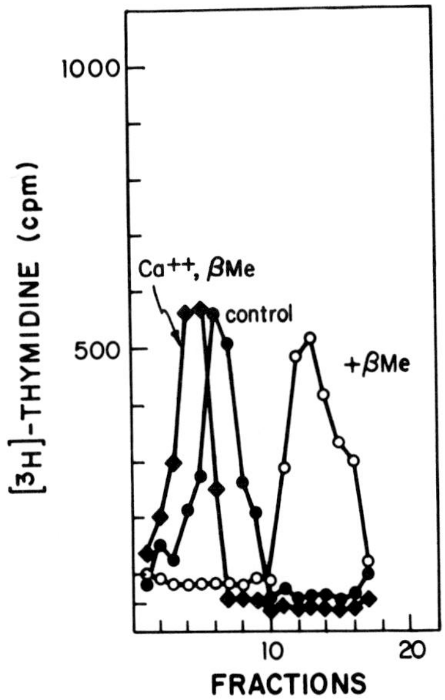

FIGURE 3. Ca^{++} can prevent the conversion of form I to II. Histone-depleted structures were studied as in the legend to figure 2, but prior to lysis nuclei were incubated with 0.5mM $CaCl_2$ at 37°C. This treatment blocks the unfolding induced by β-mercaptoethanol. (●) gradient profile of histone-depleted nuclei (form I); (O) gradient profile of histone-depleted nuclei (form II) following addition of β-mercaptoethanol to the high salt lysis mixture; (◆) $CaCl_2$ (0.5mM final concentration) was added to the nuclei and the sample incubated at 37°C for 30 min. The gradient profile shows that the histone-depleted nuclei remain in the fast form I (possibly

a little faster) despite addition of β-mercaptoethanol to the lysis mixture. Addition of $CaCl_2$ at 4^oC does not block the form I to II conversion. Centrifugation was at 3000 rpm for 90 min. A qualitatively similar observation is made with Ca^{++}-treated chromosomes.

Unfolding of histone-depleted chromosomes and nuclei: role of heavy metals.

The experiments presented above show that addition of β-mercaptoethanol dramatically reduces the sedimentation coefficient of the histone-depleted structures. To a large extent this reduction must be due to unfolding of DNA rather than loss of mass. Although precise viscoelasticity measurements have not yet been carried out, it is easily observed that histone-depleted chromosomes or nuclei are considerably more viscous in the presence of β-mercaptoethanol. This observation is consistent with the idea that the DNA is in a more unfolded conformation. An understanding of this unfolding phenomenon is of importance since it could possibly provide an insight into chromosome assembly, decondensation and the identification of the structural components involved.

Various possibilities have been considered for the effect of thiols on chromosome structure; β-mercaptoethanol could reduce disulfide bonds which are formed during isolation and which stabilize the structure, or thio-enzymes could be activated (e.g. protease or phosphatase) which in turn could damage or alter the structure. Experiments to detect important disulfide bonds or thio-enzymes were negative. Particularly telling was the finding that the reducing agent sodium borohydride was unable to bring about chromosome unfolding, suggesting that reduction was not the mode of action of β-mercaptoethanol (13,14). An alternative possibility is that β-mercaptoethanol acts as a chelator to dissociate structurally essential metalloprotein (s). Strong support for this idea comes from experiments which demonstrate that the metal chelators 1,10-phenanthroline (OP) or neocuproine (2,9-dimethyl-1,10-phenanthroline) can bring about the unfolding of the histone-depleted complexes (13,14).

The gradient profile in Figure 1 shows that OP mimicks thiols in the unfolding of the histone-depleted chromosomes. An identical observation was also made with histone-depleted nuclei showing that the form I to II conversion is brought about by these chelators (14). The chelator OP is known to degrade DNA under certain conditions requiring Cu(II), a reducing agent and oxygen (16). It is unlikely that chromosome unfolding is due to this degradative effect since it is carried out in the absence of a reducing agent, in the presence of EDTA, and under a N_2 atmosphere. EDTA as well as the specific chelator neocuproine prevent the type of DNA degradation caused by OP (16), yet neocuproine alone is sufficient to unfold these structures (13).

These observations suggest that the form I to II transition is due to the removal of a metal from a structurally important protein(s) of form I. Following removal of the metal the metalloprotein(s) loses all or part of its structural contribution, leading to the relaxation of the folded DNA. EDTA alone is not effective in this transition, providing support for the specificity of the metallo protein interaction. The pH sensitivity of this phenomenon has been studied in nuclei, and it was found that this conversion is blocked at pH 6.5 but maximal above pH8.(14).

Identification of the metal involved.

Metal-depleted chromosomes can be obtained by isolation or washing of Hex or RSB chromosomes with β-mercaptoethanol (13). Such chromosomes generate histone-depleted structures of the slow form II and thus provide a means to test whether addition of a given metal restores the fast form I. Also suitable for these experiments are a type of chromosome isolated in the presence of polyamines and EDTA, according to a modified published procedure (13). These chromosomes are stable, appear to be relatively undamaged by proteolytic enzymes and nucleases, and generate histone-depleted structures of the slow form II without the addition of mercaptoethanol. These polyamine chromosomes, as well as the β-mercaptoethanol-treated Hex and RSB chromosomes, were tested with Mn, Cu, Zn, Co and Hg in order to determine

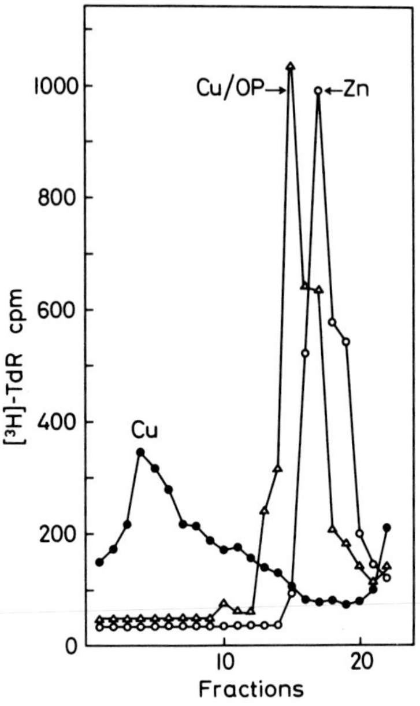

FIGURE 4. Identification of the metal involved in the maintenance of the compact form I. Chromosomes stabilized by polyamines were isolated by a modification of a published procedure (12) in the presence of EDTA. These chromosomes produce histone-depleted structures of the slow form II. To test various metals for their capacity to restore the fast sedimenting form I, the isolated chromosomes were diluted into buffers containing 0.1mM of $CuSO_4$, $ZnSO_4$, $CoCl_2$ or $Hg(C_2H_3O_2)_2$. EDTA at final concentration of 2mM was added 5 min. after dilution of the chromosomes into the metal-containing buffers. The various samples were assayed by gradient sedimentation as described in Figure 1. (●) gradient profile of histone-depleted chromosomes following treatment with $ZnSO_4$; (O) treatment of the chromosomes with $CuSO_4$; (Δ) 1,10-phenanthroline was added (3mM final concentration) to the lysis mixture to reverse the effect of $CuSO_4$.

if the fast form I could be restored bv the addition of a metal. The results show clearly that the only metal which is effective in regenerating the form I structure is Cu. Similar observations were made with nuclei. An example of such an experiment is illustrated in Figure 4. Addition of OP to copper-treated chromosomes leads to reversion and unfolding of the induced fast form I. In these experiments, all metal additions were performed under N_2 atmosphere to prevent oxidation. If this precaution is omitted, form I is still induced by Cu but is no longer reversible with OP. The fact that β-mercaptoethanol is then required for reversion to form II implies that disulfide bonds are formed during the addition of Cu under oxidizing conditions.

These experiments suggest that copper is required to stabilize the chromosomal and nuclear scaffolding by forming a metalloprotein complex. The stabilization by copper can be dissociated by thiols or by the hydrophobic chelators OP and neocuproine, but not by EDTA. The dissociation of the metal is reversible as assayed by sedimentation, indicating that copper has not caused irreversible damage.

It is important to point out that the addition of the various metals was done at the level of intact chromosomes or nuclei prior to extraction of the histones. Preliminary experiments indicate that addition of copper to the histone-depleted form II does not restore the fast form I.

Although the binding site(s) for copper has not yet been investigated in detail, there is some evidence that sulfhydryl groups are involved. If metal depleted chromosomes are treated with iodoacetate or mercury acetate to block reduced sulfhydryl groups, subsequent addition of Cu does not restore the fast form. Conversely, if chromosomes are treated with copper first and then exposed to iodoacetate or mercury acetate, the Cu-induced fast form I structure is not affected by the sulfhydryl reagents.

DISCUSSION

The initial experiments concerning the structure of histone-depleted chromosomes were carried out with chromosomes isolated in the presence of hexylene glycol and $CaCl_2$ (7, 8, 9). These chromosomes have a good morphology and can be obtained quite free of cytoplasmic contaminants (15). The use of nonphysiological media, such as the organic solvent hexylene glycol, could introduce structural alterations in chromosomes. Consequently, it was necessary to repeat the basic experiments with chromosomes isolated by an aqueous procedure. The experiments reported here confirm and extend the earlier studies. Chromosomes isolated using aqueous solutions containing divalent cations also remain in a compact and fast sedimenting form when the histones and many non-histone proteins are extracted. These histone-depleted chromosomes have a similar sedimentation profile as those isolated with hexylene glycol and also retain an expanded metaphase morphology if examined in the fluorescence or electron microscope (13). Thus, the possibility can be ruled out that hexylene glycol produces artifactual protein interactions which maintain the structure of the histone-depleted chromosomes.

Similar observations are made with histone-depleted nuclei (14). These structures display a discreet gradient profile (form I), confirming similar observations by others (17). In both chromosomes and nuclei, the histone-depleted structure is destroyed by proteolytic enzymes, but not by ribonuclease A. Therefore, scaffolding proteins must exist which maintain the DNA of both metaphase and interphase chromosomes in a compact, looped configuration. The folded DNA in the former case retains the overall morphology of expanded metaphase chromosomes (8) and in the latter the spherical shape of expanded nuclei (5).

Our experiments demonstrate that important metalloprotein interactions are involved in the stability of the fast sedimenting form I derived from nuclei and chromosomes. Addition of thiols, or of metal chelators (1, 10 phenanthroline or neocuproine) leads to unfolding of the fast sedimentation form.

Neither sodium borohydride nor EDTA can duplicate this effect. For nuclei, the sedimentation coefficient is reduced from the fast form I (18000s) to the slow form II (8500s). This slow form II has a structure which maintains an overall spherical shape of expanded nuclei. For chromosomes a similar shift is observed from an average of 5500s for the fast form I to the slow form II of about 1500s. This latter structure is extremely expanded, and its sedimentation behavior not well characterized. It is, therefore, not clear whether the DNA of the slow form II is still organized in a residual structure as in the case of nuclei.

It is unlikely that the observed unfolding is due to DNA degradation induced by 1,10-phenanthroline. Neocuproine does not have this same capacity to degrade DNA (16), yet unfolds these structures as efficiently. Furthermore, EDTA present in the lysis mixture strongly inhibits the degrading capacity of 1,10-phenanthroline.

The experiments presented here suggest that the scaffold-bound metal is copper. Addition of copper to metal-depleted chromosomes restores the fast sedimenting form I. Other metals (Zn, Mn, Co and Hg) tested clearly do not have this effect. Experiments to determine the minimum concentration of Cu needed to restore the fast sedimenting form I have not been completed. However, this concentration must be below 5×10^{-7} M since this amount is sufficient to produce the effect.

The observed selective effect of copper and the removal of the bound metal by the chelators 1,10-phenanthroline and neocuproine, but not by EDTA, suggests that this metal-protein interaction is specific and of importance for chromosome structure. It argues against the possibility that the scaffolding proteins are a random aggregation of non-histone proteins.

Identical experiments with isolated nuclei demonstrate that addition of Cu to the metal-depleted nuclei induces the fast form I as assayed by sedimentation (14). This evidence suggests that the histone-depleted chromosomes and nuclei share common structural components.

We have mentioned above that EDTA added to the lysis mixture is not able to chelate the protein-bound metal possibly because of steric hindrance. Preliminary experiments indicate that EDTA might be able to remove the protein-bound metal from chromosomes at low ionic strength. We also learned that chromosomes isolated in the presence of polyamines and EDTA without β-mercaptoethanol appear to be metal-depleted and generate the form II histone-depleted structures. These observations indicate that EDTA is able to remove the bound metal during the process of isolation or to prevent possible artifactual association in the first place. It is, of course, very difficult to prove that the copper ions are naturally associated with chromosomes, but the specificity of the interactions observed strongly favor this concept. Additional experimentation will be required to study this point.

Our observations could be related to those of Ide et al. (18). These authors have presented evidence for a supercoiled DNA-protein complex isolated from cells with a lysis procedure using SDS. Their isolated structures have a sedimentation coefficient of about 350s and must therefore be in much more expanded configuration than our histone-depleted structures which sediment with an S value of 18000s. However, they also observed a reduction of the S value from about 350 to 150s when a reducing agent was added to the lysis mixture. It is not clear at present what the relationship is between their and our observations.

Isolation of chromosomes and nuclei in the presence of 0.5 mM $CaCl_2$ or addition of $CaCl_2$ at a later stage leads to a general "toughening" of chromosomes and nuclei if incubated at 37°. This is manifested by the sedimentation behavior of the histone-depleted structures derived from these chromosomes and nuclei. After such treatment, only the fast sedimentation form I is found, irrespective of the addition of thiols. The proteins that are normally extracted by the addition of thiol during the form I to II conversion remain bound in Ca^{++}-treated chromosomes and nuclei (13, 14 and below).

We will report elsewhere our study of the non-histone proteins associated with the form I and II structures. These experiments are more advanced in the case of nuclei (14). We found that about 15 %

of (^{35}S)-labeled total nuclear proteins are associated with the fast form I. SDS gel electrophoresis shows a complex pattern of proteins. Only about 3% of the total (^{35}S)-labeled nuclear proteins remain bound to the form II structure obtained if β-mercaptoethanol or 1,10-phenanthroline is added to the lysis solution. The protein pattern of the form II structure is much less complex, consisting primarly of 3 to 4 bands in the 60 to 70 000 dalton range. Three of these major bands must correspond to the main structural components of the so-called peripheral nuclear lamina (19, 20).

In order to identify the proteins which could link the DNA to the nuclear scaffold, DNA binding studies were carried out. About 12 proteins in the type I structure and 4 proteins in the type II structure bind DNA including the 3 major lamina proteins (14). This evidence suggests that the peripheral lamina is involved in the long-range organization of the nuclear DNA. We would like to propose two levels of organization of the DNA in histone-depleted nuclei. In the type I structure, the DNA is attached at one level to the peripheral lamina of the nucleus and at a second level to a metalloprotein structure. We do not know the structural location of these metalloproteins in the nucleus but they could form an interior network. Removal of the metal dissociates this structure leading to partial unfolding of the DNA. However, attachment of the DNA to the peripheral lamina is maintained.

Our studies with the scaffolding proteins of metaphase chromosome is less complete, but again we find that the form I to II transition is accompanied by dissociation of most, if not all, of the proteins bound to the fast form I.

ACKNOWLEDGMENTS

This investigation was supported by the Swiss National Science Foundation (3.621.80) and by U.S. Public Health Service Grant GM-2511 from the National Institutes of Health.

REFERENCES

1. McGhee, J.D., and Felsenfeld, G. (1980). Ann. Rev. Biochem. 49, 1115.
2. Cole, A. (1967). Biophys. 1, 305.
3. Benyajati, C., and Worcel, A. (1976). Cell 9, 393.
4. Cook, P.R., and Brazell, I.A. (1976). J. Cell Sci. 22, 287.
5. Vogelstein, B., Pardoll, D.M., and Coffey, D.S. (1980). Cell 22, 80.
6. Igo-Kemenes, T., and Zachau, H.G. (1978). Cold Spring Harbor Symp. Quant. Biol. 42, 109.
7. Adolph, K.W., Cheng, S.M., and Laemmli, U.K. (1977). Cell 12, 805.
8. Paulson, J., and Laemmli, U.K. (1977). Cell 12, 817.
9. Adolph, K.W., Cheng, S.M., Paulson, J., and Laemmli, U.K. (1977). Proc. Nat. Acad. Sci. USA 74, 4937.
10. Laemmli, U.K., Cheng, S.M., Adolph, K.W., Paulson, J.R., Brown, J.A., and Baumbach, W.R. (1978). Cold Spring Harbor Symp. Quant. Biol. 42, 351.
11. Marsden, M.P.F., and Laemmli, U.K. (1979). Cell 17, 849.
12. Blumenthal, A.B., Dieden, J.D., Kapp, L.N., and Sedat, J.W. (1979). J. Cell Biol. 81, 255.
13. Lewis, C.D., and Laemmli, U.K., Manuscript in preparation.
14. Lebkowski, J.S., and Laemmli, U.K., Manuscript in preparation.
15. Wray, W., and Stubblefield, E. (1970). Exp. Cell Res. 59, 469.
16. Que, B.G., Downey, K.M., and So, A.G. (1980). Biochem. 19, 5987.
17. Adolph, K. (1980). J. Cell Sci. 42, 291.
18. Ide, T., Nakane, M., Anzai, K., and Andoh, T. (1975). Nature 258, 445.
19. Gerace, L., and Blobel, G. (1980). Cell 19, 277.
20. Jost, E., and Johnson, R.T. (1981). J. Cell Sci. 47, 25.

DROSOPHILA HIGH MOBILITY GROUP PROTEINS[1]

J.C. Wooley, J.S. Park, M.S. McCoy and Su-yun Chung

Department of Biochemical Sciences
Princeton University
Princeton, New Jersey 08544

ABSTRACT High mobility group proteins or HMGs are thought to be involved in modulating nucleosome and chromatin structure and may actively participate in a number of different nuclear processes including transcription. We have attempted to identify HMGs of *Drosophila melanogaster* and to ascertain the locations of HMGs along salivary gland polytene chromosomes. Following protocols developed for higher eukaryotes, we find a set of HMG-like *D. melanogaster* polypeptides ranging in size from 13,000 to 65,000 daltons in apparent molecular weight on sodium dodecyl sulfate gel electrophoresis. The most abundant polypeptide in the *Drosophila* extracts is H1. The next most abundant polypeptides have apparent molecular weights of 65,000 and 16,000 daltons, respectively. Several minor polypeptides run in the region of HMGs of higher eukaryotes. However, the salt and acid solubility of these insect polypeptides differ from HMGs of higher eukaryotes, and no clearly homologous polypeptides to HMG 1 and 2 are seen in gel electrophoretic patterns. Invertebrate HMGs may differ substantially from their mammalian and avian counterparts.

An alternate and more powerful approach to identifying invertebrate HMGs and studying their function is to utilize immunological crossreactivity. Antibody directed against mammalian HMG 14 and 17 was prepared by immunizing rabbits with mixtures of calf thymus HMG 14 and 17 (purified by sodium dodecyl sulfate gel electrophoresis). Based on indirect immunofluorescent analysis of cytological preparations of salivary glands, the HMG antisera preferentially bind to sites of current transcriptional activity or puff sites on *Drosophila melanogaster* polytene chromosomes and also to sites of previous activity or developmental puff sites. Binding of the antisera to the arms is at least several times less extensive than that observed for puff sites. More antisera is observed at active puff sites than at quiescent developmental loci, but this may reflect relative accessibility of the antigenic determinants.

[1]This work was supported by NIH Grant GM26332 and ACS Grant CD-15.

ISBN 0-12-137420-3

INTRODUCTION

Recent experimental and conceptual progress has provided a paradigm for analyzing the detailed features of the organization and function of chromatin. We are not yet able to relate the details of structure to the function of chromatin and, in particular, the functions of the nonhistone nuclear proteins remain unclarified. One partial exception is an apparent class of abundant, highly charged proteins known as high mobility group proteins or HMGs, first identified by Goodwin and Johns (1) as low molecular weight proteins rich in both basic and acidic amino acids. In avians and mammals, four major tissue-independent HMGs have been identified. These appear to fall into two families, one containing HMGs 1 and 2, the other the HMGs 14 and 17. No function for HMGs 1 and 2 has been unambiguously demonstrated, but strong, direct experimental evidence implicating the presence of HMG 14 and 17 in regions of transcriptionally active chromatin has been obtained during the two years (2,3), following many previous reports suggesting such an involvement (e.g., 4,5).

Despite the indications that HMGs are universal, conserved components of chromatin, no canonical HMGs have been unequivocally demonstrated in *Drosophila*. We wish to examine the roles of HMGs, exploiting the synergistic advantages of *Drosophila*, particularly the salivary gland polytene chromosomes. In this report, we describe our attempts to identify and characterize *D. melanogaster* polypeptides homologous to the HMGs of higher eukaryotes. We have also prepared antisera directed against mammalian HMG 14 and HMG 17 and utilized the sera to examine the distribution of HMGs in salivary gland cytological preparations, comparing their location along chromosomes and between the nucleus and the cytoplasm.

METHODS

Calf thymus HMGs were prepared according to Goodwin and Johns (1), as were HMG-like polypeptides from purified *Drosophila* nuclei. Nonionic detergent-washed, sucrose step-gradient purified nuclei were prepared from frozen *Drosophila melanogaster* embryos (0-18 hrs.) following Wu et al. (6). Sodium dodecyl sulfate gel electrophoreses were carried out following Laemmli (7). The methods of Silver and Elgin (8) were used for preparation of antisera to purified calf thymus HMG 14 and 17, and for the immunofluorescent cytological localization experiments.

RESULTS

1. Isolation of Drosophila HMG-like Polypeptides

HMGs have been operationally defined (9) for higher eukaryotes as 5% perchloric acid (PCA) soluble non-histone nuclear proteins which are also (a) co-extracted with histones by acid; (b) extracted by 0.35 M sodium chloride; (c) soluble in low amounts of trichloracetic acid (TCA); (d) capable of binding both DNA and histones. Taking this operational definition as a starting point, we have examined Drosophila chromatin for the presence of HMG-like polypeptides. A standard PCA extract of nonionic detergent-washed, sucrose step-gradient purified nuclei extracts a set of polypeptides, as shown in Figure 1; the molecular weights of the polypeptides consistently observed are given in Table 1. For the purposes of clarity at this stage of analysis, we arbitrarily identify the smaller PCA-extracted polypeptides as P1 through P8. (PCA-soluble fragments not labeled on the gel are qualitatively variable, that is, not always detectable, and appear to be H1 and D1 degradation products). H1 is the most abundant component of the extract, just as in mammals. The next most abundant polypeptides are present at roughly 5% of the mass of H1, one running at 65,000 daltons and one at 16,000. The high molecular weight polypeptide was first identified by Alfageme et al. (10) and termed D1. It has an HMG-like amino acid composition but a true molecular weight of about 50,000 daltons, which is almost twice that of the largest HMGs of higher eukaryotes. Alfageme et al. (10) have previously observed that D1 is widely distributed along polytene chromosomes with some preferential sites, but is not associated notably with active gene regions, that is, with puff sites. The 16,000 dalton polypeptide, which we term P6, is approximately present in equimolar quantities with D1. P6 is probably not a degradation product of D1 or H1 based on partial peptide mapping by the method of Cleveland et al. (11) and its stoichiometric reproducibility (data not shown).

Four less abundant polypeptides are consistently present in PCA-extracts and in the acid soluble fraction of 0.35M NaCl extracts from embryos of D. melanogaster (and from the Kc D. melanogaster cell line). Two, termed P4 and P5, migrate slightly slower on SDS gels than P6; two run slightly faster and are termed P7 and P8 (not resolved on the short, 15% acrylamide gel shown in Figure 1). Based on the previous studies of the HMGs of avians and mammals, the minor polypeptides P4, P5, P7 and P8 are present in too low abundance

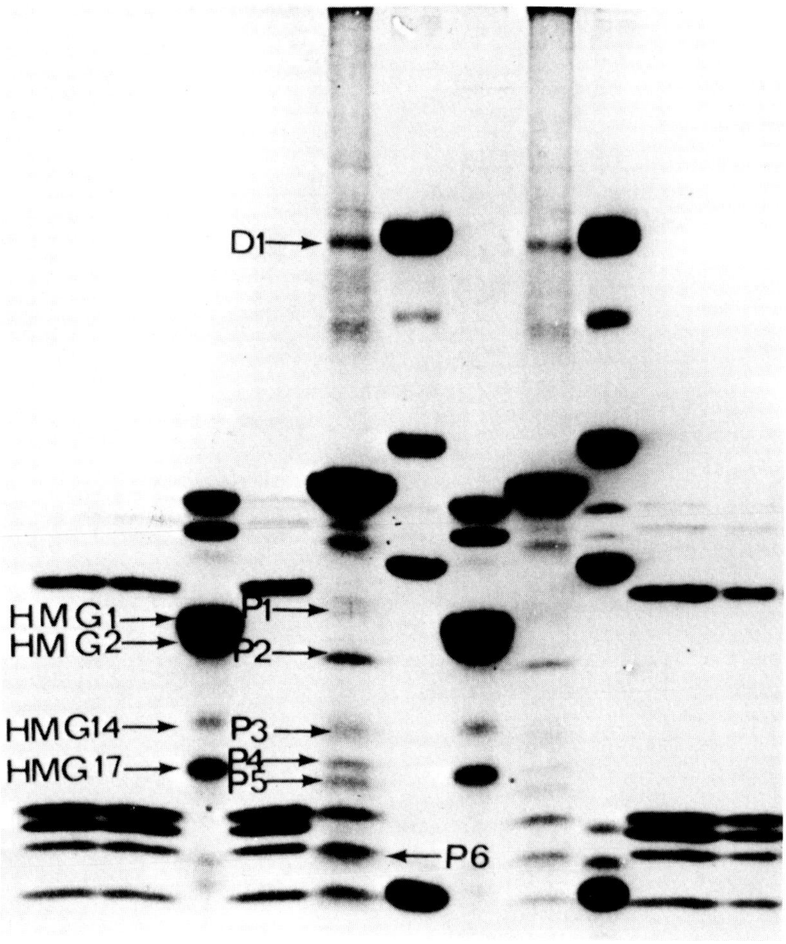

FIGURE 1. Comparison of Calf HMGs and Drosophila melanogaster HMG-like polypeptides.
Protein aliquots were electrophoresed on 15% SDS-polyacrylamide gel. Conventional protein standards for estimating molecular weight are in lanes 6 and 9. These are bovine serum albumin (68,000), aldolase (40,000), chymotrypsin (27,500) and lysozyme (14,300). Histone standards, prepared from chick erythrocytes, are in lanes 1, 2, 4, 10 and 11. HMG extract prepared from calf thymus and including H1 and HMG1, 2, 14 and 17 is in lanes 3 and 7. Lanes 5 and 8 are the 5% PCA extract of D. melanogaster embryonic nuclei, prepared from (0-18 hour) embryos using protease inhibitors.

to play general roles in gene activity analogous to HMG 14 and 17. Indeed, ignoring abundance arguments, initial candidates for invertebrate HMGs in the extract would appear to be P1, P2 (particularly) and P3, since these polypeptides have similar mobilities to HMG 1, 2 and 14, respectively. Furthermore, P2 is apparently a discrete species which can be present in substantial quantities as in the example shown. However, partial peptide mapping (11) indicates that P1, P2 and P3 are probably H1 degradation products, a result also indicated by our studies with H1 antisera; the remaining minor polypeptides are probably not H1 degradation products based on the same types of analyses and amino acid composition (data to be published elsewhere). Mammalian HMGs (and H1) are also highly sensitive to protease degradation (9).

We have examined the solubility and extraction properties of the Drosophila HMGs in terms of the known properties of HMGs of higher eukaryotes. The overall pattern we observe for the Drosophila polypeptides, outlined in part in Table 2, differs substantially from the mammalian pattern, and no clear clues emerge by comparing this pattern to that observed for avians and mammalian HMGs (1,9). For example, D1 is not a canonical HMG due to its high molecular weight, but it does follow the extraction properties of HMGs (and Alfageme et al. (10) have shown that D1 has an HMG-like amino acid composition. On the other hand, the differences in SDS mobilities are probably not significant for the low molecular weight polypeptides, since HMG 14 and 17 are highly anomalous in their migration properties (10). Final conclusions must await functional definitions, rather than operational ones, but it is clear that the HMG of Drosophila differ from their mammalian counterparts.

TABLE 1
D. melanogaster PCA-soluble Polypeptides

Protein	Apparent MW (Daltons)
D1	66,000
H1	37,000
P1	27,800
P2	24,200
P3	20,200
P4	19,000
P5	18,300
P6	16,000
P7	14,500
P8	13,500

TABLE 2

Drosophila Nonhistone Nuclear Protein Fractions

Extract	Major Polypeptides	Minor Polypeptides
1) 0.35 M NaCl followed by 2% TCA cut and acetone precipitation of 2% TCA soluble polypeptides	H1	D1, P6
2) 0.35 M NaCl extract followed by 1% TCA precipitation	50,000 dalton doublet	Many minor bands
3) 1% TCA solution followed by 10% TCA precipitation	50,000 dalton doublet and H1	Faint D1, P6 bands plus many minor bands
4) 10% TCA solution followed by 15% TCA precipitation	H1, D1, P6	Many minor bands
5) 5% TCA	D1, H1, and all 4 inner histones	50,000 dalton doublet
6) 0.4 N H_2SO_4 after 5% TCA	Inner histones	H1
7) 5% PCA	H1, D1, and P6	Many minor bands but reproducibly including only P4-P8

2. Distribution of HMG 14 and 17 in Drosophila Polytene Chromosomes

An alternative approach to identifying Drosophila HMGs is to exploit antibody cross-reactivity. We, therefore, prepared rabbit antisera to HMG 14 and 17, purified from calf thymus total HMG extract by column chromatography and gel electrophoresis. The IgG fraction was used for immunocytological localization on polytene chromosomes from late third instar larva. The unique specificity of the heat shock response provides the best test of the reactivity of antibodies with sites of transcriptional activity. The heat shock genes and their corresponding puffs were induced by

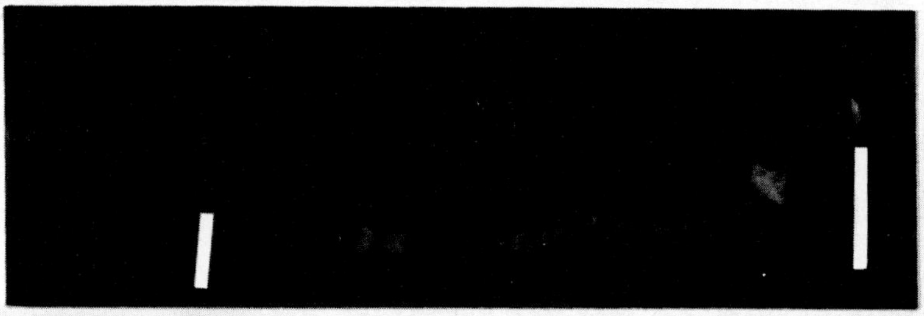

FIGURE 2. Immunofluorescent Localization of HMG 14 and 17 Determinants in D. melanogaster Polytene Chromosomes. A segment of the 3R arm, including 4 heat shock puffs, is shown. Larvae were exposed to 37°C for 20 minutes to induce the heat shock puffs and cause regression of previously active puff sites. Following Silver and Elgin (7), a squash preparation of formaldehyde pre-fixed salivary glands were incubated with calf antisera against calf HMG 14 and 17, and the binding sites visualized by incubation with rhodamine-conjugated goat anti-rabbit immunoglobin sera.

exposing larvae for 20 min to 35°C. A representative anti-(14 and 17) staining pattern (visualized by rhodamine-conjugated goat anti-rabbit sera) of the chromosomal region with the most prominent heat shock puffs (87A, 87C, 93D, 95D) is shown in Figure 2. The antisera very preferentially stain the heat shock puffs; the pre-existing developmental puff sites continue to bind antisera over the background of the arms but at a reduced level compared to the pattern observed for larvae grown at normal temperatures (for example, the prominent developmental puff 85F is still stained after heat shock; see Figure 2).

DISCUSSION

Avian and mammalian HMGs are known to be very similar in amino acid sequence, which has suggested that HMGs are conserved in evolution. However, we find that Drosophila HMG-like polypeptides do not correspond in a clear and simple fashion to the HMGs of higher eukaryotes. Similarly, yeast HMGs appear to be highly divergent from their mammalian and avian counterparts (12). Trout has two types of HMGs, one termed HMG-T, corresponding to the HMG 1 and 2 family, and one termed H6, corresponding to the HMG 14 and 17 family (13). It seems possible that the major Drosophila HMG-like polypeptides, D1 and P6, also correspond functionally to the two different major HMG families. We are currently examining the products of nuclease digestion of Drosophila chromatin as one way of ascertaining the location and role of D1 and P6. In particular, we have observed that D1 behaves like H1 in chromatin fractionation on nucleoprotein gels (14), and is probably bound to the linker region of nucleosomes. D1-containing nucleosomes can be separated from H1-containing nucleosomes by salt fractionation following Levy and Dixon (15). In this regard, it may be worth remembering that HMGs were first isolated as contaminants of H1 (1) and HMG 1 and 2 could be considered minor histones, rather than major non-histones.

The Drosophila determinants recognized by the antisera against calf HMGs 14 and 17 are very preferentially present and accessible in transcriptionally-active sites, and less so in sites of previous transcriptional activity. The experiments of Weintraub and colleagues (2,3) clearly demonstrate the preferential association of HMG 14 and 17 with potentially active genes; these and other experiments (4,16,17) argue for a 5-10 fold enrichment of HMGs in active genes. Our fluorescent distribution patterns are a direct visual confirmation of the biochemical data, although in these cytological experiments it is hard to separate the contributions of relative distribution and relative accessibility. A comparison of the developmental puff sites in control and heat-shocked larvae suggests that HMG determinants are accessible even in condensed chromatin, and that HMGs 14 and 17 are indeed preferentially associated with puff sites in polytene chromosomes.

The antisera provides us with a probe for examining changes in protein distribution on polytene chromosomes during the latter stages of larval development. The timing of HMG 14 and 17 association with a developmentally expressed gene can be determined, for example. We are currently employing the HMG antisera to determine which polypeptides in Drosophila react and are presumably homologous to HMG 14 and 17 in higher eukaryotes.

ACKNOWLEDGMENTS

We acknowledge the efforts of Margaret (Peggy) Hsieh, during her undergraduate thesis research, in initiating studies in our lab on HMGs and for preparing the first samples of purified calf thymus HMG 14 and 17. We thank Sarah Elgin, Lee Silver, Len Cohen, Candido Alfageme and Michael Bustin for advice on immunofluorescent cytological techniques and general *Drosophila* methods.

REFERENCES

1. Goodwin, G.H., Sanders, C. and Johns, E.W. (1973). Eur. J. Bioch. 38, 14.
2. Weisbrod, S. and Weintraub, H. (1979). Proc. Natl. Acad. Sci. 76, 631.
3. Weisbrod, S., Groudine, M. and Weintraub, H. (1980). Cell 19, 289.
4. Levy, W.B., Wong, N.C.W. and Dixon, G.H. (1977). Proc. Natl. Acad. Sci. 74, 2810.
5. Levy, W.B. and Dixon, G.H. (1978). Nucl. Acid Res. 5, 4155.
6. Wu, C., Bingham, P.M., Livak, K.J., Holmgren, R. and Elgin, S.C.R. (1979). Cell 16, 797.
7. Laemmli, U.K. (1970). Nature 227, 680.
8. Silver, L. and Elgin, S.C.R. (1977). Cell 11, 971.
9. Goodwin, G.H., Walker, J.M. and Johns, E.R. (1978). p. 181. In: The Cell Nucleus, Vol. VI, ed. Busch, H., Academic Press, New York.
10. Alfageme, C.R., Rudkin, G.T. and Cohen, L.H. (1980). Chromosoma 78, 1.
11. Cleveland, D.W., Fischer, S.G., Kirschner, M.W. and Laemmli, U.K. (1977). J. Biol. Chem. 252, 1102.
12. Weber, S. and Isenberg, I. (1980). Biochemistry 19, 2236.
13. Levy-Wilson, B., Kuehl, L. and Dixon, G. (1980). Nucl. Acid Res. 8, 2859.
14. Levinger, L. and Varshavsky, A. (1980). Proc. Natl. Acad. Sci. 77, 3244.
15. Levy-Wilson, B. and Dixon, G.H. (1979). Proc. Natl. Acad. Sci. 76, 1682.
16. Sandeen, G., Wood, W.I. and Felsenfeld, G. (1980). Nucl. Acid Res. 8, 3757.
17. Albanese, I. and Weintraub, H. (1980). Nucl. Acid. Res. 8, 2787.

REGULATORY FACTORS INVOLVED IN THE TRANSCRIPTION OF MOUSE RIBOSOMAL GENES

Ingrid Grummt
Gert Pflugfelder

Institut für Biochemie der Universität Würzburg, Röntgenring 11, D-8700 Würzburg, G.F.R.

ABSTRACT The transcription of ribosomal RNA correlates with the proliferation rate of the cells. In order to investigate the molecular mechanisms underlying the switch-on and switch-off of rDNA expression a cell-free system for the transcription of ribosomal genes was established. Extracts derived from cultured mouse cells contain the factor(s) required for the accurate initiation of RNA polymerase I on cloned mouse rDNA. After nutritional shift-down the ability of the cell extracts to promote specific transcription is lost. The ability of active extracts to complement inactive extracts from non-growing cells has been used to isolate the proteins which regulate the transcription of rDNA. So far, an activating factor has been purified more than 1000-fold by ion exchange and affinity chromatography. Furthermore, another protein which inhibits rDNA transcription has been identified. It is suggested that positive and negative effectors regulate the initiation frequency of RNA polymerase I.

INTRODUCTION

Ribosomal genes provide an attractive system to study control of gene expression in higher organisms since their rate of transcription is regulated in response to extracellular signals. Studies on the molecular mechanisms which correlate the rDNA transcriptional activity with the proliferation rate of the cells indicated that a more or less efficient transcription of rDNA is brought about by changes of the initiation frequency of RNA polymerase I on the ribosomal genes (1). However, the basic components are still unknown that modulate

ISBN 0-12-137420-3

the efficiency of RNA polymerase to recognize and interact with defined DNA sequences necessary for promotion of rRNA transcription.

One major approach in the investigation of the mechanisms involved in the specificity and regulation of gene expression was the development of crude cell-free systems that faithfully transcribe cloned genes (2-4). Here we show that a cytoplasmic extract (S-100) derived from rapidly growing mouse cells contain the factor(s) essential for the specific read-out of mouse ribosomal genes. Fractionation of the extracts by ion exchange and affinity chromatography revealed the existence of at least two proteins which quantitatively affect rDNA transcription. It is suggested that the ratio of activating and inhibitory factor(s) modulate the initiation frequency of RNA polymerase I both _in vitro_ and _in vivo_.

METHODS

DNA templates. The cloning and characterization of mouse rDNA fragments containing the initiation site of 45S pre-rRNA transcription has been published (5,6). To assay for specific run-off transcripts, the recombinant plasmid pMrSalB was truncated with a restriction enzyme, extracted with phenol, chloroform and ether, dissolved at 500 µg/ml in 20 mM Tris-HCl, 7.5, 0.1 mM EDTA and used at 20 µg/ml in the standard transcription assay.

In vitro transcription of rDNA. Transcriptionally active S-100 extracts, prepared as described by Weil et al. (3), were derived from Ehrlich ascites tumor cell cultured at 8 x 10^5 cells per ml in RPMI medium. The protein concentration of the S-100 extracts was 25-30 mg/ml. The standard RNA synthesis reaction mixture contained in 50 µl:1 µg template DNA, 600-750 µg S-100 proteins, 10 mM Hepes (pH 7.9), 10 % glycerol, 75 mM KCl, 0.5 mM dithiothreitol, 10 mM creatine phosphate, 0.6 mM each of ATP, CTP, UTP, 25 µM GTP and 5 µCi [^{32}P] GTP. The reaction mixture which contained 20 - 30 units RNA polymerase I was incubated at 30° C for 60 min. The samples were processed and analyzed on 4 % acrylamide gels as described (3,7). The S 1 nuclease mapping of the _in vitro_ transcripts was performed essentially according to Berk and Sharp (8).

Purification of RNA polymerase I from Ehrlich ascites cells. Polymerase I was solubilized according to Roeder and Rutter (9). The supernatant was precipitated with 20 % (w/v) polyethyleneglycol 6000. The resulting pellet was extracted with TGMED buffer containing 50 mM $(NH_4)_2SO_4$. The extract was made 200 mM $(NH_4)_2SO_4$ and run on a DEAE A25 column under ion filtration conditions (10,11). The pooled activity was further purified by phosphocellulose chromatography.

RESULTS AND DISCUSSION

In vitro transcription of mouse rDNA. The recent development of cell-free extracts which direct the specific transcription of cloned eukaryotic genes provide an *in vitro* approach to study the sequences and factors involved in the regulation of gene expression. We used the truncated template assay (3) to determine whether crude cytoplasmic extracts (S-100) contain the components required for the specific initiation of RNA polymerase I on cloned rDNA from mouse. The principle of this assay is illustrated in Fig. 1. The recombinant plasmid pMrSalB that contains the transcription initiation site for mouse ribosomal genes is digested with XhoI. This enzyme recognizes a sequence 810 nucleotides downstream the initiation site. Therefore a transcript of 810 nucleotides should be produced *in vitro* if transcription begins at the same site as *in vivo*. In fact,

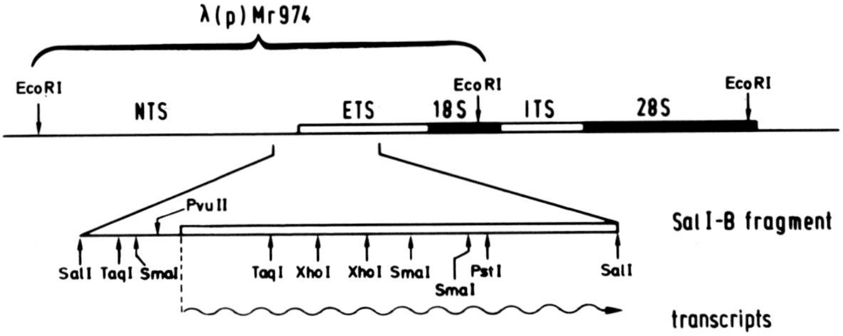

Fig. 1. Structural organization of mouse rDNA. The expanded SalI fragment B has been cloned into pBR322. The recombinant DNA pMrSalB was truncated at the XhoI site and used as template in the cell-free transcription system.

transcripts of the expected size are made if XhoI-truncated pMrSalB DNA is added to S-100 extracts (Fig. 2). Truncation of the template with PstI or SmaI results in RNA products whose lengths roughly correspond to the sizes expected if the 5'end of the *in vitro* transcript is the same as that *in vivo* (not shown).

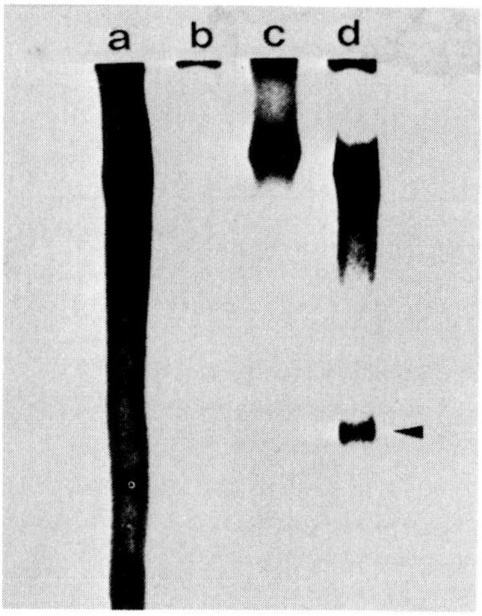

Fig. 2. Electrophoretic analysis of RNA synthesized *in vitro* by RNA polymeraseI *in the S-100* system. The autoradiogram shows RNA synthesized in the standard assay. a) pMrSalB (1 μg/50 μl) transcribed by isolated RNA polymeraseI, b) S-100 system without added DNA, c) undigested pMrSalB transcribed in the S-100 system, d) pMrSalB truncated with XhoI. The arrow indicates the position of the specific rDNA run-off transcript.

Final evidence that RNA polymerase I has initiated transcription at a unique place on the DNA that corresponds to the initiation site *in vivo* has been obtained by S1 protection experiments. For this nucleolar RNA labelled with ^{32}P *in vivo* as well as the pMrSalI-B/XhoI run-off transcript, synthesized *in vitro*, were hybridized to a 295 bp HpaII fragment which contains the initiation site of rDNA transcription (Fig. 3). In previous experiments we had determined the 5'end of 45S pre-rRNA on the cloned rDNA fragment at the nucleotide level (12). The data revealed that 85% of the pre-rRNA starts 155 bp downstream the PvuII site. As shown in Fig. 3 both the *in vivo* and the *in vitro* products exhibit a similar pattern of S1 nuclease protected DNA fragments. The major band is about 230 bp long and corresponds to the 5' end of 45S RNA estimated before. We routinely observe a certain heterogeneity in the 5' ends of pre-rRNA both *in vivo* and *in vitro*. Depending on the source of 45S RNA and the extracts used in the cell-free system additional bands, 85 and about 200 nucleotides further downstream are

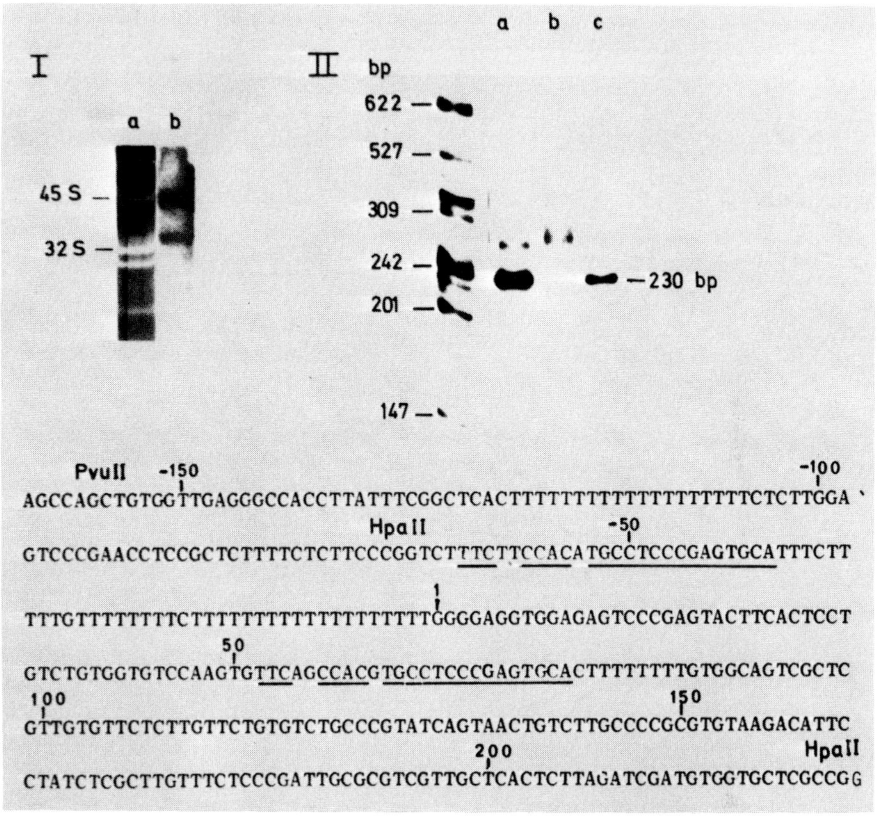

Fig. 3 5'end analysis of pre-rRNA synthesized *in vivo* and *in vitro*

A) Agarose gel analysis of nucleolar RNA labelled with ^{32}P *in vivo*.

1) ethidium bromide stain,
2) autoradiogram of ^{32}P-labelled RNA,

B) Nucleolar RNA and the pMrSalB/XhoI transcript synthesized in the S-100 system, respectively, were hybridized to a 295 bp Hpa II restriction fragment (position -68 to 227, see sequence below). After treatment with nuclease S1 the hybrids were analyzed on an 8 % acrylamide gel.

Lane a) nucleolar RNA; b) *in vitro* RNA not treated with nuclease S1, c) *in vitro* rRNA after S1 nuclease treatment.

revealed by the S1 technique. Since with this technique downstream promoters cannot be distinguished from RNA processing sites it is still uncertain whether several tandem promoters are located in front of the coding region or whether these bands represent preferred sites of 5' terminal processing. On the basis of the remarkable homology in the region flanking the major transcription initiation site with sequences located in front of another site 85 bp further downstream it is suggested that two tandem promoters are involved in the expression of mouse rDNA. In Fig. 4 common structural features of the two putative promoters are compared. Each presumptive initiation site starts with guanosine and is flanked by a cluster of thymidines in the noncoding strand. In front of the T-stretch there is a sequence homology of 25 nucleotides with only 3 bases mismatch. Whether both sites are differently used under different physiological conditions is currently being investigated in the cell-free transcription system.

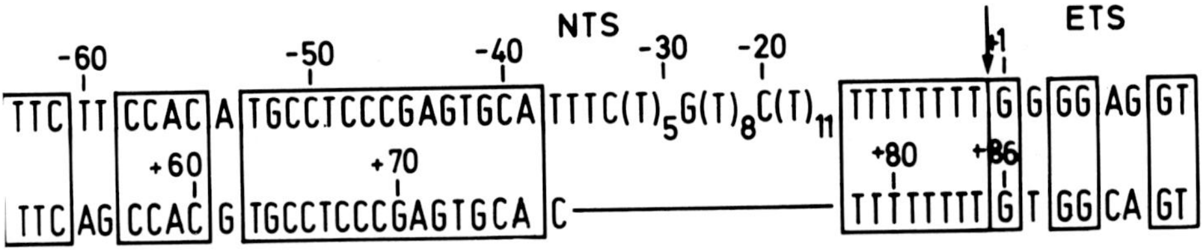

Fig. 4. Comparison of structural homologies of two putative transcription initiation sites of mouse rDNA.

Identification of factors involved in the regulation of rDNA expression. A detailed understanding of the nature and mechanism of action of the transcription factors present in the S-100 extracts requires their isolation and characterization. Fractionation of extracts that direct the faithful transcription of genes transcribed by RNA polymerase II or III, respectively, has indicated that transcription is a complex process that requires, in addition to RNA polymerase, multiple proteins for the accurate initiation reaction (13-15). The factors that promote the specific transcription of mRNA genes in cell-free systems are most likely general transcription factors which, by themselves, are insufficient for the regulation of gene expression. The mechanisms that mediate tissue-specific or developmentally regulated transcription of mRNA coding genes obviously

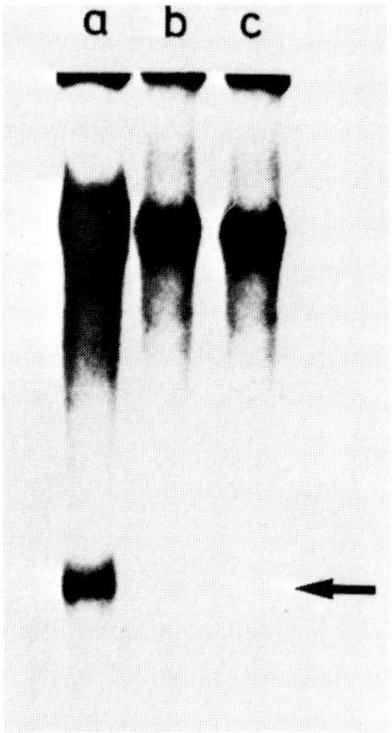

Fig. 5
Correlation of the activity of S-100 extracts with the proliferation rate of the cells. S-100 extracts were prepared from a) logarithmically growing mouse cells, b) stationary cells, and c) cells that had been starved for serum for 3 hours. They were assayed in the standard reaction mixture containing XhoI-truncated pMrSalB and 25 units of RNA polymerase I.

do not operate *in vitro* (15,16).

However, the transcription of the ribosomal genes in the S-100 system seems to reflect *in vivo* control. Only extracts prepared from rapidly growing mouse cells promote specific transcription while extracts from slowly or non-growing cells show little or no activity (Fig. 5). Thus the ability of the crude extracts to mediate faithful transcription of rDNA correlates with the proliferation rate of the

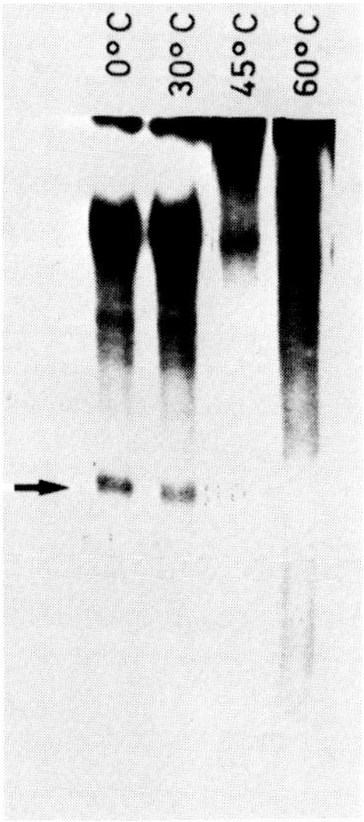

Fig. 6.
Heat inactivation of the S-100 factor(s)
1 μg pSalB DNA truncated with XhoI was used as template in the S-100 system. The S-100 extract has been incubated for 10 min at the temperature indicated before adding to the reaction. Each assay was supplemented with 25 units RNA polymerase I.

cells and reflects the cellular rRNA synthetic activity. This means, that the extracts contain in addition to proteins which suppress random transcription, factors necessary for accurate initiation, and regulatory components.

In an attempt to characterize the factors involved in the specificity and regulation of rDNA transcription we have started to purify the activating component(s). Heat inactivation experiments indicate that at least one essential factor is a rather labile protein (Fig. 6). Preincubation for 10 min at more than 30^0 C results in inactivation of the factors.

The identification and purification of the protein(s) that are involved in control of rDNA expression at the transcriptional level can be achieved by complementation of inactive extracts with active extracts (Fig. 7). This means that the inactive extracts derived from stationary cells contain all the components necessary for rDNA transcription except one (ore more) factor(s) required for the initiation reaction by RNA polymerase I.

We have started to isolate the activating factor(s) by assaying chromatographically separated S-100 fractions for their ability to promote specific rDNA transcription in the presence of extract from non-growing cells which alone is inactive. So far, we have achieved a more than 1000-fold purification of the active component(s) by ion exchange chromatography on DEAE and phosphocellulose as well

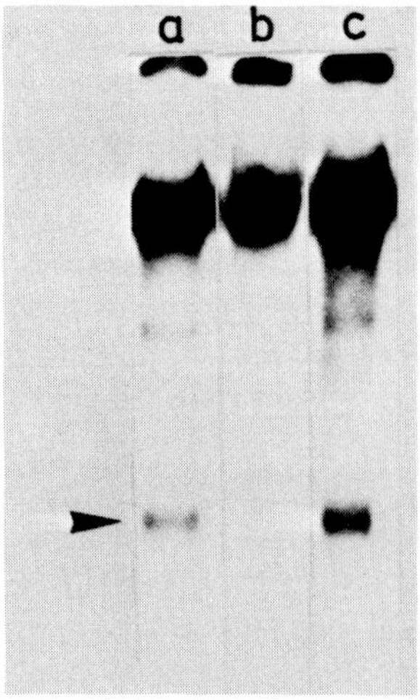

Fig. 7
Complementation of inactive extract from stationary cells with active extract from growing cells
a) S-100 from growing cells,
b) S-100 from stationary cells,
c) mixture of equal amounts of the two extracts.

as by affinity chromatography on heparin-cellulose. During the purification procedure it turned out that the extracts contain variable amounts of another factor which inhibits the production of the specific transcript drastically. Extracts prepared from slowly growing cells contain high levels of this inhibitory component while in extracts from rapidly proliferating little of this inhibitory activity is found. Thus it appears that both positive and negative effector(s) regulate the initiation frequency of RNA polymerase I. We propose that the modulation of rDNA transcription in response to nutritional and hormonal changes is brought about by variations in the amount or activity of these regulatory factors.

ACKNOWLEDGMENTS

This work was supported by the Deutsche Forschungsgemeinschaft.

REFERENCES

1. Grummt, I., Smith, V.A. and Grummt, F. (1976). Cell 7, 439-445.
2. Wu, G.J. (1978). Proc. Natl. Acad. Sci. USA 75, 2175-2179.
3. Weil, P.A., Segall, J., Harris, B., Ng, S.Y. and Roeder, R.G. (1979). Cell 18, 469-484.
4. Manley, J.L., Fire, A., Cano, A., Sharp, P.A. and Gefter, M.L. (1980). Proc. Natl. Acad. Sci. USA 77, 3855-3859.
5. Grummt, I., Soellner, C. and Scholz, I. (1979). Nucl.Acids Res. 6, 1351-1370.
6. Grummt, I. and Gross, H.J. (1980). Molec. Gen. Genet. 177, 223-229.
7. Grummt, I. (1981). Proc. Natl. Acad. Sci. USA, 78
8. Berk, A.J. and Sharp, P.A. (1977). Cell 12, 721-732.
9. Roeder, R.G. and Rutter, W.J. (1969). Nature 224, 234-237.
10. Kirkegaard, L.H., Johnson, T.J.A. and Bock, R.W. (1972). Anal. Biochem. 50, 122-138.

11. Goldberg, M.I., Perriard, J.-C. and Rutter, W.J. (1977). Biochem. 16, 1655-1665.
12. Bach, R., Grummt, I. and Allet, B. (1981). Nucl. Acids Res., in press.
13. Engelke, D.R., Ng, S.-Y., Shastry, B.S. and Roeder, R.G. (1980). Cell 19, 717-728.
14. Segall, J., Matsui, T. and Roeder, R. G. (1980). J. Biol. Chem. 255, 11986-11991.
15. Matsui, T., Segall, J., Weil, A. and Roeder,R.G. (1980). J. Biol. Chem. 255, 11992-11996.
16. Proudfoot, N.J., Shander, M.H., Manley, J.L., Gefter, M.L. and Maniatis, T. (1980). Science 209, 1329-1336.

ELEMENTS REQUIRED FOR INITIATION OF TRANSCRIPTION OF THE OVALBUMIN GENE IN VITRO

Ming-Jer Tsai
Sophia Y. Tsai
Lawrence E. Kops
Tanya Z. Schulz
Bert W. O'Malley

Department of Cell Biology
Baylor College of Medicine
Houston, Texas

I. ABSTRACT

An *in vitro* system has been used to study the initiation of transcription of the ovalbumin gene. Cloned fragments containing sequences of the natural ovalbumin gene were used as templates and a HeLa cell crude extract was used as the source of RNA polymerase and initiation factors. Correct initiation was judged by the size of RNA product, by S1 mapping and by the sizes of transcription products generated from ovalbumin DNA templates truncated at the 3' end. Recently, we have fractionated the total HeLa extracts into various components by column chromatography. Several components from different fractions are required for accurate transcription of the ovalbumin gene. A series of deletion mutants were constructed by trimming sequences which flank the 5' end of the ovalbumin gene using exonuclease III. The DNAs generated were then cloned in pBR322 and used as templates to determine which sequences were necessary for initiation of transcription. Specific transcription of the ovalbumin gene was unaffected by deletion of all but 61 nucleotides of the 5' flanking sequence. However deletion of all but 26 nucleotides of the 5' flanking sequences abolished specific initiation.

ISBN 0-12-137420-3

Thus, a region between 61 and 26 nucleotides upstream from the cap site, which includes the Hogness box (TATATAT) at position 32-26, is essential for correct initiation of the ovalbumin gene. Moreover, introduction of a point mutation in the TATA box of the ovalbumin gene which altered the DNA sequences from TATATAT to TGTATAT produced a template having a greatly reduced capacity for correct initiation of RNA synthesis. This result again demonstrates the importance of TATA box on *in vitro* transcription of ovalbumin gene. However, natural DNA fragments contain several Hogness boxes not normally located in the immediate 5' flanking region of an authentic gene. Most of these sequences did not serve as promoters for initiation of transcription. We conclude that the TATA box is necessary for correct phasing of the initiation site for transcription but that othr sequences are also involved in promoting efficient transcription of eucaryotic genes.

II. INTRODUCTION

The mechanisms which permit the transcription of eucaryotic genes are poorly understood. Development of a DNA-dependent, cell-free, *in vitro* transcription system would facilitate the delineation of the steps and possible control mechanism involved in the transcription process. In 1978, Wu demonstrated that accurate synthesis of a 5.5S RNA could be carried out using a soluble extract from KB cells containing polymerase III and adenovirus-2 (Ad-2) DNA (Wu, 1978). Subsequently, specific transcription of *Xenopus* 5S-RNA and yeast t-RNA was reported in an oocyte cell-free system (Birkenmeier *et al*., 1978; Schmidt *et al*., 1979). With the construction of 5' and 3' deletion mutants and an *in vitro* assay system, Sakonju *et al*. (1980) and Bogenhagen *et al*. (1980) established that an internal fragment between 41-87 nucleotides of the *Xenopus* 5S gene contains sufficient information to direct the initiation of specific transcription of the 5S gene by RNA polymerase III. Also, Engelke *et al*. (1980) and Pelham and Brown (1980) isolated a protein factor which specifically interacts with the internal control region (45-96) of the 5S gene and facilitates accurate transcription of that gene.

Recently, Weil *et al*. (1979a) demonstrated the selective initiation of transcription at a major late promoter of Ad-2 DNA using crude extracts from KB cells supplemented with purified RNA polymerase II. Subsequently, Manley *et al*. (1980) showed the specific initiation of transcription at late promoters of Ad-2 DNA using a whole-cell extract from HeLa cells. Modifying slightly the above polymerase II transcription system, we have demonstrated that correct initiation of RNA synthesis from ovalbumin DNA can be obtained by using HeLa crude extract (Tsai et al., 1980, Tsai et al., 1981).

Our conclusion was based on the following observations: (1) Correct RNA products were generated only from the DNA fragments containing promoter sequence. (2) The sizes of RNA products transcribed from various templates truncated at the 3' end by various restriction enzymes were of the expected size. (3) Results of S_1 mapping experiments indicated that nascent RNA chains were initiated at the cap site. (4) Finally, transcription of these RNA products was carried out by polymerase II since it was sensitive to a low concentration of α-amanitin.

In contrast to this crude cellular extract system, the purified RNA polymerase alone cannot transcribe the same DNA fragment correctly. Therefore, the crude extract must contain some transcription factor(s) other than the polymerases itself (Matsui et al., 1980). In this paper, we would like to summarize our work on the fractionation of transcription factors from both HeLa cell and oviduct tissue. Several factors which partition in various fractions were required for correct and efficient transcription of the ovalbumin RNA product. In addition, using this transcription system and a series of mutants generated by deletion or single base alteration, we have attempted to define the promoter sequences utilized in this in vitro transcription system of the ovalbumin gene.

III. RESULTS

A. Fractionation of HeLa Cell Transcription Factors

A soluble extract prepared from HeLa cell (crude HeLa cell extract) was separated into three fractions by DEAE-sephadex column chromatography as outlined in Figure 1. In general, extract prepared from 20 ml of packed HeLa cells was loaded onto a DEAE-sephadex column (30 ml) equilibrated with Buffer A + 50 mM $(NH_4)_2SO_4$ (Buffer A: 20 mM HEPES, pH 7.9, 0.2 mM EDTA, 5 mM DTT, 5 mM $MgCl_2$ and 20% glycerol). After washing the column with 2 volumes of Buffer A + 50 mM $(NH_4)_2SO_4$, proteins were first eluted

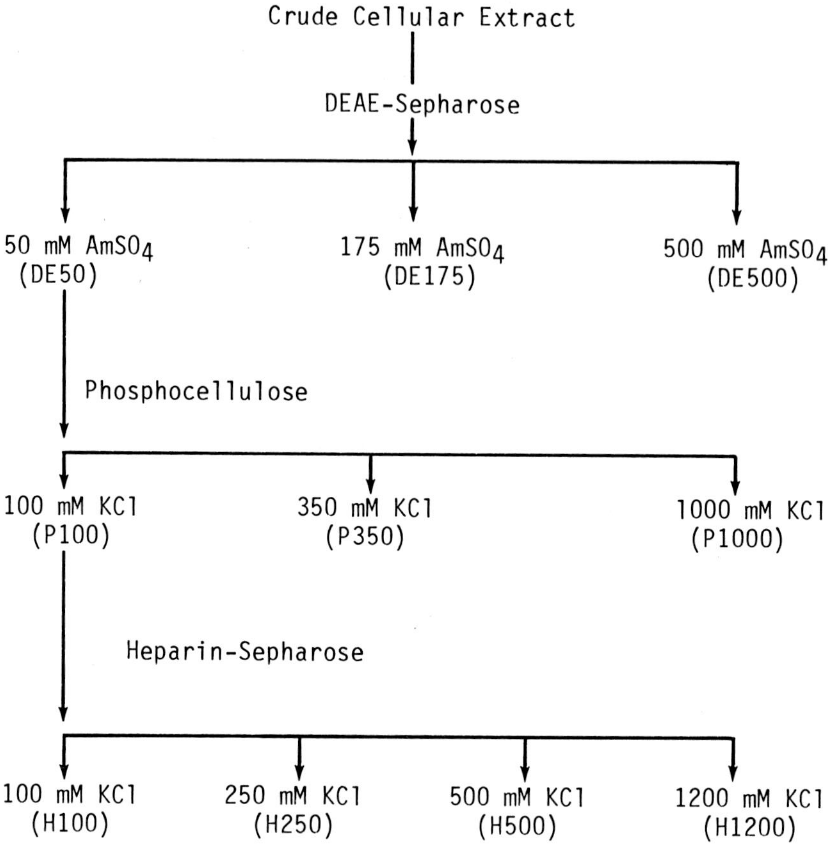

Figure 1: FRACTIONATION OF HELA AND OVIDUCT TRANSCRIPTION FACTORS

with 2 volumes of Buffer A + 175 mM $(NH_4)_2SO_4$ and then with 2 volumes of Buffer A + 500 mM $(NH_4)_2SO_4$. These three fractions were designated as DE50, DE175 and DE500 respectively. These DEAE fractions were then used to transcribe an ovalbumin DNA fragment, OV1.3 (Figure 2) which contains the 5' portion of the ovalbumin gene with presumptive promoter sequences. As shown in Figure 3, no detectable RNA product of the correct size (565 nucleotides) can be seen when individual DEAE fraction was used during transcription. When all three fractions were used, an RNA product of the right size was detected. The accuracy of initiation of RNA synthesis in such reconstituted systems can be further demonstrated by using other truncated DNA fragments as templates (data not shown). In Figure 3, we have also shown that in the absence of DE500, the correct RNA product (565 nucleotides in length) can be produced although at a slightly reduced efficiency. However, transcription of the lower molecular weight RNA bands greatly increases. In the absence of either DE175 or DE50 the 565 nucleotide RNA product is no longer detected. Therefore, all three fractions appear to be required for correct transcription of ovalbumin gene.

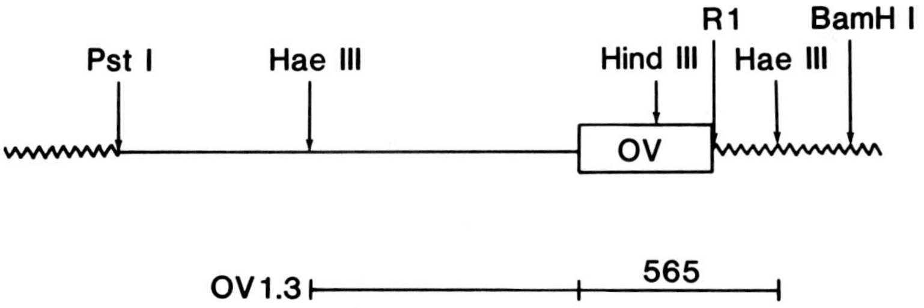

Figure 2: Restriction Map of Ovalbumin Gene at 5' End

After dialyzing against Buffer B + 100 mM KCl (Buffer B is the same as Buffer A except that 0.5 mM spermidine is used instead of 5 mM $MgCl_2$), the DE50 fraction was loaded onto 10 ml of phosphocellulose column which was preequilibrated with Buffer B + 100 mM KCl (Figure 1). After washing the column with 2 volume of Buffer B + 0.1 M KCl the protein factors were eluted with 2 volumes of Buffer B + 350 mM KCl and Buffer B + 1000 mM KCl. The unbound, 350 mM KCl, and 1000 mM KCl eluted fractions were designated as P100, P350 and P1000 respectively. When these fractions were used in combinations with DE500 and DE175 to transcribe the OV1.3 DNA fragment, we observed that both the

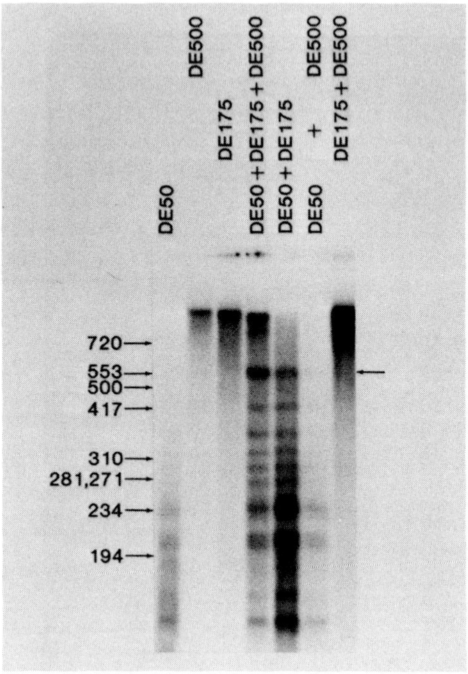

Figure 3: Reconstitution of HeLa Cell Transcription Factors.

P100 and P1000 fractions were required for correct and efficient transcription (Figure 4). In the absence of P1000, no 565 nucleotide size product could be detected. Thus, P1000 is absolutely required for the synthesis of the right RNA product. In the absence of P100, however, a very weak band at 565 nucleotides could be observed in the autoradiogram. Therefore, P100 is apparently required for maximum efficiency of transcription. Addition of P350 to the transcription mixture drastically decreased the transcription (data not shown). Therefore, a general inhibitor of transcription might be associated with this fraction. Thus, at least 4 independent factors are required for accurate initiation of RNA synthesis _in vitro_.

Preliminary results also indicated that P100 can further be separated into 4 fractions by Heparin-sepharose chromatography. Only 2 (H100 and H500) of the 4 fractions are required for accurate initiation.

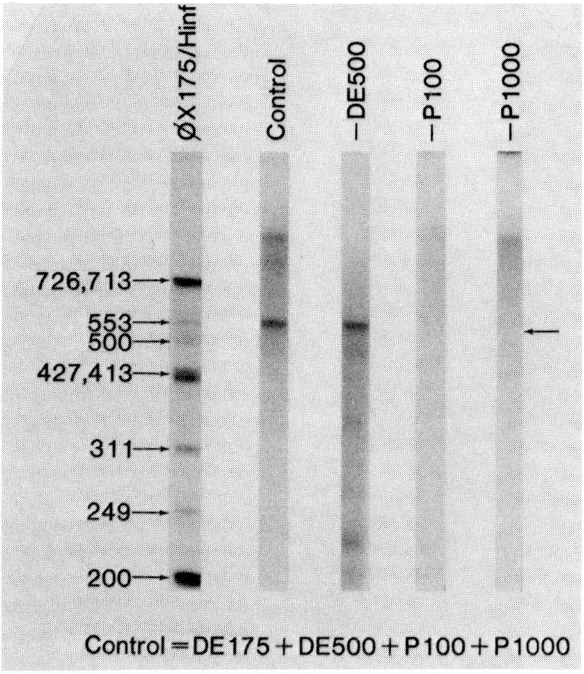

Figure 4: Reconstitution of HeLa Cell Transcription Factors

B. Fractionation of Oviduct Transcription Factors

Due to a high concentration of nuclease activity, oviduct crude extract prepared in a fashion similar to that of HeLa cell extract cannot be used to transcribe ovalbumin DNA templates. We have fractionated the oviduct crude extract according to the same scheme as that described in Figure 1 in the hope of obtaining RNase free transcription factors. When the DEAE-cellulose chromatography was carried out, we found that the majority of RNase activity was associated with DE50. Further fractionation of DE50 on phosphocellulose yielded a fraction (P350) which contained most of the ribonuclease activity. Since the P350 fraction was not required for correct initiation of transcription in the HeLa cell system, we attempted to replace the HeLa cell fraction with its oviduct counterpart which now contained low level of RNase. When this protocol was followed we found that the DE175, P100

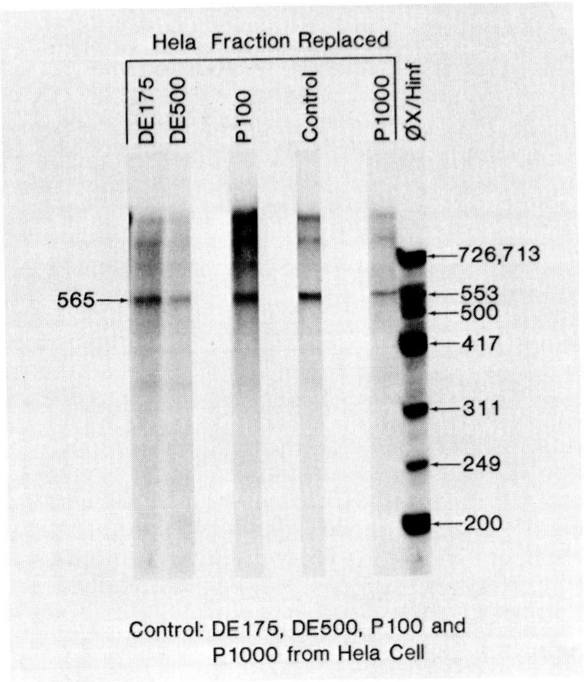

Figure 5: Exchange of Oviduct Transcription Factors For HeLa Cell Factors

or P1000 fractions could be replaced by oviduct fractions (Figure 5). Therefore, the factors required for accurate transcription of the ovalbumin gene seem to be common to both tissues.

C. DNA Sequences Required for Accurate Initiation of RNA Synthesis on the Ovalbumin Gene

In order to define the regions important for *in vitro* initiation of ovalbumin RNA we have carried out deletion experiments on cloned pOV1.7 plasmid (Figure 2). The procedure is similar to that of Sakonju *et al*. (1980) as described in Tsai *et al*. (1981). Briefly, we used exonuclease III and S_1 nuclease to digest the Ava II-Bam HI DNA fragment from both ends of the fragment (Figure 2). After ligating Bam HI linkers at the ends, the DNA was digested with Bam HI + RI. This DNA fragment was then

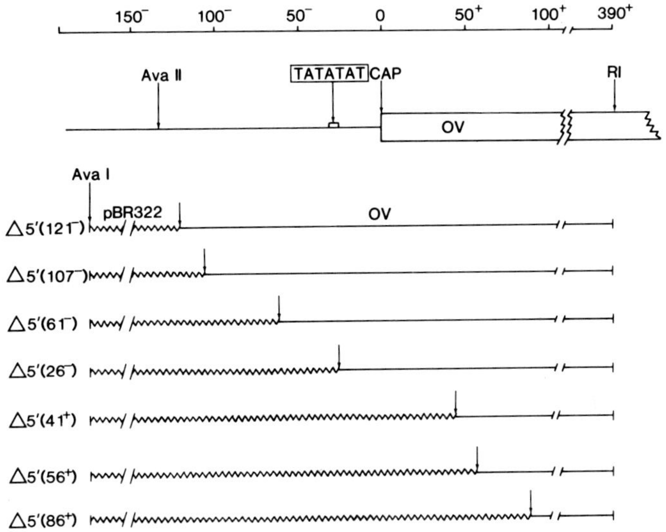

Figure 6: Ovalbumin Clones Containing 5' Deleted Sequences

cloned into Bam HI and RI sites of pBR322. In this manner, we have generated mutants containing different lengths of 5' end flanking sequences (Figure 6).

DNAs from the above 5' deletion mutants were digested with Ava I and Eco RI. The Ava I-Eco RI fragments contained the ovalbumin DNA sequences as well as 1.05 Kb of plasmid sequences located upstream from the ovalbumin DNA. These fragments were purified by agarose gel electrophoresis and then used as templates for in vitro transcription. As shown in Figure 7, the specific run-off product of the ovalbumin DNA (393 NT in length) was synthesized by deletion mutants OV Δ5'(121^-), OV Δ5'(107^-) and OV Δ5'(61^-). Thus, deletion of any DNA sequences upstream from position 61^- did not affect the initiation of transcription of the ovalbumin DNA in vitro. By contrast, transcription of the specific product was completely abolished when OV 5' (26^+), OV Δ5'(41^+), OV Δ5' (56^+) and OV Δ5' (86^+) mutants were used as templates. It should be noted that some higher molecular weight products were seen in the autoradiogram when these DNAs isolated from deletion mutants DNA were used as template. These appear to result from transcription of the plasmid DNA but no attempt was made to characterize them. The deletion of sequences

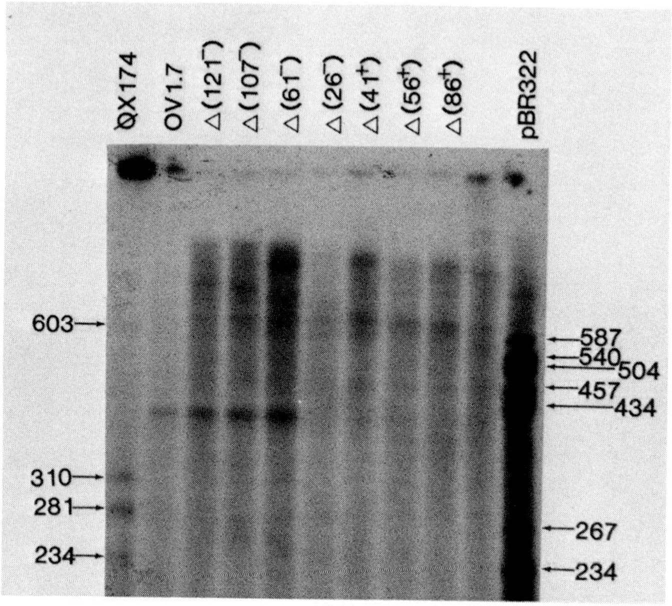

Figure 7: Transcription of Ovalbumin 5' Deletion Mutants

between 61⁻ and 26⁻ completely eliminated the initiation on ovalbumin DNA, an observation which suggests that this region of DNA is essential for initiation of transcription of ovalbumin DNA in vitro. Interestingly, OV Δ5'(26⁻) contains only the last nucleotide of the "Hogness box" (TATATAT in the case of ovalbumin) which exists in most eucaryotic genes and has been postulated to function as a possible recognition site by RNA polymerase II.

In order to define further the importance of the TATA box on the initiation of RNA synthesis, we have carried out transcription on a DNA fragment containing a single base change at the TATA box. This single base change on the TATA box was generated by using a method similar to that reported by Wallace et al. (1980) (Manuscript in preparation). Briefly, a synthetic oligonucleotide containing the base change of interest was used to prime the synthesis of DNA from a single-stranded DNA template. The end product of this synthesis was a DNA heteroduplex consisting of a parental DNA strand and a mutant complementary strand. After two transformations of E. coli, one can obtain a pure clone with the single base alteration. In this particular clone, the TATA box of the chick ovalbumin gene was changed from TATATAT to TGTATAT. If the TATA sequence is important for RNA polymerase II transcription, one would expect that transcription of the correct RNA product would be affected. As shown in Figure 8, this mutation results in a decrease in the efficiency of transcription of the correct RNA bands (234 NT for the Hae III fragment and 670 NT for the Hpa II fragment). Therefore, once again the TATA box preceding the cap site of the RNA polymerase II transcribed gene is quantitatively important for correct transcription of the mRNA genes in vitro.

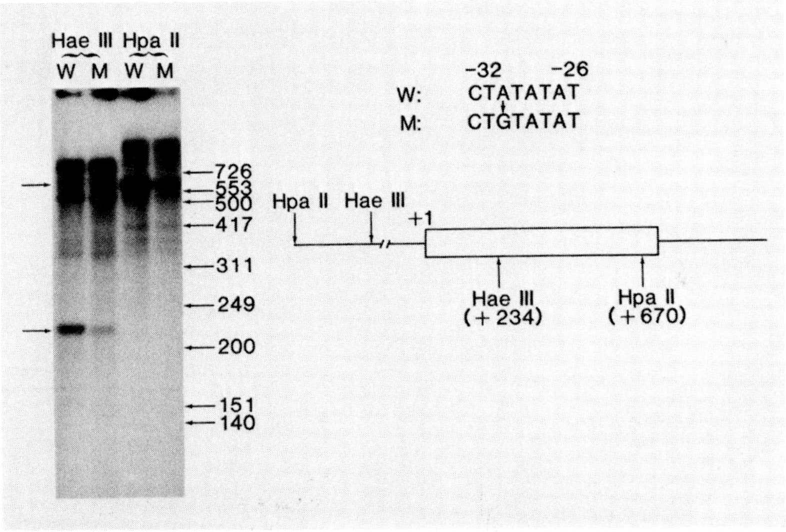

Figure 8: Transcription of DNA Fragments With Point Mutation on TATA Box

Next we examined whether the TATA sequence alone is a sufficient signal for initiation of transcription. In addition to the presumptive promoter, the Hogness box (TATATAT) at 32⁻ position of the ovalbumin DNA, there are several other similar sequences present in the OV1.7 DNA fragment (Table I). If these sequences can be recognized by RNA polymerase as promoters, we would expect to find RNA products of OV1.7 having the sizes listed in Table I.

EXISTENCE OF FALSE HOGNESS BOXES IN THE OV1.7 DNA FRAGMENT

Position	Sequences: Hogness Box		Sequences: Possible Initiation Site	Expected RNA Product
-1090	[1]TATATAT	.....	[30]AAACAAA	∿1451
- 825	TATATTA	.....	ACTCCTC	∿1186
- 624	TATAAAG	.....	GATTTAA	∿ 985
- 607	CATAAAA	.....	TATAGGA	∿ 968
*- 32	TATATAT	.....	TGTACAT	393
+ 78	TATATTA	.....	AAAAAAT	∿ 283
+ 174	TATAAAA	.....	TCTGCAC	∿ 187

TABLE I

As shown in Figure 7 and 9 the only major RNA product observed was 393 nucleotides when OV1.7 was used as a template. We conclude that specific transcription of the ovalbumin gene is directed mainly by the Hogness box situated at -32 position. One might argue that RNA polymerase or factors in our *in vitro* system may concentrate on the strong promoter (at position -32) and leave other weak ones unattended. Thus, we obtained a DNA fragment restricted by Hae III (OV0.7) which was devoid of most of the Hogness boxes including the one at position -32. With this DNA fragment we might expect that without competition from a stronger promoter, the RNA polymerase could concentrate at this promoter region if it could possibly be used. As shown in Figure 9, no major RNA band with a predicted size of 310 nucleotides could be synthesized from this DNA template (OV0.7). It should be noted that in addition to the Hogness box at position -1090, this DNA fragment (OV0.7) also contains another Hogness box at position -825. If the later Hogness box were ever used as promoter for *in vitro* transcription, we would predict an RNA product around 40 nucleotides in size. This product was too small to be analyzed in our present gel system. Therefore, at least some of the Hogness boxes cannot be used as meaningful promoters for in vitro RNA synthesis. Taken together, our entire results suggest that the Hogness box is essential but not sufficient for specific initiation of RNA synthesis of the ovalbumin gene.

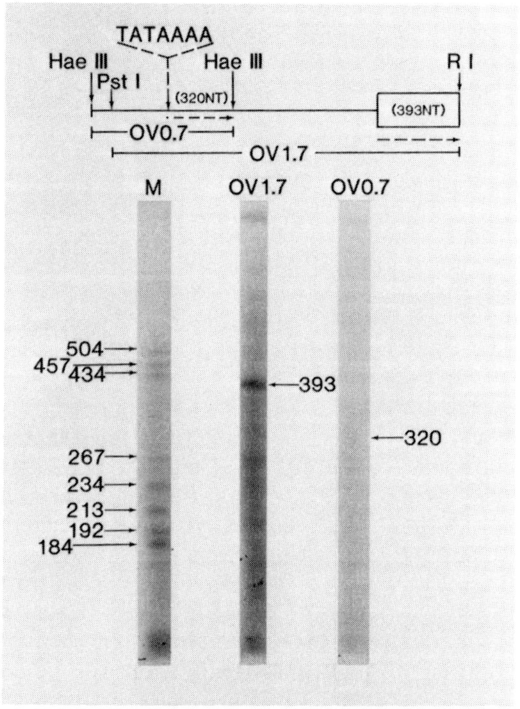

Figure 9: In Vitro Transcription of OV1.7 and OV0.7

IV. DISCUSSION

Our present study demonstrated that HeLa and oviduct crude extract can be fractionated into multiple active components. This DE175 fraction contains the majority of RNA polymerase II activity and can be replaced by partially purified oviduct RNA polymerase (30-50% purity) (data not shown). The DE500 fraction, although not required for the production of correct transcripts, reduces the appearance of low molecular weight RNA bands and increases the correct RNA product. The P100 fraction contains factor(s) necessary for efficient transcription. Deleting this fraction from the in vitro transcription system markedly decreases the efficiency of transcription. P1000 is probably the most important fraction of all; without P1000, no detectable correct transcripts can be obtained. In this fraction, one can detect multiple strong DNA binding components. It is possible that a promoter binding factor similar to that in oocyte 5S gene system exists in this fraction (Engelke et al. 1980; Pelham and Brown, 1980). Similar results were also observed by Matsui et al. (1980).

The studies on transcription of 5' deletion mutants indicates that the deletion of DNA sequences between position 61⁻ and 26⁻ completely abolishes the initiation of synthesis of the specific product. The ovalbumin gene, like many other eucaryotic genes (Ad-2, SV40, histones, conalbumin) contains a conserved DNA sequence (TATA or Hogness box) 26-32 nucleotides upstream from the cap site. Our results imply that a limited region of 5' flanking DNA is required for in vitro initiation of transcription of the ovalbumin gene. Similarly, in Ad2 (Wasylyk et al., 1980; Hu and Manley, 1981), globin (Proudfoot et al., 1980; Talkington et al., 1980), and conalbumin (Wasylyk et al., 1980a, 1980b) systems, deletionof the TATA box abolished the correct initiation of RNA synthesis. These results were further supported by generation of a point mutation on the TATA box (Schulz et al., 1981, in preparation). Taken together, these results implicate the Hogness box as the presumptive promoter which is necessary for in vitro initiation of polymerase II gene products. However, in view of the low efficiency of the in vitro transcription system, these results do not exclude the possibility that other DNA regions might also be important for specific interaction with regulatory proteins. In fact in L cell (Dierks et al., 1980), in oocyte (Grosschedl and Birnsteil, 1980a, 1980b), and in SV40 systems (Benoit and Chambon and P. Gruss and G. Khoury, personal communication) sequences upstream from the TATA box were also important for transcription. Therefore, one has to use caution in classifying a DNA sequence as unimportant if the analysis is carried out in the in vitro transcription system alone.

Finally, it should be noted that at least some of the other Hogness box type sequences at internal gene position or 5' flanking positions of the OV1.7 DNA (Pst to R1 of Figure 2) do not serve as a signals for initiation of specific RNA product in vitro. It is interesting to point out that extending or shortening the distance between

TATA box and capsite greatly affect the efficiency of transcription (Dierks et al., 1981). In addition, a natural occured mutant on the cap site of the globin gene (Talkington and Leder, 1980) cannot be transcribed by Hela cell crude extract. Thus, it is likely that other genomic sequences, in addition to the Hogness box are essential for specific initiation of gene transcription.

In summary, it appears that at least 3 separate regions of DNA may be required for correct and efficient transcription of eucaryotic genes. The cap site itself seems to be necessary for precise initiation. Second, the TATA box looks to be important in staging accurate downstream initiation (~30 bp) at the cap site. Finally, an upstream sequence at ~100⁻ ($\pm$ 30 bp) nucleotides from the cap site appears to modify the efficiency of gene transcription. Taken together these and perhaps other yet unidentified regions of genomic DNA assume important regulatory roles in expression of eucaryotic genes.

ACKNOWLEDGMENTS

We thank Dr. Charles Lawrence for providing HeLa cells and Dr. A. Dugaiczyk for his valuable advice in construction of the 5' deletion mutants. We also thank Ann Kong, Cindy Herzog, Frank Leu and B. Mathey for their expert technical assistance. This work was supported by NIH grant HD-08188 and NIH Center Grant for Population Research and Reproductive Biology, HD-07495-08.

VI. REFERENCES

Birkenmeier, E.H., Brown, D.D. and Jordan, E. (1978) Cell 15 1077-1086.

Bogenhagen, D.F., Sakonju, S. and Brown, D.D. (1980) Cell 19 27-35.

Dierks, P., Van Ooyen, A., Mantel, N. and Weissman, C. (1980) Arolla Workshop.

Engelke, D.R., Ng, S.Y., Shastry, B.S. and Roeder, R.G. (1980) 19 717-728.

Grosschedl, R. and Birnstiel, M.L. (1980a) Proc. Natl. Acad. Sci. USA 77 1432-1436.

Grosschedl, R. and Birnstiel, M.L. (1980b) Proc. Natl. Acad. Sci. USA 77 7102-7106.

Hu , S.L. and Manley, J.L. (1981) Proc. Natl. Acad. Sci. USA 78 820-824.

Manley, J.L., Fire, A., Cano, A., Sharp, P.A. and Gefter, M.L. (1980) Proc. Natl. Acad. Sci. USA 77 3855-3859.

Matsui, T., Segall, J., Weil, P.A. and Roeder, R.G. (1980) J. Biol. Chem. 255 11992-11996.

Pelham, H.R.B. and Brown, D.D. (1980) Proc. Natl. Acad. Sci. USA 77 4170-4174.

Proudfoot, N., Shander, M.H.M., Manley, J.L., Gefter, M.L. and Maniatis, T. (1980) Science 209 1329-1336.

Sakonju, S., Bogenhagen, D.F. and Brown, D.D. (1980) Cell 19 13-25.

Schmidt, O., Mao, J.L., Silverman, S., Hovemann, B. and Soll, D. (1978) Proc. Natl. Acad. Sci. USA 75 4819-4823.

Talkington, C.A., Nishiora, Y. and Leder, P. (1980) Proc. Natl. Acad. Sci. USA 77 7132-7136.

Tsai, S.Y., Tsai, M.J. and O'Malley, B.W. (1981) Proc. Natl. Acad. Sci. USA 78 879-883.

Tsai, M.J., Tsai, S.Y. and O'Malley, B.W. (1980) Sigrid Juselius Symposium on "Expression of Eucaryotic, Viral and Cellular Genes" Helsinki, Finland.

Wallace, R.B., Johnson, P.F., Tanaka, S., Schold, M., Itakura, K. and Abelson, J. (1980) Science 209 1396-1400.

Wasylyk, B., Kedinger, C., Corden, J., Brison, O. and Chambon, P. (1980a) Nature 285 367-373.

Wasylyk, B., Derbyshire, R., Guy, A., Molks, D., Roget, A., Teoule, R. and Chambon, P. (1980b) Proc. Natl. Acad. Sci. USA 77 7024-7028.

Weil, P.A., Luse, D.S., Segall, J. and Roeder, R.G. (1979a) Cell 18 469-484.

IDENTIFICATION OF A TRANSCRIPTIONAL CONTROL REGION UPSTREAM FROM THE HSV THYMIDINE KINASE GENE

Steven L. McKnight

Department of Embryology
Carnegie Institution of Washington
115 West University Parkway
Baltimore, Maryland 21210

INTRODUCTION

Transcriptional expression of eukaryotic chromosomes occurs on discrete units. This is evidenced by electron micrographs of active transcriptional complexes (Miller and Beatty, 1969). Linear regression analyses of such complexes indicate that RNA polymerase entry onto transcription units occurs at a discrete site (Laird *et al*., 1976). Biochemical studies of primary transcripts derived from specific transcription units have shown that the 5' terminus of each RNA molecule maps to a unique nucleotide or set of nucleotides (Ziff and Evans, 1978). The locus to which the 5' terminus of a primary transcript derived from a structural gene maps is referred to as the "cap site". It is believed that this locus represents the entry site for RNA polymerase (Contreras and Fiers, 1981).

I have conducted a mutagenesis study of the herpes simplex virus (HSV) thymidine kinase (tk) gene. The study was designed to resolve the identity of nucleotide sequences required for accurate expression of tk mRNA *in vivo*. The viral tk gene was chosen for this study due to the readily accessible functional assay of DNA-mediated transfection (Wigler *et al*., 1977). As it has turned out, however, I have used *Xenopus laevis* oocytes as a standard transcription assay. Frog oocytes are capable of expressing microinjected copies of the tk gene in the form of translatable mRNA that is identical at its 5' terminus to authentic tk mRNA.

The results of the present study reveal the existence of a transcriptional control region that

DEVELOPMENTAL BIOLOGY USING PURIFIED GENES

ISBN 0-12-137420-3

is located between 40 and 100 nucleotides upstream from the tk structural gene. It appears that the *in vivo* function of this control region is to both recruit entry of RNA polymerase form II, and direct the enzyme complex to initiate transcription roughly 70 nucleotides 3' to its center. It is not known whether the functional role of the control region is potentiated by transcription factors other than RNA polymerase.

RESULTS AND DISCUSSION

Structural Features of the HSV Thymidine Kinase Gene

The tk gene used in this study was derived from the macroplaque strain of HSV type I. A 3.4 kb Bam HI restriction enzyme fragment containing the tk gene was isolated from the HSV I genome and cloned in pBR-322. Several studies have shown that the tk gene is contained within a 2.1 kb Pvu II restriction fragment (Colbere-Garapin *et al.*, 1979; Wilkie *et al.*, 1979; Enquist *et al.*, 1980). I derived the nucleotide sequence of this fragment of the HSV I genome and mapped the 5' and 3' termini of tk mRNA on the DNA sequence using a hybridization/S_1 nuclease procedure (McKnight, 1980). Figure 1 is a schematic diagram showing a restriction map of the HSV tk gene superimposed on the transcript map.

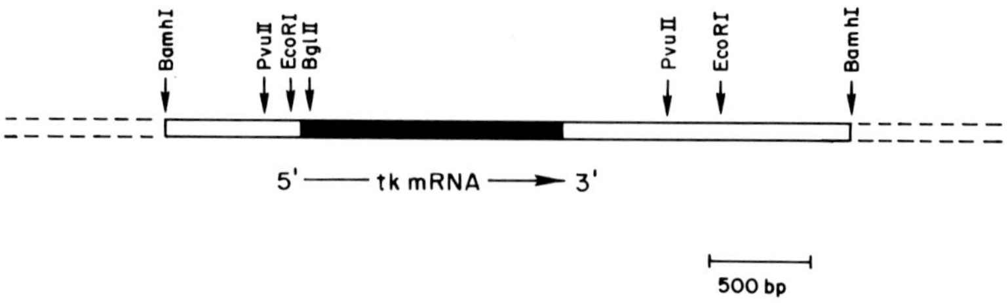

Fig. 1. Diagram shows restriction endonuclease recognition sites surrounding the herpes simplex virus thymidine kinase gene. Darkened region denotes segment of DNA complementary to the tk mRNA.

Computer-assisted sequence comparison studies have identified two elements common to the 5' flanking DNA of eukaryotic structural genes. The nucleotide sequence upstream from the HSV thymidine kinase gene exhibits conserved oligonucleotide sequences positioned roughly 25 and 80 nucleotides, respectively, from the putative transcription start site. An A-T rich heptanucleotide, ATATTAA, is located 27 nucleotides from the 5' boundary of the tk mRNA coding sequence (Figure 2). A similar sequence, TATAAA, was first recognized as an element common to eukaryotic structural genes by Hogness and Goldberg (unpublished). A second conserved sequence is located 83 nucleotides upstream from the tk gene. This -80 homology is represented by the heptanucleotide GGCGAAT (Figure 2). Conserved derivatives of this oligonucleotide sequence are commonly observed between 75 and 85 nucleotides from the 5' termini of structural genes (Benoist et al., 1980; Efstratiadis et al., 1980).

```
-200
CAGCTGCTTC ATCCCCGTGG CCCGTTGCTC GCGTTTGCTG GCGGTGTCCC CGGAAGAAAT ATATTTGCAT GTCTTTAGTT CTATGATGAC ACAAACCCCG

-100                                                                                                       -1
CCCAGCGTCT TGTCATTGGC GAATTCGAAC ACGCAGATGC AGTCGGGGCG GCGCGGTCCG AGGTCCACTT CGCATATTAA GGTGACGCGT GTGGCCTCGA
                      EcoRI
+1                                                                                                       +100
ACACCGAGCG ACCCTGCAGC GACCCGCTTA ACAGCGTCAA CAGCGTGCCG CAGATCTTGG TGGCGTGAAA CTCCCGCACC TCTTCGGCCA GCGCCTTGTA
                                                       Bgl II
                                                                                                         +200
GAAGCGCGTA TGGCTTCGTA CCCCGGCCAT CAACACGCGT CTGCGTTCGA CCAGGCTGCG CGTTCTCGCG GCCATAGCAA CCGACGTACG GCGTTGCGCC
          aug
```

Fig. 2. Nucleotide sequence surrounding the 5' terminus of the HSV thymidine kinase gene. The non-coding strand of the DNA sequence is displayed progressing from a Pvu II restriction enzyme recognition site to a position 200 nucleotides internal to the mRNA-coding component of the gene. The predominant 5' termini of tk mRNA map to adenosine residues +1 and +3. The AUG triplet located at nucleotide +110 is the translation start codon most proximal to the 5' terminus of tk mRNA. Probe DNA for quantitative S_1 mapping of tk mRNA 5' termini is made by cleaving with Bgl II, which cuts the coding strand at nucleotide +56, and Eco RI, which cuts at nucleotide -75.

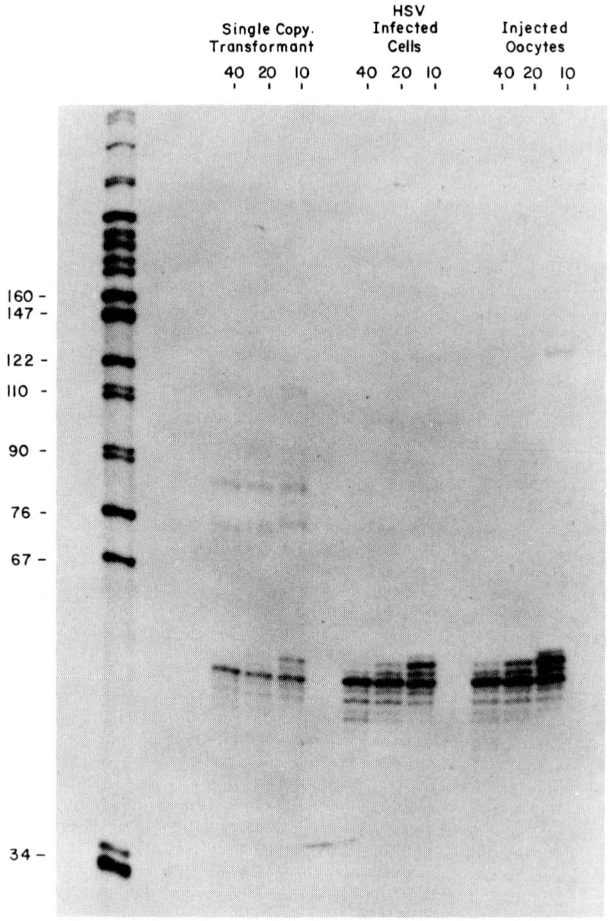

Fig. 3. S_1 nuclease maps of tk mRNA synthesized in HSV-infected cells, tk^+ transformant cells, and frog oocytes. Autoradiograph exposure of an electrophoresis gel used to size S_1 resistant tk probe DNA. RNA prepared from HSV-infected cells, tk^+ transformant cells and frog oocytes injected with tk plasmid DNA was hybridized to an excess of radiolabeled, 131 nucleotide Bgl II/Eco RI, single-stranded probe DNA. The probe DNA overlaps the 5' terminus of the tk gene (see Figure 2). RNA/DNA hybrids were digested with 10, 20 or 40 units/ml S_1 nuclease. S_1 resistant DNA was sized by electrophoresis on a 14% polyacrylamide sequencing gel. Numbers to the left denote the size of molecular weight marker DNA fragments in nucleotides.

Expression of the tk Gene in Xenopus laevis Oocyte Nuclei

When a recombinant plasmid carrying the HSV tk gene is microinjected into frog oocytes, a 100-200 fold increase in tk enzymatic activity is effected. This increase in tk enzymatic activity is a function of the synthesis of authentic HSV tk protein, and is sensitive to as little as 1 μg/ml α-Amanitin (McKnight and Gavis, 1980). In order to test the accuracy of tk mRNA synthesis in oocytes, RNA isolated from injected samples was subjected to hybridization/S_1 nuclease mapping. Figure 3 is an autoradiographic exposure of a sequencing gel used to size S_1-resistant DNA fragments derived from the annealing of a 131 nucleotide EcoRI/Bgl II probe DNA fragment to various RNA samples. As the gel exposure shows, tk mRNA synthesized in oocytes is identical at its 5' terminus to both authentic tk mRNA (synthesized in virally infected mammalian cells) and tk mRNA derived from a cell line that carries a single copy of the HSV tk gene.

Construction of Deletion Mutants of the tk Gene

As an initial method of identifying nucleotide sequences required for accurate expression of the tk gene, two systematic sets of deletion mutants were prepared and assayed by microinjection. Deletion mutagenesis was accomplished by strandard enzymatic methods (Sakonju *et al.*, 1980). A set of mutants that lack progressively greater extents of 5' flanking DNA in a 5'→ 3' direction was derived by enzymatic digestion starting at a Bam HI restriction site ∿680 nucleotides upstream from the tk structural gene. Using similar methods, a systematic set of deletion mutants progressing in a 3'→5' direction was prepared. The latter mutants were prepared by enzymatic digestion starting at a Bgl II restriction site 56 nucleotides internal to the tk structural gene. Figure 4 is a schematic diagram showing of 13 5' deletion mutants and 11 3' mutants pertinent to the results of this study.

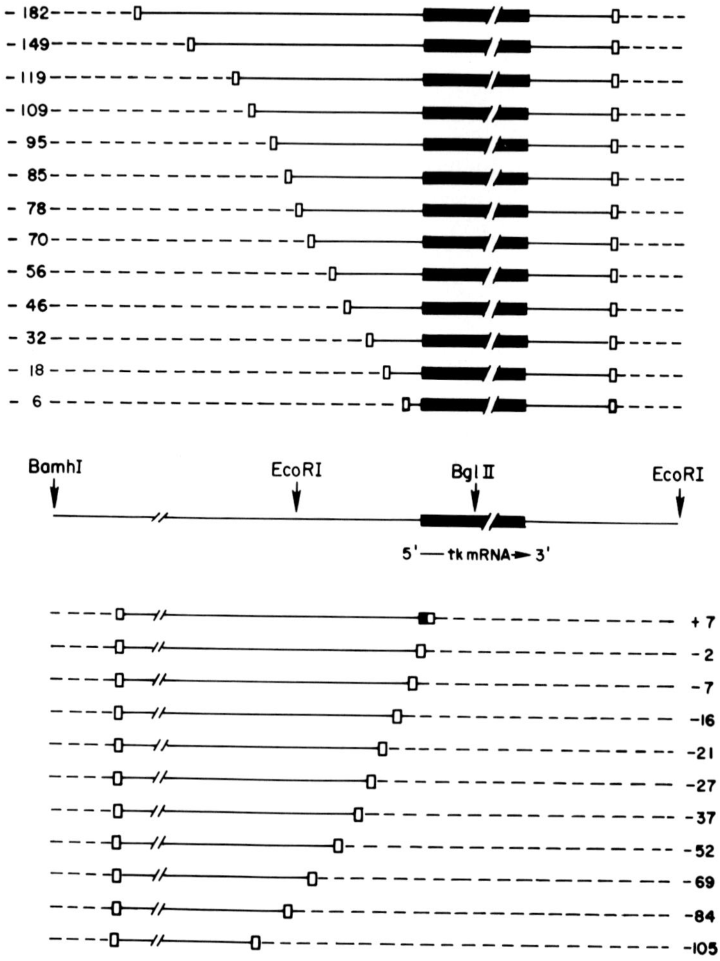

Fig. 4. Schematic diagram of 5' and 3' deletion mutants of the HSV tk gene. Restriction enzyme map of tk DNA is shown at the center of the diagram. 5' deletion mutants are displayed above restriction map. 3' deletion mutants are displayed below restriction map. Dashed lines represent pBR-322 DNA. Open boxes denote locations of synthetic oligonucleotide linker molecules. Unbroken lines denote intact 5' flanking DNA. Darkened boxes denote tk mRNA-coding DNA.

Expression of 5' Deletion Mutants in Frog Oocytes

10 ng supercoiled DNA of each 5' deletion mutant was injected into the nuclei of 100 oocytes. Following a 24 hour incubation interval at 20°C, RNA was isolated and assayed for the presence of authentic tk mRNA by quantitative hybridization/S_1 nuclease mapping. Figure 5 shows an autoradiographic exposure of a sequencing gel used to resolve transcript maps of RNA synthesized from various 5' deletion mutants. Two features of these results are noteworthy. First, the amount of authentic tk mRNA synthesized by 5' -85 is drastically reduced as compared with 5' -109 and all mutants that retain $\geq$109 nucleotides of contiguous 5' flanking DNA. Second, mutants that retain $\leq$18 nucleotides of 5' flanking DNA do not direct the synthesis of any authentic tk mRNA.

The reduction in tk mRNA expression observed at the -109/-85 boundary can be quantitated by measuring the amount of radioactive DNA in excised gel bands by scintillation spectrometry. As Table I shows, the amount of authentic tk mRNA synthesized by 5' deletion mutants that retain $\geq$109 5' flanking nucleotides is roughly 50-fold that of ones that retain between 32 and 85 contiguous 5' flanking nucleotides. These data are supported by two independent observations. First, the amount of enzymatically active tk protein made in oocytes injected with 5' -109 is 20-40 fold greater than the amount made by 5' -85 (Table I). Second, mutants that retain $\geq$109 5' flanking nucleotides transform tk^- mouse cells to the tk^+ phenotype 20-fold more efficiently than do those that retain $\leq$ 85 contiguous 5' flanking nucleotides (McKnight, Gavis, Kingsbury and Axel, manuscript submitted).

These results indicate that there is a region of DNA upstream from the tk structural gene that has a profound effect on the efficiency of transcriptional expression. The 5' boundary of this region is located somewhere between the deletion end points of 5' -109 and 5' -85. Since the deletion isolate 5' -95 retains a significant fraction of the transcription potential of 5' -109, I tentative assign the 5' boundary of this transcriptional control region to nucleotide -100.

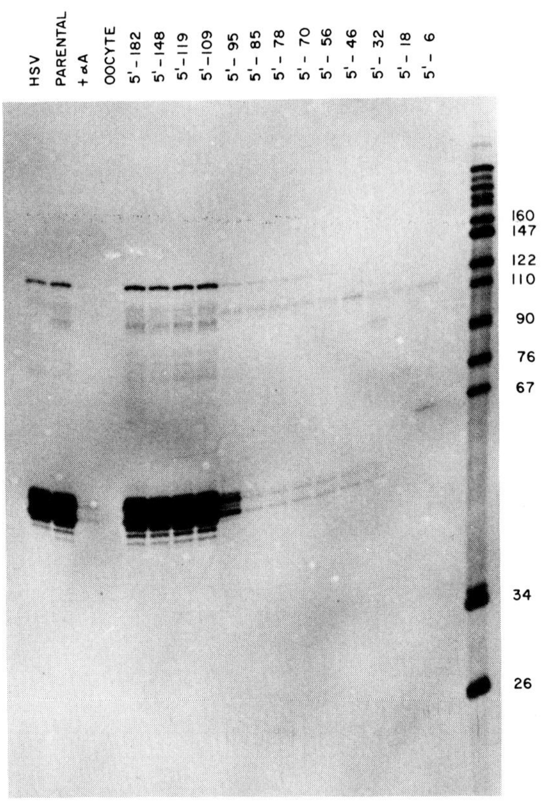

Fig. 5. S_1 nuclease maps of tk mRNA synthesized in frog oocytes injected with 5' deletion mutants. Autoradiograph exposure of a 14% polyacrylamide sequencing gel used to size S_1-resistant tk probe DNA. For each experiment 5 μg of oocyte RNA was hybridized with 5 ng radiolabeled Bgl II/Eco RI probe DNA. RNA/DNA hybrid molecules were digested with 10 units/ml S_1 nuclease. Lane labeled "HSV" shows transcript map of authentic tk mRNA. Lane labeled "parental" shows transcript map of RNA synthesized in oocytes in response to microinjection of ptk/ Δ3'-1.13, the parental tk isolate from which 5' deletion mutants were derived. Lane labeled "+αA" shows transcript map of RNA synthesized in oocytes injected with ptk/ Δ3'-1.13 and 1 μg/ml α-Amanitin. Remaining lanes show transcript maps of RNA made in oocytes in response to various 5' deletion mutants. Number to right of figure denote sizes of molcular weight marker DNA fragments in nucleotides.

TABLE I. Functional Assays of 5' Deletions in Oocytes and Fibroblasts

Deletion Mutant	Specific tk mRNA[1] [cpm x 10^{-1}]	tk Enzymatic Activity [units activity above 5 units/ml oocyte background]
ptk/Δ3'-1.13	372	389
5'-182	499	491
5'-148	312	733
5'-119	414	576
5'-109	461	681
5'-95	39	70
5'-85	7	19
5'-78	7	32
5'-70	6	20
5'-56	7	40
5'-46	5	33
5'-32	4	21
5'-18	0	13
5'-6	0	56

[1]Represents Cherenkov counts per minute above the background (45 cpm) of a gel slice corresponding to the 52-57 nucleotide range excised from a lane loaded with probe DNA that had been hybridized to RNA from uninjected oocytes prior to S_1 nuclease digestion.

Expression of 3' Deletion Mutants in Frog Oocytes

The 3' deletion mutants constructed for this study lack most, if not all, of the tk structural gene. The 5' flanking DNA retained by this series of deletions extends from a fixed site 480 nucleotides upstream from the 5' terminus of the tk structural gene to varying locations in and around the putative transcription start site (Figure 4). The nomenclature denoting 3' mutants refers to the site at which a synthetic Bam HI linker molecule was ligated to the deletion end point. For example, 3' +7 refers to a 3' mutant that retains 480 nucleotides of 5' flanking DNA as well as 7 nucleotides of tk mRNA-coding DNA.

Since these 3' deletion mutants lack the tk structural gene I have analyzed their ability to promote expression of plasmid sequences juxtaposed to each deletion mutant end point. This analysis was accomplished by a hybridization assay using the two single strands of a discrete DNA fragment derived from the tetracycline gene of pBR-322. RNA isolated from oocytes that were injected with individual 3' deletion mutants was hybridized to either the "coding" or "non-coding" strand of a radiolabeled, 53 nucleotide Hae II restriction fragment. The Hae II fragment is located within the tetracycline gene of pBR-322 some 121 nucleotides from the unique Bam HI site (Sutcliffe, 1978). "Coding" strand refers to the Hae II DNA strand that replaced the tk mRNA coding DNA strand. Similarly, "non-coding" refers to the Hae II DNA strand that replaced the non-coding strand of the tk gene.

When 3' +7 plasmid DNA is injected into oocytes, it expresses at least 10-fold more RNA complementary to the Hae II "coding" strand than to the "non-coding" strand. This transcriptional asymmetry is not observed when pBR-322 alone is introduced into oocytes, or is it observed when 3' +7 is injected in the presence of 1 μg/ml α-amanitin. These quantitations are derived from nucleic acid hybridizations performed using a vast excess of radiolabeled probe. Following annealing, RNA/DNA hybrids are exposed to S_1 nuclease, precipitated on filter discs with TCA and counted by scintillation spectrometry.

All 3' deletion mutants extending as far as 3' -52 are capable of promoting asymmetric transcription of plasmid sequenes juxtaposed to tk 5' flanking DNA (Table II). The isolate 3' -69 also directs accentuated expression of the Hae II "coding strand" sequence, yet at a level only 25% that of 3' -52. 3' deletions that remove 82 or more 5' flanking nucleotides do not promote expression of the plasmid DNA strand that replaced tk mRNA coding strand DNA (Table II).

These experiments indicate that a segment of DNA at least 52 nucleotides upstream from the putative transcription start site of the tk gene can promote transcriptional expression by RNA polymrase form II in the complete absence of mRNA-coding sequences. In concert with the analyses of 5' deletion mutants discussed earlier, these data delimit the boundaries of a transcriptional control region operative in vivo. A segment of tk 5' flanking DNA reaching from 109 to 52 nucleotides upstream from the tk structural gene appears to serve a critical role in ensuring quantitative expression of sequences that lie 3' to the region.

The quantitative hybridization assays described in the preceding paragraphs do not resolve the location of 5' termini on transcripts synthesized by the various 3' deletion mutants. I have used a primer extension assay to loate the 5' termini of such transcripts. The same 53 nucleotide Hae II "coding strand" probe used for quantitating transcripts promoted by 3' deletion mutants was annealed to RNA synthesized from different 3' deletions. RNA/DNA hybrids were then exposed to reverse transcriptase in the presence of excess deoxyribonucleoside triphosphates. Primer extension products were recovered by ethanol precipitation, boiled in formamid and sized by gel electrophoresis. Figure 6 shows an autoradiographic exposure of a sequencing gel used to map the 5' termini of transcripts synthesized in oocytes from the various 3' deletion mutants. Two aspects of these data are notable. First, the same 3' mutants that promote asymmetric expression of the pBR-322 Hae II fragment as assayed by quantitative hybridization also synthesize RNA

TABLE II. Hybridization Analysis of Transcripts Synthesized by 3' Deletion Mutants

Plasmid Sample	Hae II[1] Probe Strand	S_1-Resistant DNA[2] (cpm x 10^{-2})
3' +7	Slow	27.1
3' +7	Fast	0.6
3' +7 (+αA)[3]	Slow	2.1
pBR-322	Slow	3.1
3' -2	Slow	24.5
3' -7	Slow	25.6
3' -16	Slow	31.4
3' -21	Slow	19.9
3' -27	Slow	23.7
3' -37	Slow	28.0
3' -52	Slow	19.1
3' -69	Slow	8.4
3' -84	Slow	3.7
3' -105	Slow	4.1

[1]Hae II slow migrating single strand probe was labeled to a specific activity 3.1 x that of the fast migrating strand due to the asymmetric distribution of adenosine and cytosine residues replaced by Exo III/reverse transcriptase labeling.

[2]5 μg oocyte RNA was annealed to 1 ng single-stranded probe DNA for 2 hr at 65°C. Hybrids were diluted 10-fold into S_1 nuclease buffer, digested for 30 minutes at 20°C using 50 units/ml S_1 nuclease, and assayed for nuclease-resistance by TCA precipitation and scintillation spectrometry.

[3]α-Amanitin was co-injected at a concentration of 1 μg/ml.

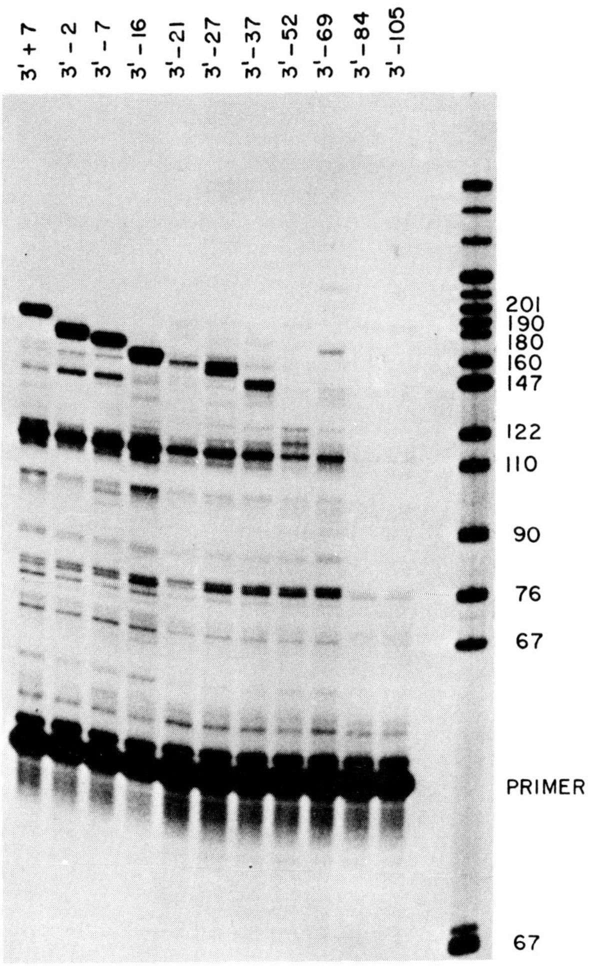

Fig. 6. Primer extension maps of RNA synthesized in frog oocytes injected with 3' deletion mutants. Autoradiographic exposure of a 10% polyacrylamide sequencing gel used to size reverse transcriptase extension products generated from various RNA samples. For each experiment 5 μg of RNA was hybridized with 1 ng of radiolabeled pBR-322 Hae II coding strand probe. RNA/DNA hybrids were extended using 100 units/ml AMV reverse transcriptase. Deletion mutants tested are indicated above gel lanes. All transcription reactions were assayed in oocytes isolated from a single female frog. Numbers to right of figure denote sizes of molecular weight markers.

capable of extending the length of the Hae II primer. Second, the largest quantitatively significant extension product derived from RNA synthesized by seven of the 3' deletion mutants (3' +7 through 3' -37) is inversely proportional to the extent of 3'→5' deletion. For example, the largest extension product derived from RNA synthesized by 3' +7 is ∿185 nucleotides in length. 3' -2, which is deleted in a 3'→5' direction 9 nucleotides beyond 3' +7, produces transcripts that extend the Hae II primer to a length of ∿176 nucleoties (Figure 6). Much the same pattern is observed for RNA synthesized by 3' -7, 3' -16, 3' -21, 3' -27 and 3' -37.

I interpret these data in the following way. When 3' +7 plasmid DNA is microinjected into oocytes it promotes synthesis of transcripts that initiate at nucleotide +1 of the tk structural gene. Elongation of the nascent RNA chain progresses through the 7 nucleotides of tk structural DNA, and continues through plasmid sequences past the 53 nucleotide Hae II fragment. Thus, when assayed by primer extension, a 186 nucleotide product is observed (7 nucleotides tk DNA +126 nucleotides pBR-322 DNA from Bam HI site to Hae II site +53 nucleotides Hae II primer = 186 nucleotides). When 3' -2, 3' -7, 3' -16, 3' -21, 3' -27 and 3' -37 are transcribed in oocyte nuclei, they all initiate transcription at or very near the position formerly occupied by nucleotide +1 of the tk structural gene. These data suggest that sequences upstream from nucleotide -37 not only promote transcriptional expression of the HSV thymidine kinase gene in a quantitative sense, but also specify the approximate start site of transcription.

CONCLUSIONS

I have shown that the HSV thymidine kinase gene is accurately expressed in *Xenopus laevis* oocytes. By constructing, characterizing and assaying two systematic sets of deletion mutants of the tk gene, I have resolved the location of a transcriptional control region. The control region reaches from a position approximately 100 nucleotides upstream from the tk gene to a position ∿40 nucleotides upstream from the gene. This 60 nucleotide segment of tk 5' flanking DNA is capable of promoting expression of sequences 3' to it in the absence of both the tk gene and the 40 nucleotides most proximal to the putative transcription start site. Moreover, expression mediated from this transcriptional control region appears to start at a specified location some 70 nucleotides 3' to its center.

REFERENCES

Benoist, C., O'Hare, K., Breathnach, R., and Chambon, P. (1980). Nucl. Acids Res. 8: 127-142.

Colbere-Garapin, F., Chousterman, S., Horodniceann, F., Kourilsky, P. and Garapin, A. (1979). Proc. Nat. Acad. Sci. USA 76: 3755-3759.

Contreras, R. and Fiers, W. (1981). Nucl. Acids Res. 9: 215-236.

Efstratiadis, A., Posakony, J.W., Maniatis, T., Lawn, R.M., O'Connell, C., Spritz, R.A., DeRiel, J.K., Forget, B.G., Weissman, S.M., Slightom, C.C. and Proudfoot, N.J. (1980). Cell: 653-668.

Enquist, L.W., Vande Woude, G.F., Wagner, M., Smiley, J. and Summers, W.C. (1980). Gene 7: 335-342.

Laird, C.D., Wilkinson, L.E., Foe, V.E. and Chooi, W.Y. (1976). Chromosoma 58: 169-192.

McKnight, S.L. (1980). Nucl. Acids Res. 8: 5949-5964.

McKnight, S.L. and Gavis, E.R. (1980). Nucl. Acids Res. 8: 5931-5948.

Miller, O.L., Jr. and Beatty, B.R. (1969). Science 164: 955-957.

Sakonju, S., Bogenhagen, D.F. and Brown, D.D. (1980). Cell 19: 13-25.

Sutcliffe, J.G. (1978). Nucl. Acids Res. 5: 2721-2728.

Wigler, M., Silerstein, S., Lee, L., Pellicer, A., Cheng. Y. and Axel, R. (1977). Cell 11: 223-232.

Wilkie, N.M., Clements, J.B., Boll, W., Mantei, N., Lonsdale. D., Weissman, C. (1979). Nucl. Acids Res. 7: 859-877.

Ziff, E.B. and Evans, R.M. (1978). Cell 15: 1463-1475.

EXPRESSION OF β-GLOBIN GENES MODIFIED BY RESTRUCTURING AND SITE-DIRECTED MUTAGENESIS

P. Dierks, B. Wieringa, D. Marti, J. Reiser, A. van Ooyen, F. Meyer, H. Weber and C. Weissmann

Institut für Molekularbiologie I
Universität Zürich
8093 Zürich, Switzerland

ABSTRACT. The effect of modifications introduced into the 5'-flanking region of the rabbit β-globin gene on correct transcription was determined in vivo and in vitro. The ATA box sequence located 25 to 31 bp upstream of the cap site was required for detectable levels of transcription; moreover, it programmed the starting points of β-globin transcripts. A region 57 to 84 bp upstream of the cap site, containing the CAAT box, had a potentiating effect on the level of transcription in vivo but was not required for correct initiation.

The large intervening sequence and its neighboring region were modified by deletions and site-directed point mutations. Most of the internal region of the large intron was not required for normal splicing of RNA transcripts. Deletion of the 5' intron-exon junction did not prevent excision of the large intron; in this case splicing occurred between the normal 3' splice site and a new (cryptic) 5' site located in the second exon. Introduction of several single purine transition mutations at the 5' splice junction had no detectable effect on splicing.

DEVELOPMENTAL BIOLOGY USING PURIFIED GENES

ISBN 0-12-137420-3

INTRODUCTION

Recombinant DNA technology has made it possible to isolate almost any gene of interest, modify it by restructuring (1,2) and/or site-directed mutagenesis (cf. Weber et al., this volume) and reinsert it into cells to investigate its functional potential (3).

In the preceding article we have described the construction of deletion, deletion/substitution, and point mutations in the globin gene and its flanking regions. One set of modifications alters the 5'-flanking region preceding the starting point of transcription ("cap site"), and a second set affects the large intron and its 5'-proximal splice junction. In this paper we report on the effect of these modifications on transcription and splicing, respectively, using both in vivo and in vitro expression systems.

I. Efficiency and accuracy of transcription of β-globin genes with deletions or deletion/substitutions in the 5'-flanking region.

The 5'-flanking region of the β-globin gene contains two regions which appear moderately conserved in a variety of eukaryotic genes. One of these, an A+T-rich region flanked by G or C residues, designated as the Goldberg-Hogness, TATA, or ATA box, is located about 31 bp upstream of the cap site (4-6) in most, but not all, genes and has been proposed as an important component of the RNA polymerase II promoter. The second conserved sequence, known as the CAAT box (5,7), has been noted in a number of eukaryotic and viral genes transcribed by polymerase II, but neither its presence nor its position (usually about 75 to 80 bp upstream of the cap site), is universal.

To test the functional role of sequences preceding the 5' terminus of the β-globin gene, three sets of experiments were performed.

a) Transcription in vivo from β-globin genes with 5'-flanking regions of 1500, 425, 76, 66, or 14 bp preceding the cap site. The hybrid plasmids containing these genes (cf. Weber et al., preceding paper) were linked to the Herpes simplex virus type I thymidine kinase (TK) gene and introduced into

TK-negative mouse L cells by the calcium phosphate method (8). Several TK^+ transformants were selected from each experiment and globin-specific RNA was analyzed by a modified S1 mapping procedure (9), using as a probe a 5'-^{32}P-labeled DNA fragment spanning the 5' end of the rabbit β-globin gene (10). In general, the 5' termini of the transcripts mapped not only to the cap site, but also to positions 42 to 48 nucleotides downstream from it or to vector sequences preceding the gene. The amount of β-globin RNA in TK^+ cell lines was similar for each of the mutants DNAs tested and ranged from 5 to 1500 copies per cell. However, the proportion of correctly initiated β-globin RNA varied with the length of the 5'-flanking region. When the gene was preceded by only 14 bp of 5'-flanking sequence, the level of β-globin-specific RNA was high, but no transcripts had the correct 5' termini; most of them originated in the vector moiety. With 66 bp preceding the gene, 5%, and with 76 bp or more, 30-85% of the transcripts were correctly initiated. We concluded that the region between 14 bp and 66 bp upstream of the cap site, which contains the ATA box, is essential for correct initiation of β-globin transcription in vivo (11). Moreover, we suggested that the region between 66 bp and 76 bp preceding the cap site, which contains the CAAT box, enhances transcription. However, since the absolute number of correctly initiated transcripts varied widely from one cell line to another, a firm conclusion regarding the role of the CAAT box could not be drawn.

b) Transcription in vivo from β-globin genes containing localized deletion/substitutions or insertions in the 5'-flanking region. To further investigate the role of the ATA and CAAT boxes, as well as the cap site, we constructed internal deletion/substitution and insertion mutants which a) removed the cap site, b) altered the distance between the ATA box and the cap site, c) removed part or all of the ATA box, d) removed a region containing the CAAT box, or e) removed regions located between these three elements. In all cases the deleted region was replaced by a 10 bp HindIII linker (Fig. 1a, and Fig. 1 in preceding paper). Because of the long time required to analyze in vivo expression using the TK-mediated expression system, we made use of the transient expression observed for

A

	Sequence (−40 … +40)	RTL
WILD-TYPE SEQUENCE	GGACTTGGGCATAAAAGGCAGAGCAGGGCAGCTGCTGCTTACACTTGCTTTTGACACAACTGTGTTTACTTGCAATCCCC	1.0
pDPVU2	acaagcttgt (inserted) GGACTTGGGCATAAAAGGCAGAGCAGGGCAGCTGCTGCTTACACTTGCTTTTGACACAACTGTGTTTACTTGCAATCCCC	0.8
pDPVU101	GGACTTGGGCATAAAAGGCAGAGCAGGGCAG— acaagcttgg —TTTTGACACAACTGTGTTTACTTGCAATCCCC	0.4
pDPVU103	GGACTTGGGCATAAAAGGCAGAGCAGGGCAG———— acaagcttg ————TGCAATCCCC	2.1

	Sequence (−50 … +1)	RTL
WILD-TYPE SEQUENCE	GGATTACATAGTTCAGGACTTGGGCATAAAAGGCAGAGCAGGGCAGCTGCTGCTTACACTT	1.0
pDPVU105	GGATTACATAGTTCAGGACTTGGGCATAAAAGGCA-ccaagcttgt-CTGCTGCTTACACTT	0.7
pDPVU1070	GGATTACA———— ccaagcttgt ————CTGCTGCTTACACTT	< 0.03
pDPVU1071	GGATTACA——— ccaagcttgt ———GAGCAGGGCAGCTGCTGCTTACACTT	< 0.03
pDPVU1072	GGATTACA—— ccaagcttgt ——AAGGCAGAGCAGGGCAGCTGCTGCTTACACTT	0.12
pDPVU1073	GGATTACA —ccaag— CTTGGGCATAAAAGGCAGAGCAGGGCAGCTGCTGCTTACACTT	0.5

	Sequence (−90 … −40)	RTL
WILD-TYPE SEQUENCE	AGCCACACCCTGGTGTTGGCCAATCTACACACGGGGTAGGGATTACATAGTTCAGGACTT	1.0
pDPVU631	AGCCACACCC——— caagcttg ———GGGATTACATAGTTCAGGACTT	0.2
pDPVU632	AGCCACACCC caagcttg TGGCCAATCTACACACGGGGTAGGGATTACATAGTTCAGGACTT	1.3

B

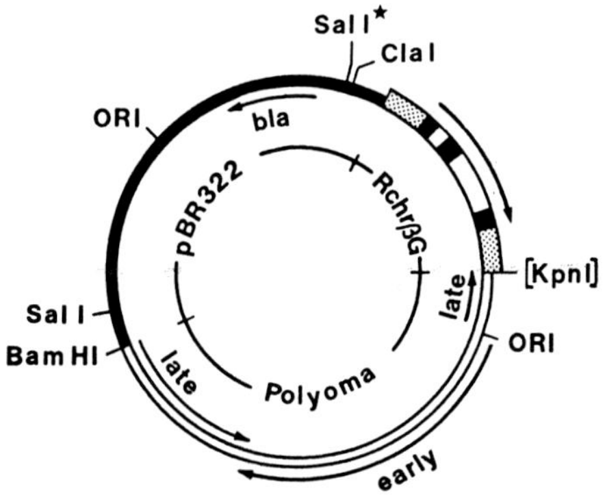

C

	100 … 150
WILD-TYPE SEQUENCE	GGGGCAAGGTGAATGTGGAAGAAGTTGGTGGTGAGGCCCTGGGCAGGTTGG
	├─ INTRON
REFERENCE GENE	----------------------------CCAAGCT---------------

genes linked to polyoma or SV40 DNA (12, 13). When the rabbit β-globin gene was inserted into the KpnI site in the late region of a polyoma-pBR322 hybrid plasmid, such that the orientation of the β-globin gene was opposite to that of the polyoma late region, high levels of correctly initiated β-globin RNA were found 30 to 42 h after transfection into mouse 3T6 cells (J. Reiser, unpublished data). Each of the modified DNAs was inserted into the vector as shown in Fig. 1b. An equimolar amount of the same vector containing a β-globin reference gene, which had a 7 bp substitution between positions 129 and 135 in the 5' exon (Fig. 1c), was added in each transformation experiment to facilitate quantitation of expression. When a uniformly labeled probe extending from the middle exon to the 5' flanking region was hybridized to correctly spliced RNA from

Fig. 1. Modified rabbit β-globin genes and expression vector used for transcription studies in vivo.

a) Nucleotide sequences in the 5'-flanking regions of the wild type and modified β-globin genes. Capital letters, nucleotides of rabbit chromosomal DNA; small letters, non-homologous nucleotides derived from a synthetic linker DNA (cf. previous paper). The ATA (positions -25 to -31) and CAAT boxes (positions -69 to -77) are underscored. Deletion/substitution mutants pdPVU101 to pdPUV632 correspond to group IV mutants 1 to 9, respectively (cf. Fig. 1, previous paper). Arrowheads indicate the approximate 5' termini of transcripts as determined by S1 mapping using 5'-^{32}P-labeled probes prepared from the homologous gene (data not shown). RTL, transcript level relative to internal reference transcript (see text).

b) pBR322-polyoma-rabbit β-globin expression plasmid. The directions of transcription are indicated by arrows. The KpnI site in brackets was obliterated by the insertion of the rabbit β-globin gene. A SalI site (SalI*) was inserted into the EcoRI site of pBR322. Stippled boxes, non-transcribed flanking sequences; black boxes, exons; open boxes, introns; ORI, origins of replication. Only restriction sites required to define the orientation and joining sites are indicated.

c) Nucleotide sequence between positions 110 and 150 of the wild type and reference rabbit β-globin genes. Only the substituted nucleotides are shown for the reference gene.

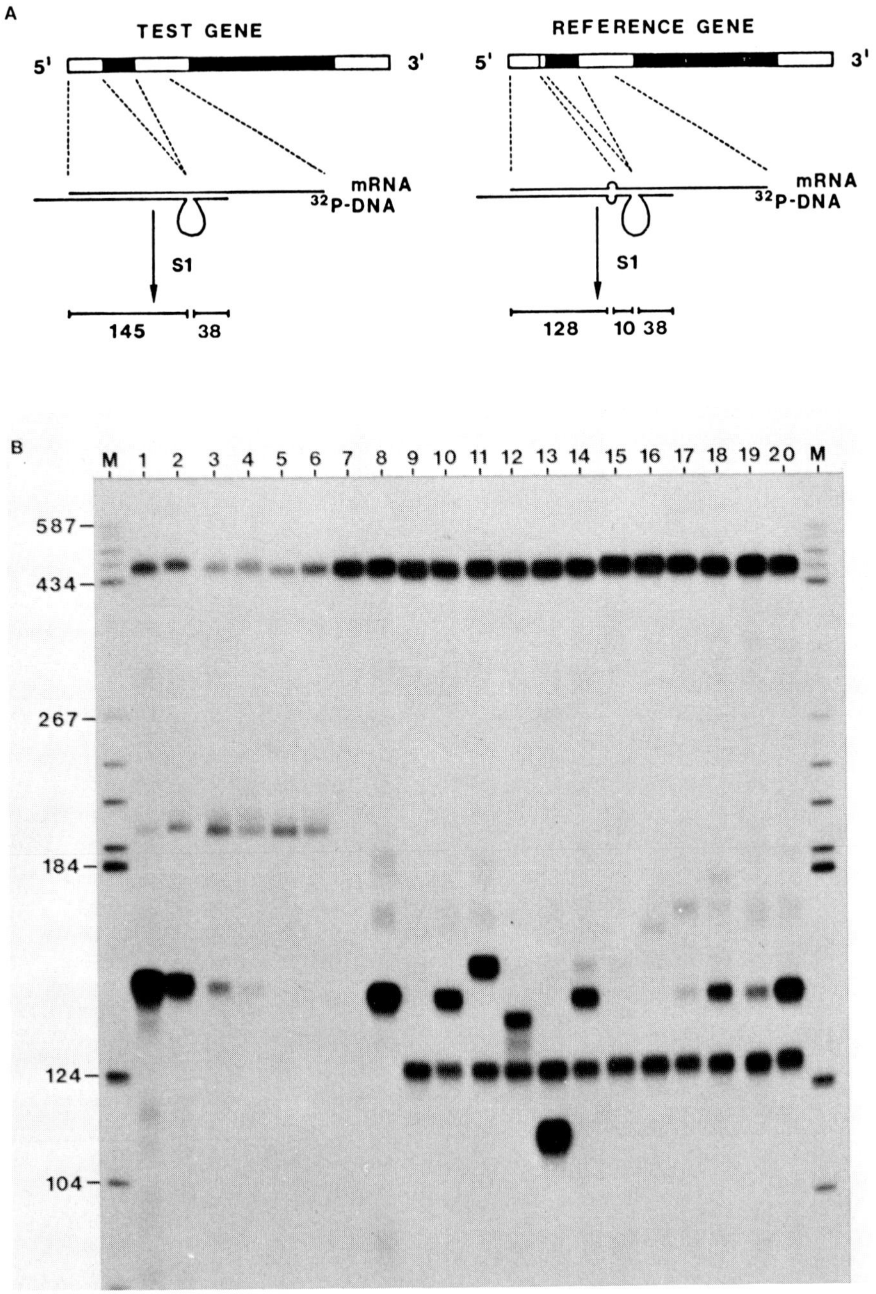
A
TEST GENE
5'
3'
mRNA
32P-DNA
S1
145
38
REFERENCE GENE
5'
3'
mRNA
32P-DNA
S1
128
10
38
B
M 1 2 3 4 5 6 7 8 9 10 11 12 13 14 15 16 17 18 19 20 M
587
434
267
184
124
104

the reference gene, S1-resistant fragments of 10, 38, and 128 nucleotides resulted, while wild type RNA yielded fragments of 38 and 145 nucleotides (cf. Fig. 2a and 2b, lanes 8 and 9). Thus, the radioactivities in the 128 and 145 nucleotide fragments are a measure of the RNA derived from the reference and test rabbit β-globin genes, respectively. When equal

Fig. 2. Determination of rabbit β-globin RNA (5' proximal region) by nuclease S1 mapping.

a) Strategy for assaying RNA from both test and reference rabbit β-globin genes. A minus strand probe uniformly ^{32}P-labeled from the 5'-flanking region to the second intron was prepared by digesting suitably linearized β-globin chromosomal DNA with exonuclease III and filling up with α-^{32}P dNTPs using E.coli DNA polymerase I (Klenow fragment). The fragment containing the 5' end of the rabbit β-globin gene was isolated after TaqYI cleavage. Hybridization of this probe to correctly initiated and spliced RNA transcribed from the wild type gene, followed by nuclease S1 digestion (14), resulted in labeled protected fragments of 145 and 38 nucleotides, which were analyzed on a denaturing polyacrylamide gel. RNA from the reference gene with a substitution in the first exon yielded fragments of 128, 10, and 38 nucleotides.

b) S1 mapping of rabbit β-globin RNAs from mouse 3T6 cells transfected with pBR322-polyoma-rabbit β-globin hybrid DNAs. 3T6 cells were transfected with wild type, reference, or a 1:1 molar ratio of reference and mutant pBR322-polyoma-rabbit β-globin DNAs and the RNA assayed as outlined above. Lanes 1 to 6, 810 pg, 270 pg, 81 pg, 27 pg, 8.1 pg, or 0 pg of authentic rabbit β-globin mRNA. Lane 7, RNA from mock transfected cells. Lanes 8 and 9, RNAs from 3T6 cells transfected with wild type and reference β-globin genes, respectively. RNAs in lanes 10 to 20 were from cells transfected with an equimolar amount of the plasmids containing the reference gene and the wild type (lane 10), pdPUV2 (lane 11), pdPUV101 (lane 12), pdPUV103 (lane 13), pdPUV105 (lane 14), pdPUV1070 (lane 15), pdPUV1071 (lane 16), or pdPUV1072 (lane 17), pdPUV1073 (lane 18), pdPVU631 (lane 19), or pdPVU632 (lane 20) rabbit β-globin genes (cf. Fig. 1a). M, ^{32}P-labeled BspI restriction fragments of pBR322.

amounts of each were used in a mixed transfection (Fig. 2b, lane 10) the ratio of radioactivities was 1.8:1; this ratio, designated the relative transcriptslevel (RTL) was assigned a value of 1.0. The deviation of this ratio from the expected equimolar value of 1.2:1 may reflect vagaries of the system or differences in the stability of the two RNAs in the cells. Substitution of the sequences between positions -9 and +10 (lane 12) or between positions -9 and +30 (lane 13) by a 10 bp linker, which removes the cap site, did not abolish β-globin transcription (RTL = 0.4 and 2.1, respectively), but resulted in the production of transcripts 4-6 and 28-31 nucleotides shorter than normal, respectively, and with new 5' termini mapping 25 to 29 nucleotides downstream of the T residue in the ATA box (cf. Fig. 1a, data not shown). Conversely, insertion of a 10 bp HindIII linker between positions -9 and -10 (lane 11), which increased the distance between the ATA box and the cap site, led to the production of β-globin transcripts (RTL = 0.8) which were initiated 11 to 13 bp upstream of the cap site, i.e. 26-28 nucleotides downstream of the ATA box (cf. Fig. 1a; data not shown). Substitution of the sequences between positions -10 and -20 with the 10 bp linker neither altered the initiation site nor significantly affected the amount of β-globin RNA (RTL = 0.7, lane 14), while complete removal of the ATA box by deletion/substitutions from positions -10 to -47 or from -21 to -47 reduced the amount of correct transcripts at least 30-fold (RTL $<$ 0.03, lanes 15 and 16, respectively). When the DNA segment between positions -27 and -47 was replaced with the linker fragment, leaving only 2 of the 3' proximal AMP residues of the ATA box, a low but significant level of correct transcripts was observed (RTL = 0.12, lane 17). A deletion/substitution between positions -38 and -47, which does not affect the ATA box, gave an RTL of 0.5 (lane 18). Removal of the CAAT box by a deletion/substitution from -57 to -84 reduced the transcript level 5-fold (RTL = 0.2, lane 19) without altering the specificity of initiation, whereas a small deletion/substitution just upstream of the CAAT box (positions -79 to -84) had no effect (RTL = 1.3, lane 20).

c) Analysis of rabbit β-globin promoter sequences by in vitro transcription. A comprehensive set of terminal deletion/substitution mutants (group II, 1-12 in Weber et al., preceding paper), which retained 12 to 178 bp of 5' flanking sequence provided templates in the in vitro transcription system developed by Manley et al. (15). This system directs specific transcription of the Adenovirus 2 major late region (15,16) and many other genes (16-19). As transcription is not accurately terminated, the templates are truncated by cleavage with a restriction enzyme and correct transcription is assayed by the production of a labeled "run-off" product whose length equals the distance between the cap site and the restriction cleavage site. When a plasmid containing the rabbit β-globin gene preceded by 425 bp of 5' flanking DNA (pΔ425) was digested with BglII or BamH1 and used as a template for in vitro transcription, the expected run-off products of 1200 and 480 nucleotides, respectively, were observed (Fig. 3a, lanes 2 and 3). Neither band was observed when template DNA was omitted from the reaction (Fig. 3a, lane 1) or when synthesis was carried out in the presence of 1 μg/ml of α-amanitin (data not shown). A mixture of the two templates gave rise to both of the expected products. Each of 14 deletion mutant DNAs was cleaved with BglII, an equal amount of BamH1-digested pΔ425 DNA was added as internal reference, and the products formed in vitro were analyzed. As shown in Fig. 3a, all reactions produced the 480 nucleotide reference RNA fragment, while the 1200 nucleotide fragment was observed only in reactions where the BglII-cleaved test DNA contained 37 bp or more of 5'-flanking DNA. As shown in Fig. 3b, the plasmid containing 37 bp of 5' flanking DNA is the smallest deletion mutant tested in which the ATA box is still entirely present. The plasmid containing 26 bp of 5'-flanking sequence retains 2 of the 3' terminal AMP residues of the ATA box, but no longer directed synthesis of the 1200 nucleotide run-off RNA.

II. Efficiency and accuracy of splicing in vivo of transcripts from β-globin genes modified by deletions or site-directed mutations.

The comparison of many intron-exon junctions of polymerase II-transcribed genes of higher eukaryotes has shown that the most common sequence is AG/GTAAG(intron)CAG/exon where the underlined nucleotides appear to be absolutely conserved in all functional splice junctions (20-22). On the

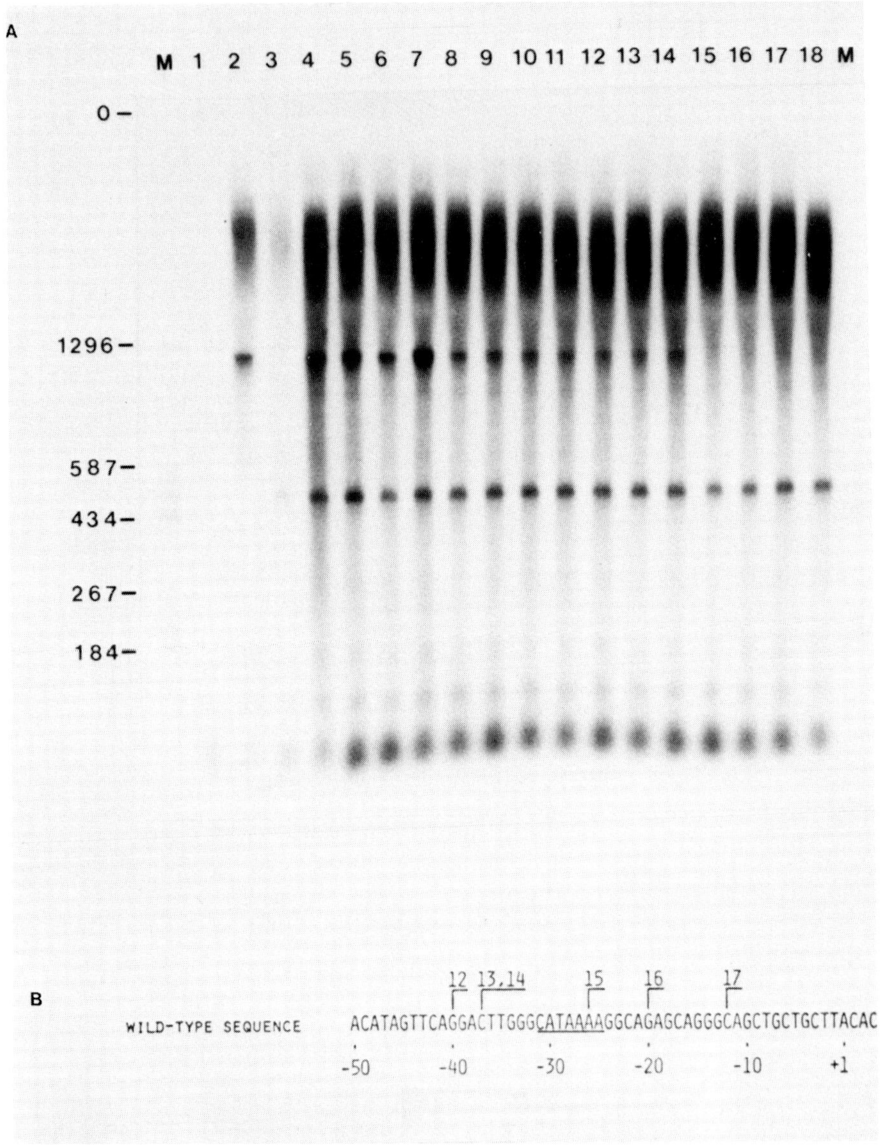

Fig. 3. Transcription in vitro of wild-type and externally deleted/substituted rabbit β-globin genes.

a) Autoradiograph showing the ^{32}P-labeled run-off products synthesized in vitro. Reactions were carried out using α-^{32}P UTP as the radioactive label in "run-off" transcription reactions, as described in the text. Hela whole cell extracts were used and transcription products were analyzed on a 2%

other hand, no other features of introns appear to be conserved, and introns of homologous genes are highly divergent in all but the regions flanking the splice sites (4,5,23).

In order to evaluate the importance of various regions of the rabbit β-globin gene for splicing, the expression of genes modified in various ways, was measured in the TK-transformation system. The genes to be tested were linked to the vector shown in Fig. 4a, namely pBR322 containing a rabbit β-globin reference gene, i.e. a gene with a small deletion in the 3rd exon. S1 mapping with a uniformly labeled β-globin chromosomal DNA probe (14) gave a 133 nucleotide fragment in the case of a normally spliced mRNA devoid of such a deletion, while the "reference mRNA" yielded fragments of 44 and 76 nucleotides instead (Fig. 4b). It is thus possible to determine the level of (normally spliced) test mRNA relative to reference mRNA. The first and second exons of the reference and test genes yielded identical signals of 145 and 222 nucleotides, respectively. As mentioned earlier (11, cf. above), the use of an internal reference is essential for

agarose gel as described by Manley et al. (15). Lanes 1 to 3, no added DNA, pΔ425 DNA truncated with BglII and with BamH1, respectively. Lanes 4 to 17, equimolar amounts of BamH1-cleaved pΔ425 DNA, as an internal control, and BglII-digested DNA from plasmids containing the rabbit β-globin gene preceded by 425 bp (lane 4), 178 bp (lane 5), 154 bp (lane 6), 109 bp (lane 7), 82 bp (lane 8), 79 bp (lane 9), 78 bp (lane 10), 56 bp (lane 11), 40 bp (lane 12), 37 bp (lane 13), 37 bp with a G ⟶ T transversion at position -33 (lane 14), 26 bp (lane 15), 20 bp (lane 16), or 12 bp (lane 17) of 5'-flanking rabbit chromosomal DNA. Lane 18, a BglII-digested plasmid in which all upstream sequences from position +31 were removed. The preparation of the DNAs is described in Weber et al. (preceding paper, Fig. 1, group II mutants 1 to 13).

b) Nucleotide sequence of the 5'-flanking region around the ATA box. The vertical line indicates the end point of the deletions; the number refers to the lane in which the corresponding transcripts were analyzed.

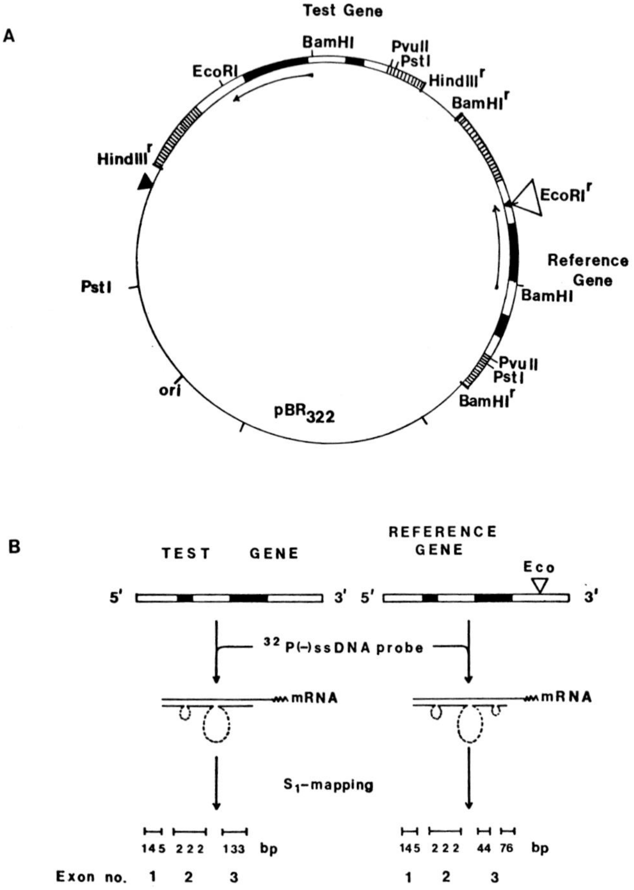

Fig. 4. Strategy for the analysis of transcripts of rabbit β-globin genes with modification in or around the large intron.

A) The modified chromosomal β-globin DNA (test gene) and a β-globin reference gene with a 13 bp deletion in the third exon were linked in tandem into pBR322. The linearized hybrid DNA was concatenated with a HSV-1 TK gene hybrid, introduced into TK^- mouse L-cells as described (8,10,11), and β-globin specific RNA was analyzed as outlined below.

B) Simultaneous analysis of β-globin-specific transcripts from test and reference gene. The structure and relative abundance of RNAs transcribed from the test and reference quantitation of transcripts, since distinct cell lines resulting from one and the same transformation experiment show very variable levels of globin gene expression; however, the ratio of reference and wild type test signals was found to be quite constant, with values slightly above one.

a) Deletions and deletion/substitutions of the large intron and its flanking region. A 439 bp MboII segment was removed from within the large 573 bp intron of the rabbit β-globin gene, leaving only 110 bp of the 5' and 24 bp of the 3' part of the intron. No effect was seen on the level or the structure of the mRNA derived from the modified gene. This result makes it unlikely that any particular internal sequence or the secondary structure of the intron is required for splicing.

A globin gene with a 38 bp deletion in the second exon, 18 bp upstream from the large intron (Fig. 5, no. 2), gave rise to a mRNA level about half that of wild type; as expected, the signal corresponding to the second exon had only 165 rather than 222 nucleotides. Two constructions had deletions removing the junction between the second exon and the large intron; no. 3 in Fig. 5 lacked 51 bp of the exon and 8 bp of the intron, while no. 4 lacked 13 and 8 bp, respectively. Surprisingly, the reduction in mRNA level was marginal, and S1 mapping showed that while a normal signal resulted from the first and third exons, the middle exon gave a fragment of about 90 nucleotides in both

genes were determined by the S1 mapping procedure (14), using a uniformly ^{32}P-labeled chromosomal β-globin DNA minus strand extending from position -98 (PstI) to +1198 (BglII) as hybridization probe.

The RNA segments corresponding to the first and second exons of normally spliced transcripts gave rise to probe fragments of 145 and 222 nucleotides, respectively. The RNA segment corresponding to the third exon yielded a fragment of 133 nucleotides in the case of the test gene, whereas in the case of the reference gene two fragments, of 76 and 44 nucleotides, were generated because the probe loops out in the region of the deletion.

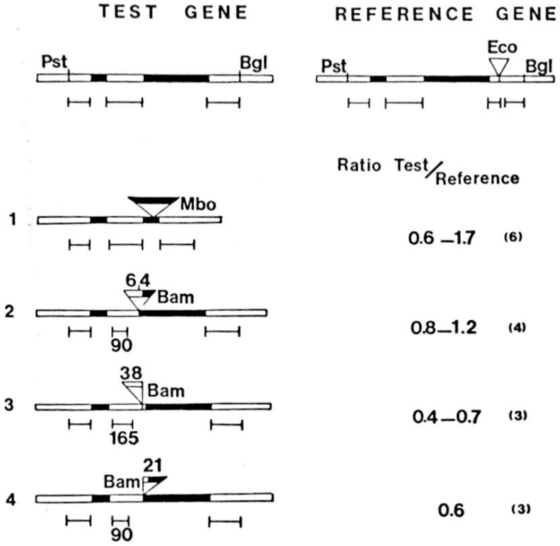

Fig. 5. Analysis of transcripts of rabbit β-globin genes with deletions within and bordering the second intron.

The transcripts of reference and test genes were analyzed by S1 mapping, as described in the legend to Fig. 4. The value "test/reference" is the ratio of the radioactivity in the 133 nucleotide fragment (derived from the test gene) to the sum of radioactivities of the 76 and 44 nucleotide fragments (derived from the reference gene). The numbers given are the maximal and minimal values determined; the number of individual L-cell lines examined is given in parentheses. The test genes are: 1) RβG Mbo 439 (439 bp MboII fragment removed from within large intron); 2) RβG Bam 38 (38 bp deleted upstream from the Bam site); 3) RβG Bam 59 (38 bp deleted upstream and 21 bp downstream); 4) RβG Bam 21 (21 bp deleted downstream from the Bam site) (cf. van Ooyen et al. (4) for nucleotide sequence of the rabbit β-globin gene).

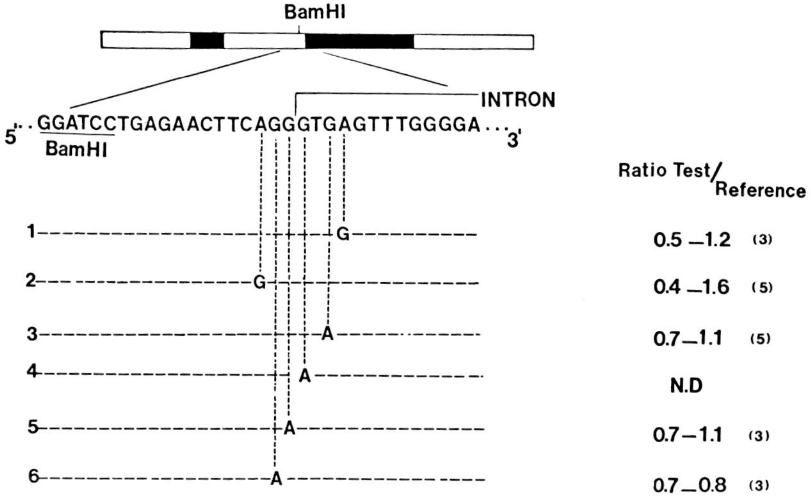

Fig. 6. Analysis of transcripts of rabbit β-globin genes with point mutations in the region of the 5'-proximal splice junction of the second intron.

The analysis was carried out as in Fig. 5. The mutated rabbit β-globin DNAs were prepared as described by Weber et al. (preceding paper). Only the base in which the test gene sequence differs from the wild type rabbit β-globin sequence is given. The range of the ratio of test and reference gene transcripts is given for the number of independent transformed L-cell lines indicated in parentheses, as described in Fig. 5. N.D., not done.

cases. This suggests that splicing resulted in correct excision of the large intron at its 3' junction, but that, in absence of the normal 5' junction, a normally unrecognized (cryptic) splice signal located in the middle exon was used. Inspection of the middle exon shows the existence of a sequence AAG/GTGA (positions 357 to 363, cf. sequence in ref. 4), which is very similar to the normal 5' splice junction AGG/GTGA (positions 492 to 498). If splicing occurred at the position indicated by the slash, the middle exon would yield an 88 nucleotide fragment.

b) Point mutations around the 5' junction of the large intron. Fig. 6 shows that transition mutations affecting the three nucleotides immediately preceding the splice site had no effect on the β-globin mRNA level in the cell, or did mutations in the 3rd and 4th positions after the splice site. Unfortunately, the experiment with the gene carrying a mutation at the position immediately following the splice site, which converts the G residue of the "Chambon sequence" (21) into an A was unsuccessful; it is currently being repeated.

DISCUSSION

The expression of a variety of modified rabbit β-globin genes (cf. Weber et al., preceding paper) has been studied in vivo and in vitro. The quantitation in vivo was carried out by the S1 mapping technique (14), using β-globin genes with small deletions or substitutions as internal references. S1 mapping measures the level of transcripts at a given time, and not the rate of synthesis or the efficiency of splicing. In the case of mutations in the 5'-flanking region which do not change the structure of the primary transcript, it is reasonable to correlate the mRNA content of a cell with the rate of transcription, and in the following discussion we shall assume this to be the case. When the structure of the mRNA or its precursor is modified, however, other phenomena, such as transport to the cytoplasm or stability of the RNA may affect the transcript level.

Our studies on transcription show that the ATA box plays a dual role in the transcription of the rabbit β-globin gene, in that it is required to obtain significant levels of transcription in vivo and in vitro and to localize the starting point of the transcripts, regardless of whether the normal cap site is present or not. Furthermore, we confirm our previous suggestion (11) that the DNA segment between -57 and -84, which contains the CAAT box, has an important potentiating effect on the level of transcription in vivo, but is not required for correct initiation of the transcripts. Studies of

ATA box function in vivo for the SV40 early region transcription unit (24-26) and the sea urchin H2A histone gene (27) have demonstrated the critial role of the ATA box in positioning the 5' termini of the transcripts from these genes. In contrast to our findings with the β-globin gene however, significant levels of transcripts (albeit with variant 5' termini) from the SV40 early region (24-26, 28) and the sea urchin H2A gene (27) were found even after deletion of the ATA box. Moreover, sequences more than 100 bp upstream of the cap sites of these genes had a dramatic effect on the level of transcription (24, 26,28,29) while deletion of the CAAT-like sequence of the sea urchin H2A gene did not adversly affect production of H2A mRNA (29). Thus, although the initiation site "fixing" function of the ATA box regions of these three genes appears to be similar, the nature and interplay of upstream elements involved in transcription may be fundamentally different. The essential role of the ATA box for transcription has been confirmed in vitro for the Adenovirus 2 late transcription unit (6,30,31), the conalbumin gene (30,32,33) and now for the rabbit β-globin gene. In contrast, the effect of the CAAT box was not detected in the in vitro systems. In view of the fact that in vitro transcription proceeds at a very low level compared to that in vivo, it could be that whatever mechanism is mediated by the CAAT box is not operative in vitro, perhaps because the DNA is not present as a native nucleoprotein, or because one or more factor(s) are not present in sufficient level.

Our studies on the structural requirements for splicing have shown that removal of 77% of the large intron has no effect on either the accuracy of splicing nor on the level of spliced transcripts. Khoury et al. (34) have reported that natural deletions within the large early intron of SV40 DNA do not prevent slicing, provided that the splice junctions remain intact, and Chu and Sharp (35) found that hybrid introns are spliced normally. All of these data suggest that the internal part of an intron plays no role in the splicing mechanism.

It seems that there is no very elaborate sequence requirement even at the splice junction, since a number of transition mutations around the splice point did not diminish the level of correctly spliced transcripts, even though two of these mutations would disrupt one putative base pair interaction with U1snRNA (36,37).

Our finding that deletions removing the 5' splice junction cause splicing to take place at a similar sequence 135 bp further upstream can be explained if splicing occurrs in a processive mode, from 3' to 5' (22,38) or from within the intron both upstream and downstream. Alternatively, the cryptic splice site may have such a low affinity for whatever initiates the splicing event that it is very rarely used. The fact that the cryptic site is so efficient in the deletion mutants makes the latter explanation less likely.

ACKNOWLEDGMENTS

P.D. was supported by a fellowship of the American Cancer Society, and B.W. of EMBO. This work was supported by the Schweizerische Nationalfonds and the Kanton of Zürich.

REFERENCES

1. Nathans, D., and Smith, H.O. (1975). Annu. Rev. Biochem. 44, 273-293.
2. Methods in Enzymology, vol. 68 (R. Wu, ed.), Academic Press (1979).
3. Weissmann, C. (1978). TIBS 3, N109-N111.
4. van Ooyen, A., van den Berg, J., Mantei, N., and Weissmann, C. (1979). Science 206, 337-344.
5. Efstratiadis, A., Posakony, J.W., Maniatis, T., Lawn, R.M., O'Connell, C., Spritz, R.A., DeRiel, J.K., Forget, B.G., Weissman, S.M., Slightom, J.L., Blechl, A.E., Smithies, O., Baralle, F.E., Shoulders, C.C., and Proudfoot, N.J. (1980). Cell 21, 653-668.
6. Corden, J., Wasylyk, B., Buchwalder, A., Sassone-Corsi, P., Kedinger, C., and Chambon, P. (1980). Science 209, 1406-1414.
7. Benoist, C., O'Hare, K., Breathnach, R., and Chambon, P. (1980). Nucl. Acids Res. 8, 127-142.
8. Wigler, M., Pellicer, A., Silverstein, S., and Axel, R. (1978). Cell 14, 725-731.

9. Weaver, R., and Weissmann, C. (1979). Nucl. Acids Res. 7, 1175-1193.
10. Mantei, N., Boll, W., and Weissmann, C. (1979). Nature 281, 40-46.
11. Dierks, P., van Ooyen, A., Mantei, N., and W Weissmann, C. (1981). Proc. Natl. Acad. Sci. USA 78, 1411-1415.
12. Banerji, J., Rusconi, S., and Schaffner, W. (1981). Experientia 37, in press.
13. Weidle, U., Henco, K., Schmid, J., Fujisawa, J., Schamböck, A., Nagata, S., and Weissmann, C. (1981). Experientia 37, in press.
14. Berk, A.J., and Sharp, P.A. (1977). Cell 12, 721-732.
15. Manley, J.L., Fire, A., Cano, A., Sharp, P.A., and Gefter, M.L. (1980). Proc. Natl. Acad. Sci. USA 77, 3855-3859.
16. Hagenbüchle, O., and Schibler, U. (1981). Proc. Natl. Acad. Sci. USA 78, 2283-2286.
17. Handa, H., Kaufman, R.J., Manley, J., Gefter, M., and Sharp, P.A. (1981). J. Biol. Chem. 256, 478-482.
18. Proudfoot, N.J., Shander, M.H.M., Manley, J.L., Gefter, M.L., and Maniatis, T. (1980). Science 209, 1329-1336.
19. Tsai, S.Y., Tasi, M.-J., and O'Malley, B.W. (1981). Proc. Natl. Acad. Sci. USA 78, 879-883.
20. Seif, I., Khoury, G., and Dhar, R. (1979). Nucl. Acids Res. 6, 3387-3398.
21. Breathnach, R., Benoist, C., O'Hare, K., Gannon, G., and Chambon, P. (1978). Proc. Natl. Acad. Sci. USA 75, 4853-4857.
22. Sharp, P.A. (1981). Cell 23, 643-646.
23. van den Berg, J., van Ooyen, A., Mantei, N., Schamböck, A., Grosveld, G., Flavell, R.A., and Weissmann, C. (1978). Nature 275, 37-44.
24. Gluzman, Y., Sambrook, J.F., and Frisque, R.J. (1980). Proc. Natl. Acad. Sci. USA 77, 3898-3902.
25. Ghosh, P.K., Lebowitz, P., Frisque, R.J., and Gluzman, Y. (1981). Proc. Natl. Acad. Sci. USA 78, 100-104.
26. Benoist, C., and Chambon, P. (1981). Nature 290, 304-310.
27. Grosschedl, R., and Birnstiel, M.L. (1980). Proc. Natl. Acad. Sci. USA 77, 1432-1436.
28. Benoist, C., and Chambon, P. (1980). Proc. Natl. Acad. Sci. USA 77, 3865-3869.

29. Grosschedl, R., and Birnstiel, M.L. (1980). Proc. Natl. Acad. Sci. USA 77, 7102-7106.
30. Wasylyk, B. Kedinger, C., Corden, J., Brison, O., and Chambon, P. (1980). Nature 285, 367-373.
31. Hu, S.-L., and Manley, J.L. (1981). Proc. Natl. Acad. Sci. USA 78, 820-824.
32. Wasylyk, B., and Chambon, P. (1981). Nucl. Acids Res. 9, 1813-1824.
33. Wasylyk, B., Derbyshire, R., Guy, A., Molko, D., Roget, A., Téoule, R., and Chambon, P. (1980). Proc. Natl. Acad. Sci. USA 77, 7024-7028.
34. Khoury, G., Gruss, P., Dhar, R., and Lai, C.-J. (1979). Cell 18, 85-92.
35. Chu, G., and Sharp, P.A. (1981). Nature 289, 378-382.
36. Rogers, J., and Wall, R. (1980). Proc. Natl. Acad. Sci. USA 77, 1877-1879.
37. Lerner, M.R., Boyle, J.A., Mount, S.M., Wolin, S.L., and Steitz, J.A. (1980). Nature 283, 220-224.
38. Avvedimento, V.E., Vogeli, G., Yamada, Y., Maizel, J.V., Pastan, I., and de Crombrugghe, B. (1980). Cell 21, 689-696.

MODIFICATION OF THE RABBIT CHROMOSOMAL β-GLOBIN GENE BY RESTRUCTURING AND SITE-DIRECTED MUTAGENESIS

H. Weber, P. Dierks, F. Meyer, A. van Ooyen,
C. Dobkin, P. Abrescia, M. Kappeler, B. Meyhack,
A. Zeltner, E.E. Mullen and C. Weissmann

Institut für Molekularbiologie I
Universität Zürich
8093 Zürich, Switzerland

ABSTRACT. Modifications were introduced into the rabbit chromosomal β-globin gene by 1) gene restructuring techniques, leading to a variety of internal and external deletions and deletion/substitutions, 2) site-directed mutagenesis involving the incorporation of a mutagenic deoxynucleoside triphosphate into specific sites of a DNA. The modifications affect the 5'-flanking sequences including putative promoter sites, the 5'-non-coding region of the mRNA sequence, introns and splicing sites, and the termination codon.

INTRODUCTION

We have prepared a large number of variants of the cloned chromosomal β-globin gene of the rabbit in order to investigate the effect of specific modifications on the expression of this gene, both in vivo and in vitro (1). The methods used fall into two classes: (I) Restructuring of cloned DNA, using nucleases, polymerases and DNA ligase to generate defined terminal and internal deletions, and (II) site-directed mutagenesis, to obtain point

ISBN 0-12-137420-3

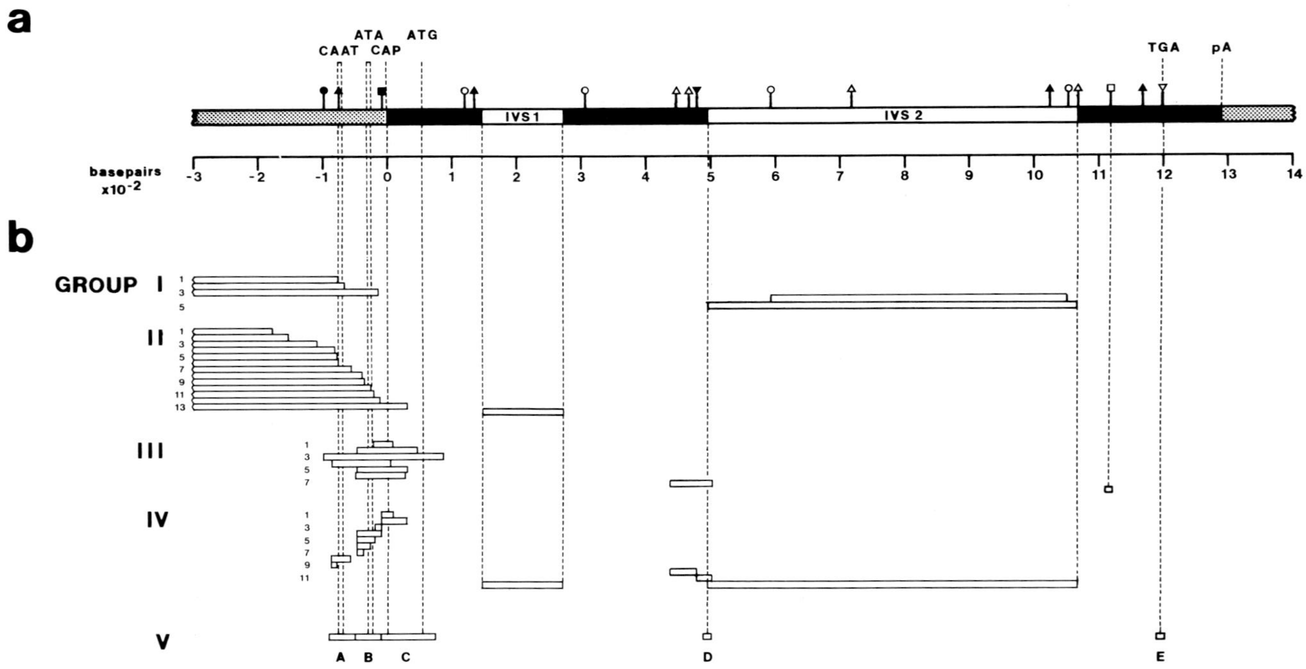

a
CAAT
ATA
CAP
ATG
TGA
pA
IVS1
IVS 2
basepairs
x10-2
-3
-2
-1
0
1
2
3
4
5
6
7
8
9
10
11
12
13
14
b
GROUP I
II
III
IV
V
A
B
C
D
E

mutations, both transitions and transversions, at specific positions. The alterations we describe are located in the nontranscribed 5'-flanking region, the 5'-non-coding region of the mRNA, the introns and splicing sites, and near restriction sites within the coding region including the termination codon (the latter on cDNA only), as shown in the map of Fig. 1.

I. Gene restructuring

1) Excision of restriction fragments and religation.
The simplest way to produce a defined deletion is to remove an entire restriction fragment from a

Fig. 1. Physical map of the rabbit β-globin gene and summary of mutant DNAs generated in vitro.

A: Physical map of the rabbit β-globin gene. Transcription is from left to right. Stippled boxes, flanking rabbit chromosomal DNA; solid boxes, exons; open boxes, introns. "CAAT box" and "ATA box" denote the sequences (-77) GGCCAATCT (-69) and (-31) CATAAAA (-25), respectively; "cap site", transcription initiation site; "ATG", translation initiation codon; "IVS1" and "IVS2", first and second introns, respecitvely; "TGA", translation termination codon; "pA", polyadenylation site of the transcript. Restriction sites: † ,PstI; †, BspI; † ,PvuII; †, MboII; †, AluI; †, BamH1; † , EcoRI; †, BglII. Positions are numbered relative to the cap site.
B: Summary of mutants generated in vitro. Group I deletions were constructed by digesting the DNA with restriction enzymes and religating the desired fragments, group II deletions and deletion/substitutions by primer extension methods (see section I-3), group III deletion/substitutions by the exonuclease III/nuclease S1 method, group IV deletions and deletion/substitutions by simple restriction enzyme digestion and fragment ligation as for group I, but the starting DNAs were mutants from group II or III. Group V point mutations were generated in vitro by site-directed mutagenesis. Regions A-E refer to segments selected for individual mutagenesis experiments and are shown in detail in Fig. 7.

given DNA and to rejoin the termini by ligation. Depending on the restriction sites available, such a procedure may require one or several fragment isolation and ligation steps. Since these techniques are standard in recombinant DNA methodology (2), they will not be described in detail. The plasmids prepared in this way are listed in Fig. 1 under group I. They include deletions in the 5' flanking region (I-1 to I-3) and in the large intron (I-4). Using similar techniques, the BamH1-BglII fragment spanning the large intron was replaced by the corresponding fragment from a cDNA plasmid, resulting in a β-globin plasmid lacking the entire large intron but retaining the 5'- and 3'-flanking sequences and the small intron (I-5).

2) Restriction cleavage followed by exonuclease trimming and religation.
The potential of the "cut and join" approach described above is limited since the end points of the deletions are determined by existing restriction sites. This limitation can be overcome by cleaving a cloned DNA segment at a unique restriction site (either a natural one, or one generated by insertion of a linker (3)), within or near the region of interest, and trimming back the newly created ends with an exonuclease. The digestion may be controlled kinetically when using exonuclease Bal31 (4) or E.coli exonuclease III followed by nuclease S1 (5), or controlled by nucleoside triphosphate addition, when using the 3'-exonuclease activity of T_4 DNA polymerase (6) followed by nuclease S1 digestion. The ends resulting from the digestion can be rejoined directly by flush-end ligation to give pure deletions, or, in the more versatile approach mostly used here, indirectly, with use of a synthetic "linker" oligonucleotide containing the restriction cleavage site used as starting point for the trimming. The latter procedure gives rise to deletion/substitutions. DNAs with deletions of about the desired length were selected on the basis of restriction analysis and sequenced around the deletion end points to yield a collection of precisely characterized mutations. The rabbit β-globin deletions prepared by this method are listed in Fig. 1, group III. The restriction site linker made it possible to cleave the DNA at the deletion site and, after cleavage

at a second unique restriction site, to physically separate the two plasmid fragments carrying the deletion end points. Appropriate fragments from the collection of deletion plasmids could then be combined to generate a variety of additional deletion/substitution mutants. Alternatively, a fragment with a deletion terminating with a restriction linker was joined to a fragment carrying the original restriction end; in this way, deletions extending unidirectionally from a given restriction site to various upstream and downstream positions were constructed. Such "second generation" deletion mutants are shown in Fig. 1, group IV.

3) Preparation of deletion/substitution mutants by limited primer extension using AMV reverse transcriptase and suboptimal concentrations of dCTP. As an alternative to the methods described above we have used limited primer extension on a single-stranded template as a method of constructing a comprehensive set of unidirectional deletion/substitutions. The principle of the procedure (cf. Fig. 2) is the following. The primer DNA strand contains a unique internal restriction recognition site and its 3' end defines the outer limit of the set of deletions to be generated. The primer is hybridized to the single-stranded template and elongated in a substrate-controlled reaction to generate a set of molecules with increasing lengths of double-stranded segment. After removal of single-stranded DNA by S1 nuclease, an appropriate linker is added to the termini, and the DNA segment is cleaved at the unique restriction site. This fragment is then joined to a matching receptor plasmid.

Fig. 2 shows how a set of deletion/substitutions was generated extending into the 5'-flanking segment of the rabbit β-globin gene and deleting the CAAT box and/or the ATA box (7). The PvuII-BglII fragment (1209 bp) containing most of the rabbit β-globin gene was isolated, denatured and hybridized to single-stranded phage fd 107/RchrβG·27(+) DNA, which contains the plus strand of a rabbit chromosomal β-globin DNA insert. The 5' terminus of the minus strand primer thus corresponded to the BglII cleavage site at position 1200, and the 3' end to the PvuII site at position -9. The template-primer complex was isolated by sucrose gradient centrifugation and the primer elongated with reverse transcriptase in the presence of 10 μM

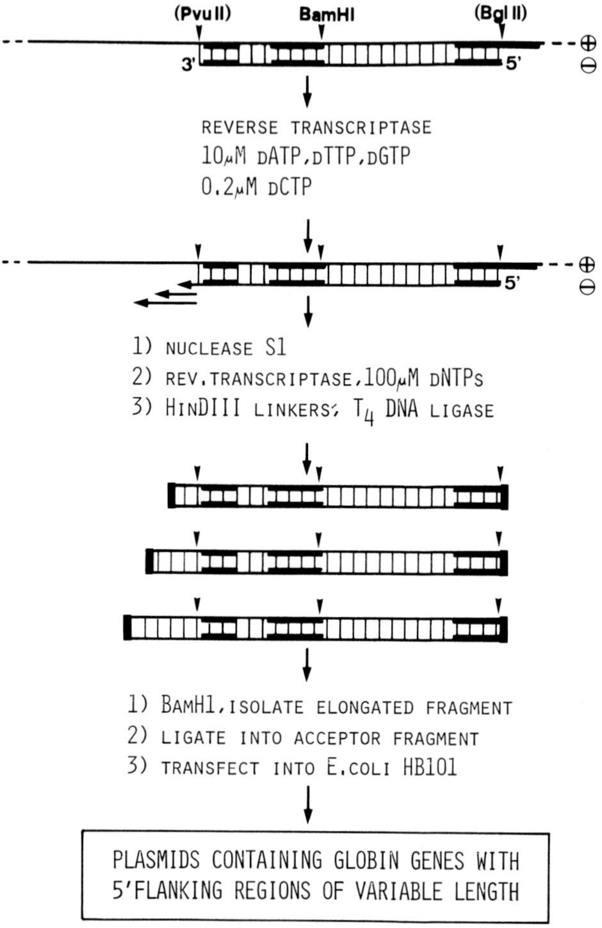

Fig. 2. Construction of β-globin genes with 5'-flanking regions of variable length by the primer extension method.

The primer, the 1209 bp PvuII-BglII minus strand fragment (corresponding to most of the chromosomal β-globin gene) was annealed to the plus strand of the gene cloned in a single-stranded fd phage vector (the plus strand is the strand carrying the mRNA sequence). Thick horizontal lines indicate exon sequences. The HindIII linkers are symbolized by vertical black bars. The procedure is described in Section I-3.

dGTP, dATP, dTTP and 0.2 μM dCTP for 15' at 41°C (Fig. 2). When suboptimal concentrations of dCTP are used, DNA synthesis terminates prematurely at multiple sites which correspond roughly to positions where multiple dGMP residues are present in the template (8). In the 5'-flanking sequence of the rabbit β-globin gene there are 8 locations in the first 100 nucleotides of the plus strand at which there are 2 or more dGMP residues: two of these locations flank the ATA box located between positions -25 and -31. Analysis of the extended primer showed multiple discrete terminations corresponding roughly to the expected positions. The portions of the template DNA remaining single-stranded were digested with nuclease S1 under mild conditions and the duplex DNA was isolated by gradient centrifugation. To insure a high frequency of flush-ended molecules, the DNA was again incubated with reverse transcriptase and 100 μM concentrations of each of the four dNTPs, and HindIII linkers were added (3) with T4 DNA ligase. The fragment was digested with BamH1, which cleaves in the middle of the globin gene, and HindIII, and the fragment corresponding to the 5' half of the globin gene was ligated into a HindIII/BamH1 acceptor fragment supplying the 3' half of the gene as well as most of the vector pBR322 sequences. Fig. 1 (group II) shows twelve cloned plasmids containing 5'-flanking sequences of 12 to 178 bp, as determined by nucleotide sequence analysis. It should be noted that each of the deletion/substitution mutants constructed above has the same pBR322 sequences flanking the rabbit chromosomal DNA insert, thereby providing a constant vector environment, which is important for evaluation of the biological activity of in vitro mutagenized genes.

4) Construction of genes lacking introns by primer extension on a heteroduplex template-primer. The removal of introns from a chromosomal gene by simple restriction and ligation techniques is usually not possible due to the absence of suitable restriction sites. We have generated a rabbit β-globin DNA insert lacking the small intron, but retaining the large intron as well as the 5'- and the 3'-flanking sequences by a variation of the primer extension technique. A plus strand chromosomal globin template was prepared by exonuclease

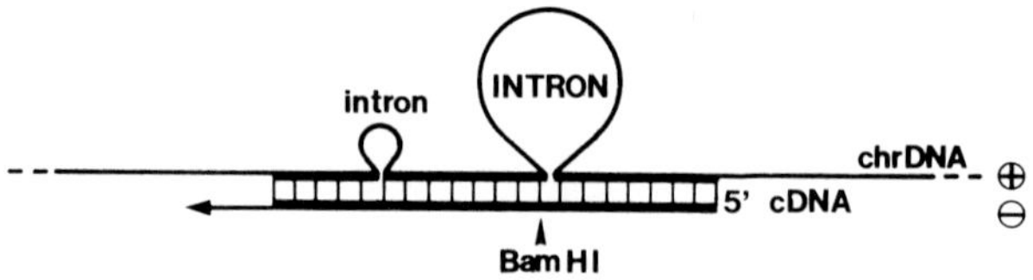

Fig. 3. The use of heteroduplex template-primer systems to construct β-globin genes with 5'-flanking regions but lacking introns.

The minus strand of a β-globin complementary DNA (cDNA) segment (obtained by excision and slight trimming of the β-globin insert of plasmid PβG (13) followed by cloning via HindIII linkers (9)) was used as a primer; it was annealed to a chromosomal β-globin DNA (chr DNA) segment from which the minus strand had been removed with exonuclease III. Primer extension led to the attachment of the 5'-flanking sequences to the cDNA moiety. Isolation of the elongated primer, its conversion to double-stranded DNA and cloning of the relevant DNA segment is illustrated in Fig. 4. The 5' half of the globin gene lacking the small intron was joined via the BamH1 site with 3' halves either possessing or lacking the large intron.

III digestion of a suitably linearized chromosomal globin DNA plasmid (but a single-stranded template prepared by cloning in fd or M13 is equally suitable), and a primer fragment consisting of an almost complete copy of rabbit β-globin complementary DNA (cDNA), flanked by HindIII linkers (9) was annealed to it. As illustrated in Fig. 3, the cDNA primer and the chromosomal DNA template form a heteroduplex structure, with the introns on the chromosomal DNA strand remaining in the form of single-stranded loops. Elongation of the cDNA primer in vitro with E.coli DNA polymerase I and the four deoxynucleoside triphosphates resulted in the formation of a minus strand segment containing the 5'-flanking sequences derived from the chromosomal template but lacking introns. The elongated primer was isolated by alkaline sucrose gradient sedimentation and converted into a double-stranded DNA segment by in vitro synthesis using E.coli DNA polymerase I and a suitable restriction fragment serving as a plus strand primer. The PstI-BamH1 fragment of this segment was used to replace the corresponding piece of a plasmid containing the chromosomal β-globin gene, resulting in an otherwise complete gene lacking the small intron (II-14). This modified gene was subsequently converted into one lacking both introns (IV-12) by replacing its BamH1-BglII segment with the corresponding fragment from a cDNA insert, as described in Section 1. Thus, four different forms of the β-globin gene including its flanking sequences became available: the "natural" gene containing both introns, and three artificial genes lacking either the small or the large or both introns.

II. Site-directed mutagenesis.

The principle of incorporating analog nucleoside triphosphates at selected sites of a nucleic acid in order to achieve specific point mutations was discussed previously both for RNA and DNA systems (10-12). In earlier work mutations were introduced into the single EcoRI recognition site of the β-globin cDNA plasmid PβG (13) by nicking the plasmid with EcoRI and carrying out limited "nick-translation" using E.coli DNA polymerase and two or three deoxynucleoside triphosphates, one of them being the analog N^4-hydroxy-deoxycytidine triphosphate (11). More recently, this method was applied to the mutagenesis of the BglII site at the end of the β-globin coding region, resulting in a 2-4% yield of BglII-resistant mutant plasmids, several of which carried mutations in the termination codon (Fig. 7E).

Modifications of this approach were developed in order to extend the use of the method to regions devoid of restriction sites and to increase the rate of mutagenesis. Satisfactory results were obtained by using template-primer systems of the kind described in the last section and illustrated in Fig. 4. Single-stranded (plus or minus strand) β-globin DNA, mostly integrated in circular fd phage vector DNA, was used as template. The primers which were annealed to it were derived from

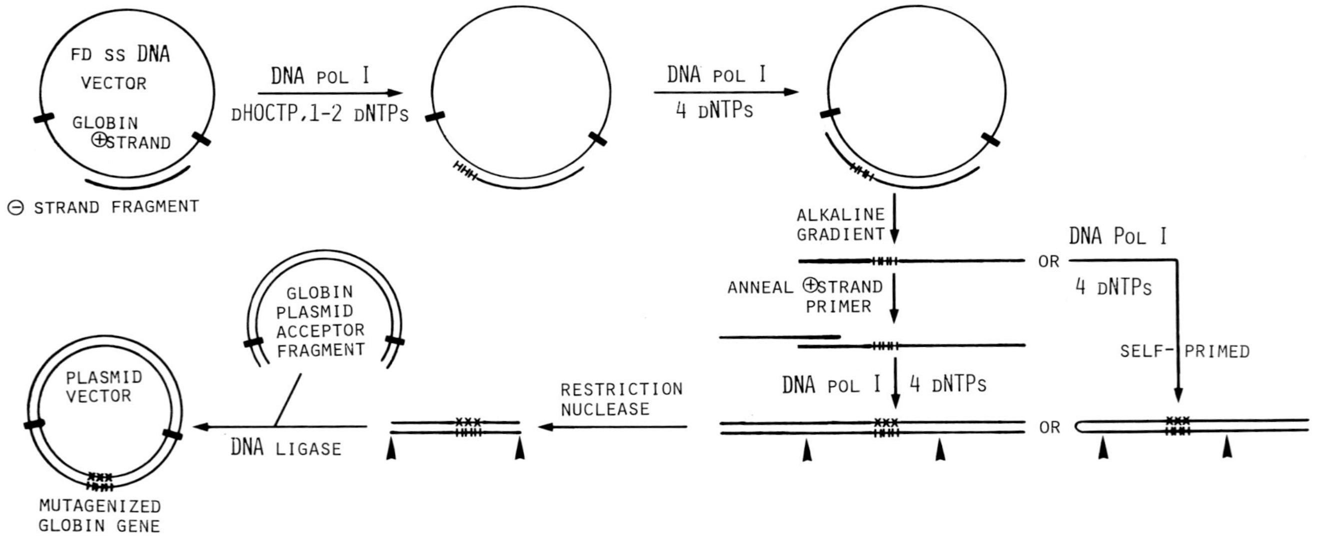

Fig. 4. General scheme of site-directed mutagenesis of the globin gene.

Globin plus strand DNA cloned in single-stranded fd DNA served as template in this example; globin minus strand in fd DNA and globin DNA segments made single-stranded by exonuclease III have also been used. The primer is a fragment of globin DNA with an appropriate 3' end (see section II). After incorporation of mutagenic hydroxydCMP (HOC) residues (H) and further elongation (see Fig. 5), the extended primer was isolated and made double-stranded by DNA polymerase I either in a self-primed reaction or using a suitable plus strand primer. The mutagenized segment was cleaved out by restriction nucleases (arrow heads) and ligated into an acceptor fragment supplying the missing parts of the globin gene as well as the vector (pBR322) sequences. X denotes sites at which point mutations may arise in vitro due to ambiguous basepairing with HOC.

restriction fragments whose ends had been trimmed back or extended, fitted with linkers and cloned in an appropriate plasmid. In the subsequent DNA polymerase I reaction, the non-base paired linker residues were removed by the 3'⟶5' exonuclease activity of the enzyme. The mutagenic nucleotide dHOCTP was incorporated in various substrate combinations depending on the desired site of mutation(s). The analog was found to substitute for both dCTP and dTTP, but in the presence of one pyrimidine triphosphate it substituted specifically for the other due to competition. Thus, mutations could be limited to either G⟶A or A⟶G transitions. The length of the mutagenized region could be varied by including one, two or three triphosphates in the reaction mixtures. Sites somewhat removed from a primer terminus could be reached in a two step reaction, the first step serving to elongate the primer by a defined number of nucleotides in presence of less than four of the normal triphosphates. Longer stretches were mutagenized by elongating in the presence of dHOCTP and both dGTP and dATP, under which conditions the reaction is not strictly substrate limited but incorporation goes on at a low rate and terminations are frequent. A further variation could be introduced by using either a plus or a minus strand primer for a particular region to be mutagenized, which results in either pyrimidine or purine transitions in the plus strand, respectively. After the analog incorporation step, the primers were extended for several hundred nucleotides in presence of high concentrations of the 4 normal triphosphates, separated from the template DNA by alkaline sucrose gradient sedimentation and used in turn as templates for the synthesis of a complementary strand, either by self-priming (14) or with use of a suitable restriction fragment as primer. This step was thought to be important for increasing the efficiency of mutagenesis for two reasons: 1) It avoids the dilution of mutated progeny plasmids by descendants of the unmutated template strand; 2) mispair-correcting repair systems which might preferentially remove the analog after transfection into E.coli will no longer reduce the mutation frequency significantly, since the base change on the complementary strand is carried out in vitro. The mutagenized portion of the globin gene

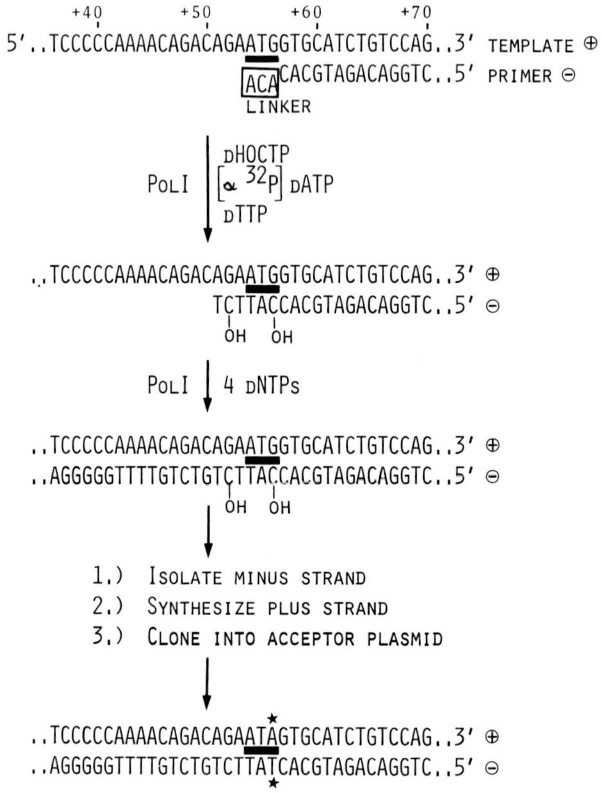

Fig. 5. Site-directed mutagenesis: Incorporation of mutagenic HOC residues resulting in a point mutation in the initiation codon of the β-globin gene.

The 3 nucleotides (boxed) at the 3' end of the minus strand primer are from the synthetic linker oligonucleotide and do not base-pair with the template. They are removed by 3'⟶5' exonuclease activity of E.coli DNA polymerase I which subsequently elongates the primer by 6 residues, 2 of them HOC. The primer is then extended with the 4 standard triphosphates and processed as depicted in Fig. 4, to yield progeny sequences, some of which have the initiator codon of the β-globin gene ATG mutated to ATA (mutant C-10). The initiation codon is indicated by a horizontal black bar, the mutated base pair by asterisks. Nucleotides are numbered from the cap site.

was recovered as a defined fragment by restriction cleavage and ligated into an acceptor fragment containing the missing parts of the gene and the vector sequences. After transfection, a number of progeny plasmids were tested for mutations by partial sequence analysis, usually limited to one or two sequencing lanes. The mutagenization efficiencies obtained ranged between 8 and 20% per mutable site. However, in one experiment which was expected to yield two distinct nucleotide transitions, only one of the sites was mutated (4 out of 23 isolates), while no mutations were found in the other position. Fig. 5 shows, as an example, the conversion of the initiation codon ATG to ATA. Out of 6 clones analyzed, one was found to be the desired mutant.

The point mutations were prepared by the general procedures outlined above in the five regions indicated A-E in Fig. 1. Groups A and B are located in the 5'-flanking region in two sites considered important for correct initiation of transcription: Group A in the -70 to -80 region (the "CAAT-site"), group B in the -25 to -35 region (the "ATA-site"). The mutants are listed in Fig. 7A and B. Group A was obtained using a plus strand primer terminating at position -86 whereas for the three mutants of group B two different minus strand primers were used. One, terminating at the wild-type PvuII site (position -9) gave rise to the two single mutants (B-1 and B-2). The other, terminating at position -26, was used in a modified procedure designed to yield not only transition but also transversion mutations (Fig. 6). The elongation of this primer was carried out with the four normal triphosphates (and no nucleotide analog), but with AMV reverse transcriptase replacing E.coli DNA polymerase I. Since reverse transcriptase does not possess any 3'⟶5' exonucleolytic activity, the primer was elongated without prior removal of the three 3'-terminal unpaired nucleotides derived from the linker. This left two mismatched nucleotides which subsequently gave rise to an A⟶T transversion and an A⟶G transition in mutant B-3. This approach obviously has considerable potential for the preparation of further transversions; it is not yet clear whether the 3'-terminal nucleotide of the linker residue has to be able to base pair, as in the present case, in order to

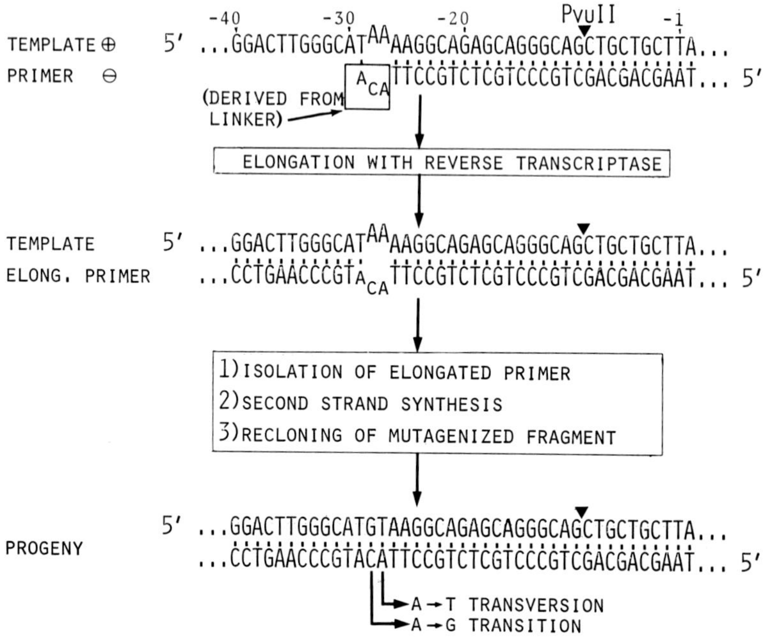

Fig. 6. A transversion mutation in the ATA site obtained by elongation of a mismatched primer terminus with reverse transcriptase.

Procedures are generally similar to the ones shown in Fig. 4 and Fig. 5, except that reverse transcriptase and the 4 standard triphosphates are used to elongate the primer. Since reverse transcriptase has no 3'⟶5' exonuclease activity, the linker-derived 3'-terminal residues (boxed) which are not (or only partially) base-paired with the template are not removed; their elongation leads to mispairs resulting in both transition and transversion mutations in the progeny plasmid (mutant B-3).

allow the fragment to be elongated efficiently.

The point mutations mapping in region C (Fig. 1) and listed in Fig. 7C were obtained in an effort to investigate the role of the 5'-extracistronic region of the β-globin mRNA for the initiation of protein synthesis. A collection of terminally trimmed β-globin cDNA segments were prepared by exonuclease III and SI digestion and cloned using HindIII linkers in pBR322 (9). Two such cloned fragments, terminating in positions +57 and +25, were used as minus strand primers. They were annealed to chromosomal β-globin plus strands to form template-primer heteroduplex structures as described above. Another minus strand primer, terminating at position +51, was used on a cDNA template. The mutated DNA segments resulting from the heteroduplex structures contained 5'- flanking sequences but lacked the small intron. They were linked via the BamH1 site to a 3' half of the globin gene either containing or lacking the large intron, depending on which constructions were desired. This group of mutants includes mutations of the capping site (C-5), of the initiation codon (C-10) and of the PvuII site (C-8). The A⟶G transition at position +19 generates a new PvuII site (C-3). A double mutant that has lost the old and acquired the new PvuII site was also obtained (C-9).

The group D mutations were made from a minus strand primer terminating at position 505 (Fig. 1 and Fig. 7D). An almost complete set of single purine transitions at the 5'-splicing site of the large intron could be isolated.

DISCUSSION

We have outlined a number of approaches which we have used to restructure the rabbit β-globin gene; several of these are based on conventional methodology, while others represent interesting variations of general applicability. In many cases, the modifications were deletion/substitutions, i.e., a DNA segment was removed from the gene or its flanking region and replaced by a (mostly substantially shorter) DNA linker fragment.

```
REGION A
                      -90       -80       -70       -60       -50
WILD TYPE SEQUENCE    ACACCCTGGTGTTGGCCAATCTACACACGGGGTAGGGATTA
          MUTANT 1    ---------------T-------------------------
                 2    ------------C---T------------------------
                 3    ------C----------------------------------
                 4    ------------C----G-----------------------
                 5    ------C-----C----G-----------------------
                 6    -----------CC-------------------------C--
                 7    ---------------------C-------------------
                 8    ---------C--C----------------------------
                 9    ------C--------------C-------------------
                10    -----------------------T-----------------

REGION B
                      -50       -40       -30       -20       -10
WILD TYPE SEQUENCE    ACATAGTTCAGGACTTGGGCATAAAAGGCAGAGCAGGGGAG
          MUTANT 1    ----------------------------------G------
                 2    ------------------------G----------------
                 3    ----------------------GT-----------------

REGION C
                      -20       -10       -1+1      +10       +20
WILD TYPE SEQUENCE    GAGCAGGGCAGCTGCTGCTTACACTTGCTTTTGACACAACTGTGTTTAC..
                                PvuII       Cap
          MUTANT 1    --------------------------A----------------------
                 2    -------------------------------------G-----------
                 3    --------------------------------------G----------
                 4    -------------------------------------------A-----
                 5    ----------------A---G----------------------------
                 6    ----------------------G----------G---G-----------
                 7    ----------------------G---A--------G-G-----------
                 8    -------- PvuII^R --------------------------------
                 9    -------- PvuII^R ---------------------G----------

                      +30       +40       +50       +60       +70
WILD TYPE SEQUENCE    ..TTGCAATCCCCCAAAACAGACAGAATGGTGCATCTGTCCAG
                MUTANT 10 ----------------A--------------
                       11* -----G-G-----------------------

REGION D
                      470       480       490       500       510
WILD TYPE SEQUENCE    GCACGTGGATCCTGAGAACTTCAGGGTGAGTTTGGGGACCC
                             BamHI                splice
          MUTANT 1    ----------------------------G------------
                 2    ----------------------G------------------
                 3    ---------------------------A-------------
                 4    -------------------------A---------------
                 5    ------------------------A----------------
                 6    -----------------------A-----------------

REGION E
                      1180      1190      1200      1210
WILD TYPE SEQUENCE    CTCACAAATACCACTGAGATCTTTTTCCCTC
                                      BglII
          MUTANT 1*   ----------------G--------------
                 2*   ----------------G-G------------
                 3*   ------------G---G--------------
```

The generation of site-directed point mutations, first described for the Qβ RNA genome (10), can be carried out by a number of different approaches, other than the mutagenic nucleotide technique described by us. Shortle and Nathans (15) introduced nicks into circular DNA using an appropriate restriction enzyme, enlarged the nicks to gaps using DNA polymerase I from Micrococcus luteus and mutagenized the exposed single-stranded DNA by the bisulfite method, which yields C to U transitions. This approach was recently extended to DNA regions not adjacent to restriction sites (16). A very versatile approach is that of using synthetic oligonucleotides containing a desired mutation as a primer on a single-stranded wild-type template; such primers can be prepared enzymatically (17) or by organic synthesis (18-20). Each of these methods has some disadvantages and some advantages. The mutagenic nucleotide approach yields only transition mutations; however, the primer extension with reverse transcriptase now opens the way to generate at least some transversions as well. On the other hand, a variety of mutations can be prepared using only few incubation reactions, and no special technology is required. Mutagenesis of gapped molecules with bisulfite only yields transitions, and the mutations cannot be introduced into precisely predetermined positions. The use of mutagenic primers is the most versatile in regard to the positioning and the nature of the

Fig. 7. List of point mutants obtained by site-directed mutagenesis in regions A-E of the β-globin gene (cf. Fig. 1).

Numbering is from the cap site. The heavy horizontal bars indicate the CAAT box (A), the ATA box (B), the initiation codon (C) and the termination codon (E), respectively. Mutants A4 and A5 each contain an unexpected A⟶G transition which might be due to occasional T-G pairing in the absence of dATP. Mutants C3 and C9 contain a new PvuII site (position +17 to +22). Mutants C8 and C9 have lost the original PvuII site by uncharacterized mutations (marked PvuIIR). Mutants which were made on cDNA only are marked by asterisks.

mutation, and its major drawback is that synthetic primers with a desired sequence are at present not equally available to all laboratories. In all methods, mutant DNAs must usually be sought for by sequencing a number of plasmids and identifying the desired individual. The primary yield of mutants is therefore an important consideration; in our experiments recoveries of 8-20% per mutable site were usually found, while values ranging from 1% (19) to over 30% (17,20) have been reported for the oligonucleotide method. A procedure for enriching a population for mutants with use of specific primers has been described (21), and has been reported to lead to preparations of almost pure mutant DNA. The expression of several of the mutant β-globin genomes described in this paper is presented in the accompanying report.

ACKNOWLEDGMENTS

P.D. was supported by a fellowship of the American Cancer Society, and C.D. of the Damon Runyon-Walter Winchell Cancer Fund. This work was supported by the Schweizerische Nationalfonds and the Kanton of Zürich.

REFERENCES

1. Weissmann, C. (1978). TIBS 3, N109-N111.
2. Methods in Enzymology, vol. 68 (R. Wu, ed.), Academic Press (1979).
3. Heffron, F., So, M., and McCarthy, B.J. (1978). Proc. Natl. Acad. Sci. USA 75, 6012-6016.
4. Gray, H.B. Jr., Ostrander, D.A., Hodnett, J.L., Legerski, R.J., and Robberson, D.L. (1975). Nucl. Acids Res. 2, 1459-1492.
5. Roberts, T.M., and Lauer, G.D. (1979). Methods in Enzymology, vol. 68 (R. Wu, ed.), Academic Press, pp. 473-482.
6. Englund, P.T. (1971). J. Biol. Chem. 246, 3269-3276.

7. Benoist, C., O'Hare, K., Breathnach, R., and Chambon, P. (1980). Nucl. Acids Res. 8, 127-142.
8. Devos, R., van Emmelo, J., Celen, P., Gillis, E., and Fiers, W. (1977). Eur. J. Biochem. 79, 419-432.
9. Meyer, F., Heijneker, H., Weber, H., and Weissmann, C. (1979). Experientia 35, 972.
10. Flavell, R.A., Sabo, D.L., Bandle, E.F., and Weissmann, C. (1974). J. Mol. Biol. 89, 255-272.
11. Müller, W., Weber, H., Meyer, F., and Weissmann, C. (1978). J. Mol. Biol. 124, 343-358.
12. Weber, H., Taniguchi, T., Müller, W., Meyer, F., and Weissmann, C. (1979). Cold Spring Harbor Quant. Biol. 43, 669-677.
13. Efstratiadis, A., Kafatos, F.C., and Maniatis, T. (1977). Cell 10, 571-585.
14. Efstratiadis, A., Kafatos, F.C., Maxam, A.M., and Maniatis, T. (1976). Cell 7, 279-288.
15. Shortle, D., and Nathans, D. (1978). Proc. Natl. Acad. Sci. USA 75, 2170-2174.
16. Shortle, D., Pipas, J., Lazarowitz, S., DiMaio, D., and Nathans, D. (1979). In Genetic Engineering, Principles and Methods, vol. 1 (J.K. Setlow & A. Hollaender, eds.) pp. 73-92.
17. Hutchison, C.A.,III, Phillips, S., Edgell, M.H., Gillam, S., Jahnke, P., and Smith, M. (1978). J. Biol. Chem. 253, 6551-6560.
18. Razin, A., Hirose, T., Itakura, K., and Riggs, A.D. (1978). Proc. Natl. Acad. Sci. USA 75, 4268-4270.
19. Kössel, H., Buyer, R., Morioka, S., and Schott, H. (1978). Nucleic Acids Res. (special publication) 4, S91-S94.
20. Bhanot, O.S., Saleem, A.K., and Chambers, R.W. (1979). J. Biol. Chem. 254, 12684-12693.
21. Smith, M., and Gillam, S. (1981). In Genetic Engineering, Principles and Methods, vol. 3 (J.K. Setlow and A. Hollaender, eds.), in press.

TRANSCRIPTION OF ADENOVIRUS DNA IN INFECTED CELL EXTRACTS[1]

Andrew Fire[+], Carl C. Baker[≠], Edward B. Ziff[≠] and Phillip A. Sharp[+]

[+]Center for Cancer Research and Department of Biology
Massachusetts Institute of Technology
Cambridge, Massachusetts 02139

and

[≠]The Rockefeller University
New York, New York 10021

ABSTRACT The activity of the nine known adenovirus promoter sites has been studied *in vitro* in a whole cell extract system. Six sites (four early, one intermediate (PIX) and the major late promoters) functioned with comparable efficiency in uninfected extracts. The other early promoter (for early region II) was utilized only poorly. Two intermediate promoters (leftward at 15.9 and 72.0 map units) which function *in vivo* only at intermediate or late stages in the lytic cycle, were inactive in uninfected extracts. Although extracts prepared from late infected cells did not show transcription of these two promoters, these extracts did show an early to late shift *in vitro*. In late infected extracts the late and PIX promoters were enhanced five- to ten-fold relative to early promoters. DNA titration and extract mixing experiments indicated the presence in uninfected and infected extracts of factors which could distinguish between early and late promoter classes.

INTRODUCTION

The lytic cycle of a human adenovirus has been extensively studied (1,2). At the moment, expression of viral gene products is thought to result from initiation at nine distinct promoter sites (3). RNA from each transcription unit is processed along varying pathways to yield multiple mRNAs, each

[1]This work was supported by Grant PCM78-23230 from NSF, Grant CA-26717 (Program Project Grant) from NIH to P.A.S., by Grant GM21779 from NIH and Grant MV75 from ACS to E.B.Z. A.F. is an NSF pregraduate fellow and C.C.B. is an NIH trainee.

ISBN 0-12-137420-3

coding for a single protein. The primary level of gene regulation during the lytic cycle is control of initiation of transcription. Little is known about the molecular processes involved in this control. In one case it has been established that a pre-early gene product enhances transcription of early promoters (4,5). A second viral early gene product seems to be responsible for shutting off transcription from other early regions (6,7,8). Finally, expression from the late transcriptional "operon" and perhaps other intermediate promoters seems to be coupled to viral DNA replication (9). Whether the interdependence of viral DNA replication and late transcription is a gene dosage effect or results mechanistically from DNA replication is not clear.

Preparation and properties of the whole cell extract system. Several extracts have been developed that are capable of faithfully initiating RNA polymerase II transcription on exogenously added DNA (10). A convenient and reasonably efficient extract from whole cells was developed by Manley *et al*. (11). Hypotonically swollen cells are dounced to break the plasma membrane, nuclei lysed by addition of 0.43 M $(NH_4)_2SO_4$ and the resulting viscous gel freed of DNA and membranes by ultracentrifugation. The supernatant is concentrated to original cell volume by $(NH_4)_2SO_4$ precipitation, dialyzed and used for transcription. A typical 20 µl reaction contains 2×10^6 cell equivalents of whole cell extract (WCE), 0.2-1.0 µg of DNA, 5-50 µM of CTP, UTP and GTP, 50 µM ATP and 10 mM creatine phosphate to regenerate triphosphate pools (12).

Despite the high specific activity and copious amounts of radioactivity obtainable in specific RNAs *in vitro*, the system is only a shadow of its *in vivo* counterpart in terms of efficiency. While estimates of *in vivo* RNA polymerase II initiation rates center on 10^4/cell per minute (13), the equivalent amount of WCE protein yields only about 100 initiations per minute even under the most favorable conditions. Likewise, the maximal ratio moles RNA product to moles DNA template, even for the strongest *in vitro* promoters, is no more than 0.1.

It was of great interest to test the WCE for other components involved in RNA metabolism. Specifically, one could examine transcripts produced *in vitro* for termination of transcription, cleavage and polyadenylation, RNA splicing or degradation. Termination, polyadenylation and cleavage have been assayed by using templates (linear or circular) which retain an intact transcription unit; the well defined SV40 genome and the Hind III G clone encompassing the EIa region of Ad5 have allowed the most critical evaluation. When a hybridization probe spanning the in vivo 3' end was used, no RNA products were found which terminated at the predicted position. A sensitive test for polyadenylation involves chromatography of in vitro labeled RNA on oligo dT cellulose columns; less than 1% of the in vitro label binds (12 and unpublished results). In addition, the small amount of radioactivity that does bind to the oligo dT column, does not electrophorese as a unique band in an agarose gel after treatment with glyoxal. Similarly, a sensitive test for splicing of in vitro transcribed RNA involves the in vitro translation of RNA transcribed from the early region of SV40. Synthesis of the large T polypeptide (94,000 m. wt.) is dependent on the removal of intervening sequences containing termination codons by RNA splicing. Synthesis of the small t polypeptide, which shares amino terminal sequence with large T, is an internal control for the presence of translatable RNA. These experiments showed that less than 1/500th of in vitro transcripts could have been modified by RNA splicing (C. Cepko, U. Hansen, H. Handa and P. Sharp, submitted). Thus, as the WCE system is now constituted, post-transcriptional processing rarely if ever occurs.

The only post-transcriptional modification of RNA that efficiently occurs in the WCE is cap synthesis. Cap synthesis on RNA initiated at late and early promoter sites of Adenovirus 2 and on promoter sites for α-globin, β-globin and a murine retrovirus have been demonstrated (10,11,14-17). It is also likely that a large fraction of the RNA polymerase II non-specific initiation events are also modified by capping (18).

Transcription from adenovirus promoters. As mentioned before, the strongest adenovirus promoter and indeed strongest known promoter in the WCE system is the major late Ad2 promoter site. We have also examined the transcriptional activity from the other promoter sites from adenovirus 2 or 5. Shown in Figure 1 is the list of the nucleotide sequences of the known adenovirus promoters (3). The sites of initiation of transcription at these promoters in vivo are indicated by an underline. Mixtures of DNA segments containing each of these promoters have been transcribed in a standard reaction. The level of initiation from a site is reflected in the amount of radioactivity found in a run-off transcript originating at the initiation site and extending to the end of the fragment. By determining the relative amounts of run-off transcripts generated by two DNA segments in the same reaction mix, it is possible to calculate the relative strength of the two promoters. The absolute concentration of

DNA Sequences Preceding mRNA Cap Sites

"TATA" BOX / CAPS	Promoter (Coordinate)	Relative Efficiency Mock	Relative Efficiency Late
GTGTATTTATACCCGGTGAGTTCCTCAAGAGGCCACTCTTGAGTG	Ad5 EIa (1.4)	0.2	0.03
GGGTATATAATGCGCCGTGGGCTAATCTTGGTTACATCTGACCTC	Ad5 EIb (4.7)	0.35	0.05
GAATATATAAGGTGGGGTCTTATGTAGTTTTGTATCTGTTTTGC	Ad5 Protein IX (9.8)	0.35	0.6
GGCTATAAAAGGGGGTGGGGGCGCGTTCGTCCTCACTCTCTTCCG	Ad2 Major Late (16.4)	1.0	1.0
TCCTTCGTGCTGGCCTGGACGCGAGCCTTCGTCTCAGAGTGGTCC	Ad2 IVa$_2$ (15.9)	<0.005	<0.005
TAGTCCTTAAGAGTCAGCGCGCAGTATTTGCTGAAGAGAGCCTCC	Ad2 EII (75)	0.04	0.005
AGGTACAAATTTGCGAAGGTAAGCCGACGTCCACAGCCCCGGAGT	Ad2 EIIa Late (72)	<0.005	<0.005
——— Not Sequenced ———	Ad5 EII (75)	0.02	0.005
GGGTATAACTCACCTGAAAATCAGAGGGCGAGGTATTCAGCTCAA	Ad2 EIII (76.6)	0.3	0.05
——— Not Sequenced ———	Ad5 EIII (76.6)	0.3	0.05
TCCTATATATACTCGCTCTGCACTTGGCCCTTTTTTACACTGTGA	Ad5 EIV (99.1)	0.3	0.05

FIGURE 1. *Sites for initiation of in vivo transcription on adenovirus DNA and their relative activities in vitro. At left are the sequences surrounding initiation sites (3, 19–25). The Goldberg-Hogness or "TATA" consensus sequence is boxed. Cap sites in vivo for each region are underlined. The "G-string" homology between the PL and PIX promoters is indicated. At right are the activities of the different promoters in vitro normalized to an activity of 1.0 for the late promoter.*

DNA in a WCE reaction mix can have a significant effect on the rate of transcription at a promoter site (10,11). Thus, an optimal DNA concentration must be established for each segment and with each new WCE preparation.

Figure 1 also shows the relative strengths of each of the promoter sites of Adenovirus 2 or 5 in the WCE prepared from uninfected cells. Note that there is little correlation between the *in vitro* transcriptional activity from a promoter site and the *in vivo* transcriptional activity observed at any period during the lytic cycle. For example, polypeptide IX is actively transcribed in an extract prepared from uninfected cells while this promoter is thought to be inactive during the early stage of infection. All the early promoter sites, with the exception of region II, are transcribed *in vitro* but at an efficiency 1/2 to 1/5 that of the late promoter. The promoter for early region II is poorly transcribed in the *in vitro* reaction, 1/25 that of the late promoter. This promoter is very active during the intermediate and late stage of adenovirus infection (2). Two intermediate promoters show no *in vitro* activity, the IVa_2 promoter and the late region II promoter.

While there is little correlation between *in vivo* and *in vitro* response, there is a strong correlation between the presence of a Goldberg-Hogness box or "TATA" sequence and *in vitro* activity. Early region II and IVa_2 promoter sites have no obvious "TATA" consensus sequence and have little transcriptional activity. Surprisingly, late region II promoter is inactive *in vitro* even though it has a TACAAA sequence -25 to -31 nucleotides from the cap site. S. L. C. Woo, R. Snead, T. Chandra, D. W. Bullock (these proceedings) have shown that this same sequence appears upstream from the rabbit α-uteroglobin gene and that this segment is actively transcribed *in vitro*.

A few general points about the relationship of promoter structure and *in vitro* activity are worth recalling: (a) Deletions of sequence in the vicinity of the "TATA" sequence eliminate *in vitro* transcription. (b) This or other *in vitro* systems do not appear to require sequences further than 50 nucleotides from the cap site (26-31). In particular, *in vitro* activity has been shown to be independent of sequences in the -70 to -80 region where the GGCCAAT sequence is found (Breathnach and Chambon, Ann. Rev. Biochem., in press). (c) The *in vitro* system must recognize more than just "TATA" consensus sequences as perfect prototype sequences are found at many positions in DNA that are not utilized as promoters. In fact, in the segments of Ad2 DNA studied, only two sites other than reported *in vivo* promoter sites are utilized efficiently *in vitro*. One such site is 300 nucleotides upstream from the EIa promoter. It is possible that this

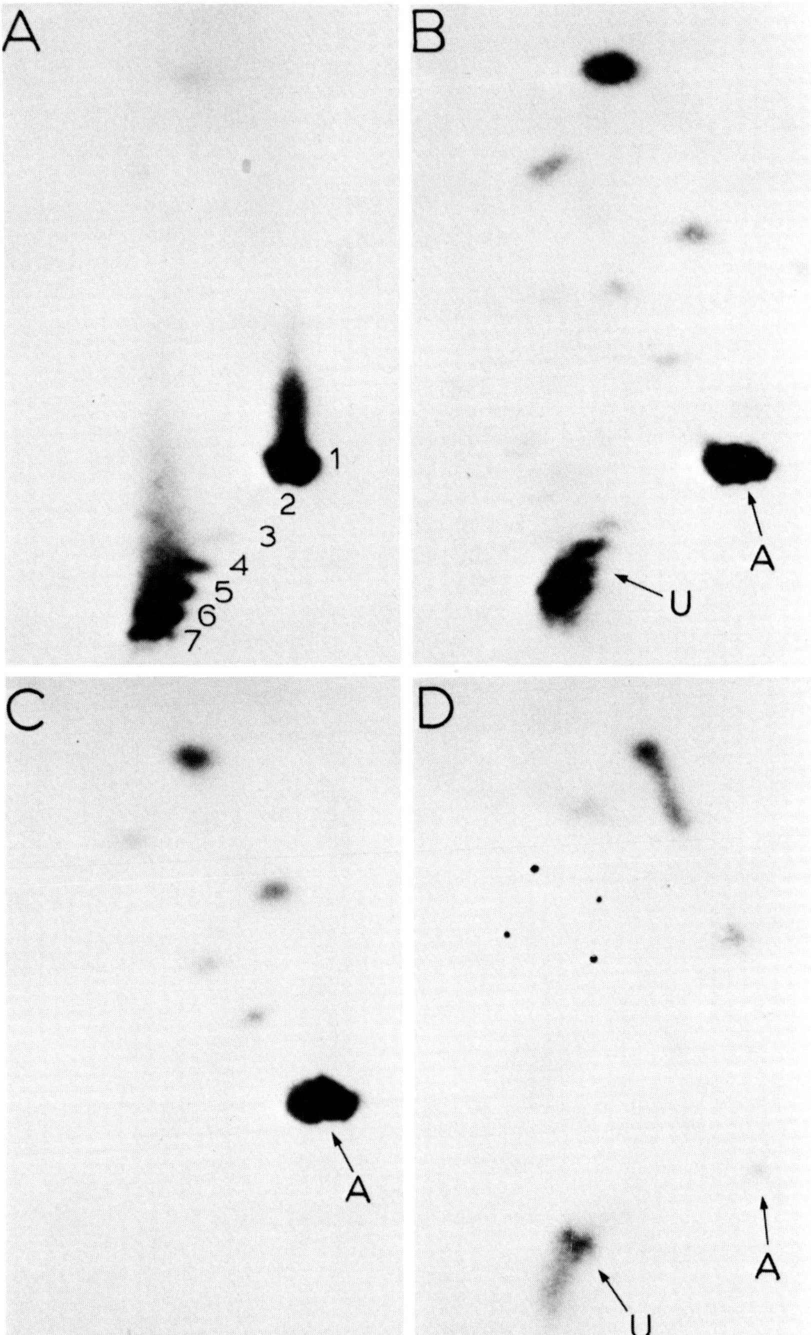
A
1
2
3
4
5
6
7
B
U
A
C
A
D
U
A

FIGURE 2. *Analysis of 5' termini from the EIV region. RNA samples were 5' labeled by decapping and kinasing using a previously described procedure (3). The RNA was then selected by hybridization to Ad2 Sma K fragment (98.3-100 mu). The sequence around the initiation site and the positions of the cap sites for PEIV are shown in Figure 1. Spot 1 is derived from the major A terminus and spots 2-7 from the minor U termini. Panel A: Analysis of cytoplasmic RNA prepared from cycloheximide treated Ad5 infected HeLa cells at 5 hr p.i. The identification of the 5' terminal oligonucleotides in vivo has been described by Baker and Ziff (Fig. 9 of reference 3). Panel B: Analysis of in vitro transcribed RNA; conditions are as described (18) except that nucleotide concentrations were 50 μM of ATP, CTP, GTP and UTP; template was Hind III cleaved Ad5 Eco RIB (4 μg) in a 0.2 ml reaction mix. Panel C: Conditions were as in panel B except that nucleotide concentrations were 500 μM ATP and 1 μM UTP (endogenous pool). Panel D: Conditions were as in panel B except that nucleotide concentrations were 10 μM ATP and 500 μM UTP. Total transcription was much lower in this case due to limiting ATP. The identity of spots in panels A, B, and C was confirmed by redigestion with ribonucleases T_2 and A, and nuclease P_1 and subsequent electrophoresis on DEAE or Whatman 540 paper.*

site is an in vivo promoter site, since a possible counterpart mRNA is found in Ad2, Ad5, Ad7 and Ad12 (4,32-34). The second site of in vitro initiation not seen in vivo is on the ℓ strand at position 96.3 mu. (d) The fidelity of the RNA polymerase II-DNA in vitro interaction is recorded in the specificity of the nucleotide sequence of initiation. Initiation at early region IV of adenovirus 5 is heterogeneous in vivo, distributed over six contiguous uridine residues and an adjacent adenine (see Figure 2; reference 3). RNA initiates in vitro in the WCE with an identical pattern of sites. The observation that changes in the ratio of ATP and UTP in the reaction mix are specifically reflected in the ratio of uridine and adenine caps from the EIV promoter is indicative of an initiation rather than a 5' processing event (Figure 2).

Although it now seems clear that only part of the machinery directing recognition of promoters in vivo is duplicated in the in vitro reaction, the WCE system does have high specificity and fidelity and thus certainly warrants further study.

Searching for regulation of adenovirus promoters with in vitro transcription. To test whether some of the regulatory changes in transcription observed *in vivo* could be duplicated in an *in vitro* reaction, WCEs were prepared from cells infected with 50 pfu/cell of Ad2 at both early (6 hrs p.i.) and late (21 hrs p.i.) stages. The relative rate of transcription of different viral promoters in these two extracts was compared to that observed with extracts from mock infected cells. An example of the most pronounced effect observed is shown in Figure 3. Here the run-off transcripts from an equal molar mixture of different fragments are displayed for mock, early, and late infected extracts.

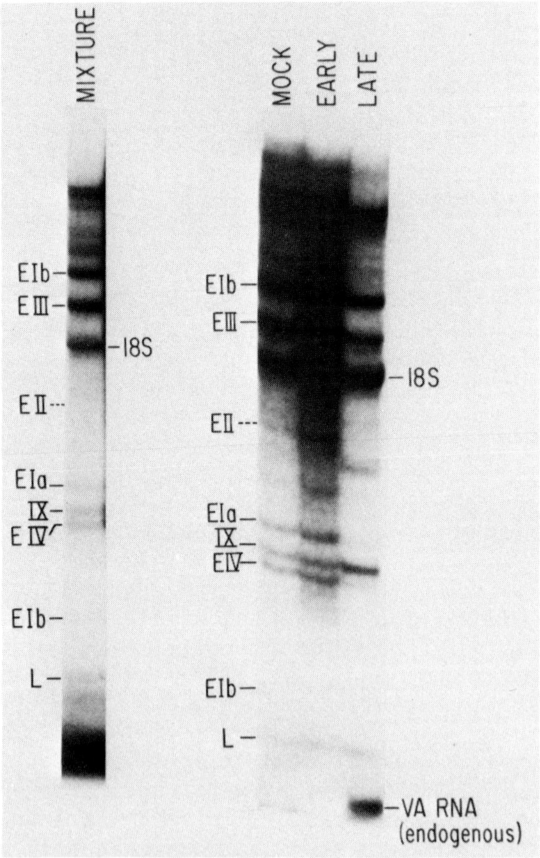

FIGURE 3. *Mixture of templates from different regions. An equimolar mixture of the following fragment was prepared. Ad5 Hind III G cut with Xba I and Kpn I, which generates run-*

offs of 340 nucleotides (n) from PEIb, 840 n from PEIa, and 1165 n from the initiation site at 0.7 mu. Ad5 Sma IF cut with Bam HI, which generates run-offs of 2575 n from PEIb and 715 n from PIX. Ad2 Bal (14.7) - Hind III (17.0) cloned from Bam HI to the Hind III site of pBR322; this recombinant cut with Eco RI, to generate a run-off of 225 n from the major late promoter. Ad5 Hind III B cut with Xho I and Sal I, which generates run-offs of 1375 n from PEII and 2180 n from PEIII. Ad5 Eco RI B cut with Hind III, which generates a 680 n run-off from PEIV. 10 μg/ml of the equimolar mixture was supplemented with 15 μg/ml of poly (dIC:dIC) carrier (U. Hansen, personal communication). Omission of any of the above DNAs removed the corresponding run-off bands from the labeled RNA pattern.

Panel A: Mixture transcribed in an uninfected extract in the presence of ^{32}P-UTP. Products were resolved on denaturing glyoxal gels (35) and autoradiogram is shown. Due to differences in run-off length and promoter strength, it was often necessary to examine several exposures to accurately compare activity.

Panel B: Comparison of extracts from mock, early, and late infected cells. Cells were harvested and extracts prepared from mock-infected cells and cells at 6 and 21 hr p.i. with Ad2. Extract concentrations were titrated and the reactions giving peak transcription are shown. The relative ratios of bands from an extract did not change over the range of the titrations. Protein concentrations were: mock, 6.3 mg/ml; early, 6.1 mg/ml; late, 5.8 mg/ml. A 150 n band in the late extract results from endogenous V.A. RNA synthesis. The level of V.A. transcription is comparable to that obtained with 1.0 μg/ml of exogenously added Ad2 DNA. Two run-off bands of 2575 n and 340 n from EIb have been included. Lighter and darker exposures allow comparison of these two bands in the different extracts. The ratio of radioactivity in these two EIb run-offs did not vary. The band of 1140 n which is strongly enhanced in the late extract corresponds to the rightward promoter at 0.7 mu.

These and other experiments suggest that mock and early infected extracts have the same transcription capacities. Disappointingly, extracts from both early and late infected cells fail to stimulate the synthesis of RNAs from the early promoter for region II, late promoter for region II and the IVa_2 promoter. As can clearly be seen in Figure 3, extracts prepared from late infected cells showed enhanced transcriptional activity on both the late and IX promoters, relative to the other active early promoters. Quantitation shows a five-to ten-fold shift in relative yield between the two extracts.

A number of effects could explain the increased transcription from the late and IX promoters relative to early promoters in extracts from late infected cells. For example, optimal DNA concentrations could vary between promoters in these two extracts. When the relative transcription of early promoters and the late promoter is compared in either mock or late infected extracts, the late promoter increases in relative activity as the DNA concentration is raised. However, at all total DNA concentrations the ratio of late to early promoter activity is five to ten times greater in late extracts as compared to mock extracts. As has been described previously, WCEs are dependent on bulk DNA concentration for specific transcription and are inhibited by too high a total exogenously added DNA concentration (11). Extracts prepared from late infected cells also show a similar bell shape DNA titration curve, suggesting that endogenous viral DNA is not significantly contributing to the total DNA concentration. When late WCEs are examined for free viral DNA by electrophoresis of extracted nucleic acids, no viral DNA is observed. In addition, the level of V.A. RNA transcription by RNA polymerase III from endogenous DNA suggested each 50 μl reaction mix contained less than 50 ng of viral DNA.

Since it is difficult to prepare WCEs with identical transcription activities, the relative ratio of the early to late promoters was determined in reaction mixes with varying concentrations of WCE protein. The ratio did not vary between 4.5 mg/ml and 11 mg/ml of extract protein. Similarly, varying the ionic strength of reaction mixes prepared with either infected or mock WCE did not affect the relative transcription activity from viral promoters.

Although *in vitro*-labeled RNAs have always appeared to be stable when incubated in a WCE, it was possible that degradation could have affected relative yields of transcripts. However, *in vitro* pulse chase experiments showed that all RNAs were stable in both mock and late infected extracts for over two hours.

It was theoretically possible that late infected cells could accumulate a specific factor for suppression of initiation at early promoter sites. In order to test for such an effect, the relative activity of non-adenovirus promoters was compared in mock and late extracts. Three of the four prominent *in vitro* initiation sites on SV40 DNA respond to late extracts in a way similar to early adenovirus promoters. One late SV40 promoter (68.5 mu) actually behaves like late and pIX adenovirus promoters (12). *In vitro* transcription from human and mouse β-globin genes is similarly suppressed relative to the late promoter in extracts from late infected cells. Thus, the alterations that occur between mock and late extracts are not specific for early viral promoters.

The differential activity between early and late adenovirus promoters in late extracts must reflect some difference in sequence recognition of these promoters. Hu and Manley (26) have recently reported the construction of a series of deletions removing sequences either 5' or 3' to the "TATA" or Goldberg-Hogness box of the late Ad2 promoter site and replacing these sequences with pBR322 sequences. Mutants with deletions ending 15 nucleotides from the 5' end of the "TATA" box and 12 nucleotides from the 3' end of the "TATA" box are specifically transcribed in the _in vitro_ system (26). These same mutants also show enhanced transcription in extracts prepared from late infected cells relative to early Ad2 promoters. Thus the sequence recognition for the enhanced activity in late extracts maps to within about 15 nucleotides of the "TATA" box. The G-rich sequence or "G string" present between -18 to -25 in all of the promoters enhanced in the late infected extract [adenovirus PL and PIX, and SV40 L (68.5)] may be the recognition sequence. This sequence is also found in many strong _in vitro_ promoters.

The enhanced relative activity on late and IX promoter sites in late extracts must be the result of a change in the number or activity of factors in the late versus the mock extract. This shift in activities from various promoters could constitute an assay for purification of the relevant factors. As a first step in this direction, the activities of various mixtures of extracts were compared. An equal mixture of late infected and mock extract gave an intermediate shift in the relative activities of the late promoter as compared to the EIa promoter. This suggests that any factors that are altered between the two extracts are not present at an excess level in either extract.

Weil _et al_. (10) have demonstrated that a mixture of purified RNA polymerase II and an S100 fraction from uninfected HeLa cells specifically initiates transcription on a variety of promoter sites. These two components can be prepared and added in varying amounts to any _in vitro_ system. Addition of RNA polymerase II purified from uninfected cells (a kind gift from M. Samuels) to late extracts stimulated the transcription of the late promoter but did not restore transcription of early promoters. This suggests that a modification of RNA polymerase II is not responsible for the enhanced transcription of the late promoter. A different result was obtained with supplementation by S100 fraction from uninfected cells. Supplementation of late extracts with small amounts of S100 fraction preferentially stimulated transcription of early promoters. However, supplementation of mock extracts with S100 fraction also stimulated transcription of early promoters. This suggests that both extracts are limiting in some factor found in the S100. Since this factor decreases the ratio of late to early promoter transcription, it is possible that its absence or modification in the late extract could yield the observed shift. These results suggest that fractionation of the S100 (36) fraction might be a good starting point for identifying factors altered in the late extract.

ACKNOWLEDGMENTS

We would like to thank K. Berkner for the gift of the terminal region recombinants; S. L. Hu and J. Manley for the late promoter deletions; G. Chu and F. Laski for other recombinants; M. Samuels for RNA polymerase II; C. Cepko, U. Hansen, R. Kaufman and S. Woo for communicating unpublished procedures and results; L. Spencer for technical assistance; M. Siafaca for preparation of the manuscript; and finally our colleagues at MIT and Rockefeller for stimulating discussions.

REFERENCES

1. Tooze, J. (1980). Tumor Viruses. Cold Spring Harbor Laboratory, Cold Spring Harbor, New York.
2. Ziff, E. B. (1980). Nature *287*, 491.
3. Baker, C. and Ziff, E. B. (1981). J. Mol. Biol. *148*, in press.
4. Berk, A., Lee, F., Harrison, T., Williams, J. and Sharp, P. A. (1979). Cell *17*, 935.
5. Jones, N. and Shenk, T. (1979). Proc. Natl. Acad. Sci. USA *76*, 3665.
6. Blanton, R. and Carter, T. (1979). J. Virol. *29*, 458.
7. Carter, T. and Blanton, R. (1978). J. Virol. *28*, 450.
8. Nevins, J. R. and Winkler, J. (1980). Proc. Natl. Acad. Sci. USA *77*, 1893.
9. Thomas, G. and Mathews, M. (1980). Cell *77*, 523.
10. Weil, P. A., Luse, D., Segall, J. and Roeder, R. (1979). Cell *8*, 469.
11. Manley, J., Fire, A., Cano, A., Sharp, P. A. and Gefter, M. (1980). Proc. Natl. Acad. Sci. USA *77*, 3855.
12. Handa, H., Kaufman, R., Manley, J., Gefter, M. and Sharp, P. A. (1981). J. Biol. Chem. *256*, 478.
13. Cox, P. (1976). Cell *7*, 455.
14. Talkington, C., Nishioka, Y. and Leder, P. (1980). Proc. Natl. Acad. Sci. USA *77*, 7132.
15. Luse, D. and Roeder, R. (1980). Cell *20*, 691.

16. Proudfoot, N., Shander, M., Manley, J., Gefter, M. and Maniatis, T. (1980). Science *209*, 1329.
17. Yamamoto, T., de Crombrugghe, B. and Pastan, I. (1980). Cell *22*, 787.
18. Fire, A., Baker, C., Manley, J., Ziff, E. and Sharp, P. A. (1981). J. Virology, submitted.
19. Baker, C., Herisse, J., Courtois, G., Galibert, R. and Ziff, E. B. (1979). Cell *18*, 569.
20. Maat, J., Van Beveren, C. P., and Van Ormondt, H. (1980). Gene *10*, 27.
21. Maat, J. and Van Ormondt, H. (1979). Gene *6*, 75.
22. Steenbergh, P. and Sussenbach, J. (1979). Gene *6*, 307.
23. Van Ormondt, H., Maat, J., DeWaasd, A. and van der Eb, A. (1978). Gene *4*, 309.
24. Ziff, E. and Evans, R. (1978). Cell *15*, 1463.
25. Alestrom, P., Akusjarvi, G., Perricaudet, M., Mathews, M., Klessig, D. and Pettersson, U. (1980). Cell *19*, 671.
26. Hu, S.-L. and Manley, J. (1981). Proc. Natl. Acad. Sci. USA *78*, 820.
27. Corden, J., Wasylyk, B., Buchwalder, A., Sassone-Corsi, P., Kedinger, C. and Chambon, P. (1980). Science *209*, 1406.
28. Tsai, S., Tsai, M. J. and O'Malley, B. (1981). Proc. Natl. Acad. Sci. USA *78*, 879.
29. Wasylyk, B., Kedinger, C., Corden, J., Brison, O. and Chambon, P. (1980). Nature *285*, 367.
30. Wasylyk, B., Derbyshire, R., Guy, A., Molko, D., Roget, A., Teoule, R. and Chambon, P. (1980). Proc. Natl. Acad. Sci. USA *77*, 7024.
31. Mathis, D. and Chambon, P. (1981). Nature *290*, 310.
32. Sambrook, J., Greene, R., Stringer, J., Mitchison, T., Hu, S.-L. and Botchan, M. (1980). Cold Spring Harbor Symp. Quant. Biol. *44*, 569.
33. Yoshida, K. and Fujinaga, K. (1980). J. Virol. *36*, 337.
34. Fujinaga, K., Sawada, Y., Uemizu, Y., Yamashita, T., Shimojo, H., Shiroki, K., Sugisaki, H., Sugimoto, K. and Takanami, M. (1980). Cold Spring Harbor Symp. Quant. Biol. *44*, 519.
35. McMasters, G. and Carmichael, G. (1977). Proc. Natl. Acad. Sci. USA *79*, 4835.
36. Matsui, T., Segall, J., Weil, P. and Roeder, R. (1980). J. Biol. Chem. *255*, 11992.

CHARACTERIZATION OF FACTORS THAT IMPART SELECTIVITY TO RNA POLYMERASE II IN A RECONSTITUTED SYSTEM

William S. Dynan
Robert Tjian

Department of Biochemistry
University of California
Berkeley, California 94720

ABSTRACT Fractionation of a HeLa cell extract reveals the presence of factors that do not copurify with RNA polymerase II but that are necessary for selective transcription *in vitro*. The transcription activity in the extract can be separated into a polymerase component, which binds to DEAE-cellulose, and a selectivity component, which flows through. The two components must be mixed to restore the ability to selectively transcribe the adenovirus 2 major late promoter. The selectivity component from the DEAE flow-through binds to and appears to elute as a single peak from phosphocellulose. Further purification of this selectivity factor shows that it is made up of at least two and more likely three separate components, all of which must be present to restore selectivity. The factors must be present simultaneously with RNA polymerase II in order for selective transcription to occur. Thus, RNA transcribed in the absence of selectivity factors is not processed into the selective product by subsequent addition of factor.

INTRODUCTION

Although purified RNA polymerase II from a variety of eukaryotic sources has been available for some years, these purified preparations appear to be deficient in their ability to initiate RNA synthesis at physiologically relevant promoter sites (1,2). Recently, however, it has been discovered that relatively crude human cell extracts are capable of recognizing adenovirus and other promoters *in vitro* and carrying out selective transcription beginning at or near the *in vivo* start sites(1,3). This finding suggested

ISBN 0-12-137420-3

that by using an appropriate assay it would be possible to purify the factor or factors that were needed for selectivity.

Previous attempts to purify polymerase have used as an assay the enzyme's ability to incorporate labeled nucleoside triphosphates into TCA precipitable material. In the experiments described here, a "run-off" transcription assay was used instead. Although substantially more work, this assay yields direct information about the enzyme's ability to transcribe selectively.

The starting material for the purification was a HeLa cell extract prepared by a modification of the method of Sugden and Keller (4). This extract contains all the components needed to direct selective transcription (3). A cloned fragment of adenovirus 2 containing the major late promoter was used as template. The template DNA was cut at a restriction site located 533 nucleotides downstream from the promoter. Selective transcription of this truncated template yielded an RNA run-off product of discrete length that could be detected by gel electrophoresis.

This assay was used to determine at what stage of purification of the extract selectivity was lost, and also whether selectivity could be restored by mixing back material that had been fractionated away. There was reason to believe that selectivity, once lost, could be restored, since Weil and coworkers (1) have shown that a crude cytoplasmic extract of KB cells imparts selectivity to purified mammalian polymerase II. This selectivity appears to depend on the presence of at least four separate factors in the extract (5). In the experiments described here, we use a different approach and a different starting material, but come to a similar conclusion: there are multiple factors needed for selective transcription.

MATERIALS AND METHODS

Purification of Polymerase and Selectivity Factors

A cell-free extract was prepared from exponentially growing HeLa cells as described by Manley et al. (3), except that the final dialysis buffer contained 50 mM Tris-HCl, pH 7.9, in place of 20 mM Hepes. A typical prep used 16 L of HeLa cell spinner culture (30 g of cells), which yielded 35 ml of extract. Extracts were stored at -70°C until needed for chromatography.

All chromatography was performed at 4°C, in TGED buffer, which consisted of 50 mM Tris-HCl (pH 7.9 at 25°C), 1 mM EDTA, 1 mM dithiothreitol, 20% glycerol, and KCl as indicated. After thawing, 35 ml of extract was passed over a 40 ml DEAE-cellulose column (Whatman DE52) equilibrated with TGED containing 0.1 M KCl (TGED 0.1 M). The flow-through was collected, pooled, and saved for subsequent phosphocellulose chromatography. The column was washed with 2 volumes TGED 0.1 M and 3 volumes TGED 0.18 M, and was eluted with 3 volumes TGED 0.35 M. Eluate was pooled narrowly on the basis of protein content, dialyzed 3-4 h against two changes of TGED 0.1 M, aliquotted, frozen in liquid nitrogen, and stored at -70°C. This fraction, referred to subsequently as D.35, contained most of the RNA polymerase II activity but did not transcribe selectively; it was used to assay for selectivity factors.

The DEAE-cellulose flow-through was adjusted to 0.175 M KCl and was passed over a 10 ml phosphocellulose column (Whatman P11) equilibrated with TGED 0.2 M. The column was washed with 2 volumes TGED 0.2 M and 3 volumes TGED 0.3 M, and was eluted with a 7 column volume linear gradient of TGED 0.3 M-TGED 0.65 M. Fractions were collected and portions of them were individually dialyzed 3 h against 2 changes of TGED 0.1 M. The dialyzed fractions were assayed for their ability to restore selectivity of transcription to D.35 or purified RNA polymerase II.

Fractions from the phosphocellulose column were pooled broadly on the basis of selectivity factor activity. To the pooled fractions was added 31.5 g/100 ml solid ammonium sulfate (45% saturation). After the ammonium sulfate had dissolved, the sample was stirred at 0°C for an additional 30 min, centrifuged 30 min at 10,000 g, and the pellet resuspended in 6 ml of TGED 0.1 M. The resuspended material was dialyzed 3 h against 2 changes of TGED 0.1 M, aliquotted, frozen in liquid nitrogen, and stored at -70°C. This fraction, which contained factors needed for selective transcription, is referred to subsequently as PC1.

Transcription Reactions

A 6.8 kb plasmid, pAL, which consists of the Sma I F fragment of adenovirus 2 inserted in the Bam HI site of pBR322 using synthetic linkers, was used as template in our experiments. Plasmid was grown in *E. coli* strain 294. After overnight amplification with chloramphenicol, cells were incubated with lysozyme and EDTA, and lysed with SDS. Plasmid DNA was purified by the method of Hirt (6), followed by isopycnic centrifugation in cesium chloride/ethidium bromide gradients. Supercoiled DNA was collected by side puncture, passed over Dowex AG50W-X8 (200-400 mesh) to remove ethidium bromide, and dialyzed extensively against TE buffer (10 mM Tris-HCl, pH 7.9, 1 mM EDTA). Plasmid was digested with restriction endonuclease Bam HI (gift of T. Burgess) in buffer containing 20 mM Tris-HCl pH 7.9, 6 mM $MgCl_2$, 1 mM dithiothreitol, 100 mM NaCl, and 100 µg/ml bovine serum albumin. The reaction was carried out for a length of time and with an amount of enzyme just sufficient for complete digestion. The cut DNA was twice extracted with 1:1 phenol:chloroform, precipitated with ethanol, and dried under vacuum. The pellet was resuspended to 500 µg/ml in TE buffer, and stored frozen or at 4^oC. Each batch of template was analyzed by agarose gel electrophoresis after denaturation in alkali, and was not used unless it appeared to be free of single-strand nicks.

Transcription reactions were generally carried out in 100 µl, and contained 50% (vol/vol) transcription factors. Thus, the reaction buffer consisted of 0.5X TGED 0.1 M. Reactions also contained 6.25 mM $MgCl_2$, 500 nmol creatine phosphate, 2.5 nmol ATP, GTP, CTP and UTP, and 10 µCi [α-^{32}P] GTP (25 Ci/mmol, ICN). Reactions contained 1.0 µg Bam HI cut pAL DNA, and did not contain any synthetic poly dG dC. Transcription was carried out for 30 min at 30^oC, and reactions were terminated by addition of 200 µl of a solution containing 0.15 M NaCl, 0.015 M EDTA, 0.75% SDS, and 150 µg/ml tRNA. To each reaction was added 150 µl phenol (containing 0.1% 8-hydroxyquinoline). The tubes were shaken vigorously, 150 µl of chloroform was added, and the tubes but were shaken again. The tubes were then centrifuged for 5 min in an Eppendorf microcentrifuge. The aqueous phase was drawn off and added to 0.75 ml of 100% ethanol. After three successive ethanol precipitations, transcripts were glyoxalated and analyzed by electrophoresis on a 1.9% agarose gel (3). Transcription reactions were not treated with DNase during the workup, because we found that this treatment had no effect.

RESULTS AND DISCUSSION

Fractionation of the Extract

The scheme adopted for partial purification of transcription factors is shown in figure 1. When the extract was passed over a DEAE-cellulose column, it separated into a polymerase component, that bound, and a selectivity component, that flowed through. The selectivity component, when added back to polymerase, restored the ability to make a selective transcript from the adenovirus 2 late promoter (figure 2, lanes 10 and 11).

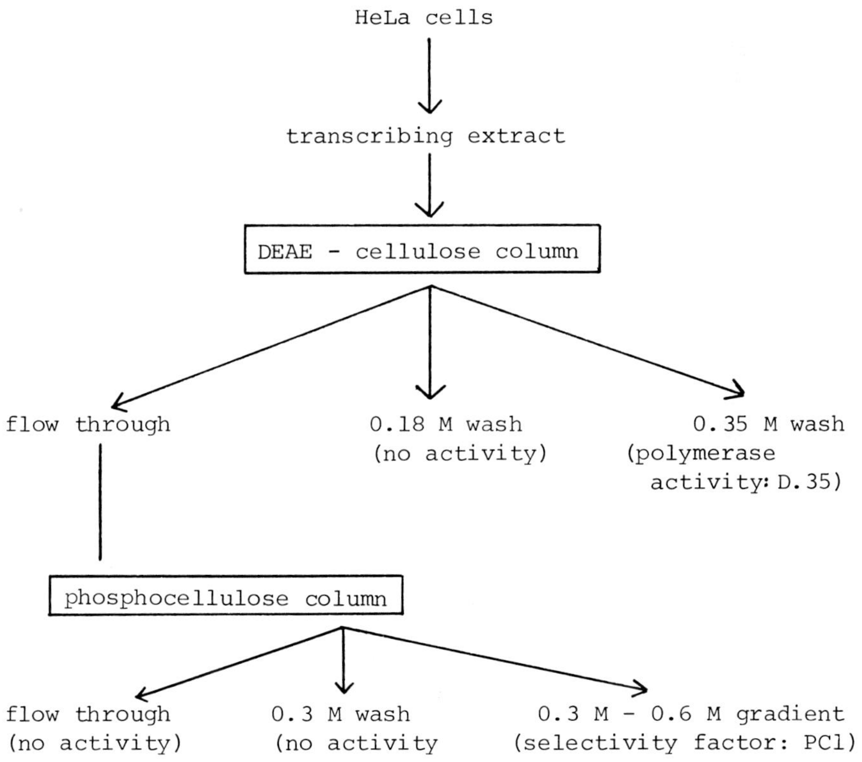

Figure 1. Scheme for purification of polymerase and selectivity factors. Selective transcription of the adenovirus 2 major late promoter was observed when the D.35 fraction was mixed either with the DEAE flow-through or with the PC1 fraction (see text). Other fractions had no effect on selective transcription.

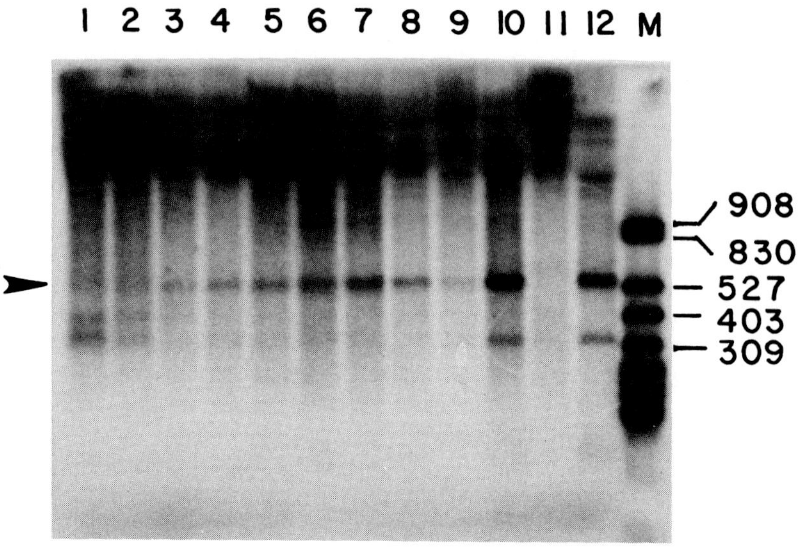

Figure 2. Gradient elution from phosphocellulose. Reactions 1-11 contained 25 µl of the D.35 polymerase fraction, complemented with dialyzed phosphocellulose column fractions (lanes 1-9), or with phosphocellulose column onput (lane 10). Lane 11 shows transcription by uncomplemented D.35. Lane 12 is a control showing transcription by unfractionated extract. Lane M contains end-labeled DNA restriction fragments whose size, in nucleotides, was determined from the DNA sequence and is indicated to the right of the figure. RNA migrates about 5% more slowly than DNA in this gel system, and thus the 533 nucleotide selective run-off transcript (arrowhead) migrates visibly behind the 527 nucleotide DNA marker.

TABLE I. Purification of Transcription Factors[a]

fraction	volume	protein conc	total protein	A_{260}/A_{280}
extract	28.5 ml	23.4 mg/ml	667 mg	1.12
PC1	4.0 ml	3.8 mg/ml	15 mg	1.05
D.35	8.0 ml	2.6 mg/ml	21 mg	1.48

[a]Results are from a typical purification starting with 21 g of cells. Protein concentration was determined by the Lowry method using bovine serum albumin as a standard.

The selectivity component was further purified on phosphocellulose. The factors needed for selectivity bound to phosphocellulose, while the bulk of the protein flowed through. Selectivity activity eluted as a single peak between 0.35 and 0.55 M KCl (Figure 2, lanes 1-9).

Table I shows that the D.35 polymerase and PC1 selectivity components each represent only a few percent of the protein in the original extract. Thus, the degree of purification appears to be substantial. The ratios of absorbance at 260 and 280 nm (table I) suggest that there is nucleic acid present in both the D.35 and PC1 fractions. This finding is not surprising in the case of D.35, since nucleic acid may begin to elute from the DEAE-cellulose column at 0.35 M KCl, but it was unexpected to find nucleic acid tightly bound to phosphocellulose and eluting with the PC1 activity.

It is evident from Figure 2 that the amount of high molecular weight background RNA was higher in reconstituted reactions than in reactions with unfractionated extract. One possible explanation is that there was an insufficient quantity of the selectivity component. Varying the amount of selectivity component in the reaction showed that within a certain range the amount of selective transcript was proportional to the amount of PC1 present, but also that increasing the amount of PC1 was in general not a very effective way of suppressing the nonselective background (data not shown). Above a certain level, the amount of selective transcript actually declined, with a concomitant increase in certain background bands. The cause of this complex behavior is unknown.

Selective transcription in the reconstituted reaction was completely inhibited by 2 µg/ml amanitin (figure 3A) as expected for RNA synthesized by RNA polymerase II. Synthesis of two of the high molecular weight background bands was inhibited by 10 µg/ml amanitin but not by 2 µg/ml. Therefore, these are probably polymerase III transcripts. Another large species, present in figure 3A, lane 3, and very prominent in figure 2, appears even in the absence of template DNA (data not shown) and may result from labeling of endogenous RNA in the extract. The presence of high levels of amanitin-insensitive products was unexpected, because essentially all transcription by unfractionated extract was amanitin sensitive (figure 3A, lanes 4 and 5). Evidently there are inhibitors of these amanitin insensitive activities that are present in the extract but lost during the

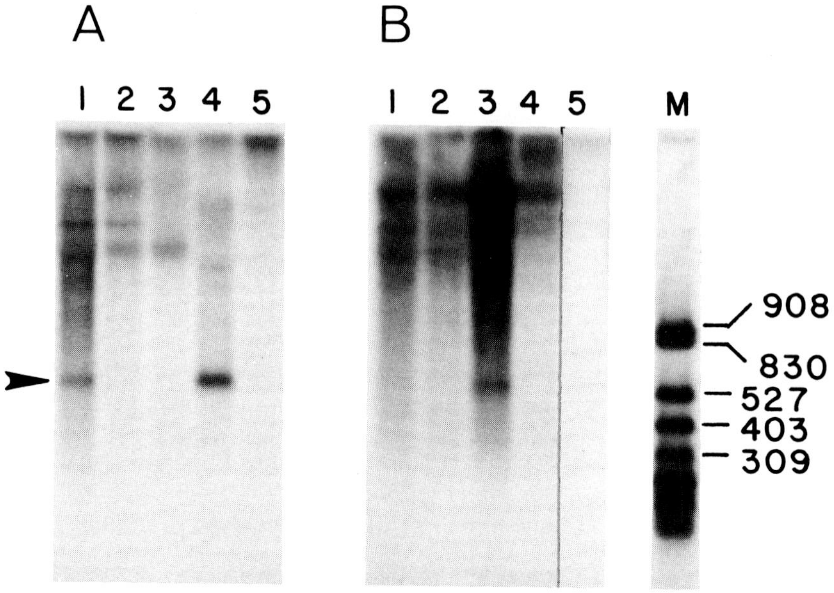

Figure 3A. Amanitin sensitivity of transcription in the reconstituted extract. Reactions in lanes 1-3 contained 25 µl of PC1 and 25 µl of D.35. Lanes 4-5 contained 25 µl of unfractionated extract. Reactions contained no amanitin (lanes 1 and 4), 2 µg/ml amanitin (lanes 2 and 5), or 10 µg/ml amanitin (lane 3). All reactions were carried out in a 100 µl volume. The position of the 533 nucleotide selective runoff transcript is marked with an arrowhead.

Figure 3B. Selectivity factor exerts its effect only in the presence of active RNA polymerase II. Each transcription reaction was incubated for 30 min at 30°C, then additions were made and the reaction incubated a further 30 min at 30°C. Contents of the reaction were as follows:

	present in first incubation	added for second incubation
1.	D.35, amanitin	PC1
2.	D.35	PC1, amanitin
3.	D.35	PC1
4.	D.35, amanitin	buffer
5.	buffer	PC1

The final reaction volume was 100 µl, and 25 µl of each transcription factor (or buffer) was used. When amanitin was present, 200 ng was used. Lane M contains DNA markers.

purification procedure.

How Does the Selectivity Factor Work?

One question that arises, given the finding that RNA polymerase II and selectivity factor are separable, is whether the so-called selectivity factor is involved in directing *de novo* initiation of RNA synthesis, or whether it merely processes previously-synthesized RNA. In order to distinguish between these possibilities, an experiment was carried out in which the D.35 fraction, containing RNA polymerase II, was incubated in a complete transcription reaction and RNA synthesis was allowed to proceed. After 30 min, PC1 fraction was added, with or without amanitin, and the reaction was allowed to continue for an additional 30 min. Figure 3B shows that selective transcription occurs only when functional polymerase and factor are present in the same reaction at the same time (lane 3), and that addition of PC1 to previously synthesized transcripts does not result in processing into the 533 nucleotide selective species (lane 2).

The PC1 fraction has no polymerase activity of its own, as shown in figure 3B, lane 5. Moreover, it has little effect on the amanitin insensitive activity in the D.35 fraction, except that it promotes labeling of the template-independent species discussed previously (compare figure 3B, lanes 2 and 4). Finally, the presence of amanitin in the initial 30 min incubation makes little difference in the total amount of RNA synthesized, indicating that in the absence of PC1 the amount of polymerase II activity expressed by D.35 on this template is relatively small.

Our findings suggest that the PC1 selectivity factor exerts its effect by directly modifying the action of RNA polymerase II, and that the factor appears neither to process nor to synthesize RNA on its own. The PC1 fraction may also have other effects on RNA polymerase II, such as stimulating overall transcription or decreasing the relative amount of nonselective transcription, but these effects are hard to sort out because of the non-polymerase II background.

How Many Selectivity Factors are Needed?

Neither the D.35 nor the PC1 fractions are pure, and both give a complex pattern of polypeptides when analyzed on SDS gels. Thus, it is not certain whether these fractions each contain one or more than one active component. For example, would purified RNA polymerase II transcribe selectively in the presence of the PC1 fraction, or does the D.35 fraction contribute some essential factor in addition to the polymerase?

To answer this question, transcription was carried out with HeLa RNA polymerase II purified nearly to homogeneity by a modification of the method of Hodo and Blatti (7). It is unlikely that a transcription factor not actually bound to RNA polymerase II would copurify so extensively as to be present in this preparation. Figure 4 shows that purified RNA polymerase II was able to carry out selective transcription in the presence of the PC1 fraction. Although the overall level of RNA synthesis with purified RNA polymerase II was lower than with the D.35 fraction, the signal-to-noise ratio was significantly improved (Figure 4, compare lanes 1 and 2). Thus, it appears that the PC1 fraction alone is sufficient to impart selectivity to polymerase, and that there are no essential selectivity factors in the D.35 other than the RNA polymerase II.

The amount of selective transcript synthesized by purified RNA polymerase II could not be increased by adding more enzyme; doing so only increased the nonselective background without increasing the amount of the selective transcript (results not shown). It is possible that the purified polymerase has lost a stimulatory factor that, although not strictly required for selectivity, acts to increase the rate of transcription. An alternative possibility is that a population of damaged polymerase molecules has accumulated during purification, and that these inhibit promoter recognition by active molecules.

Attempts were also made to further purify the PC1 selectivity activity. Ammonium sulfate precipitation and heparin-agarose chromatography both gave good recovery of activity, but little purification. Chromatography on DEAE-Sepharose, phenyl-Sepharose, and single stranded DNA-cellulose gave substantial fractionation of protein, but in each case two column fractions had to be mixed back together to restore the selectivity factor activity (data to be presented elsewhere).

If there were just two factors required for selectivity, then it should have been possible to mix together pairs of fractions from different columns and obtain a simple, self-consistent pattern of in vitro complementation. Such experiments were carried out, but the results were complex, and suggested that there are three factors, rather than two, needed for selective transcription of the adenovirus late promoter. We are currently attempting to separate all three factors.

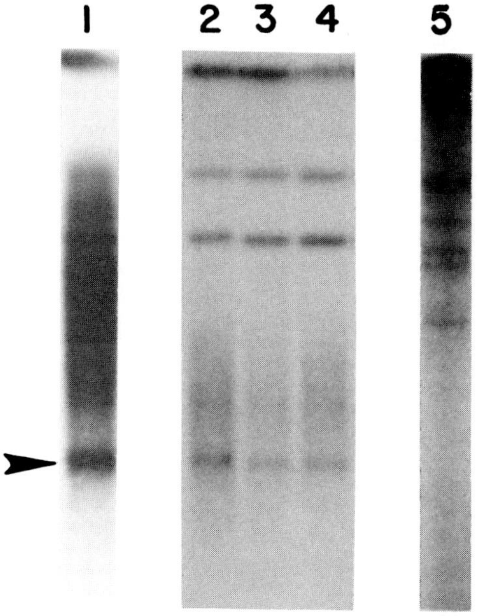

Figure 4. Addition of selectivity factor to purified HeLa RNA polymerase II. The reactions in lanes 1-4 contained 15 µl of PC1 selectivity factor. Lane 1 contained 25 µl of D.35, and lanes 2-4 contained 1.0, 0.5, and 0.25 µl, respectively of purified HeLa RNA polymerase II. Lane 5 contained 2.0 µl of HeLa polymerase II without selectivity factor. The concentration of the HeLa polymerase preparation was estimated to be 20-40 µg/ml by comparing its activity on calf thymus DNA template with that of RNA polymerase II of known concentration. The arrowhead to the left of the figure marks the position of the 533 nucleotide selective runoff transcript.

The mode of action of the selectivity factors is presently unknown. From their chromatographic behavior it appears that they may be DNA-binding proteins, but whether this binding is promoter-specific is not yet known. It is possible that one or more of the factors is an artifact resulting from the way the experiments have been done. For example, changing the ratios of two of the factors, or simply increasing their concentration, might eliminate the need for a third factor. Clearly, much work remains to be done to rule out trivial mechanisms and to establish that all three factors play a functional role.

Matsui et al. (5) have presented evidence that in reactions containing exogenous, purified RNA polymerase II, four separate factors are needed to impart selectivity of transcription. They begin their purification with a cytoplasmic extract, pass it over a phosphocellulose column, and find a component that flows through, one that elutes between 0.35 M and 0.6 M KCl, and one that elutes above 0.6 M KCl. The component eluting between .35 M and 0.6 M does not bind DEAE-cellulose and can be further fractionated on single-stranded DNA-cellulose into at least two activities, both required for selectivity. On the basis of its behavior on all three resins, our PC1 fraction appears to be analogous to the .35-.6 M component of Matsui et al.

Whether the other two components described by Matsui are functional in our system is uncertain. Occasionally an activity (PC2) is present that elutes above 0.6 M KCl on phosphocellulose, and that stimulates specific and non-specific transcription, but this result is not reproducible. There is no evidence in our system for the fourth factor that flows through phosphocellulose. One significant difference between our experiments and those of Matsui et al. that may account for some of the differences in our results is that our reactions probably contain less RNA polymerase II (3).

It is not clear why the biological regulation of adenovirus transcription requires so many factors in addition to RNA polymerase II. It is possible that different promoters will require different sets of factors, and that the reason multiple factors are involved is to permit coordinate regulation of overlapping classes of genes. It now appears that not all eukaryotic promoters contain the same regulatory elements. For example, fine structure mapping of the adenovirus 2 major late promoter shows that sequences more than 32 nucleotides upstream from the mRNA cap site are dispensable for *in vitro* transcription, but an AT rich sequence 25-31 nucleotides upstream is required (8,9). In contrast, mapping of the SV40 early promoter shows that a region 70 to 155 nucleotides upstream from the mRNA cap site is required, but that an AT rich region approximately 25 nucleotides upstream is dispensable (10;11; Myers, Rio, Robbins, and Tjian, unpublished results). In other words, the structure of these two promoters appears to be very different. Thus, the adenovirus major late promoter and the SV40 early promoter would be good candidates to test the idea that not all promoters interact with the same selectivity factors. The SV40 early promoter is also of considerable interest for its own sake, since it has been shown that the SV40 A gene product, large T antigen, binds to the template and represses transcription *in vitro* (11; Myers, Rio, Robbins, and Tjian, unpublished results).

It should be feasible to determine what factors are needed for recognition of the SV40 promoters using the same approach used here for the adenovirus major late promoter, since the unfractionated HeLa extract recognizes both early and late SV40 promoters (11,12). Experiments to find out what is needed for transcription in a reconstituted system are now underway.

ACKNOWLEDGMENTS

We thank Alan Robbins for purified RNA polymerase and Taffy Mullenbach for HeLa cells. This work was supported by Damon Runyon Fellowship DRG 480F (to W.S.D.) and by National Institutes of Health grant no. CA 25417.

REFERENCES

1. Weil, P.A., Luse, D.S., Segall, J., and Roeder, R.G. (1979) *Cell* *18*, 469-484.
2. Roeder, R.G. (1976) *in* RNA Polymerase (Losick, R., and Chamberlin, M., eds.) Cold Spring Harbor Laboratory, Cold Spring Harbor, New York, pp. 285-329.
3. Manley, J.L., Fire, A., Cano, A., Sharp, P.A., and Gefter, M.L. (1980) *Proc. Natl. Acad. Sci. USA* *77*, 3855-3859.
4. Sugden, B. and Keller, W. (1973) *J. Biol. Chem.* *248*, 3777-3788.

5. Matsui, T., Segall, J., Weil, P.A., and Roeder, R.G. (1980) J. Biol. Chem. 255, 11992-11996.
6. Hirt, B. (1967) J. Mol. Biol. 26, 365-369.
7. Hodo, H.G., and Blatti, S.P. (1977) Biochemistry 16, 2334-2343.
8. Corden, J., Wasylyk, B., Buchwalder, A., Sassone-Corsi, P., Kedinger, C., and Chambon, P. (1980) Science 209, 1406-1414.
9. Hu, S-L., and Manley, J.L. (1981) Proc. Natl. Acad. Sci. USA 78, 820-824.
10. Benoist, C., and Chambon, P. (1980) Proc. Natl. Acad. Sci. USA 77, 3865-3869.
11. Rio, D., Robbins, A., Myers, R., and Tjian, R. (1980) Proc. Natl. Acad. Sci. USA 77, 5706-5710.
12. Handa, H., Kaufman, R.J., Manley, J., Gefter, M., and Sharp, P.A. (1981) J. Biol. Chem. 256, 478-482.

EXPRESSION OF THE HEAT SHOCK GENES IN *DROSOPHILA MELANOGASTER*[1]

Mary Lou Pardue
Dennis G. Ballinger
Matthew P. Scott[2]

Department of Biology
Massachusetts Institute of Technology
Cambridge, Massachusetts

INTRODUCTION

In attempts to understand the mechanisms of gene expression in higher organisms a number of workers have been studying a response in *Drosophila* called the heat shock response (see Ashburner and Bonner, 1979, for review). The response is induced by subjecting *Drosophila* or their cells to a heat shock (usually a shift to 37°) or to a number of other environmental stresses. Removal of the inducing agent terminates the response. During the heat shock response *Drosophila* cells produce a very limited set of proteins, primarily a set of some seven polypeptides which have been named the heat shock proteins (HSPs). The heat shock response appears to be directed by a coordinately controlled set of genes and can, we believe, serve as a model system for study of the kinds of changes in genetic activity that occur during cell determination and differentiation in higher organisms. In addition, the heat shock response is of interest in its own right. What is known about the biology of the heat shock response suggests it might be a mechanism that enables the insect to cope with

[1]Supported by NIH grant 2 R01 GM21874-06
[2]Present address: Department of Biology, Indiana University, Bloomington, Indiana.

ISBN 0-12-137420-3

environmental stress. Unfortunately no function has been identified for any of the heat shock proteins so this suggestion remains speculative. It is interesting, however, that responses similar to the heat shock response have been reported in a variety of organisms, including some that are very distantly related to Drosophila.

Expression of the Heat Shock Genes Is Controlled at the Level of Transcription

During the heat shock response a very small number of newly synthesized RNA species is found in the cytoplasm, in contrast to the extremely diverse set of newly synthesized RNAs found in the cytoplasm of control cells (Spradling et al., 1977). The heat shock-induced change in the RNA population is largely due to control at the level of transcription. It has been assumed since the discovery of the heat shock response (Ritossa, 1962) that the inducer was acting to cause changes in gene activity because the induction produced specific changes in the pattern of polytene puffing. Puffing in polytene chromosomes is generally thought to be a morphological manifestation of transcription at the puff site. Work on the heat shock response has provided the first complete proof of this assumption. RNA complementary to the sequences in the heat shock puffs is transcribed in salivary gland cells only when these regions are puffed (Bonner and Pardue, 1976). Later studies show a similar correlation between puffing and appearance of RNA complementary to sequences in ecdysone-induced puffs (Bonner and Pardue, 1977). Although the change in chromatin structure that produces a puff is apparently necessary for transcription of RNAs such as the major heat shock RNAs, there are other genes (for example those coding for histone mRNA) that do not seem to undergo this chromatin change when they are being transcribed. An interesting question that remains to be answered is whether the genes that puff when they are being transcribed undergo an analogous change when they are transcribed in non-polytene cells.

Because there are only a few newly synthesized RNAs entering the cytoplasm during heat shock it has been possible to isolate the major species and to show by in situ hybridization that the major poly(A)$^+$ RNA species are complementary to DNA sequences within the major heat shock puffs (Spradling et al., 1977). The isolated RNA species have also been shown to code for the major heat shock-induced polypeptides (Mirault, et al., 1978). In addition the poly(A)$^-$ fraction of newly synthesized RNA in the cytoplasm contains histone mRNA. The histone genes were not originally considered to be members of the heat shock set because no heat shock puff is induced at the histone locus. However histone mRNA and histone protein is synthesized during the heat shock (Spradling et al, 1977). This newly made histone becomes associated with 11s particles in chromatin during the heat shock (M. Sanders, personal communication).

Expression of the Heat Shock Genes Is Also Controlled at the Level of Translation

As first shown by Tissieres et al. (1974), heat shocked Drosophila cells synthesize a very limited set of polypeptides (Fig. 1). (We will refer to these polypeptides as heat shock

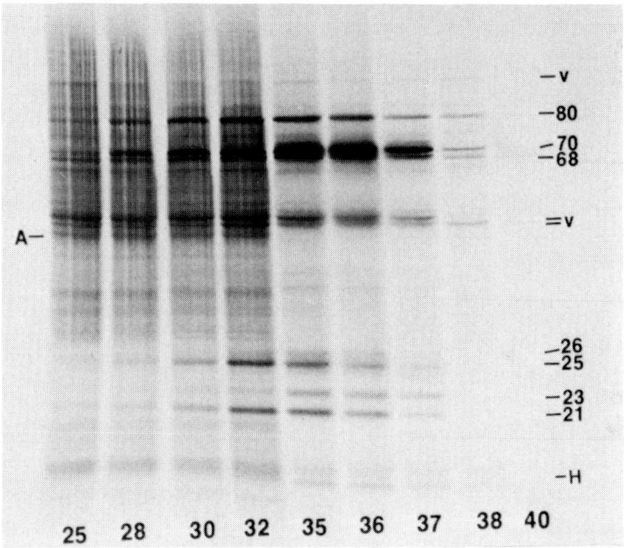

FIGURE 1. Autoradiogram showing proteins synthesized in cultured Drosophila cells at different temperatures. For each lane, $6x10^5$ cells were incubated at the indicated temperature for 30 min with 80 μCi/ml of ^{35}S-methionine. The cells were washed in saline and lysed in sample buffer. After a brief sonication, the labeled proteins were separated on a 9-14% SDS-polyacrylamide gel. Equal numbers of cells were loaded on each lane. The TCA-precipitable cpm loaded on each lane were, from right to left, 36,000, 38,000, 44,000, 78,000, 34,000, 30,000, 13,500, 5900 and 1600. Autoradiograph exposed 2d. A=actin, v=putative viral proteins, H=histones. Heat shock proteins are indicated by molecular weight in daltons x 10^{-3}. From Storti et al., 1980. Reprinted with permission.

proteins (HSPs), although some, notably histones and the 83k HSP, are also synthesized in control cells.) As mentioned earlier, the heat shock polypeptides are encoded by RNAs transcribed during the heat shock so it is perhaps not surprising that they are the major products of protein synthesis. What is surprising is the speed with which translation of most of the pre-existing mRNAs (which we will call 25° mRNAs) is stopped. Although most eukaryotic mRNA has a halflife of several hours in the cytoplasm (Perry and Kelley, 1973), translation of the pre-existing mRNAs declines rapidly after induction of the heat shock (McKenzie et al., 1975) as a result of the heat shock-induced translational control. The resulting rapid shift in protein synthesis would have an obvious advantage to the animal if it is important to concentrate available resources on the synthesis of the heat shock polypeptides. In addition, the 25° mRNAs, though undertranslated, are not degraded during heat shock; instead they are sequestered in such a way that they become available for translation after the cell recovers. This protection of the 25° mRNA would appear to be advantageous to the organism, especially during development when it might be difficult to recapitulate a transcriptional program once the mRNA was destroyed.

The first evidence that the 25° mRNA is not degraded came from experiments by McKenzie (1976). Actinomycin D was added to cells that were in heat shock and the cells were returned to their normal growth temperature. At 25° cells resumed synthesis of the control set of polypeptides, although the actinomycin D prevented renewed transcription of 25° mRNA. Thus the cells must have been utilizing 25° mRNA that had been preserved during the heat shock. In recovery experiments cells growing at 25° synthesize both 25° and heat shock polypeptides. The heat shock transcripts persist in the cell and continue to be translated for several hours (Storti et al., 1980).

It is also possible to detect intact 25° mRNAs in heat shocked cells by extracting RNA from such cells and translating the RNA in cell-free protein synthesizing systems made from rabbit reticulocytes or wheat germ. In such experiments the translation products include both heat shock and 25° proteins, showing that functional mRNAs for both groups of polypeptides can be extracted from heat shocked cells even though the 25° mRNAs and not being translated in vivo (Mirault et al., 1978; Storti et al., 1980).

Because the heterologous cell-free translation systems that we have used do not show preferential translation of heat shock mRNA, we have developed a translation system from lysates of *Drosophila* cells (Scott et al., 1979). Briefly, cells of a cultured Drosophila cell line are homogenized and nuclei and mitochondria are removed by centrifugation. Endogenous mRNA is digested with micrococcal nuclease which is then inactivated by chelating Ca^{++} with EGTA before the lysate is used for translation of added mRNA. These lysates appear to reproduce completely the translational discrimination found in the intact cells from which the lysates were made (Fig. 2). Lysates made from cells growing at 25° translate both 25° and heat shock mRNAs efficiently. Lysates made from cells that have been heat shocked at 36° for 1 hour before the lysate is made translate only heat shock mRNA.

Cell-free translation in the Drosophila lysates allows us to draw some preliminary conclusions about the mechanism of the heat shocked translational control. First, although induced by heat shock, the discriminating mechanism is not dependent on high temperature for its activity. The optimal temperature for translation in both the heat shock and the control lysates is 28°, which is also the optimal temperature for protein synthesis in intact cells. As a result, all of our translation experiments have been done at 28°, which permits direct comparison of the two lysates. A second conclusion from the cell-free translation experiments is that the lysate is discriminating mRNAs on the basis of a structural feature of the RNA. It does not seem to be the case that heat shock mRNAs are selected on the basis of a non-structural feature such as a special protein(s) that might be added during heat shock to all newly transcribed RNA before it is exported to the cytoplasm. The RNA that is added to the lysate in our experiments has been treated with Proteinase K and extracted with phenol and chloroform. Therefore it is unlikely that any protein of the original RNP remains associated with this RNA ;selection must be on the basis of an intrinsic structural feature.

What components of the lysate are responsible for the translational discrimination? We have begun to study the ability of different fractions of the Drosophila cell-free translation system to rescue translation of 25° mRNA in heat shock lysates (Scott and Pardue, 1981). In our initial experiments lysates of both heat shocked and control cells (after digestion with micrococcal nuclease) were fractionated by centrifugation into a crude ribosomal pellet and a supernatent soluble fraction. The fractions were then used to supplement translation reactions. As shown in Figure 2, the crude ribosomal pellet from 25° cells had a dramatic effect in rescuing the translation of 25° mRNA in heat shock lysates. None of the other lysate fractions affected translation in either the heterologous or the homologous lysate. These results argue strongly that at least part of the discrimination system is

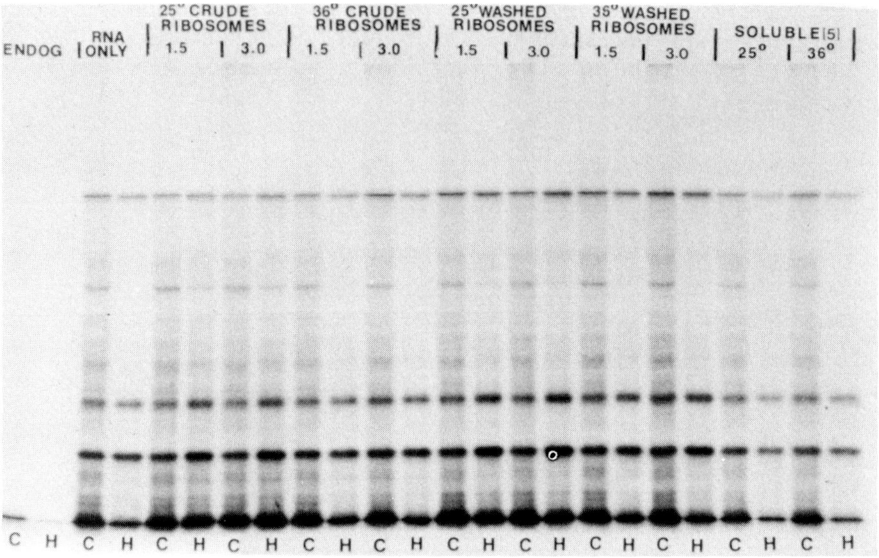

FIGURE 2. Translation products of lysates from control and heat shocked cultured Drosophila cells: effects of supplements on translational control. Lysates were prepared as described by Scott et al.(1979). The same 1:1 mixture of poly(A)$^+$ RNA from 22° and from heat shocked (1 hr at 36°) Drosophila embryos was translated in all reactions. Equal amounts of RNA were added to each reaction and equal amounts of each reaction were loaded on the gel. Supplements were prepared by fractionating micrococcal nuclease-digested lysates as described by Scott and Pardue (1981). Lysates were centrifuged for 30 min at 222,600 x g in a Beckman Ti50 rotor. The supernatent is designated as the "soluble" fraction. The pellet is "crude ribosomes". The crude ribosome fraction was treated with 0.5M KCl and centrifuged as before to yield the "washed ribosomes". The supplements added to lysates are indicated above the gel lanes. 1.5, 3.0 and 5.0 refer to the μl of fraction added. (2 μl of washed ribosomes would equal the number of ribosomes already in each reaction). "C" and "H" identify lanes containing products of control and heat shock lysates, respectively. The gel is SDS-10% polyacrylamide. Autoradiographic exposure was 3 d. Heat shock proteins (H) and actin (A) are indicated. ENDOG: synthesis without added mRNA. RNA ONLY: RNA but no supplements added. From Scott and Pardue (1981). Reprinted with permission.

either on the ribosomes or pellets with them in the fractionation. There is one obvious potential artifact in this experiment. If the 25° crude ribosomal pellet contained polysomes, the apparent recovery of 25° translation in the heat shock lysate might be spurious, simply representing run-off from the polysomes. We have tested the crude ribosome pellets for translation activity in rabbit reticulocyte lysates, which are very sensitive to the addition of *Drosophila* mRNAs. These experiments give no evidence of mRNA in the crude ribosome pellets.

In other experiments the crude ribosome pellets were washed with high salt to study the association between the discriminating element(s) and the ribosomes. Salt washed ribosomal pellets from 25° cells are somewhat less active than the corresponding crude ribosomal pellets in rescuing translation of 25° mRNAs in heat shock lysates. On the other hand, salt washed 36° ribosomal pellets are slightly active in rescuing translation of 25° mRNAs in heat shock lysates while crude ribosomal pellets from 36° cells have no detectable activity. At present the most attractive possible explanation for these results is that the salt wash might damage ribosomes and thus decrease the rescuing activity of 25° ribosomes; at the same time the salt wash might alter the ability of the heat shocked ribosomes to discriminate mRNAs, thus allowing them to increase the translation of 25° mRNA in a heat shock lysate.

What structural features allow mRNA selection in heat shocked lysates? Large portions of several of the heat-shock induced genes have now been sequenced. There are sequences within the transcribed portions of these genes that are common to all of the heat shock genes that have been studied (Holmgren et al., 1981; Ingolia and Craig, 1981). Such common sequences are attractive candidates for "translation selector" sequences but this remains to be demonstrated directly.

It appears that the feature(s) by which heat shock mRNA is recognized may not be unique to Drosophila heat shock mRNA. A virus, HPS-1, found in the cultured cell line, Schneider 2-L, also produces mRNA that is translated by heat shocked cells (Scott et al., 1980). Another virus, vesicular stomatitis virus (VSV), isolated from Chinese Hamster cells, also produces mRNAs that are discriminated by the Drosophila lysates. The RNAs encoding the N and NS viral proteins are translated by both heat shock and control lysates. The RNA coding for the M polypeptide, on the other hand, is translated efficiently only by the control lysate. The translation of the M polypeptide can be rescued by supplementing the heat shock *Drosophila* lysates with crude ribosomes from the 25°

lysate. Thus the VSV RNA coding for the M polypeptide shows the translational properties of a Drosophila 25° mRNA while the RNAs encoding the N and NS polypeptides are treated like heat shock mRNAs. Comparison of the mRNA sequences of these viruses with Drosophila mRNA sequences should give additional insight into the mechanism of mRNA discrimination.

What is the subcellular localization of 25° mRNA in heat shocked cells? The change in protein synthesis during heat shock is accompanied by a striking alteration in the size distribution of polyribosomes (McKenzie et al., 1975). Instead of the broad distribution of polyribosomes of many sizes found in 25° cells, the polyribosome profile in heat shocked cells has two predominant peaks of the size expected to translate the two major sizes of HSPs (Fig.3). The

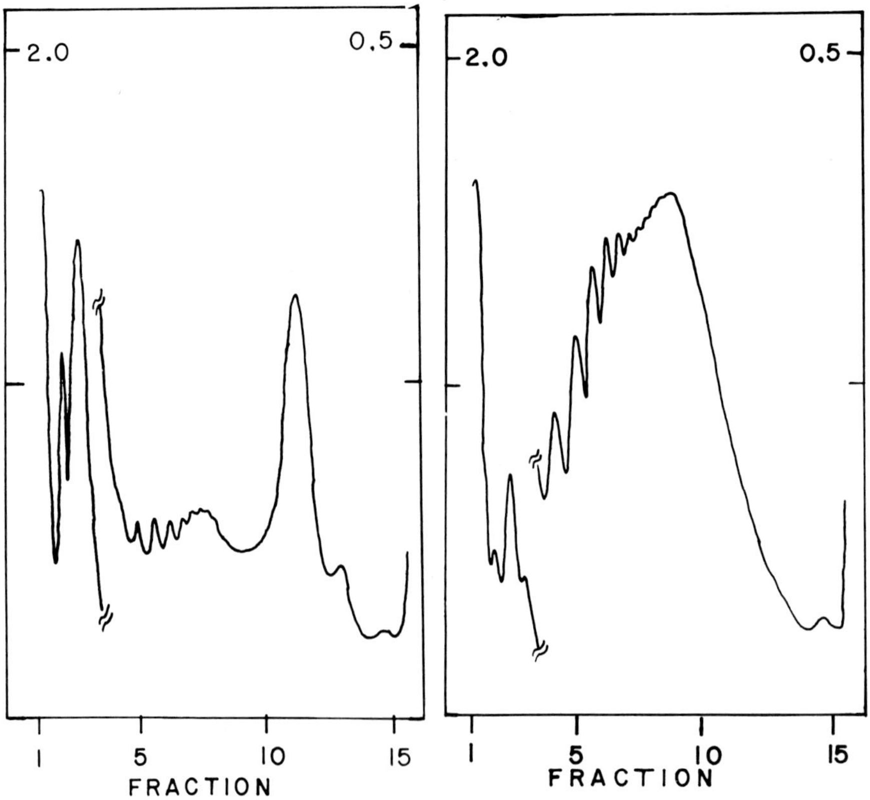

FIGURE 3. Optical density profile of sucrose gradients of polysomes from heat shocked (left) and control (right) cells. Exponentially growing cultures of Schneider 2-L cells (6-7 x 10^6 cells/ml) were fed with 1/20 vol of fresh growth

decrease in the number of ribosomes in the polysomal gradients from heat shocked cells is paralleled by an increase in in the size of the 40s and 60s ribosomal subunit peaks, indicating that ribosomes are recovered with equal efficiency from heat shocked and control cells.

To evaluate the status of 25° mRNAs in heat shocked cells we collected fractions from sucrose gradients such as those shown in Figure 3. The RNA was extracted from these fractions with phenol and chloroform and translated in a rabbit reticulocyte cell-free system. In spite of the dramatic change in the polyribosomal profile in heat shocked cells, we find that many 25° mRNAs are present in the same fractions of polyribosome gradients from either heat shocked or control cells (Fig. 4). The proteins encoded by these RNAs are barely detectable among the translation products of heat shocked cells (Fig. 1), yet a significant amount of these RNAs can be detected in polyribosome fractions by in vitro translation analysis.

We see some non-specific association of mRNA in polyribosome regions when gradients are run in low salt conditions. However under the high salt conditions used here (0.5 M KCl), only specific messages remain in rapidly sedimenting material. None of the mRNAs detected in rapidly sedimenting material appear to be sedimenting as large RNPs

---

medium 2 hr before the culture was split and equal portions incubated for 1 hr at 36.5° (heat shock) or at 25° (control). Cells were harvested by centrifugation and washed once with ice cold physiological saline. 2×10^9 cells were then resuspended in 3 ml polysome buffer (250 mM KCl, 2.5 mM $MgCl_2$, 10 mM EGTA, 20 mM Hepes, pH 7.2 with KOH), containing 5% rat liver supernatent (Scott et al., 1979) and 1% Triton X-100 detergent. After 10 min on ice with occasional vortexing, lysates were spun at 12,000 x g for 15 min and the resulting supernatent was frozen in liquid nitrogen and stored at -70°. 400λ aliquots (2×10^7 cells) were thawed, adjusted to gradient buffer (0.5 M KCl, 5 mM $MgCl_2$, 10 mM EGTA, and 20 mM Hepes, pH 7.2 with KOH) by the addition of 1 vol of equilibration buffer, and layered onto 36 ml 0.5 M-1.5 M linear sucrose gradients. Gradients were centrifuged 3 hr at 23,000 rpm in a Beckman SW27 rotor at 1° and analyzed for A_{254} with an ISCO density gradient fractionator. The vertical axis indicates full scale deflection for the top (left) and bottom (right) of the gradients. The point at which scale was changed is indicated by a break in the tracing. The horizontal axis indicates the numbers of the fractions used for in vitro translation (see Fig. 4).

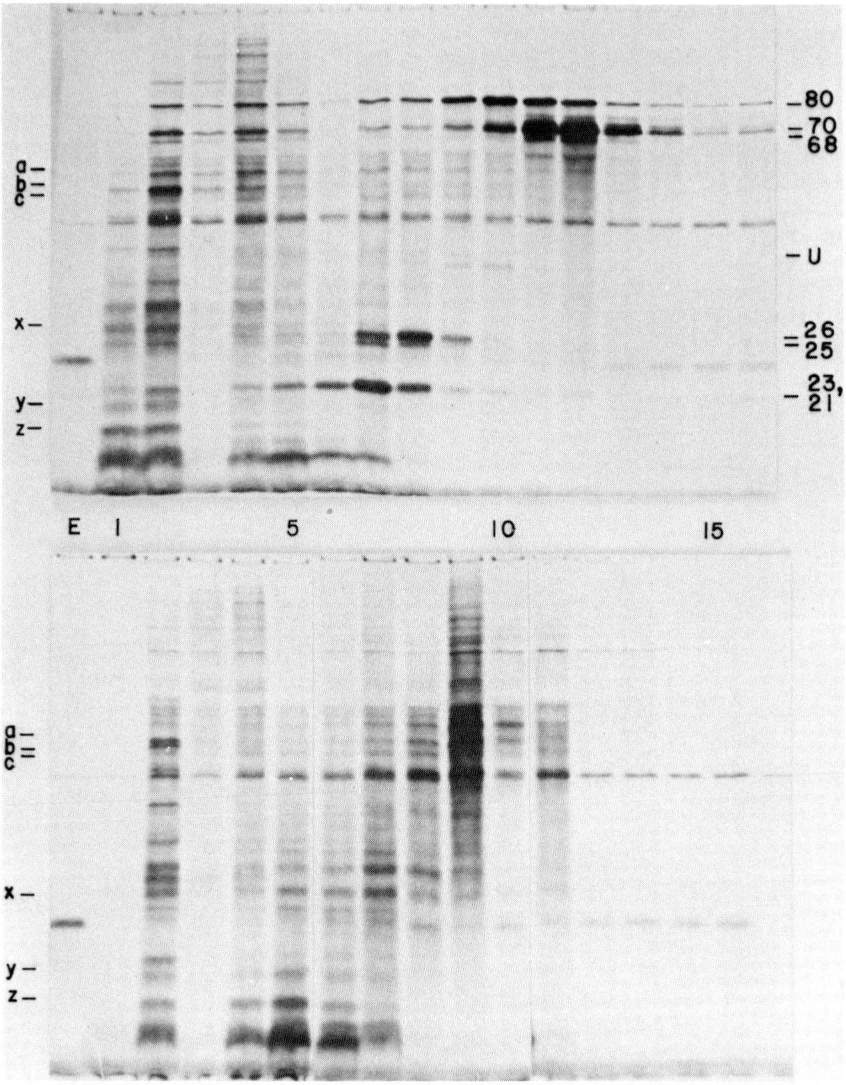

FIGURE 4. In vitro translation products of RNA extracted from polysomal gradients from heat shock (top) and control (bottom) cells. 2.4 ml fractions from the gradients in Fig.3 were collected into 5 ml 100% ethanol (at -70°) containing 20 µg Proteinase K and held at -70° overnight. The precipitates were collected by centrifugation, resuspended in 0.5 ml of RNA buffer (100 mM NaCl, 1 mM EDTA, 10 mM Tris HCl, pH 7.4,

not associated with ribosomes; all of the polyribosomes in gradients from both control and heat shocked cells are dissociated into subunits by 25 mM EDTA and most of the functional mRNAs are released by this treatment into the subunit region of the gradients.

In sucrose gradients from heat shocked cells, a number of RNAs coding for smaller 25° proteins (less than 40,000 M_r) are recovered in fractions where one would expect RNPs (and naked RNAs under the salt conditions used here) to sediment (e.g. x, y, and z in Fig. 4). Many RNAs, on the other hand, sediment in the polysomal region of the gradients, as well as in the subunit region (a, b, and c in Fig. 4). This second class of RNAs sediments as though each mRNA was associated with approximately its normal number of ribosomes in the heat shocked cells. A third class of RNA molecules (primarily HSPs, Fig.4) sediment exclusively in the polysomal region of the gradients. Thus it appears that, even though undertranslated in vivo, a significant subset of the 25° mRNA is found on polysomes in heat shocked cells.

---

containing 0.25% SDS and 0.625% Sarkosyl) and 0.5 ml phenol on ice. One vol of chloroform: isoamyl alcohol (48:2) was added and the aqueous layer was extracted 2 more times with phenol: chloroform:isoamyl alcohol (50:48:2) and 3 times with chloroform:isoamyl alcohol. The aqueous phase was brought to 0.16 M NaAcetate and precipitated twice with 2.5 vol ethanol. The final ethanol precipitate was resuspended in 60 µl H_2O, lyophilized, and resuspended in 20 µl H_2O (except fractions 1, 2, and 3 which were resuspended in 40 µl of H_2O) 1/10 vol of this RNA was added to 25 µl of rabbit reticulocyte translation mix (Pelham and Jackson, 1976) and incubated for 60 min at 30°. The translation reaction was stopped by the addition of 5 vol of sample buffer. Equal vols of each reaction were analyzed on parallel lanes of an SDS-polyacrylamide gel containing 15% acrylamide and 0.086% bis-acrylamide in the discontinuous buffer system of Laemmli (1970). The figure shows autoradiographs of dried gels exposed for 1 d at room temperature. All translations were shown by serial dilution to be within the linear range of the RNA concentration. E: endogenous synthesis without added RNA. U: unidentified HSP. HSPs are identified by molecular weight in daltons x10^{-3}. Fraction numbers refer to fractions of the gradient shown in Figure 3. Other labels are explained in the text.

SUMMARY

Gene expression is controlled at several levels in heat shocked cells. Transcriptional control results in the production of a small set of RNAs. In the cytoplasm, a translational control produces a rapid switch in the proteins being synthesized. This translational control may involve several components. At least one of the components implicated in mRNA discrimination appears to be associated with the ribosomes. A second component of the control appears to be a structural feature of the mRNA. Finally, a specific set of RNAs coding for proteins which are undertranslated in heat shocked cells remains associated with polysomes in those cells. This association seems normal by several criteria but the proteins encoded by the mRNAs are not among the translation products of the cells in vivo. The exact nature of the association is under investigation , but the data indicate that two classes of non-translated messages exist in heat shocked cells. Some messages are not associated with ribosomes, and other messages appear to be associated with ribosomes but to have their expression blocked at some stage of protein synthesis other than initiation.

REFERENCES

Ashburner, M., and Bonner, J.J. (1979). Cell 17, 241.
Bonner, J.J., and Pardue, M.L. (1976). Cell 8, 43.
Bonner, J.J., and Pardue, M.L. (1977). Cell 12, 219.
Holmgren, R., Croces, V., Morimoto, R., Blackman, R., and Meselson, M. (1981). Proc. Nat. Acad. Sci.USA (in press).
Ingolia, T.D., and Craig, E.C. (1981). Nuc. Acids Res. (in press).
Laemmli, U.K. (1970). Nature 227, 680.
McKenzie, S.L. (1976). Ph.D. Thesis, Harvard University, Cambridge, Massachusetts.
McKenzie, S.L., Henikoff, S., and Meselson, M. (1975). Proc. Nat. Acad. Sci. USA 72, 1117.
Mirault, M.-E., Goldschmidt-Clermont,M., Moran, L, Arrigo, A. P., and Tissieres, A. (1978). Cold Spring Harb. Symp. Quant. Biol. 42, 819.
Pelham, H.R.B., and Jackson, R.J. (1976). Eur. J. Biochem. 67, 247.
Perry, R.P., and Kelley, D.E. (1973). J. Mol. Biol. 79, 681.
Ritossa, F.M. (1962). Experentia 18, 571.

Scott, M.P., and Pardue, M.L. (1981). Proc. Nat. Acad. Sci. USA (in press).

Scott, M.P., Storti, R.V., Pardue, M.L., and Rich, A. (1979). Biochemistry 18, 1588.

Spradling, A., Pardue, M.L., and Penman, S. (1977). J. Mol. Biol. 109, 559.

Storti, R.V., Scott, M.P., Rich, A., and Pardue, M.L. (1980). Cell 22, 825.

Tissieres, A., Mitchell, H.K., and Tracey, U.M. (1974). J. Mol. Biol. 84, 389.

ANALYSIS OF EUKARYOTIC GENE TRANSCRIPTION IN VITRO

Robert G. Roeder
David C. Lee
Barry M. Honda
B.S. Shastry

Departments of Biological Chemistry and Genetics
Division of Biology and Biomedical Sciences
Washington University
St. Louis, Missouri

ABSTRACT

We have begun an analysis of the factors and events governing the specific initiation of transcription *in vitro* at both class II and class III genes. In the former case, adenovirus DNA fragments have been transcribed in a cell-free system comprising RNA polymerase II and a soluble extract (S100) from uninfected human cells. Transcription initiation has been observed for at least 6 temporally regulated promoters, including some which may require prior viral gene expression for maximal activation *in vivo*. Active initiation at these promoters correlates with the presence of an upstream "TATA" sequence and has been shown to be accurate even for those transcription units whose *in vivo* transcripts contain a heterogeneous array of 5' termini. For example, transcription from the EIV promoter utilizes the same 6 start sites *in vivo* and *in vitro*.

Fractionation of the cell-free extract has revealed a multiplicity of factors required for reconstitution of transcription initiation at the major late adenovirus promoter. At least 4 distinct factors have now been implicated, and these are distinct from at least 3 additional factors required for reconstitution of transcription of class III genes by RNA polymerase III; 2 of these latter factors may be commonly required for vertebrate tRNA and 5S RNA genes, whereas the third is specific for 5S RNA.

Soluble extracts prepared from *Xenopus* tissues also contain at least 3 factors, in addition to RNA polymerase III,

DEVELOPMENTAL BIOLOGY USING PURIFIED GENES

ISBN 0-12-137420-3

required for transcription of homologous 5S RNA genes. One of these factors is the previously described 5S gene-specific initiation factor TFIIIA. Immunological analyses have indicated the presence of high levels of TFIIIA in immature oocytes, reduced levels in large mature oocytes, and still lower levels in unfertilized eggs and early embryos. An immunoreactive protein equivalent in size to TFIIIA persists, albeit at much lower levels, in late embryos and in adult-derived cells and tissues. In addition, somatic cells also contain a slightly larger immunoreactive protein which could be a new factor associated with the reduced level of 5S RNA synthesis and/or selective transcription of somatic-type genes in these cells. Chromatographic fractions containing the TFIIIA-sized protein, but not those containing the larger TFIIIA-related protein, complement TFIIIA-deficient extracts for 5S gene transcription. The implications of these findings are discussed.

INTRODUCTION

A detailed understanding of how transcriptional processes are regulated must ultimately require the ability to duplicate natural transcription events in cell-free systems reconstituted with purified components. Major advances toward this goal have followed the recent development of soluble systems in which exogenous purified genes are accurately transcribed. These have now been described for genes transcribed by RNA polymerase III (reviewed in 1), for RNA polymerase II (reviewed in 2) and for RNA polymerase I (3). These systems have allowed the analysis of mutated genes and the identification of DNA sequences critical for accurate transcription (e.g. 4,5) and in one instance, an analysis of the mechanism of action of a known viral transcription repressor (6). Equally important, however, is an understanding of the factors which are required for mediating accurate transcription by the complex RNA polymerases; these latter enzymes have already been characterized (7) and shown not to mediate accurate or selective transcription in the absence of other factors (7,8,9). Initial experiments with crude soluble systems (cell extracts) and a variety of class II and III genes suggested that the factors which are necessary for the observed transcription are neither tissue nor species specific (9,10,11). At the same time, the ability to separate these factors from the RNA polymerases (8-10,12) suggested that they might be purified and analyzed (cf. 12) and ultimately used to construct a more purified system for the analysis of other regulatory factors/controls.

In the following, we summarize studies relevant to the further analysis of both class II and class III genes. First we describe recent studies on the accuracy and extent of transcription from a group of regulated adenovirus promoters and we summarize our present understanding of the complexity of factors required by one of these promoters *in vitro*. Second, we review the complexity of those factors required for class III genes and we present more recent information on the 5S gene-specific factor in *Xenopus* oocytic and somatic tissues.

RESULTS AND DISCUSSION

TRANSCRIPTION OF CLASS II GENES

Adenovirus 2 Promoters

Our analysis of the *in vitro* transcription of a variety of class II genes (those transcribed by RNA polymerase II) is exemplified by the results obtained with promoters present in the genome of adenovirus type-2 (Ad2). The Ad2 chromosome contains at least 8 transcription units, and their activities are temporally regulated during the lytic cycle (reviewed in 14). In order to examine the activity of these promoters *in vitro*, we have transcribed appropriate cloned viral fragments (14) in a crude cell-free system derived from uninfected human KB cells and supplemented with purified RNA polymerase II from the same source (9); specific transcription initiation at a defined site is demonstrated using the so-called "run-off" assay as initially described for the major late promoter of Ad2 (9).

As shown in Fig. 1, most Ad2 promoters are active *in vitro* and generate run-off transcripts of the expected length (14). These include even the promoter for the polypeptide IX gene, which appears not to be expressed *in vivo* until several hours after infection (16). Other active promoters include those of the EIB, EIII and EIV regions, whose activation *in vivo* may require expression of a very early viral product (17,18) and which, in any case, are not expressed for 1-3 hr following infection (19). [Analogous results have been obtained by others for some of these promoters using similar (5) and different (15) extracts.] However, the transcriptional activity of these promoters varies considerably; thus the transcriptional efficiencies of the EIA and major late transcription units (the weakest and strongest, respectively) differ by as much as 10-fold. We do not yet know whether these differences reflect variations in initiation or elongation efficiencies, and we have thus far been unable to demonstrate competition between ostensibly weak and strong promoters. In addition two of the Ad2 promoters, those of the early EII (map position

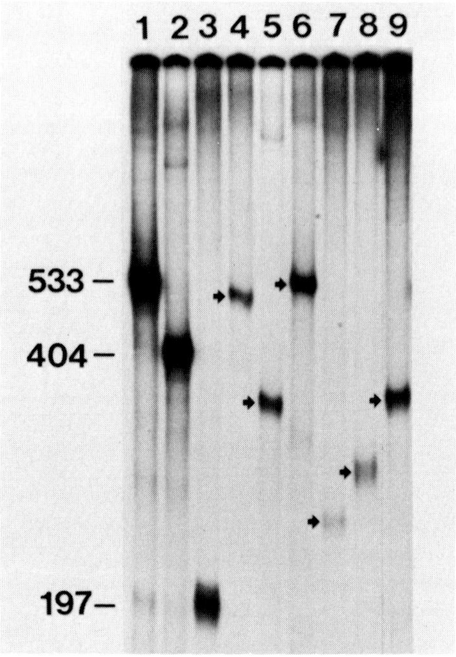

FIGURE 1. Run-off RNAs synthesized in vitro from Ad2 promoters. The autoradiogram shows a polyacrylamide gel analysis of glyoxal-denatured run-off RNAs synthesized in vitro from cloned plasmids containing various Ad2 promoters. The indicated transcripts originate from the following promoters: (lanes 1-3) major late; (lane 4) EIA; (lane 5) EIB; (lane 6) EIII: (lanes 7 and 8) EIV; (lane 9) polypeptide IX. The length of each of the major late run-off transcripts as deduced from sequence information is shown on the left and corresponds to the distance from the initiation site to the downstream termination site generated by restriction endonuclease cleavage. [For further details see ref. 14.]

(75) and IVa2 transcription units, are not actively transcribed in vitro (data not shown, see 14). As indicated in Fig. 2, which lists the nucleotide sequences surrounding the Ad2 cap sites, both of these latter promoters lack a recognizable TATAA sequence upstream from the cap site. Since others have shown that deletion or modification of the conalbumin and major late TATAA sequences dramatically reduces their transcriptional efficiency in vitro (5,20,21), and since we have shown that specific deletion of the TATAA sequence from the EIII promoter abolishes transcription (D.C. Lee, R.G. Roeder, and W.S.M. Wold, manuscript in preparation), it is tempting to correlate the lack of transcription in vitro with the lack of a TATAA box. Specific transcription of the early EII and IVa2 pro-

	TATAA
	-30 -20 -10 CAP
EIA	GTGTATTTATACCCGGTGAGTTCCTCAAGAGGGCACTCTT
EIB	AGGGTATATAATGCGCCGTGGGCTAATCTTGGTTACATCT
EII	TAGTCCTTAAGAGTCAGCGCGCAGTATTTGCTGAAGAGAG
EIII	CAGGGTATAACTCACCTGAAAATCAGAGGGCGAGGTATTC
EIV	CCTATATATACTCGCTCTGTACTTGGCCCTTTTTACACTG
PIX	GAATATATAAGGTGGGGGTCTCATGTAGTTTTGTATCTGT
MAJOR LATE	GGCTATAAAAGGGGGTGGGGGCGCGTTCGTCCTCACTCTC
IVA_2	GTGCTGGCCTGGCCTGGACGCGAGCCTTCGTCTCAGAGTG

FIGURE 2. DNA sequence upstream from Ad2 cap sites. The diagram lists DNA sequences upstream from the cap sites of the various Ad2 promoters. The putative TATAA sequences, as well as the location of the cap sites, are indicated by underlining. This data is derived from ref. 22.

moters could also require host (or viral) factors not present or active in our cell-free system.

We have previously shown by a detailed fingerprint analysis that initiation *in vitro* at the adenovirus major late promoter and at the mouse β-globin promoter is accurate in our cell-free system (9,11). In order to further document the fidelity, and hence the utility, of the cell-free system, we have characterized the 5' terminii of the transcripts generated in response to the other adenovirus promoters. In this case, we have used a reverse transcriptase-mediated primer extension method to analyze the 5' terminii. As shown in Fig. 3A, examination on high-resolution sequencing gels of the extension products obtained with *in vivo* and *in vitro* major late RNAs reveals identical 5' ends; similar results have been obtained with *in vivo* and *in vitro* transcripts from each of the active Ad2 promoters (14). As shown in Fig. 2, several of the Ad2 promoters display microheterogeneity of 5' ends *in vivo* (22). This is dramatically evident, for example, in the case of the EIV promoter where initiation occurs primarily with the A, but also to a significant extent with the 5 T's immediately preceding the A. As indicated in Fig. 3B, primer extension analysis of EIV *in vivo* RNAs reflects this kind of microheterogeneity; significantly, the analysis of the 5' ends of EIV *in*

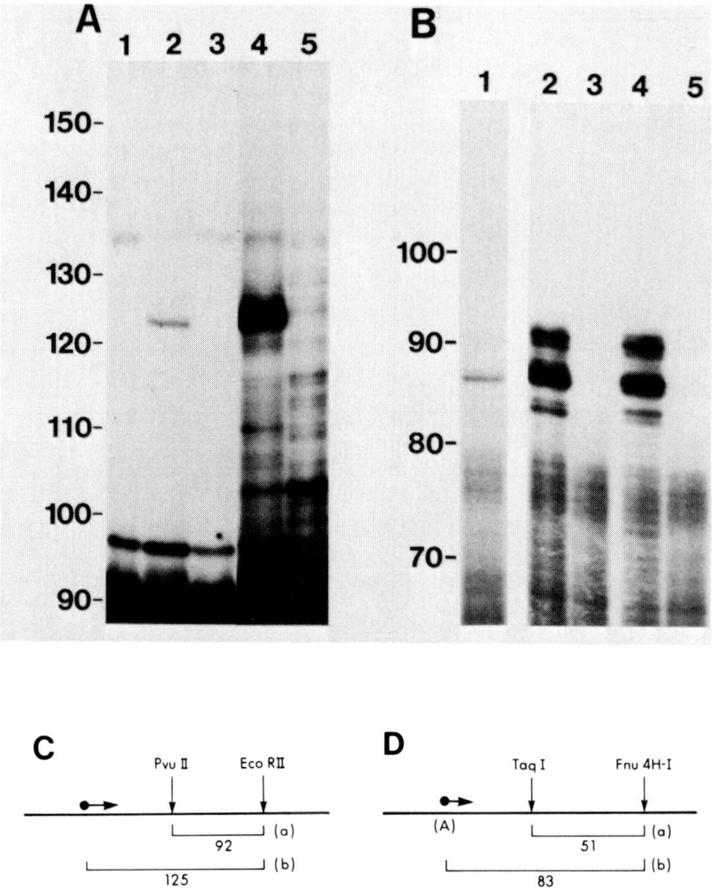

FIGURE 3. 5' End analysis of major late and EIV RNAs. The autoradiograms show the gel analyses of primer extension products obtained with probes specific for major late (A) and EIV (B) transcripts. The details of the preparation of the probes and RNAs, as well as of the hybridization, extension reaction and electrophoretic conditions, are given in ref. 14. For (A), the major late probe was hybridized to either E. coli tRNA (lane 1), poly (A+) late RNA from Ad2-infected KB cells (lane 2), poly (A+) RNA from uninfected KB cells (lane 3) or total RNA from an in vitro transcription reaction of Ad2 late promoter DNA carried out in the absence (lane 4) or presence (lane 5) of 0.5 μg/ml α-amanitin. For (B), the EIV probe was hybridized to poly(A)-containing early RNA from Ad2-infected cells (lanes 1 and 2; lane 1 is a relatively short exposure

from a different experiment) and uninfected cell RNA (lane 3), or to gel-purified RNAs from in vitro transcription reactions carried out in the absence (lane 4) or presence (lane 5) of 0.5 μg/ml α-amanitin. For both (A) and (B), the numbered lines denote sizes determined from marker sequence reactions run in adjacent lanes (not shown). The diagrams below show the restriction enzyme sites used to generate the probes for the 5' end analysis and indicate the sizes of the probes (a), as well as the sizes of the expected extension products (b). The locations of the cap sites are shown as closed circles (●), and the direction of transcription is indicated. In the case of EIV, only the predominant cap site is shown.

As shown in (A), hybridization of the 92 base-long PvuII-EcoRII major late primer to either late-infected in vivo RNA or RNA derived from an in vitro transcription reaction results in an extension product of approximately 125 nucleotides in length, as expected. This extension product is not present when the hybridization is carried out with either tRNA, uninfected-cell RNA or in vitro transcripts synthesized in the presence of α-amanatin. Hybridization of the 51 base-long EIV primer (B) to either early-infected in vivo RNA or in vitro transcripts yields a complex pattern of 6 extension products not present when the hybridization is carried out with either uninfected-cell RNA or the products of in vitro transcription carried out in the presence of α-amanatin. Examination of the extension products reveals a prominent band whose size is consistent with the expected size of 86 nucleotides, followed by 5 consecutive bands of higher molecular weight; the relative intensities of these 5 bands are greatest for the largest and smallest. These results are consistent both qualitatively and quantitatively with the multiple initiation events described by Baker and Ziff (ref. 22) and indicated in Fig. 2. The 2 relatively prominent bands immediately below the principal set of extension products are artifactual. For a more extensive discussion of these results, see ref. 14.

---

vitro transcripts demonstrates precisely the same kind of heterogeneity. We have similarly observed identical microheterogeneity in the 5' ends of in vivo and in vitro polypeptide IX and EIII transcripts. Thus, initiation of transcription in vitro is accurate from promoters which display not only a wide range of transcriptional efficiency but also microheterogeneity of cap sites.

As indicated previously, the above results were obtained using cell-free extracts derived from uninfected KB cells. We have thus far failed to observe any significant or selective differences when transcribing Ad2 promoters in extracts derived from virus-infected cells (data not shown). In particular, we have failed to note any selective enhancement of transcription of any of the early promoters, or even those of the polypeptide IX and major late genes whose activities in vivo are clearly turned on or enhanced at later times of infection. In addition, the use of infected-cell extracts does not appear to facilitate the expression of the inactive early EII or IVa_2 promoters. These results point to the generalization that we have failed to observe any indication of positive regulation of transcription in vitro of any of the class II genes thus far examined in our laboratory. This includes cellular (11) as well as viral genes. As indicated earlier, the cell-free system appears to be permissive in nature (both with respect to tissue and species specificity), at least when naked DNA templates are employed. Whether this indicates a critical role for chromatin structure in transcriptional regulation or reflects, instead, the inadequacy or inappropriateness of the run-off assay and the particular cell-free system is not known (discussed further in 9,11,14).

Factors Involved in RNA Polymerase II-Mediated Transcription

The permissiveness of the transcription system from human KB cells suggests that the factors active in the extract are general in nature. To investigate the complexity, and ultimately the specificity and mechanism of action, of these factors, we have begun to fractionate the KB extract. The results of chromatography on phosphocellulose, DEAE-cellulose, and DNA cellulose are summarized in Fig. 4; in this particular analysis we have identified at least four components which are required for active and selective transcription initiation at the Ad2 major late promoter (2). More recently we have resolved one of these fractions (TFIIB) into two separable components, thus indicating at least 5 distinct factors (D. Dignam and R.G. Roeder, unpublished). One of these factors (TFIIC) appears to function by suppressing random transcription; it has been purified to homogeneity and consists of a single 130,000 dalton polypeptide (T. Matsui and R.G. Roeder, unpublished). Whether this factor functions as a transcription factor in vivo is not established, since it might simply block initiation in vitro at non-physiological sites (E. Slattery and R.G. Roeder, unpublished). The roles of the other factors are not yet established. However it is clear that they are distinct from those employed by RNA polymerase III (Fig. 4 and below). Thus, changes in the cellular levels or activities of any of these factors might effect an overall regulation of RNA polymerase II transcription or even differential rates of transcription from subsets of class II genes if the relative affinities for different genes vary. It will be important in this regard to determine whether the factors identified are indeed commonly required for different class II genes; it does not seem unreasonable to expect that different subclasses of genes/promoter structures might utilize partially distinct sets of "general" factors.

TRANSCRIPTION OF CLASS III GENES

Factors Involved in RNA Polymerase III-Mediated Transcription

As described for class II genes, a variety of class III genes (transcribed by RNA polymerase III) are accurately transcribed in both heterologous and homologous extracts; these include various tRNA and 5S RNA genes, as well as genes encoding the Ad2 VA RNAs (reviewed in 1). From these results it was suggested (10) that the factors active in these extracts are "general" factors which are necessary and sufficient for transcription of purified genes but not sufficient for effecting all of the regulation observed *in vivo*.

To gain an understanding of these factors, we have focused our attention on the fractionation of extracts derived from human cells and from amphibian tissues. As summarized in Fig. 4, human cell extracts contain at least three separable factors required for accurate transcription (initiation and termination) by RNA polymerase III. Two of these fractions, IIIB and IIIC, are required for all genes tested (tRNA, VA RNA, 5S RNA), while a third, IIIA, is specifically required for 5S RNA genes (1). Since none of the human factors are purified to homogeneity, it remains an open issue whether there are factors common to all class III genes and whether other subclasses such as the tRNA genes have specific factors. As indicated in Fig. 4 and elsewhere (1,2) the RNA polymerase III factors appear distinct from the RNA polymerase II factors. Thus a general modulation of classes or subclasses of genes appears possible via alteration in the activities and/or levels of a given factor. The general complexity of the group of polypeptides (a large number of RNA polymerase III subunits and at least 2-3 separable factors) required for transcription of these simple genes is also somewhat surprising. At present we have no firm indication as to the involvements of these factors/subunits in initiation versus elongation or termination events (see also below).

We have also begun to define the factors required for 5S gene transcription in *Xenopus* tissues. This system is of particular interest because of the presence of two classes of 5S genes: oocyte-type genes are expressed uniquely in oocytes while somatic-type genes are expressed in oocytes and in somat-

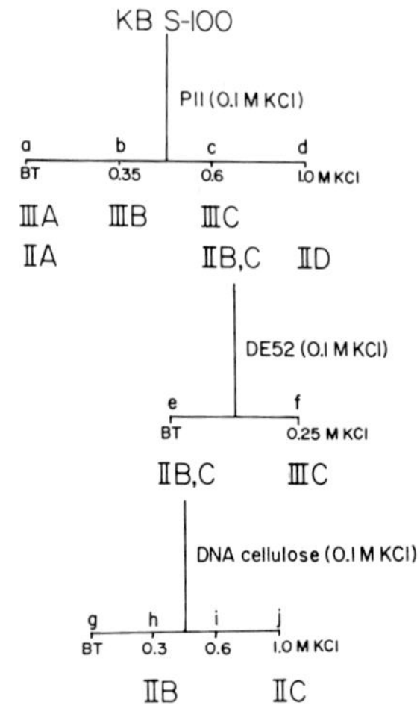

FIGURE 4. Separation of RNA polymerase II and RNA polymerase III transcription factors. A human KB cell S100 fraction was subjected to the chromatographic fractionation indicated as detailed elsewhere (1,2). Accurate initiation of transcription by RNA polymerase II at the major late promoter of Ad2 requires the fractions containing the "factors" IIA, IIB, IIC, and IID. The fractions containing the factors IIIA, IIIB, and IIIC are required for selective transcription by RNA polymerase III (see text). BT, P11, and DE52 denote breakthrough, phosphocellulose, and DEAE-cellulose, respectively. The KCl concentrations indicate either the salt concentration at which protein was loaded onto or eluted from a column.

ic tissues (reviewed in 23). Both types of genes were shown to be transcribed in germinal vesicle (23) and oocyte (8) extracts as expected. However, both genes were also shown to be transcribed in extracts from cultured kidney cells expressing only the somatic genes (10), suggesting that the endogenous factors mediating the transcription are not tissue specific and not those responsible for the selective transcription in vivo (discussed further in 10,25, and below). Fractionation of oocyte, egg, embryonic, and kidney cell extracts has now revealed the presence of at least three factors needed for 5S gene transcription in vitro in addition to RNA polymerase III. One of these (designated TF IIIA) was purified to homogeneity

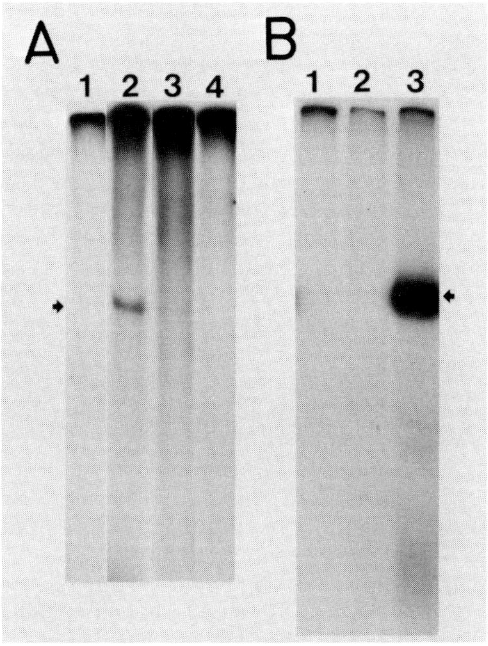

FIGURE 5. Chromatographic separation of 5S gene transcription factors in oocyte and egg extracts. S100 extracts from ovarian tissue (panel B) or unfertilized eggs (panel A) were subjected to chromatography on phosphocellulose (P-11) under standard conditions to yield flow-through (0.21 M KCl) and step-eluted (1.0 M KCl) fractions. Transcription of somatic-type 5S genes in the presence of these fractions was analyzed as described (12). Panel A shows the RNAs synthesized with exogenous RNA polymerase III and the following: lane 1, P-11 flow-through and P-11 step fractions; lane 2, P-11 flow-through fraction, P-11 step fraction, and TFIIIA; lane 3, P-11 flow-through fraction and TFIIIA; lane 4, P-11 step fraction and TFIIIA. Panel B shows the RNAs synthesized in the presence of saturating levels of exogenous RNA polymerase III and oocyte TFIIIA and the following: lane 1, P-11 flow-through fraction; lane 2, P-11 step fraction; lane 3, P-11 flow-through plus step fractions.

from oocytes (12) and is considered further below. A demonstration of the other activities is presented in Fig. 5. The fractionation of egg extracts (deficient in TFIIIA; see 12) on phosphocellulose yields two fractions which are jointly required, along with TFIIIA, for 5S gene transcription in the presence of saturating levels of exogenous RNA polymerase III (Fig. 5A). A similar fractionation of oocyte extracts (containing abundant TFIIIA) yields two fractions which are both required for 5S gene transcription in the presence of saturating amounts of TFIIIA and RNA polymerase III (Fig. 5B). It has also been possible to completely remove these latter two components from both egg and oocyte extracts and to show directly a requirement for the three factors in addition to the RNA polymerase; similar results have been obtained with extracts from cultured kidney cells (data to be reported elsewhere). In general, under comparable conditions, the chromatographic properties of the IIIA, IIIB and IIIC activities are similar to those indicated for the human cell factors; moreover, the Xenopus IIIB and IIIC fractions suffice for transcription of the tRNA genes, whereas the IIIA activity is uniquely required for somatic- and oocyte-type 5S genes.

Xenopus 5S Gene-Specific Transcription Factors

The oocyte factor TFIIIA (see above) is of particular interest because it is specific for 5S genes and represents the first specific eukaryotic transcription factor purified to homogeneity (12). This factor was originally shown to bind independently to a specific intragenic region on both oocyte and somatic genes (12); since this region was independently shown to be necessary and sufficient for transcription initiation (4), this factor clearly represents an initiation factor. More recently, it was shown (24,25) that this factor is identical to the protein associated with oocyte-type 5S RNA as a 7S-ribonucleoprotein (RNP) complex in oocytes (26); this confirmed a dual function for this protein and more interestingly, the possibility that 5S RNA might inhibit its own synthesis *in vivo* when TFIIIA is limiting. It has in fact been shown that 5S RNA can inhibit its own synthesis *in vitro* (24) and that the levels of TFIIIA in somatic tissues are dramatically lower than in oocytes (25).

A further immunological analysis of the level of TFIIIA during oogenesis is shown in Fig. 6. The results indicate a near constant cellular level of TFIIIA in earlier stages of oogenesis and a much lower level in mature oocytes. Other more quantitative analyses indicate at least a 20-30 fold decline in the amount per oocyte (from roughly 50 to 2 ng). This is compatible with the notion that TFIIIA is overproduced at earlier stages to ensure both maximal transcription and 5S RNA stabilization, and that most of the protein complexed with 5S RNA is destroyed late in oogenesis as 5S RNA is transferred to ribosomes. The analysis presented in Fig. 7 (lanes 1 and 2) shows more clearly the presence of residual TFIIIA in stage 6 oocytes and, in addition, a reduction of the TFIIIA level in unfertilized eggs. This reduction is estimated to be in the

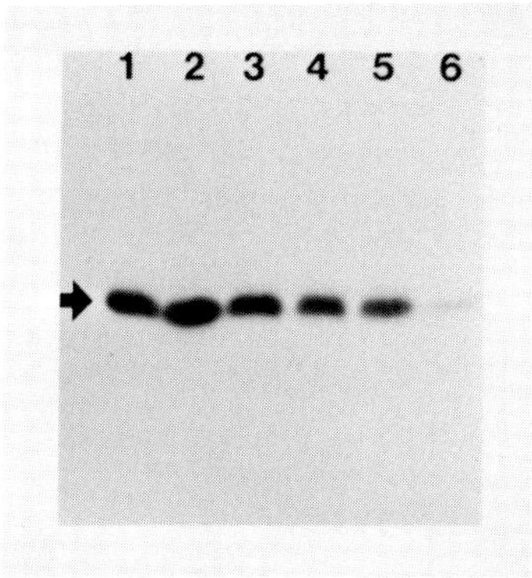

FIGURE 6. Analysis of TFIIIA in staged Xenopus oocytes. Total homogenates of Xenopus oocytes were subjected to electrophoresis in the presence of SDS on 12% polyacrylamide gels. The proteins were transferred electrophoretically to DPT-paper which was then treated with anti-TFIIIA antiserum and ^{125}I-labelled S. aureus protein A (cf. 25) and subjected to autoradiography. The figure shows the analyses of 0.25 oocyte equivalents for stage I-VI oocytes (in lanes 1-6, respectively). The radiolabeled bands comigrate with authentic TFIIIA.

range of 20-50 fold, and is compatible with the supposed decreased level of 5S RNA synthesis in eggs and early embryos. However, it is not possible to say with any certainty that the reduced level of 5S RNA accumulation is directly related to the loss of TFIIIA since neither the levels of 5S synthesis nor the levels of TFIIIA are accurately quantitated; moreover the intracellular localization/state of the TFIIIA detected is unknown. The additional analyses in Fig. 6 also show an immunologically cross-reactive protein, equivalent in size (40,000 daltons) to oocyte TFIIIA, in gastrula stage embryos. Preliminary peptide analyses (CNBr fragments) suggest that this protein is similar or identical to oocyte TFIIIA (data not shown); it is not clear whether this represents newly synthesized versus oocyte-derived protein.

As shown in Fig. 8, a similar immunological analysis of kidney cell and liver tissue extracts reveals two immunoreactive protein bands; the lower band is equivalent in size to oocyte TFIIIA and the more predominant upper band appears larger by 2000-3000 daltons. The upper band has never been observed in oocyte, egg or early embryo extracts (cf. Fig. 7),

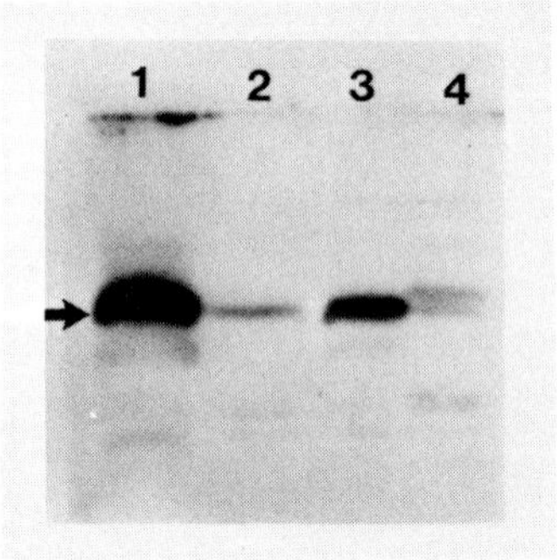

FIGURE 7. Analysis of TFIIIA in stage VI oocytes, eggs, early embryos and cultured kidney cells. Extracts prepared from the various cell and tissue types were analyzed for the presence of proteins cross-reacting with anti-TFIIIA antibody, as described in Figure 6. The resulting autoradiogram shows the TFIIIA-related proteins in extracts from: lane 1, two stage VI oocytes; lane 2, 10 unfertilized eggs; lane 3, 20 early gastrula embryos; lane 4, 40 µl of a kidney cell S100 extract (10). The arrow indicates the migration position of authentic TFIIIA.

although it is visible in later stage embryos (data not shown). This result raises the intriguing possibility that the upper band might reflect the presence of a distinct (but related) TFIIIA in somatic cells, whose synthesis and function might be better adapted to the somatic cell requirements for 5S RNA synthesis (see discussion in 25 and below). Obviously such a factor might either reflect an altered product of the same gene (as that which encodes oocyte TFIIIA) or the product of a distinct gene.

To further assess the nature and function of the two immunoreactive protein bands in the somatic cell extracts, we have subjected kidney cell extracts to standard fractionation techniques (Fig. 9A). The analysis in Fig. 9B shows that the phosphocellulose flow-through fraction contains the IIIA activity, as assayed by its ability to complement the phosphocellulose step fraction (this contains both the IIIB and IIIC activities; data not shown) or an unfertilized egg extract (data not shown). Similarly, the blot immunoassay in Fig. 9C indicates that the phosphocellulose flow-through fraction contains both of the immunoreactive polypeptides. Subsequent chromatography of this fraction on DEAE-cellulose (Fig. 9A) resolves these two proteins (Fig. 9C). Significantly, only that fraction (DEAE-cellulose bound) which has the TFIIIA-sized protein complements the phosphocellulose step fraction

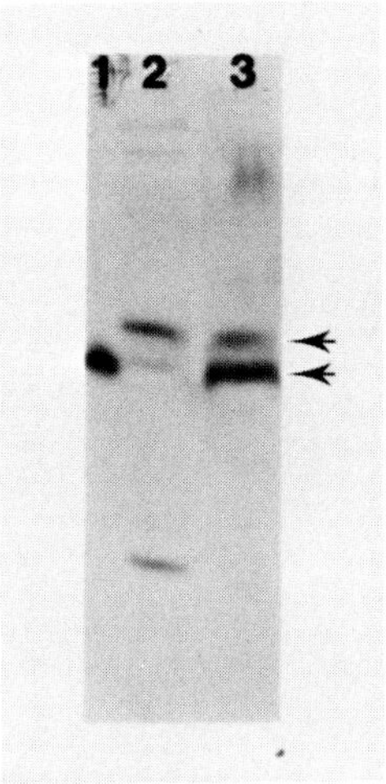

FIGURE 8. Analysis of TFIIIA-related proteins in cultured kidney and juvenile liver cell extracts. Extracts from these cell types were examined as described in Figure 6. The autoradiogram shows: lane 1, purified oocyte TFIIIA marker; lane 2, kidney cell extract (40 μl); lane 3, liver extract (40 μl). The consistently observed bands in liver and kidney extracts are indicated with arrows. The bands of greater mobility apparently represent degradation products.

(Fig. 9B) or unfertilized egg extracts (data not shown); the fraction (DEAE-cellulose flow-through) containing the upper band appears inactive in this regard. These results indicate, therefore, that the TFIIIA-sized protein _is_ active in somatic cells; although not yet established, this protein could be identical to oocyte TFIIIA. The negative result raises the possibility that the upper cross-reactive band is not active in 5S synthesis in somatic cells; but such a conclusion would be premature because of the possibility that the _in vitro_ manipulations (eg. chromatography) may functionally inactivate the protein. Nonetheless, these results ostensibly suggest that the lower band may be identical to or function in a similar fashion as oocyte TFIIIA and that the upper band may have another role, perhaps as a modulator of 5S gene activity. (A possible role in the repression of oocyte-type genes might even be considered.) The resolution of these questions will require the further purification and analysis of these components, including their interactions with oocyte-type versus somatic-type genes.

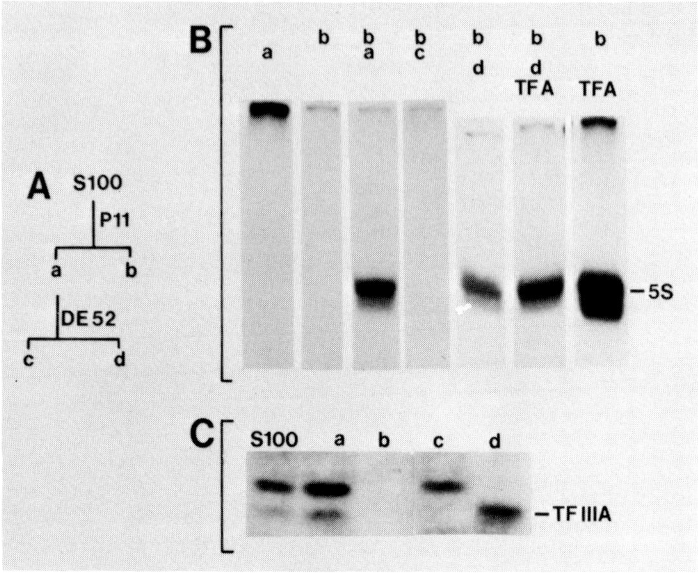

FIGURE 9. Functional and immunological analysis of TFIII-related proteins in Xenopus kidney cell extracts.

Panel A. Flow chart for the fractionation of a kidney cell S100 on phosphocellulose to yield a flow-through (a) and a step fraction (b), and the subsequent fractionation of the (a) flow-through fraction on DEAE-cellulose to yield a flow-through (c) and a step (d) fraction.

Panel B. Autoradiogram showing the RNAs synthesized in the presence of somatic 5S DNA, exogenous RNA polymerase III and the chromatographic fractions (refer to Panel A) indicated above each lane. Purified oocyte TFIIIA was added in the experiments analyzed in the last two lanes.

Panel C. Autoradiogram showing an immunological analysis of TFIIIA-related proteins. Aliquots of the S100 and the resulting chromatographic (indicated above each lane; refer to Panel A) fractions were analyzed as described in Figure 6. The lower band comigrates with oocyte TFIIIA. The same relative amounts of fractions c and d were analyzed in the functional analysis (Panel B) and in the immunological analysis.

These studies still leave unanswered the basis for the selective transcription of somatic-type 5S RNA genes in somatic cells. While it has been tempting to speculate that the somatic cell-specific TFIIIA-related protein (upper band) might be directly responsible (as a new positively-acting factor) for this selective transcription, several facts argue against such a simple explanation. First, we have not been able yet to show an activity for this factor on somatic genes. Second, it is not clear why somatic cells would continue to produce the other "oocyte type" factor. Third, the "somatic" protein has not been detected in gastrula stage embryos which have already begun to selectively accumulate somatic 5S RNA (although this could be a matter of sensitivity of the immunoassay). Finally, somatic cell extracts efficiently transcribe oocyte-type genes (10). Thus it seems most likely that there are additional factors/mechanisms which control the differential expression of these genes *in vivo*. Such controls might include modifications of complex (chromatin) templates.

CONCLUSIONS

The above-described results provide evidence that, for most promoters examined, the basic events involved in the initiation of transcription are faithfully duplicated *in vitro*. The challenge which now remains is to superimpose known *in vivo* regulation on these initiation events in the cell-free systems. The demonstrated ability to isolate and characterize initiation factors, as well as to define those DNA sequences involved in the primary initiation event, indicate the feasibility of these attempts. Indeed, the power of these analyses is best exemplified by our present understanding of the role of TF IIIA in the regulation of 5S gene transcription. Similar studies with other class II and class III genes should greatly aid our understanding of the control of gene expression during differentiation and development.

ACKNOWLEDGMENTS

This work was supported by research grants CA 16640 and CA 23615 from the NCI; by research grant NP 284A from the American Cancer Society; and in part by Cancer Center Support Grant CA 16217 (to Washington University) from the NCI. R.G.R is a Camille and Henry Dreyfus Teacher-Scholar awardee, and D.C.L. is a fellow of the Helen Hay Whitney Foundation.

REFERENCES

1. Segall, J., Matsui, T., and Roeder, R.G. J. Biol. Chem. 255, 11986-11991 (1981).

2. Matsui, T., Segall, J., Weil, P.A., and Roeder, R.G. J. Biol. Chem. 255, 11992-11996 (1981).
3. Grummt, I. Proc. Natl. Acad. Sci. USA 78, 727-731 (1981).
4. Bogenhagen, D.F., Sakonju, S., and Brown, D.D. Cell 19, 27-35 (1980).
5. Corden, J., Wasylyk, B., Buchwalder, A., Sassone-Corsi, P., Kedinger, C., and Chambon, P. Science 209, 1406-1413 (1980).
6. Rio, D., Robbins, A., Myers, R., and Tjian, R. Proc. Natl. Acad. Sci. USA 77, 5706-5710 (1980).
7. Roeder, R.G. in RNA Polymerases (Losick, R. and Chamberlin, M., eds) pp. 285-329, Cold Spring Harbor Laboratory, New York (1976).
8. Ng, S.-Y., Parker, C.S., and Roeder, R.G. Proc. Natl. Acad. Sci. USA 76, 136-140 (1979).
9. Weil, P.A., Luse, D.S., Segall, J., and Roeder, R.G. Cell 18, 469-484 (1980).
10. Weil, P.A., Segall, J., Harris, B., Ng, S.-Y., and Roeder, R.G. J. Biol. Chem. 254, 6163-6173 (1979).
11. Luse, D.S., and Roeder, R.G. Cell 20, 691-699 (1980).
12. Engelke, D.R., Ng, S.-Y., Shastry, B.S., and Roeder, R.G. Cell 19, 717-728 (1980).
13. Sakonju, S., Bogenhagen, D.F., and Brown, D.D. Cell 19, 13-25 (1980).
14. Lee, D.C., and Roeder, R.G. Mol. Cell. Biol., in press (1981).
15. Manley, J.L., Fire, A., Cano, A., Sharp, P.A., and Gefter, M.L. Proc. Natl. Acad. Sci. USA 77, 3855-3859 (1980).
16. Wilson, M.C., Fraser, N.W., Darnell, J.E. Virology 94, 175-184 (1979).
17. Jones, N., and Shenk, T. Proc. Natl. Acad. Sci. USA 76, 3665-3669 (1979).
18. Berk, A.J., Lee, F., Harrison, T., Williams, J., and Sharp, P.A. Cell 17, 935-944 (1979).
19. Nevins, J.R., Ginsberg, H.S., Blanchard, J.-M., Wilson, M.C., and Darnell, J.E. J. Virol. 32, 727-733 (1979).
20. Wasylyk, B., Derbyshire, R., Guy, A., Molko, D., Roget, A., Teoule, R., and Chambon, P. Proc. Natl. Acad. Sci. USA 77, 7024-7028 (1980).
21. Hu, S.-L., and Manley, J.L. Proc. Natl. Acad. Sci. USA 78, 820-824 (1981).
22. Baker, C.C., and Ziff, E.B. J. Mol. Biol., in press (1981).
23. Korn, L.J., and Brown, D.D. Cell 15, 1145-1156 (1978).
24. Pelham, H.R.B., and Brown, D.D. (1980) Proc. Natl. Acad. Sci. USA 77, 4170-4174 (1980).
25. Honda, B.M., and Roeder, R.G. Cell 22, 119-126 (1980).
26. Picard, B., and Wegnez, M. Proc. Natl. Acad. Sci. USA 76, 241-245 (1979).

STUDIES ON THE DEVELOPMENTAL CONTROL OF 5S RNA GENE EXPRESSION

Hugh R.B.Pelham
Daniel F.Bogenhagen
Shigeru Sakonju
W.Michael Wormington
Donald D. Brown

Department of Embryology
Carnegie Institution of Washington
Baltimore, Maryland

ABSTRACT

There are two types of 5S RNA genes in *Xenopus*: oocyte-type genes are transcribed only in oocytes, whereas somatic-type genes are transcribed in all tissues. Transcription of both types of genes in a cell-free system derived from oocytes requires several protein factors, one of which interacts specifically with a control region in the center of the gene. This specific factor mediates the formation of a stable transcription complex in which one or more essential factors remain bound for multiple rounds of transcription.

Somatic cells contain much lower levels of the specific factor than do oocytes. Nevertheless, they contain all the factors necessary for transcription of oocyte-type genes present in genomic DNA purified from somatic cells. However, chromatin isolated from somatic cells directs synthesis of only somatic-type 5S RNA when added to a cell-free system. The repressed oocyte genes in chromatin can be activated by washing the chromatin with 0.6 M NaCl.

Transcription of 5S RNA genes in extracts prepared from *Xenopus* oocytes or somatic cells or

ISBN 0-12-137420-3

from human tissue-culture cells is inhibited by added 5S RNA. This inhibition is due to interaction of 5S RNA with the specific transcription factor. This interaction may be the basis of a feedback regulatory mechanism for 5S RNA synthesis.

INTRODUCTION

Oocytes of *Xenopus* and other amphibia accumulate ribosomes at least a thousand times more rapidly than the most active somatic cell. These ribosomes are stored for use during embryogenesis. A high rate of synthesis of 18S and 28S rRNA is made possible by a thousand-fold amplification of their structural genes in oocytes, so that each oocyte contains about two million copies of these genes (1,2). The 5S rRNA genes are not amplified. Instead, the oocytes transcribe a large family of genomic 5S RNA genes that are not expressed in somatic cells (3); these are the so-called oocyte-specific or oocyte-type 5S RNA genes. Even with these extra copies, however, the 5S RNA genes are outnumbered twenty to one by the amplified 18S and 28S rRNA genes. This imbalance is compensated by an unusual mechanism. Immature oocytes synthesize and store 5S RNA and tRNA for about two months prior to the onset of 18S and 28S rRNA synthesis. Ribosomes are then assembled using both newly-synthesized and stored 5S RNA (4). Much of the 5S RNA in immature oocytes is stored in the form of a 7S cytoplasmic particle which contains one molecule of 5S RNA bound to a single protein with a molecular weight of about 40,000 (5). This protein comprises 10-15% of the soluble protein in immature ovaries, and plays a major role in 5S RNA metabolism.

The oocyte and somatic 5S RNA genes from two species of *Xenopus*, *X. laevis*, and *X. borealis*, have been studied in detail (6-9). Both types of genes are found in long tandem arrays, separated by several hundred base pairs of spacer DNA. A somatic cell of *X. laevis* contains some 40,000 oocyte-type genes, compared to about 800 somatic-type genes; yet at least 95% of the RNA made in these cells is of the somatic type. Thus during development of an oocyte into an embryo, the oocyte

5S RNA genes are repressed; their level of expression becomes, on average, less than one thousandth of that of the somatic genes. To understand the mechanism of this specific developmental control is the ultimate aim of our research.

Somatic and oocyte genes from both species of _Xenopus_ have been cloned and sequenced (6-9). The genes differ by a few base changes, which allow the different RNAs to be distinguished by fingerprinting. The most striking differences, however, lie in the spacer regions, which are GC-rich in the somatic families and very AT-rich in the oocyte families. It is possible that this difference is an essential part of the developmental control mechanism.

Cloned copies of each type of _Xenopus_ 5S RNA gene are faithfully and efficiently transcribed in a cell-free extract prepared from the nuclei of mature oocytes (10). To define the DNA sequences that are essential for transcription, deletion mutants of one repeating unit of _X. borealis_ somatic 5S DNA were constructed and assayed for template activity. The only deletions that abolish transcription completely are those that extend from either side into a region in the center of the gene (between residues 50 and 83) which we term the control region (11,12). Provided this region is intact, RNA polymerase III will initiate transcription 50 ± 5 bases to the 5' side. Similar deletion analysis has defined the sequence responsible for termination of transcription at the 3' end of the gene (13).

Engelke _et al_. (14) have isolated a protein from _Xenopus_ ovary that is required for 5S RNA synthesis in a partially fractionated system. This protein binds to the control region in 5S DNA. Collaborative experiments have shown a correlation between the ability of various deletion mutants to bind this protein, and their ability to be transcribed _in vitro_ (15). This correlation suggests that binding of the protein to the control region is an essential step in transcription.

This protein which acts as a transcription factor is identical to the protein in immature oocytes which is bound to 5S RNA in the 7S storage particle (16, 17). We were originally led to this finding by the observation that addition of 5S RNA to the cell-free extract specifically inhibits transcription of 5S RNA genes, but not of other genes transcribed by polymerase III such as tRNA genes. This inhibition is due to interaction of the RNA with the transcription factor, which prevents it from binding to 5S DNA (16). Transcription of other genes is unaffected, because the factor does not bind to them and is not required for their activity. We suggested (16) that binding of 5S RNA to this factor, which is specifically required for its synthesis, may be the basis of a mechanism for feedback regulation of 5S RNA gene expression.

RESULTS AND DISCUSSION

A Difference in Transcription of Oocyte and Somatic Genes

Although the control region was first identified in a somatic gene, the same basic mode of expression applies also to the oocyte genes. However, a more quantitative assay has revealed a difference between the oocyte and somatic genes.

The relative "strength" of different 5S genes can be estimated from their ability to outcompete each other when added simultaneously to the transcription system (18). In order to distinguish the transcripts from the two genes, it is convenient to insert extra DNA into one of the genes outside the control region, so that its transcript migrates differently during gel electrophoresis. Fig. 1 shows such a competition experiment between oocyte and somatic genes. It can be seen that 5 μg/ml of a cloned somatic gene was sufficient to reduce transcription of the same amount of a cloned oocyte gene by 50%, whereas 40 μg/ml of cloned oocyte gene was required to inhibit transcription of a somatic gene (at 5 μg/ml) by 50%. To determine which part of the DNA sequence is responsible for this relative weakness of the oocyte genes, we constructed hybrid genes in which the control region of an oocyte gene was replaced by the corresponding region from a somatic gene and *vice versa* (18). The competitive strength of these genes is determined entirely by their control regions. Thus as few as four base changes within the control region can have a quantitative effect on the transcription of a 5S RNA gene. Direct binding studies with purified factor and DNA have confirmed that the factor binds more tightly to somatic than to oocyte genes, although this difference is small compared to the observed specificity of transcription in somatic cells.

Formation of a Stable Transcription Complex

Despite the obvious importance of the control region and the factor that binds to it, the relative activity of two 5S genes can be affected by other variables. In the competition experiments described above, the two DNAs were added simultaneously to the transcription system. A very different result is obtained if one DNA is

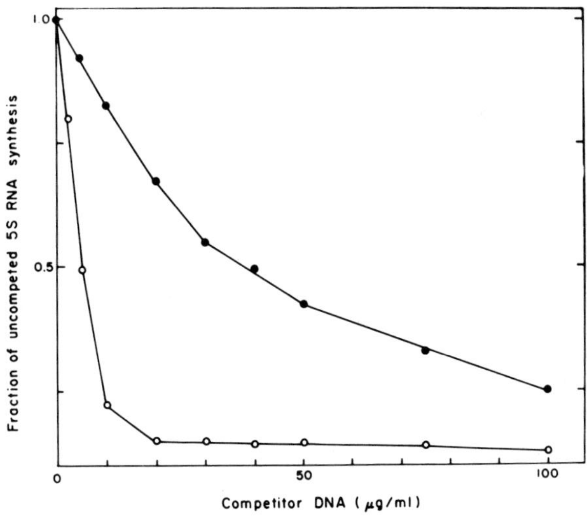

FIGURE 1. Competition between oocyte and somatic genes. Mixtures of plasmids were transcribed in the oocyte extract. Solid symbols: effect of increasing concentrations of oocyte gene on transcription of 5 μg/ml somatic gene. Open symbols: effect of increasing concentrations of somatic gene on transcription of 5 μg/ml oocyte gene.

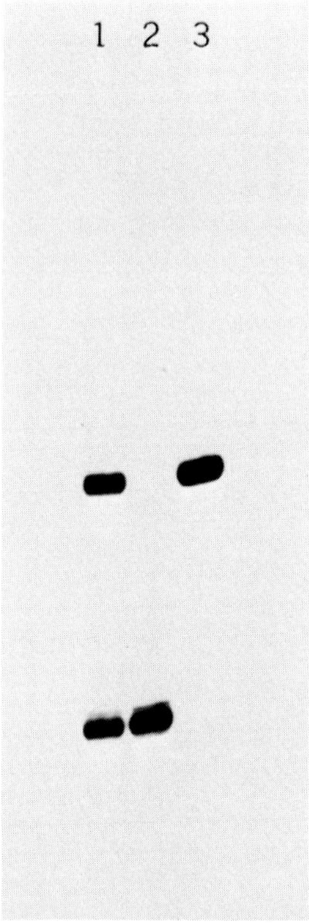

FIGURE 2. Formation of a stable transcription complex. Transcription products were separated on a gel and detected by autoradiography. Templates were as follows: **lane 1**, equal amounts of a somatic gene and a somatic "maxigene", containing an insert at position 115, mixed before transcription; **lane 2**, maxigene added 15 minutes after the other gene; **lane 3**, maxigene alone. Final DNA concentrations were 20 μg/ml for each template. In each case label was added 30 minutes after the start of the incubation.

preincubated in the extract before the second DNA is added (fig. 2). In this case, the first DNA is preferentially transcribed. Addition of a sufficient quantity of the first DNA (15 μg/ml or more) can completely prevent transcription of the second DNA. This effect persists for several hours. The second DNA remains inactive even after each original gene has been transcribed, on average, over thirty times. This implies that a stable transcription complex is formed on the first DNA and persists through many rounds of transcription. Evidently some component of this complex is limiting in the oocyte extract, and thus is not available to the second DNA.

Formation of a transcription complex on a 5S RNA gene requires binding of the 5S-specific factor as an initial step. This conclusion is derived from experiments in which the factor is first removed from the cell-free system with Sepharose-bound antibodies, and subsequently added back. Preincubation of DNA in the extract without factor does not increase its subsequent transcription activity. A requirement for the specific factor is also suggested by the observation that oocyte-type genes form complexes less efficiently than somatic genes. The complex contains more than just the specific factor, however, since addition of this purified factor is not sufficient to induce efficient transcription of the second DNA. The complex probably contains factors that are not gene-specific, because competition between 5S genes and tRNA genes is also affected by preincubation of one of the DNAs. At least two such factors, in addition to polymerase III, are known to be required for accurate transcription (19).

Transcription of preincubated 5S RNA genes is resistant to inhibition by 5S RNA (16), which binds to free factor and prevents its interaction with DNA. This implies that once the transcription complex is formed, the 5S-specific factor either remains bound in the complex (despite transcription through the control region), or possibly is no longer required for transcription.

The existence of stable transcription complexes has several implications regarding possible control mechanisms. For example, feedback regulation by 5S RNA, if it occurs *in vivo*, would not be a minute-by-minute control of the *level* of expression of each gene, but rather a long-term control of the number of genes that are active. This type of control may be very useful for a multigene family in which the total number of genes varies between individuals (and perhaps between cells) due to unequal crossing-over, since the amount of transcription would be independent of the total number of genes present.

It is also clear that discrimination between oocyte and somatic genes in somatic cells does not have to occur at each round of transcription -- once established, the control could be quite stable. Furthermore, the unidentified factors that must be present in the transcription complex could accentuate the discrimination between oocyte and somatic genes by interacting with sequences outside the control region. It is not surprising that such discrimination does not show up in the oocyte cell-free system, since both types of genes are transcribed in oocytes in vivo. We therefore prepared extracts from somatic (tissue-culture) cells, and investigated their ability to support transcription of 5S RNA genes in vitro.

Transcription of 5S RNA Genes in Extracts from Somatic Cells

Extracts were prepared from Xenopus tissue-culture cells by the method of Manley et al. (20). Transcription of partially deleted genes showed that these extracts had the same requirement for the control region as the oocyte extract. Similar results were obtained with extracts of human KB cells, suggesting a conserved transcription mechanism in vertebrates. Both cloned oocyte and

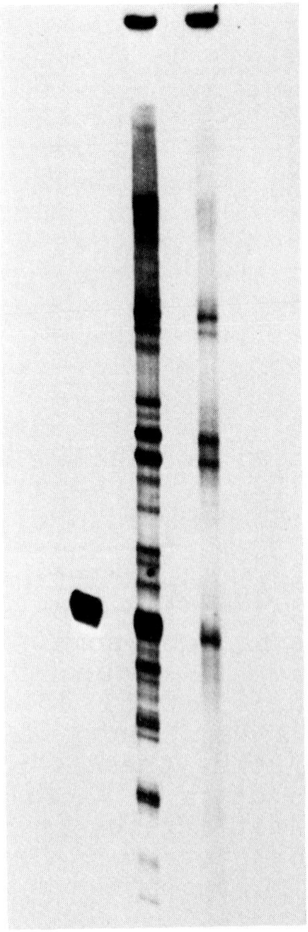

FIGURE 3. Transcription of genomic DNA. Radioactive transcripts were separated on a polyacrylamide gel and detected by autoradiography. Total erythrocyte DNA was transcribed in the oocyte extract (center lane) or in the somatic cell extract (right hand lane). The left hand lane contained a 5S RNA marker.

somatic genes were transcribed in somatic extracts, and their relative efficiencies were similar in the oocyte and somatic systems, although the overall efficiency of transcription in the latter was considerably lower. However, one could argue that discrimination between oocyte and somatic genes requires that they be present in tandem arrays, or that the DNA is modified in some way, for example by methylation, or that DNA sequences adjacent to the tandem arrays are involved. We therefore prepared genomic DNA from nucleated erythrocytes and transcribed it in both cell-free extracts (fig. 3). In both oocyte and somatic extracts 5S RNA was a prominent product, and fingerprinting showed it to be principally oocyte-type RNA in both cases. We conclude that the oocyte-type genes are not irreversibly inactivated by covalent modification in somatic cells, and that these cells contain all the factors necessary for transcription of these genes. Thus control *in vivo* must involve repression of the oocyte genes. If there are specific signals such as methyl groups on the DNA, they are not recognized by our assay system. In fact, both oocyte and somatic 5S RNA genes are methylated in somatic cells, and we have found no significant differences in the extent of their modification (unpublished observations).

The requirement for the control region in somatic extracts suggested that somatic cells contain a factor functionally equivalent to the one identified in oocytes. This equivalence apparently extends to the RNA binding property of the factor, since 5S RNA specifically inhibits transcription of 5S RNA genes in extracts prepared either from *Xenopus* tissue-culture cells or from human KB cells (Table 1). The existence of this phenomenon in cells which are not known to store 5S RNA in particle form supports the idea that interaction of 5S RNA with the factor has some function of general importance, such as in feedback regulation, and is not merely an adaptation of the factor for its unique storage function in oocytes. One might imagine that binding of the factor to 5S RNA is an essential step in the transcription process, facilitating release of the product, for example. This does not seem to be the case, however, because several altered 5S RNA genes are transcribed and compete with normal efficiency even though their

TABLE 1. Inhibition of transcription by 5S RNA[a]

Extract	Addition	Synthesis(% control) 5S RNA	tRNA
Xenopus somatic	none	(100)	(100)
	10 μg/ml 5S RNA	13	73
	10 μg/ml tRNA	63	44
Human KB cell	none	(100)	(100)
	10 μg/ml 5S RNA	21	112
	10 μg/ml tRNA	69	68

[a]Mixtures of cloned 5S RNA and tRNA genes were transcribed, total DNA concentration being 5 μg/ml.

transcripts are not able to bind to the factor. This can be assayed conveniently by immunoprecipitation of the endogenous factor in the oocyte cell-free system at the end of a transcription reaction. Typically, 20% of the transcripts from either oocyte or somatic genes are bound to the endogenous factor and hence are precipitated by antibody. Transcripts from a cloned oocyte pseudogene (7,18) or a somatic gene with a 20-base insertion after residue 40 (11) remain entirely in the supernate after immunoprecipitation (unpublished observations).

Related Transcription Factors in Somatic Cells and Oocytes

As a further comparison between the oocyte and somatic cell-free systems, we tested the effects of antibodies raised against highly-purified oocyte factor on transcription in the two systems. Fig. 4 shows that for both extracts, preincubation with the antibody caused a significant inhibition of 5S transcription, while synthesis of tRNA in the same reaction was actually stimulated (due to relief of competition from 5S RNA genes). Antibodies prepared from control serum had no effect. These results confirm the importance and specificity of the oocyte factor, and imply that the somatic extract contains a factor that is antigenically related to the oocyte one. We have looked in somatic tissues for proteins that crossreact with the antibodies using a gel transfer technique (21,22). The most prominent such protein in tissue-culture cells and liver migrates more slowly than the oocyte factor on both SDS and acid-urea gels, but like the oocyte factor binds tightly to heparin-agarose. Somatic cells also contain a smaller amount (about one tenth as much) of a protein that comigrates with the oocyte factor. It is not clear whether there is enough of this protein to account for the observed level of 5S RNA synthesis in the somatic cell-free system (22). The larger protein is thus a good candidate for a second type of 5S-specific transcription factor.

The abundance of 5S transcription factor in oocytes and somatic cells is dramatically different (17,22). A mature oocyte nucleus contains about 2 ng of factor, approximately one million times as much as a somatic cell. Thus, a somatic cell contains about ten thousand molecules of factor, or about ten molecules per somatic 5S RNA gene. Much of the oocyte factor (about 95%) is lost during

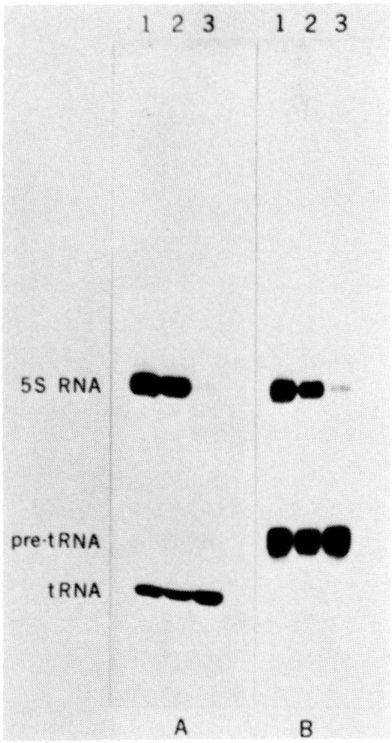

FIGURE 4. Effect of antibodies on transcription. Mixtures of cloned 5S RNA and tRNA genes were transcribed in the oocyte extract (A) or the somatic cell extract (B). The IgG fraction from antiserum raised against the purified oocyte transcription factor was added to the reactions 15 minutes before the template. **Lane 1**, no IgG; **lane 2**, preimmune IgG; **lane 3**, immune IgG.

maturation of the oocyte into an egg, the rest being further diluted out during development (22). In our somatic cell extracts, this factor is limiting, since addition of purified oocyte factor greatly stimulates transcription of added 5S DNA.

There is a general correlation between the amount of 5S-specific factor present in oocytes and somatic cells and the amount of 5S RNA synthesized. However, the factors present in somatic cells appear to be functionally equivalent (though perhaps not identical) to those in oocytes, so the repression of oocyte-type genes in somatic cells cannot be explained merely by a specificity of such factors for the somatic genes.

Transcription of Chromatin

Since we found no more than a four-fold discrimination between oocyte and somatic 5S RNA genes using purified DNA and soluble factors, we have begun to study the template activity of these genes in chromatin form. Chromatin prepared from *Xenopus* tissue-culture cells has no transcription activity on its own, but when added to a cell-free extract from either oocytes or somatic cells, it directs the synthesis of a small amount of 5S RNA (fig. 5). Fingerprinting shows this RNA to be principally of the somatic type. The oocyte genes remain inactive even in the presence of a large excess of soluble factors. Thus we have an assay which preserves the repression of oocyte genes that is observed *in vivo*.

The oocyte genes remain repressed even after the chromatin has been washed with 0.35 M NaCl (fig. 5), a treatment that removes many non-histone proteins and has been shown to abolish the difference between DNaseI-sensitive and -insensitive genes in other systems (23). Washing the chromatin for ten minutes with 0.6M NaCl or more activates all the genes that are transcribed from purified DNA. These *in vitro* results are basically in agreement with those obtained by Korn and Gurdon (24), who found that NaCl treatment could activate synthesis of oocyte-type 5S RNA by somatic cell nuclei injected into mature oocytes. These experiments are the first steps toward identifying the nature of oocyte-type 5S RNA gene repression *in vivo*.

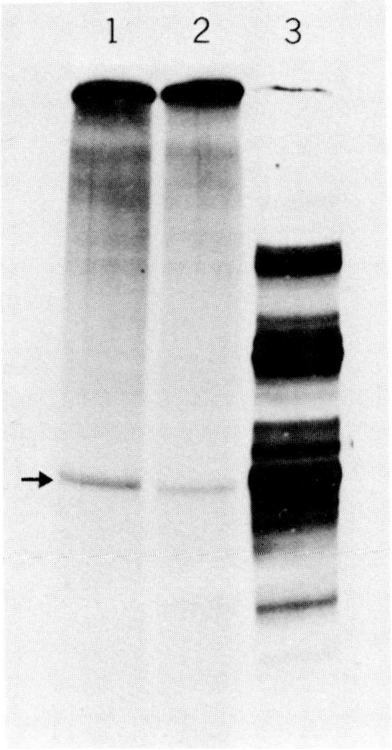

FIGURE 5. Transcription of chromatin. Intact chromatin was prepared from Xenopus tissue-culture cells (23) and added to the oocyte cell-free system. **Lane 1**, chromatin prepared in low salt; **lane 2**, chromatin washed with 0.35 M NaCl; **lane 3**, chromatin washed with 2 M NaCl (the same pattern was observed when chromatin was washed for ten minutes with 0.6 M NaCl). The arrow indicates the position of 5S RNA.

CONCLUSIONS

The dual 5S RNA gene system in Xenopus provides a well-defined transcriptional control mechanism that is relatively easy to study. One specific transcription factor has already been identified, and its role is being elucidated. Specific interaction of 5S RNA with this protein factor suggests a possible mechanism for quantitative control of 5S RNA synthesis.

It is becoming clear from this work that stable complexes of DNA and protein are a key feature of 5S RNA gene expression. The transcription complex stably activates a gene for many rounds of transcription. Formation of this complex is dependent on binding of the specific transcription factor to the center of the 5S RNA gene. The complex probably contains other proteins, although some factors, such as RNA polymerase III, may dissociate from the complex between rounds of transcription. The transcription complex can be assembled from its components *in vitro*, and is amenable to purification and further study.

Another kind of stable DNA-protein complex is responsible for repressing the oocyte-type genes in somatic cell chromatin. We do not know whether this complex contains only common chromatin proteins such as histones, or whether a specific repressor is present. We have not yet been able to assemble this type of complex in vitro, but these preliminary experiments provide a functional assay for its controlled disassembly.

The existence of these stable complexes does not in itself account for the discrimination between the oocyte and somatic 5S RNA genes. However, the stability of the active and repressed states suggests that such discrimination need not be continuous. If daughter cells can somehow inherit the chromatin structure of their parent, the discrimination may only have to occur at a single brief moment during development. Perhaps when we can reassemble a specifically repressed gene we will begin to understand this process.

ACKNOWLEDGMENTS

We thank E. Jordan for expert technical assistance. This work was supported in part by a grant from the National Institutes of Health. D.F.B. is a fellow of the Helen Hay Whitney Foundation; W.M.W. is an NIH postdoctoral fellow.

REFERENCES

1. Gall, J.G. (1968). Proc. Natl. Acad. Sci. USA **60**, 553-560.
2. Brown, D.D. and Dawid, I.B. (1968). Science **160**, 272-280.
3. Brown, D.D. (1979), in "Mechanisms of Cell Change" (J.D.Ebert and T.S.Okada,eds. John Wiley, New York.
4. Mairy, M. and Denis, H. (1972). Eur. J. Biochem. **25**, 535-543.
5. Picard, B. and Wegnez, M. (1979). Proc. Natl. Acad. Sci. USA **76**, 241-245.
6. Fedoroff, N.V. and Brown, D.D. (1978). Cell **13**, 701-716.
7. Miller, J.R., Cartwright, E.M., Brownlee, G.G., Fedoroff, N.V. and Brown, D.D. (1978). Cell **13**, 717-725.

8. Korn, L.J. and Brown, D.D. (1978). Cell **15**, 1145-1156.
9. Peterson, R.C., Doering, J.L. and Brown, D.D. (1980). Cell **20**, 131-141.
10. Birkenmeier, E.H., Brown, D.D. and Jordan, E. (1978). Cell **15**, 1077-1086.
11. Sakonju, S., Bogenhagen, D.F. and Brown, D.D. (1980). Cell **19**, 13-25.
12. Bogenhagen, D.F., Sakonju, S. and Brown, D.D. (1980). Cell **19**, 27-35.
13. Bogenhagen, D.F. and Brown, D.D. (1981). Cell **24**, 261-270.
14. Engelke, D.R., Ng, S.Y., Shastry, B.S. and Roeder, R.G. (1980). Cell **19**, 717-728.
15. Sakonju, S., Brown, D.D., Engelke, D., Ng, S-Y, Shastry, B.S. and Roeder, R.G. (1981). Cell **23**, 665-669.
16. Pelham, H.R.B. and Brown, D.D. (1980). Proc. Natl. Acad. Sci. USA **77**, 4170-4174.
17. Honda, B.M. and Roeder, R.G. (1980). Cell **22**, 119-126.
18. Wormington, W.M., Bogenhagen, D.F., Jordan, E. and Brown, D.D. (1981). Cell **25**, in press.
19. Segall, J., Matsui, T. and Roeder, R.G. (1980). J. Biol. Chem. **255**, 11986-11991.
20. Manley, J.L., Fire, A., Cano, A., Sharp, P.A. and Gefter, M.L. (1980). Proc. Natl. Acad. Sci. USA **77**, 3855-3859.
21. Towbin, H., Staehelin, T. and Gordon, J. (1979). Proc.Natl.Acad.Sci. USA **76**, 4350-4354.
22. Pelham, H.R.B., Wormington, W.M. and Brown, D.D. (1981). Proc. Natl. Acad. Sci. USA **78**, in press.
23. Weisbrod, S., Groudine, M. and Weintraub, H. (1980). Cell **19**, 289-301.
24. Korn, L.J. and Gurdon, J.B. (1981). Nature **289**, 461-465.

INFLUENCE OF 5' FLANKING SEQUENCES ON tRNA TRANSCRIPTION *IN VITRO*

Stuart G. Clarkson[1]
Raymond A. Koski[2]
Janine Corlet
Robert A. Hipskind[1]

Department of Microbiology
University of Geneva Medical School
Geneva, Switzerland

ABSTRACT

Two *Xenopus laevis* methionyl-tRNA genes, which differ at a single position intragenically and at several positions in their flanking regions, are transcribed with very different efficiencies in a homologous cell-free system. By sequence manipulation, we show that this is due to their 5' flanking regions and, in particular, that sequences somehow inhibitory for transcription normally precede the relatively inactive gene. Possible candidates for such inhibitory sequences are discussed.

---

[1]Supported by Swiss National Science Foundation Grant 3-418.78.
[2]Supported by NIH postdoctoral fellowship 1 F32 GM06510-01.
Present address: Department of Genetics, University of Washington, Seattle, WA 98195, USA.

ISBN 0-12-137420-3

INTRODUCTION

One of our prejudices is that the organization of genes in multigene families is important for their expression and is not merely an evolutionary accident. An intriguing example of a complex multigene family is provided by a 3.18kb fragment of *X.laevis* DNA that is tandemly repeated about 150-fold. This repeat unit was originally identified by its hybridization to initiator methionyl-tRNA (1,2) but more recent sequence analysis has revealed that it potentially contains 8 tRNA genes (3,4). These are distributed over both DNA strands in no obviously ordered way (Fig. 1). Two of the genes are very similar but not identical: the Met-A gene codes for the $tRNA^{met}_{f}$ found in oocytes and somatic cells of *X.laevis* (5) and indeed in all vertebrates so far examined (6). In contrast, the Met-B gene differs at a single position intragenically (Fig. 2) and at many positions in its flanking sequences (4; Fig. 6). Met-B transcripts have not been detected *in vivo* (5), suggesting that this gene may not normally be functional. We have investigated this question with the aid of a homologous cell-free transcription system (7). Here we show that the Met-A gene is transcribed *in vitro* more efficiently than the Met-B gene, and we describe our preliminary attempts to characterize the DNA sequences that must be in part responsible for the differential expression of these two genes.

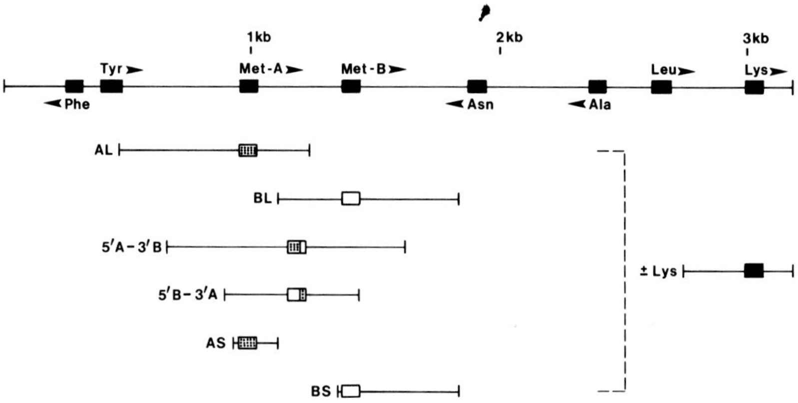

Figure 1. Organization and subclones of the 3.18kb *X.laevis* tDNA fragment.

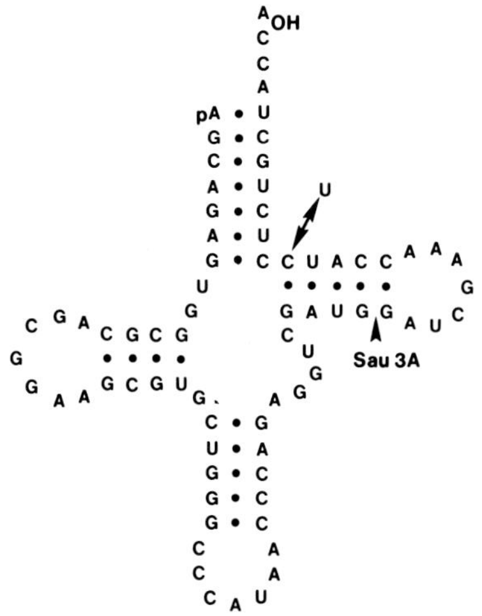

Figure 2. The unmodified sequence of *X.laevis* $tRNA^{met}_{f}$. Nucleotide 64 is a C in the Met-A gene and a T in the Met-B gene. The indicated *Sau*3A sites (GATC in the DNA) were used to construct the hybrid genes shown in Fig. 1.

METHODS

The subclones indicated in Figure 1 were prepared from the following restriction fragments, making use of their known map positions (2). Subclone AL contains the Met-A gene within the 773bp *Hpa*II/A fragment, and BL the Met-B gene within the 721bp *Hinf*I/A fragment. Each of these fragments was cleaved at its single intragenic *Sau*3A site (Fig. 2). The resulting subfragments were then exchanged and ligated to yield the hybrids 5'A-3'B and 5'B-3'A. AS comprises the 181bp *Hinf*I/G fragment and BS a 472bp *Pvu*II digestion product of BL. All 6 fragments were inserted, via *Eco*RI linkers, into the *Eco*RI site of the plasmid pBR322 (8). They were also inserted into the *Eco*RI site of the Lys subclone (Fig. 1), which comprises the 442bp *Eco*RI/B fragment inserted between the *Eco*RI and *Hin*dIII sites of pBR322. In all cases, the direction of transcription of the inserted genes was clockwise according to the conventional pBR322 map (8).

Supercoiled plasmid DNAs were transcribed with a cytoplasmic extract (7) of *X.laevis* tissue culture cells as previously described (9). The RNA, labelled with α-^{32}P-GTP, was fractionated at 4°C on 10% acrylamide, 7M urea gels (9).

RESULTS

The Met-A Gene Transcriptional Unit

The *in vitro* initiation and termination sites of transcription of the Met-A gene have been determined with subclone AL, which contains this gene together with 489bp of its 5' and 213bp of its 3' flanking sequences. It yields three major RNA products, p1, p2 and m, none of which are found after a control reaction with pBR322 DNA (cf. lanes b and a, Fig. 3). These RNAs have been sequenced and shown to result from the selective transcription of the Met-A gene by RNA polymerase III (10).

p1 is a tRNA precursor of 86-87 nucleotides which starts with a pppG 7 nucleotides upstream of the 5' end of the mature tRNA sequence. It terminates within a $(dT)_5$ tract just after the gene, yielding molecules with mostly 3 or 4 residues at their 3' ends. p2 is a second primary transcript of 83-84 nucleotides that terminates at the same points but initiates with a pppA 4 nucleotides upstream of the mature 5' end. These two initiation sites are used with approximately equal frequency with subclone AL as template. This is not obvious from

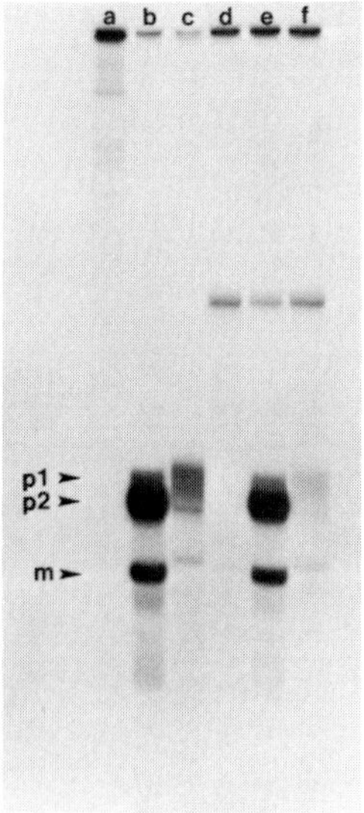

Figure 3. *In vitro* transcription products of the Met-A and Met-B genes. p1 and p2, two precursors to m, mature $tRNA^{met}_{f}$, synthesized from subclone AL. Lanes: a, pBR322; b, AL; c, BL; d, Lys; e, AL+Lys; f, BL+Lys.

Figure 3 which shows the result of a 2hr *in vitro* reaction. Within this time some p1 molecules have been partially processed so that they co-migrate with the p2 precursor. In addition, some molecules of p1 and p2 have been completely matured by 5' and 3' processing enzymes, nucleotide modification enzymes, and tRNA nucleotidyl transferase to yield m, the 75 nucleotide mature $tRNA^{met}_{f}$.

The Met-B Gene Is Transcribed Less Efficiently

The *in vitro* transcription products of subclone BL, which contains the Met-B gene together with 275bp of its 5' and 395bp of its 3' flanking sequences, are shown in Figure 3. The transcripts migrate in the 4-5S RNA region of the gel but the bands are much less intense and apparently are longer than the Met-A transcripts. Sequence analysis of these RNAs has not yet been completed but the available evidence suggests that they comprise tRNA precursors, and their processing products, that were initiated and terminated within a few nucleotides of the Met-B gene sequence.

The reduced transcription of the Met-B gene is not due to inhibitory contaminants of the DNA because a double recombinant of the Met-B and Lys genes still produces a $tRNA^{lys}$ precursor of ~160 nucleotides (cf. lanes d and f, Fig. 3). Two further points emerge from *in vitro* reactions with such double recombinants. Transcription of both the Met-A and Lys genes is reduced with the two genes on the same plasmid (cf. lanes b, d and e, Fig. 3), suggesting that they compete for a limiting factor in the transcription extracts. In contrast, the Met-B + Lys subclone yields normal levels of pre-$tRNA^{lys}$ whereas transcription of the Met-B gene is reduced even further (cf. lanes c, d and f, Fig. 3), thereby suggesting that this gene is a less effective competitor for this factor(s).

5' Flanking Sequences Influence Transcription Efficiency

If the reduced intensity of Met-B transcripts is indeed due to less efficient transcription of this gene, rather than instability of the transcripts, then this must reflect the single intragenic base change and/or the different sequences flanking the Met-A and Met-B genes. To distinguish between these possibilities, the two genes were first cleaved at a *Sau*3A site which is 5' to the intragenic base change (Fig. 2). The flanking regions and half-genes were exchanged, recloned in both pBR322 and the Lys subclone (Fig. 1), and were then tested in the cell-free reactions.

The hybrid recombinant 5'A-3'B is transcribed as efficiently as the Met-A subclone AL (cf. lanes d and b, Fig. 4). Hence the presence of the C→T change within the gene does not impair its transcription nor does it lead to instability of the transcripts. Similarly, the 5'B-3'A hybrid is transcribed no better than the Met-B gene (cf. lanes e and c, Fig. 4), confirming that the presence of the normal C at position 64 of the gene does not ensure its efficient transcription. More importantly, the 5'B-3'A hybrid is transcribed much less efficiently than the Met-A gene (cf. lanes e and b, Fig. 4). Their sole difference is the nature of their 5' flanking sequences. We conclude that it is this region that brings about the differential expression *in vitro* of the Met-A and Met-B genes.

Although the intragenic base change has no effect on transcription, it does appear to induce a conformational change in the transcripts. Thus the 5'B-3'A hybrid yields a RNA that co-

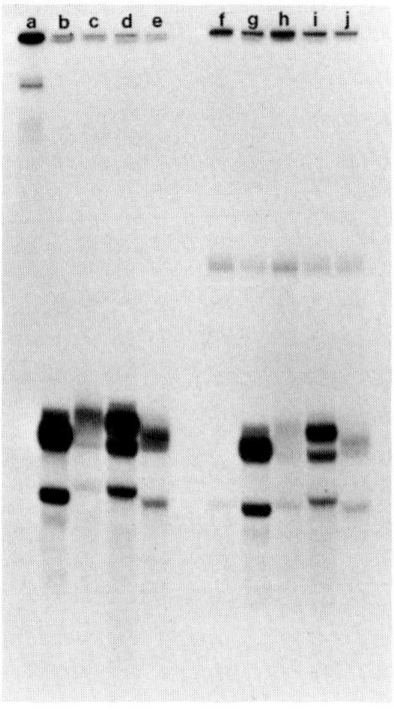

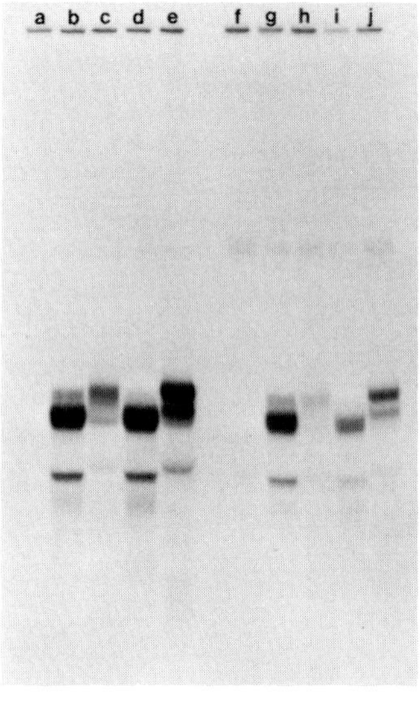

Figure 4 (above left). *In vitro* transcription products of the hybrid genes. Lanes: a, pBR322; b, AL; c, BL; d, 5'A-3'B; e, 5'B-3'A; f, Lys; g, AL+Lys; h, BL+Lys; i, 5'A-3'B+Lys; j, 5'B-3'A+Lys.

Figure 5 (above right). *In vitro* transcription products of the Met-A and Met-B genes with deleted 5' flanking sequences. Lanes: a, pBR322; b, AL; c, BL; d, AS; e, BS; f, Lys; g, AL+Lys; h, BL+Lys; i, AS+Lys; j, BS+Lys.

migrates with mature $tRNA^{met}_{f}$ whereas the corresponding band from the 5'A-3'B hybrid and from the Met-B gene has slightly slower mobility (Fig. 4).

Sequences Inhibitory For Transcription Precede The Met-B Gene

The 5' flanking regions presumably exert their influence in one of two ways: either sequences essential for transcription precede the Met-A gene, or sequences inhibitory for transcription precede the Met-B gene. A partial answer to this question has already been provided by the demonstration that the Met-A gene can still be efficiently transcribed after its micro-injection into *X.laevis* oocyte nuclei when it contains only 22bp of its 5' flanking sequence (11). The same result is obtained with the homologous *in vitro* transcription system: the characteristic Met-A transcripts are produced from subclone AS (Fig. 1), although the G start at -7 appears to be used less frequently (cf. lanes b and d, Fig. 5). We conclude that the sequences upstream of -22 influence the choice of initiation site but are not essential for efficient transcription of the Met-A gene.

To test the corollary, a subclone was prepared that contains the Met-B gene with just 8bp of its 5' flanking sequence (subclone BS, Fig. 1). This elimination of 249bp of upstream sequences results in a stimulation of transcription of the Met-B gene (cf. lanes c and e, Fig. 5). We conclude that sequences somehow inhibitory for transcription normally precede the Met-B gene.

DISCUSSION

The efficient *in vitro* transcription of the Met-B gene from a clone containing only 8bp of its 5' flanking sequence adds to the growing body of evidence suggesting that eukaryotic tRNA genes contain sequences essential for their own expression (9,11-16). This is emerging as a general feature of transcription by RNA polymerase III for such "intragenic control regions" have been well documented for *X.laevis* 5S RNA genes (17,18) and for those encoding adenovirus VA RNAs (19). Indeed, some weak sequence homologies can be detected within these three classes of genes (9,19). While their transcription appears to be promoted by different protein factors (20,21), it is possible that these intragenic homologies have transcriptional significance in reflecting, for example, a common protein-DNA recognition mechanism.

The boundaries of the control region within tRNA genes have not yet been determined but two observations suggest the particular importance of sequences in the 3' portions of these genes. First, while neither the 5' nor 3' half of the *X.laevis* Met-A gene is capable of being transcribed, only the 3' half can effectively compete out transcription of the intact gene (14). Second, the only point mutations so far found to reduce *in vitro* transcription of a yeast *SUP*4 $tRNA^{tyr}$ gene are located in its 3' half (9).

The G→T difference between the *X.laevis* Met-A and Met-B genes is clearly not another example of a mutation within an intragenic control region. Instead, the differential expression of these two genes is due to their 5' flanking regions and, in particular, is due to sequences lying upstream of the Met-B gene that exert a negative control over its expression. Removal of all the sequences preceding a *Drosophila melanogaster* $tRNA^{lys}$ gene similarly leads to more efficient *in vitro* transcription than if they are present (15). However, these two inhibitory 5' flanking regions contain no obvious sequence homologies (15; Fig. 6).

```
-70       -60       -50       -40       -30       -20       -10  ▼  ▼-1
ATGTTGGAAACATTAGCTGCATGTCTTAGAAGCAACGTAAAGGAAGAGAATCACCCAAAGGACGGCAATC:  AL
                                                ↑
                                              Hinf I

CCTATAAAAATAGGCGTATCACGAGGCCCTTTCGTCTTCAAGAATTCcAATCACCCAAAGGACGGCAATC:  AS
                                        ↑
                                      EcoRI

GTTATTGCCCACCCCCACCCCCATACAATGCACAGCGCAGCAAGAATACCTGCGCGTGCCAGCTGCGATT:  BL
                                                              ↑
                                                            Pvu II

TATCATGACATTAACCTATAAAAATAGGCGTATCACGAGGCCCTTTCGTCTTCAAGAATTCcCTGCGATT:  BS
                                                       ↑
                                                     EcoRI
```

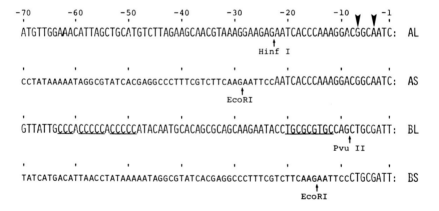

Figure 6. 5' flanking regions of normal and truncated Met-A and Met-B genes. Sequences are numbered negatively from the 5' nucleotide of mature $tRNA^{met}_{f}$. The non-coding strand is shown in each case. pBR322 and *Eco*RI linker sequences are set in lower case letters for subclones AS and BS; the restriction sites used in their construction are shown in AL and BL. The two possible *in vitro* transcription initiation sites are arrowed above the AL sequence. An extensive stretch of dC residues and a 9bp alternating purine-pyrimidine tract are underlined in the BL sequence.

At least one case is known of positive control: transcription of a *Bombyx mori* $tRNA^{ala}_{2}$ gene requires the presence of the normal 5' flanking sequences in a homologous cell-free system (16). In addition, there exist several examples of transcription efficiency being unimpaired when 5' flanking regions are replaced by various plasmid sequences (11-13,17). Together, these results suggest that a variety of DNA sequences are inhibitory for transcription, a second collection is merely compatible, while a third is actually essential for gene expression.

Without knowledge of the protein-nucleic acid interactions that presumably cause these effects, it may be futile to try to pick out the features of the Met-B 5' flanking sequence that make it inhibitory. Nevertheless, it contains two regions of potential significance (Fig. 6). One is a stretch of 13/15 dC residues located 48bp upstream of the mature 5' end which could hamper the unwinding of the DNA prior to transcription. The second is a 9bp alternating purine-pyrimidine tract, comprising mostly d(GC) pairs, that is even closer to the gene. Oligonucleotides with these features can crystallize as left-handed DNA helices (22-25). It is conceivable, therefore, that such a tract has an inhibitory effect on transcription by locally altering the DNA conformation. Clearly it will be of interest to determine the transcriptional effects of deleting these regions.

ACKNOWLEDGMENTS

We are indebted to Andrew Lassar for his gift of the AL subclone, and to David Galas for many stimulating discussions.

REFERENCES

1. Clarkson, S.G., and Kurer, V. (1976). *Cell 8*, 183.
2. Clarkson, S.G., Kurer, V., and Smith, H.O. (1978). *Cell 14*, 713.
3. Müller, F., and Clarkson, S.G. (1980). *Cell 19*, 345.
4. Müller, F., Clarkson, S.G., and Galas, D.J. Manuscript submitted.
5. Wegnez, M., Mazabraud, A., Denis, H., Petrissant, G., and Boisnard, M. (1975). *Eur.J.Biochem. 60*, 295.
6. Gauss, D.H., and Sprinzl, M. (1981). *Nucleic Acis Res. 9*, rl.
7. Weil, P.A., Segall, J., Harris, B., Ng, S-Y., and Roeder, R.G. (1979). *J.Biol.Chem. 254*, 6163.

8. Bolivar, F., Rodriguez, R.L., Greene, P.J., Betlach, M., Heynecker, H.L., Boyer, H.W., Crosa, J.H., and Falkow, S. (1977). *Gene 2*, 95.
9. Koski, R.A., Clarkson, S.G., Kurjan, J., Hall, B.D., and Smith, M. (1980). *Cell 22*, 415.
10. Koski, R.A., and Clarkson, S.G. Manuscript submitted.
11. Telford, J.L., Kressmann, A., Koski, R.A., Grosschedl, R. Müller, F., Clarkson, S.G., and Birnstiel, M.L. (1979). *Proc.Nat.Acad.Sci.USA 76*, 2590.
12. Garber, R.L., and Gage, L.P. (1979). *Cell 18*, 817.
13. Hagenbüchle, O., Larson, D., Hall, G.I., and Sprague, K. (1979). *Cell 18*, 1217.
14. Kressmann, A., Hofstetter, H., DiCapua, E., Grosschedl, R., and Birnstiel, M.L. (1979). *Nucleic Acids Res 7,* 1749.
15. DeFranco, D., Schmidt, O., and Söll, D. (1980). *Proc.Nat. Acad.Sci.USA 77*, 3365.
16. Sprague, K.U., Larson, D., and Morton, D. (1980). *Cell 22*, 171.
17. Sakonju, S., Bogenhagen, D.F., and Brown, D.D. (1980). *Cell 19*, 13.
18. Bogenhagen, D.F., Sakonju, S., and Brown, D.D. (1980). *Cell 19*, 27.
19. Fowlkes, D.M., and Shenk, T. (1980). *Cell 22*, 405.
20. Engelke, D.R., Ng, S-Y., Shastry, B.S., and Roeder, R.G. (1980). *Cell 19*, 717.
21. Pelham, H.R.B., and Brown, D.D. (1980). *Proc.Nat.Acad.Sci. USA 77*, 4170.
22. Wang, A.H-J., Quigley, G.J., Kolpack, F.J., Crawford, J., van Boom, J.H., van der Marel, G., and Rich, A. (1979). *Nature 282*, 680-
23. Arnott, S., Chandraskaran, R., Birdsall, D.L., Leslie, A. G.W., and Ratliff, R.L. (1980). *Nature 283*, 743.
24. Crawford, J.L., Kolpak, F.J., Wang, A.H-J., Quigley, G., van Boom, J.H., van der Marel, G., and Rich, A. (1980). *Proc.Nat.Acad.Sci.USA 77*, 4016.
25. Drew, H., Takano, T., Tanaka, S., Itakura, K. and Dickerson, R.E. (1980). *Nature 286*, 567.

TRANSCRIPTION INITIATION AND TERMINATION SIGNALS IN THE YEAST *SUP*4 tRNA^{Tyr} GENE[1]

Raymond Koski[2]
Stuart Clarkson

Department of Microbiology
University of Geneva
Geneva, Switzerland

Janet Kurjan[3]
Benjamin Hall

Department of Genetics
University of Washington
Seattle, Washington

Shirley Gillam
Michael Smith

Department of Biochemistry
University of British Columbia
Vancouver, British Columbia

The nucleic acid-protein interactions that mediate transcription initiation are a fundamental aspect of gene regulation in both prokaryotes and eukaryotes. E. coli RNA polymerase initiates transcription by interacting with specific

[1]*This research was supported by grants from the National Institutes of Health, the Swiss National Science Foundation and the Medical Research Council of Canada.*
[2]*Present address: University of Washington, Seattle, Washington.*
[3]*Present address: University of Oregon, Eugene, Oregon.*

ISBN 0-12-137420-3

promoter sequences located upstream from transcription initiation sites (reviewed in ref. 1). These promoter regions generally contain variations of the consensus sequences TATAAT TTGACA about 10 and 35 base pairs upstream from the transcription start site, respectively. Promoter mutations that affect transcription efficiency usually alter one of the base pairs within these homologous regions.

Eukaryotic genes transcribed by RNA polymerase II are also flanked by homologous sequences that are involved in directing efficient and accurate transcription initiation (reviewed in ref. 2). In contrast to both prokaryotic genes and eukaryotic genes encoding mRNA, eukaryotic tRNA and 5 S rRNA genes possess no common sequences upstream from transcription start sites. Although short regions of sequence homology flank related Drosophila tRNA genes (3, 4), few, if any, of the sequences upstream from eukaryotic tRNA coding sequences are required for transcription (5-8). These data suggest that the RNA polymerase III transcription complex may recognize promoter elements within the coding regions of eukaryotic tRNA genes that are analogous to the intragenic RNA polymerase III promoter regions identified in Xenopus 5 S rRNA genes (9-11) and adenovirus VA genes (12).

In a search for individual base pairs that control expression of a eukaryotic tRNA gene, we have isolated several mutant yeast *SUP*4 $tRNA^{Tyr}$ genes that are inactive in vivo and tested them for their ability to support RNA polymerase III transcription in vitro.

ISOLATION OF MUTANT *SUP*4 $tRNA^{Tyr}$ GENES

The yeast Saccharomyces cerevisiae contains eight unlinked tyrosine tRNA genes that encode identical $tRNA^{Tyr}$ sequences, but differ considerably in their 5'-flanking sequences (13-15). Point mutations in the anticodon regions of these genes can convert the codon recognition specificity of the $tRNA^{Tyr}$ from UA^{C}_{U} to UAA, thereby allowing it to suppress nonsense mutations within structural genes. By appropriate genetic selection, yeast mutants carrying suppressor-inactivating mutations can be isolated in a $SUP4_{UAA}$ strain (16). The loss-of-suppression mutations fall into several categories, including reversion of the suppressor tRNA genes to wild type, second-site mutations within the suppressor tRNA gene, and mutations in unlinked genes that are essential for suppressor tRNA biosynthesis or function.

Within a collection of such mutations, we identified 69 spontaneous second-site *SUP*4 alterations that block efficient suppression of the *can*1-100, *ade*2-1, *lys*2-1, *met*4-1 and *trp*5-2 ochre nonsense alleles (17).

The *SUP*4 mutations were positioned with respect to the anticodon and each other in a fine structure map by meiotic recombination (18). The mutations were grouped into ten clusters, such that within any one cluster recombination frequencies were low.

MUTANT *SUP*4 GENE SEQUENCES

Fine structure map position and differences in phenotype enabled us to choose a variety of mutant *SUP*4 $tRNA^{Tyr}$ genes for molecular cloning and DNA sequence analysis. Rapid determination of the mutant gene sequences was achieved with a synthetic oligonucleotide primer, d(pAAAAACAAA), that hybridized to a specific site downstream from the tRNA gene. The sequence changes found in 40 mutant genes are presented in Figure 1 and Table 1. Thirty-four different mutations were identified. Most of the sequence changes were point mutations distributed throughout the coding regions of the tRNA gene. Three independent mutations were identified in the intervening sequence. All three were identical A→T transversions in the fourth base pair of the intervening sequence. No mutations analogous to the promoter mutations found upstream from an E. coli $tRNA^{Tyr}$ gene (19) were found in the 5'-flanking sequences.

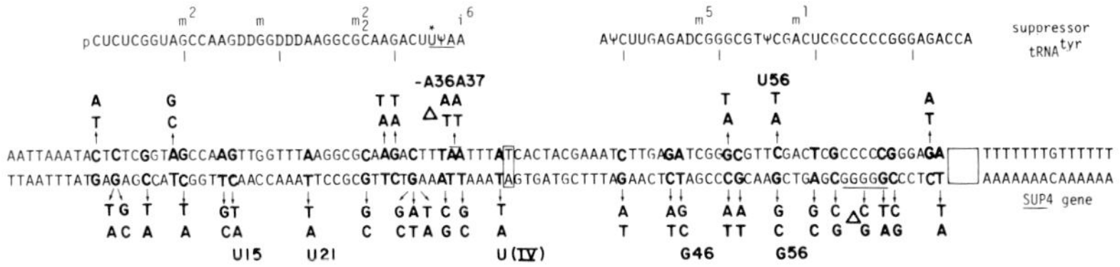

Figure 1. Sequence alterations in *SUP*4 genes with second-site mutations. Positions at which mutants have been isolated are shown in boldface type. Only the names of the mutations known to significantly affect transcription are given. Each mutation is named according to the position in the tRNA and the nucleotide alteration, as described by Celis (20). [Adapted from Koski et al. (21)]

TABLE I. Mutant Alterations

Mutant[a]	Original Name	Nucleotide Change	tRNA Location	Conserved Nucleotides
A1	M145	C → A	aa stem	
U3	M52, M107	C → U	aa stem	
G3	M123	C → G	aa stem	
U6	M69	G → U	aa stem	
G9	M105	A → G	aa stem-D stem junction	
U10	M91	m^2G → U	D stem	usually G or m^2G
G14	M60	A → G	D loop	A
U15	M83	G → U	D loop	purine
U21	M112	A → U	D loop	usually A
G27	M94	C → G	AC stem	
U29	M106	A → U	AC stem	
U30	M98	A → U	AC stem	
G32	M50	C → G	AC loop	pyrimidine
A32	M90	C → A	AC loop	pyrimidine
U32	M125	C → U	AC loop	pyrimidine
C35	M128	Ψ → C	anticodon	
-A36A37	M119	$\Delta A i^6A$	AC loop	modified purine
G37	M54, M109	i^6A → G	AC loop	modified purine
U(IV)	M55, M57, M100	A → U	Intervening Sequence	
A40	M99, M113	C → A	AC stem	
A45	M110	G → A	extra arm	
G46	M51	A → G	extra arm	
A51	M108	G → A	TΨ stem	
U51	M95	G → U	TΨ stem	
A52	M88	C → A	TΨ stem	
G56	M70	C → G	TΨ loop	C
U56	M137	C → U	TΨ loop	C
G60	M49	U → G	TΨ loop	usually pyrimidine
C62	M79, M129	G → C	TΨ stem	
-C(63-67)	M77	ΔC	TΨ stem-aa stem	
U67	M56	C → U	aa stem	
C68	M59	G → C	aa stem	
A72	M127	G → A	aa stem	
U73	M102	A → U	aa stem	

[a]*Mutants are named according to the position in the tRNA and the nucleotide alteration as described by Celis (20).*

TRANSCRIPTION OF THE MUTANT *SUP*4 GENES IN VITRO

Since all of the *SUP*4 mutations that we isolated block *SUP*4 $tRNA^{Tyr}$ gene expression in vivo, we expected some of the mutations to inactivate the gene at the transcriptional level. An in vitro transcription system derived from Xenopus tissue culture cells (22) provided a convenient means of testing the

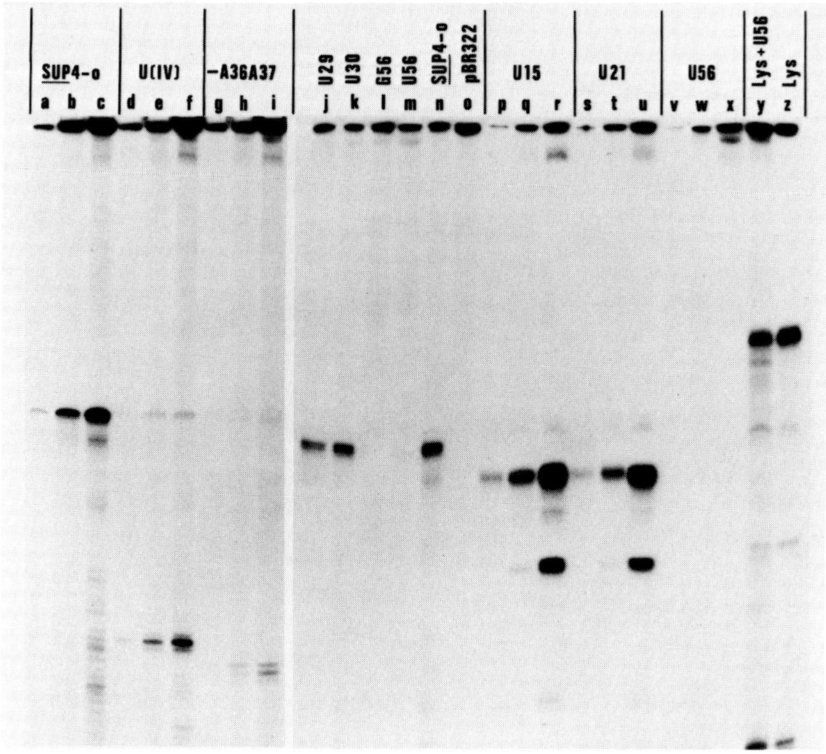

Figure 2. Transcription time course experiments. Each transcription reaction contained the indicated gene linked to pBR322 vector DNA, X. laevis tissue culture cell S100 extract, ATP, CTP, UTP and α-^{32}P-GTP. (Lys), X. laevis DNA containing a $tRNA^{Lys}$ gene. Reactions were terminated after 20 min (lanes a, d, g, p, s, v), 60 min (b, e, h, q, t, w), 120 min (lanes j-o, y-z) or 180 min (lanes c, f, i, r, u, x). The RNAs synthesized in each reaction were extracted and analyzed in a 7 M urea, 10% polyacrylamide gel.

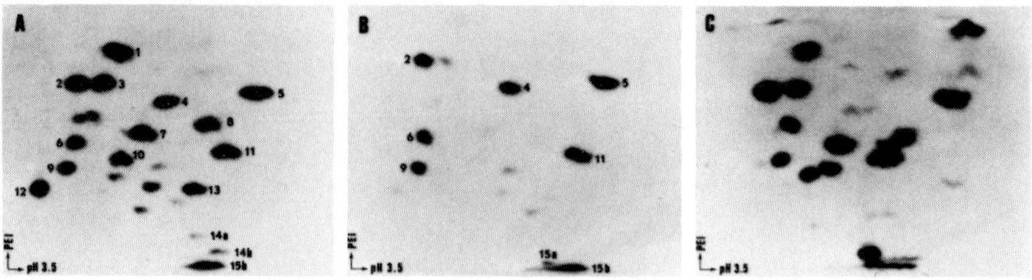

Figure 3. RNAase T1 oligonucleotide fingerprints of the major *SUP*4-o, U(IV) and U15 RNAs synthesized in vitro. (A) *SUP*4-o RNA fingerprint. (B) -A36A37 RNA fingerprint. (C) U15 RNA fingerprint. Characteristic *SUP*4-o RNA oligonucleotides are numbered as in Koski et al. (21).

interaction of RNA polymerase III with the mutant $tRNA^{Tyr}$ genes. The major transcripts generated from the *SUP*4 ochre suppressor $tRNA^{Tyr}$ (*SUP*4-o) gene contain approximately 100 nucleotides (Figure 2, lanes a-c). RNA sequence analyses of the ~100 nucleotide transcripts (Figure 3 and ref. 21) determined that these transcripts initiate with pppA five nucleotides before the beginning of the *SUP*4 gene and terminate almost immediately after the coding region. Most of the *SUP*4 genes with second-site mutations yield transcripts in vitro that correspond in size and yield to those produced from the *SUP*4-o gene (Figure 4). In these cases, the spontaneous mutations within the *SUP*4 locus probably inactivate expression of the $tRNA^{Tyr}$ gene through posttranscriptional events. However, a few mutations do have significant effects on *SUP*4 gene transcription in vitro, providing indications as to the nature of RNA polymerase III transcription initiation and termination signals.

THE -A36A37 AND U(IV) MUTATIONS GENERATE TRANSCRIPTION TERMINATION SIGNALS

In contrast to the ~100 nucleotide major transcripts produced from the *SUP*4 gene, the -A36A37 and U(IV) genes accumulate primary transcripts with approximately 45 and 50 nucleotides, respectively (Figure 2, lanes d-i). RNAase T1 fingerprints of the -A36A37 RNA contain oligonucleotides characteristic of the 5' end of the normal *SUP*4 gene transcript (Figure 3). Secondary analyses of these oligonucleotides established that -A36A37 gene transcription initiates at the same site as for the *SUP*4-o gene and terminates near the site of the mutation. Sequence studies of the ~50 nucleotide U(IV) transcript provided a second example of a mutation that alters the transcription termination site without affecting the transcription start site (21). Both mutations

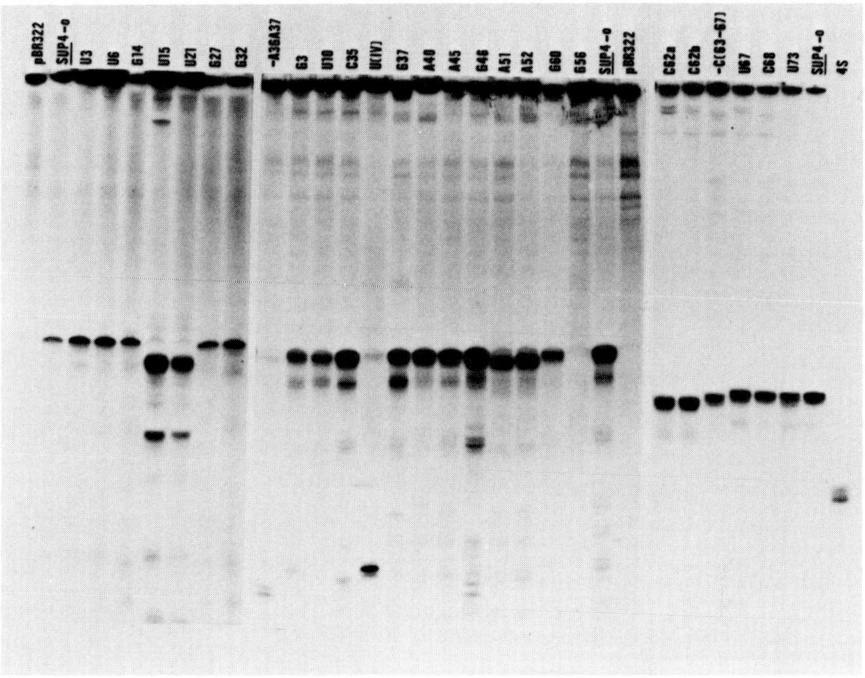

Figure 4. RNAs synthesized from *SUP4*-o genes with second-site mutations. pBR322 DNA, or recombinant DNAs with the *SUP4*-o gene or one of its mutant variants, were incubated with X. laevis S100 extract, ATP, UTP, CTP and α-^{32}P-GTP. The reaction products were analyzed in 7 M urea, 10% polyacrylamide gels. C62a and C62b are recombinant DNAs from two independently isolated mutants. (4S) ^{32}P-labeled tRNA. [Adapted from Koski et al. (21)].

cause premature transcription termination near the site of the mutation, and both mutations create new T clusters in the noncoding strand of the gene. The -A36A37 mutation alters the sequence TTTAATTTAT to TTTTTTAT through a two base pair deletion, and the U(IV) mutation changes the same sequence to TTTAATTTTT (Figure 1). The new T clusters generated by these mutations resemble the T tracts found after the *SUP*4 tRNA coding region and after other genes transcribed by RNA polymerase III (3-6, 23). These data demonstrate that a T cluster in the noncoding strand is the most important sequence element of RNA polymerase III terminator signals. The structure of the nascent transcript may also influence the RNA polymerase termination site, since three mutations that disrupt the GC-rich TΨ stem of the tRNA structure (A51, A52, C62) apparently induce slightly shortened primary transcripts (Figure 4).

THE U15 AND U21 MUTATIONS ENHANCE TRANSCRIPTION AND ALTER THE TRANSCRIPTION TERMINATION SITE

The U15 and U21 mutations both enhance *SUP*4 gene transcription approximately five-fold in vitro and generate abnormally short transcripts (Figures 2 and 4). For both mutant genes, a ~90 nucleotide RNA accumulates as a primary transcript, and during longer incubations a ~75 nucleotide RNA is produced. RNase T1 fingerprints of the shortened transcripts all lack the CCCCCG spot characteristic of the 3' end of the normal *SUP*4 gene transcript (oligonucleotide 12, Figure 3; and R. K., unpublished data). The altered positions of other fingerprint spots indicate that the shortened U15 and U21 transcripts also have extensive nucleotide modifications not present in the in vitro *SUP*4-o gene transcripts. Both mutations alter base pairs encoding conserved nucleotides within the tRNA D loop that probably interact with other regions of the tRNA precursor structure (17). Perturbations of the tRNA precursor structure may allow the GC-rich TΨ stem, which is followed by a C_5 tract in the noncoding strand, to function as a transcription termination signal. Loss of tertiary structure may also account for the enhanced rate of nucleotide modification. More extensive sequence analyses of the aberrant oligonucleotides are required to determine the exact sequence of the U15 and U21 transcripts.

THE G56 and U56 MUTATIONS REDUCE *SUP*4 GENE TRANSCRIPTION

The G56 and U56 mutations alter the same GC base pair in the 3' half of the *SUP*4 gene and both drastically reduce transcription of the tRNA gene (Figures 1 and 2). The U56 mutation reduces transcription approximately ten-fold and the G56 mutant gene yields no detectable *SUP*4 transcripts. Enhanced susceptibility to degradative nucleases is not likely to account for the reduced levels of G56 and U56 transcripts. A wide variety of RNA molecules, including prematurely terminated *SUP*4 gene transcripts, are not degraded in the in vitro reactions. Short reaction times fail to reveal G56 and U56 transcripts that are unusually susceptible to nuceases. Inhibitory components in the U56 and G56 DNA preparations are not responsible for reduced transcription since Xenopus laevis lysine tRNA precursors are produced in transcription reactions containing the inactive *SUP*4 genes (Figure 2, lanes y-z, and data not shown). The reduced levels of transcription from the G56 and U56 genes suggest that the transcription promoter for the *SUP*4 $tRNA^{Tyr}$ gene encompasses the tRNA coding sequences surrounding these two mutations. Both mutations alter the first C within the highly conserved 5'GTΨCRANYC3' tRNA sequence. This C is absolutely invariant in all known prokaryotic and eukaryotic tRNAs (24). Thus, the Xenopus RNA polymerase III transcription initiation complex can recognize a base pair within the *SUP*4 $tRNA^{Tyr}$ gene that has been conserved for translational functions and utilize it as a transcription initiation signal.

CONCLUSIONS AND PROSPECTS

The accessibility of the yeast *SUP*4 $tRNA^{Tyr}$ gene to both formal genetics and molecular genetics allowed us to analyze many mutations that block *SUP*4 gene expression at the molecular level. We began studying the biochemical implications of these mutations by testing the mutant genes in cell-free transcription reactions containing Xenopus laevis RNA polymerase III. Two mutations prematurely terminate transcription through the creation of new noncoding strand T clusters. Two mutations that alter conserved nucleotides in the tRNA D loop cause enhanced transcription and premature termination. Two mutations that alter an invariant C in the tRNA TΨ loop drastically reduce *SUP*4 gene transcription. These results suggest that Xenopus RNA polymerase III utilizes conserved features of tRNA genes and tRNA precursor structure as transcription initiation and termination signals. Initial experiments with a homologous yeast RNA polymerase III in vitro transcription system suggest that these transcription signals are highly conserved throughout evolution (R.K., unpublished data). In both the yeast and Xenopus in vitro transcription reactions, the -A36A37, U(IV), U15 and U21 mutations cause premature transcription termination, and the G56 and U56 mutations reduce transcription. We are now extending our studies of *SUP*4 gene expression by isolating additional mutant genes and characterizing the effects of the mutations on homologous transcription and processing reactions.

ACKNOWLEDGMENTS

We thank S. Goh for sequencing one of the mutant *SUP*4 genes, and M. Worthington for excellent technical assistance.

REFERENCES

1. Siebenlist, U., Simpson, R. B., and Gilbert, W., *Cell 20*, 269 (1980).
2. Minty, A., and Newmark, P., *Nature 288*, 210 (1980).
3. Hovemann, B., Sharp, S., Yamada, H., and Söll, D., *Cell 19*, 889 (1980).
4. Robinson, R. R., and Davidson, N., *Cell 23*, 251 (1981).
5. Telford, J. L., Kressmann, A., Koski, R. A., Grosschedl, R., Müller, F., Clarkson, S. G., and Birnstiel, M. L., *Proc. Nat. Acad. Sci. USA 76*, 2590 (1979).
6. Garber, R. L., and Gage, L. P., *Cell 18*, 817 (1979).
7. De Franco, D., Schmidt, O., and Söll, D., *Proc. Nat. Acad. Sci. USA 77*, 3365 (1980).
8. Sprague, K. U., Larson, D., and Morton, D., *Cell 22*, 171 (1980).
9. Bogenhagen, D. F., Sakonju, S., and Brown, D. D., *Cell 19*, 27-35 (1980).
10. Sakonju, S., Bogenhagen, D. F., and Brown, D. D., *Cell 19*, 13-25 (1980).
11. Engelke, D. R., Ng, S.-Y., Shastry, B. S., and Roeder, R. G., *Cell 19*, 717 (1980).
12. Fowlkes, D. M., and Shenk, T., *Cell 22*, 405 (1980).
13. Olson, M. V., Hall, B. D., Cameron, J. R., and Davis, R. W., *J. Mol. Biol. 127*, 285 (1979).
14. Goodman, H. M., Olson, M. V., and Hall, B. D., *Proc. Nat. Acad. Sci. USA 74*, 5453 (1977).
15. Goh, S., and Smith, M., personal communication.
16. Rothstein, R., *Genetics 85*, 55 (1977).
17. Kurjan, J., Hall, B. D., Gillam, S., and Smith, M., *Cell 20*, 701 (1980).
18. Kurjan, J., and Hall, B. D., Manuscript in preparation.
19. Berman, M. L., and Landy, A., *Proc. Nat. Acad. Sci. USA 76*, 4303 (1979).
20. Celis, J. E., *Nucl. Acids Res. 8*, r23 (1980).
21. Koski, R. A., Clarkson, S. G., Kurjan, J., Hall, B. D., and Smith, M., *Cell 22*, 415 (1980).
22. Weil, P. A., Segall, J., Harris, B., Ng, S.-Y., and Roeder, R. G., *J. Biol. Chem. 254*, 6163 (1979).
23. Korn, L. J., and Brown, D. D., *Cell 15*, 1145 (1978).
24. Sprinzl, M., Grueter, F., Spelzhaus, A., and Gauss, D. H., *Nucl. Acids Res. 8*, r1 (1980).

PROCESSING OF YEAST tRNA^{TYR} IN *XENOPUS* OOCYTES MICROINJECTED WITH CLONED GENES

Kazuko Nishikura[1]
Eddy M. De Robertis[2]

MRC Laboratory of Molecular Biology
Cambridge, England

ABSTRACT

Microinjected tRNA genes are transcribed, and then the transcripts are processed in a precise sequential order by frog oocytes (De Robertis and Olson 1979, Nature 278, 137-143). In this study, we have investigated further details of the RNA processing of transcripts copied from yeast tRNA^{Tyr} genes following microinjection into *Xenopus* oocytes. 1) We analyzed base modifications of ^{32}P-labelled 104 and 92 nucleotide precursors and 78 nucleotide mature tRNA^{Tyr} separately and found that all the base modifications occur in a strict order which correlates with the size alteration of the tRNA^{Tyr} precursors. 2) We have studied the effect of single nucleotide changes on the processing of tRNA gene transcripts, and found that mutations which affect splicing of the 92 precursor are mainly located in the anticodon loop and the intervening sequence. 3) We analyzed localization of the splicing enzymes in frog oocytes by manually separating the nuclear envelope, the nuclear contents, and the cytoplasm and then using these cell components in an in vitro assay system for splicing. All the splicing activity was found to be located in the nuclear contents and not associated with the nuclear membrane.

[1]*Present address: Department of Structural Biology, Stanford University School of Medicine, Stanford, California.*
[2]*Present address: Biozentrum University of Basel, Basel, Switzerland.*

ISBN 0-12-137420-3

INTRODUCTION

Since the presence of intervening sequences in many eukaryotic genes and splicing of precursor RNAs were found in 1977, it has become increasingly clear that the post-transcriptional events play important roles in eukaryotic gene expression (1-2). However, the post-transcriptional events such as splicing, nucleoside modification, and transport of RNA from nucleus to cytoplasm may involve a great number of different enzymatic reactions, and could be very difficult to reproduce in vitro. The combination of DNA cloning in bacterial plasmids and DNA injection into living frog oocytes nuclei provides opportunities for the study of eukaryotic gene expression (3-5). Since genes transcribed by RNA polymerase III are expressed very efficiently and with high fidelity in frog oocytes, as shown for 5S ribosomal RNA genes (6-7) and also for tRNA genes (8-9), we have chosen yeast $tRNA^{Tyr}$ genes and studied in some detail the tRNA processing in *Xenopus* oocyte microinjected with the cloned genes.

RESULTS

A. *Sequential Addition of Base Modifications to tRNA Precursors*

1. *$tRNA^{Tyr}$ Precursors in Xenopus Oocytes.* The injection of many copies of a pure gene (10^8 to 10^9) together with radioactive ribonucleotides into each oocyte allows one to label the transcripts after short incubation times and thus to isolate the early transcripts before extensive processing has occurred. When a yeast $tRNA^{Tyr}$ gene, cloned in plasmid pBR322 and known to contain a 14-nucleotide intervening sequence (10), was injected into frog oocyte nuclei, three main bands (102, 92 and 78 nucleotides length) of newly synthesized RNAs in oocytes were detectable by acrylamide-gel electrophoresis (Fig. 1). Using standard RNA fingerprinting methods (11-12) it was found that the early transcript contains, in addition to the intervening sequence, extra segments of RNA at the 5' end (5' leader) and 2 or 3 U residues at the 3' end (13-14). After maturation of the termini, including removal of 5' leader and 3' trailer sequences and addition of the CCA end, the molecule is converted into a 92 nucleotide precursor which still retains the intervening sequence. Finally, the intervening sequence is excised and the molecule religated giving rise to a mature 78 nucleotide-long $tRNA^{Tyr}$ (13). One possi-

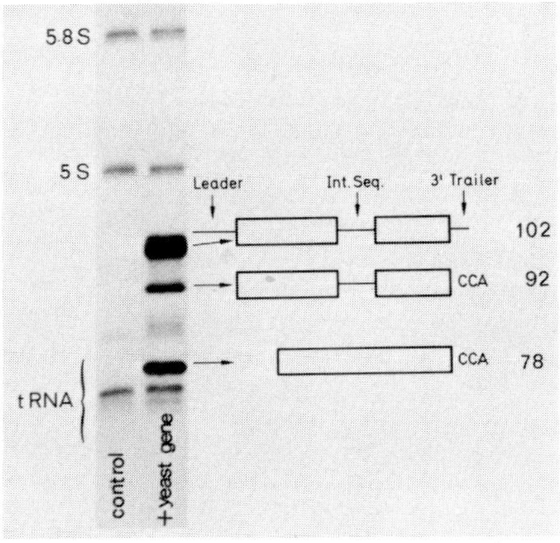

FIGURE 1. Transcription products of tRNA^{Tyr} genes in injected oocytes. Oocytes were injected with a cloned tRNA^{Tyr} gene (plasmid SUP4) and α-^{32}P GTP, incubated for 24h and the ^{32}P RNA was electrophoresed in a 12% acrylamide gel as described (13, 16). The molecular structure of the different precursor bands is indicated (13-14).

ble explanation why these extra segments, 5' leader, 3' trailer and intervening sequence, are transcribed and processed in such a complicated order instead of being simply deleted all at once is suggested in the following section.

2. Base Modifications Are Added in a Sequential Order. Most tRNAs have a set of modified bases at certain positions, and the types of modification and the places are well conserved among species (15). When we analyzed the different tRNA^{Tyr} precursors for base modifications to test whether frog oocytes injected with tRNA genes would introduce base modifications on a yeast tRNA, we found not only that oocytes can almost perfectly achieve the modification, but also that they are added in a stepwise fashion (16). Figure 2 shows the results of analysing 104, 92 and 78-nucleotide RNAs for modified bases by two-dimensional cellulose thin-layer chromatography (16-17). As can be seen in Fig. 2, m^{5}C is present in the 104 molecule (d), but the m^2_2G, m^{2}G and D modifications are introduced in the 92 precursor after maturation of the 5' and 3' termini (b and e). The modification Q and an unidentified Q* derivative modification were found in the 78 tRNA^{Tyr}. All the base modifications added to tRNA^{Tyr} transcripts in oocytes are summarized in Fig. 3. The most important point shown in Fig. 3 is that base modifications are introduced in a sequential order. Each type of precursor seems to be a substrate

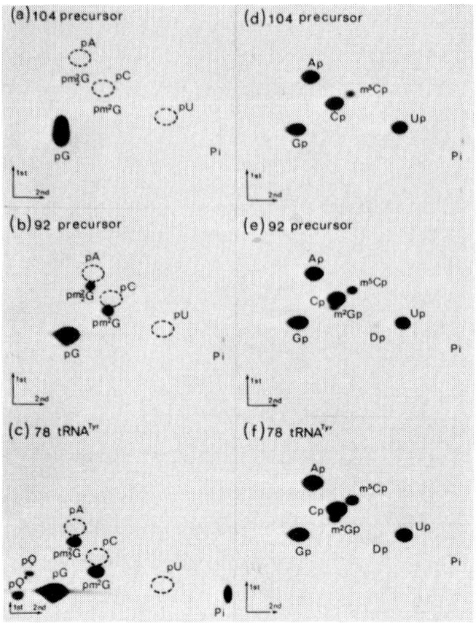

FIGURE 2. Modified base analysis on tRNA^{Tyr} gene transcripts. The RNAs labelled with α-^{32}P GTP in frog oocytes were digested with nuclease P_1 into 5'-mononucleotide (a, 104; b, 92; c, 78 tRNA^{Tyr}), and with RNase T_2 into 3'-mononucleotide (d, 104; e, 92; f, 78 tRNA^{Tyr}). Two-dimensional chromatography on a cellulose thin-layer plate was performed as described previously (16).

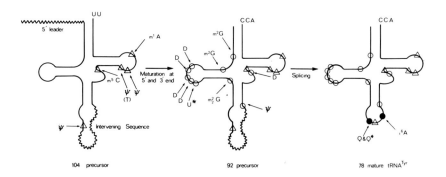

FIGURE 3. Sequential addition of base modifications to tRNA^{Tyr} precursors. Δ, modified bases added at the 104 precursor stage; o, modified bases detected at the 92 precursor stage; •, base modifications added in the 78 mature tRNA^{Tyr} after splicing.

only for a specific subset of base modification enzymes. This sequence of events can be correlated with the presence of extra RNA segments in the molecule, which could enforce this sequential order perhaps by altering the general conformation of the tRNA precursors.

B. Genetic Analysis of tRNA Processing

The detailed processing pathway for yeast tRNA^{Tyr} we have described in the previous section was useful for a genetic analysis of RNA processing. Kurjan *et al.* have recently prepared a collection of point mutations in 26 different locations of the SUP4 gene (an ochre suppressor tRNA^{Tyr} gene derived from the Gene A) isolated by selecting for loss of suppressor activity (18). We analyzed the effect of 18 different single nucleotide changes on the RNA processing by frog oocytes injected with cloned tRNA^{Tyr} mutant genes. The main conclusion is that the maturation of the 5' and 3' ends (102 to 92 precursor) is the step most sensitive to mutations. These studies will be reported elsewhere (Nishikura, Kurjan, Hall and De Robertis, in preparation), but here we would like to discuss those nucleotide changes that can produce quantitative changes in the splicing efficiency of tRNA^{Tyr} precursors in the frog oocyte system.

1. Single Nucleotide Changes Can Affect Splicing Efficiency. When we compared the transcripts from various tRNA^{Tyr} genes which have one single nucleotide change in the vicinity of the intervening sequence and the anticodon (Fig. 4A), we noticed quantitative changes in the efficiency of splicing by frog oocytes as shown in Fig. 4B. Each radioactive band was extracted from the gel, and counted for quantitation. The result is summarized in Table I. It seems that the SUP4 precursor is spliced twice as efficiently as its wild-type counterpart. The substrate for the splicing enzymes, the 92 precursor, differs only by a single G to U change in the anticodon between Gene A and SUP4 (Fig. 4A). In all intervening sequence-containing wild-type yeast tRNA genes studied so far the anticodon region can be base paired with a part of the intervening sequence (19). In the *S. cerevisiae* tRNA^{Tyr} only three base pairs can be formed between the anticodon region and the intervening sequence (Fig. 4A, wild-type Genes A and C). The corresponding gene for *Xenopus laevis* tRNA^{Tyr} has been sequenced, and found to contain the 13-nucleotide intervening sequence, but the sequence, unlike yeast, can not base pair with the anticodon region (20). Thus, it seems as if the SUP4 mutations, which do not have the G-C base pair in the anticodon-intervening sequence region of complementarity, make

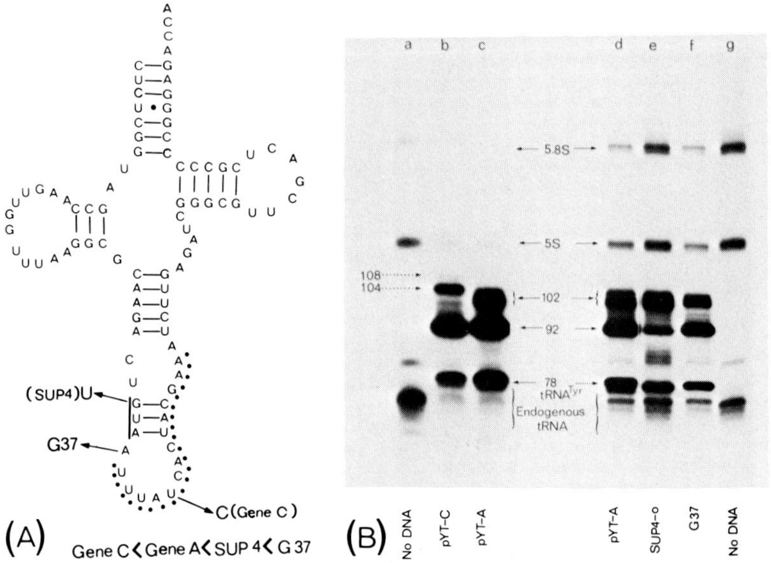

FIGURE (4A). The location of the nucleotide changes in different tRNA^{Tyr} gene transcripts. The nucleotide changes of two wild-type (Gene A and Gene C), an ochre suppressor (SUP4) and one mutant (G37) which was derived from SUP4 are shown. The secondary structure shown is that of the 92 precursor for Gene A transcripts. The order of splicing efficiency is indicated (see Fig 4B.).

(4B) Efficiency of splicing in various tRNA^{Tyr} gene transcripts. Splicing activity can be assesed by the ratio of 92 precursor to mature 78 tRNA. Oocytes were incubated at 19°C for 24h and analyzed in 8% polyacrylamide gels. Note that the early transcripts of Gene C (which comes from a different yeast tRNA^{Tyr} gene locus) are different from the rest: this is because they have a 5' leader of different length and composition (13).

the yeast precursor more similar to the frog precursor RNA, which is spliced more efficiently by the oocytes. Other single nucleotide changes that also affect splicing efficiency (Gene C and G37) are shown in Fig. 4 and Table I. Mutant G37 (18) differs from its parental strain SUP4 only by a single A to G change at the splicing point (Fig. 4A), and the splicing efficiency is decreased (Fig. 4B lane f, Table I). G37 decreases splicing to about the level of Gene A and, interestingly, the mutation also restores a possible 3 base pair complementarity between the anticodon region and the

TABLE I. Different Efficiency of tRNA Splicing among Various Gene Transcripts

	RNA precursor	^{32}P cpm[a]	Percentage of total $tRNA^{Tyr}$ transcripts(%)
Gene C	104	204	14
(Fig. 4B,	92	1,090	73
lane b)	78	201	13
Gene A	102-100	337	11
(Fig. 4B,	92	2,074	70
lane c)	78	562	19
SUP4	102-100	2,242	48
(Fig. 4B,	92	830	18
lane e)	78	1,578	34
G37	102-100	2,006	16
(Fig. 4B,	92	8,290	66
lane f)	78	2,306	18

[a]*Corrected counts. Background counts were subtracted. The tRNA transcripts were eluted from polyacrylamide gels.*

intervening sequence, which was not possible in the SUP4 parental strain. Gene C (Fig. 4B, lane b) is reproducibly spliced less efficiently than Gene A from which it differs by only a single nucleotide change within the intervening sequence in the 92 precursor (Gene C transcripts also differ in the length and composition of the 5' leader, in the early precursors). In most of our previous work on tRNA^{Tyr} processing we had injected Gene C DNA, and this explains the lower splicing efficiency reported earlier (13-14).

The relative efficiency of splicing by frog oocytes is: Gene C < Gene A < SUP4 > G37 (Fig. 4A). The quantitative differences in the splicing efficiency of various tRNA^{Tyr} genes are very reproducible, and all were observed using oocytes from at least 4 different frogs. Different frogs, however, may vary in the maximum extent of splicing (some frogs may have up to 80% of spliced transcripts 24 hours after injection of SUP4). The reason for this variability between oocytes is not understood, but results are constant when oocytes from the same frog are utilized (up to 4 times over a period of several months). We concluded from these observations that single nucleotide changes in the vicinity of the intervening sequence and anticodon region affect the efficiency of splicing. The secondary structure of the precursor seems to play an important role in the tRNA splicing reaction.

C. Intranuclear Location of the tRNA Splicing Enzymes

Finally we asked where within the cell does RNA splicing take place. We had previously found that the splicing activity was in some way associated with the cell nucleus when tested by microinjecting the 92 precursor $tRNA^{Tyr}$ into the nucleus or cytoplasm of living oocytes (14). More recently, the localization was studied further to decide whether tRNA splicing enzymes are confined to the nuclear envelope.

1. Two Enzymatic Steps of Splicing Can Be Distinguished in Vitro. The studies by Abelson and his colleagues have shown, using an in vitro yeast extract system, that tRNA splicing is a two-stage reaction. The excision of the intervening sequence, producing half-tRNA molecules, is followed by the ligation of the two half-molecules to generate a mature tRNA (22-23). When the radioactive 92 precursor $tRNA^{Tyr}$ was incubated with the *Xenopus* oocyte nucleus (the oocyte has a large nucleus or germinal vesicle of 0.4 mm in diameter), which can be isolated manually under the microscope, it was possible to distinguish the two splicing activities (24).

2. Intranuclear Location of the Excision and Ligation of Splicing Activities. The nuclear membrane can be isolated manually from the nuclear contents, which do not become dis-

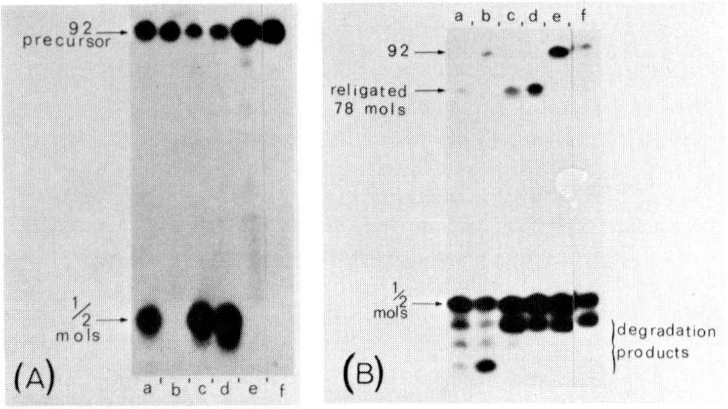

FIGURE 5A. Excision of the intervening sequence from a 92 precursor by various oocyte fractions. FIGURE 5B. Religation of $tRNA^{Tyr}$ half-molecules by various fractions. ^{32}P-labelled 92 precursor (A) or ^{32}P-labelled half-molecules (B) were incubated with: a) one whole oocyte; b) one cytoplasm; c) five oocyte nuclei; d) 5 demembranated nuclear contents; e) 5 nuclear membranes; f) buffer under two different conditions (24).

persed if Mg^{++} is present in the isolation medium because the nucleoplasm forms a gel (24-25). Using assays for both the excision and the ligation reaction and the manually isolated cell components, we investigated the intracellular locaton of the splicing enzymes. Fig. 5A shows that the intervening sequence excision activity can be detected in intact oocyte nuclei (lane c) and in demembranated nuclear contents (lane d), but not in the cytoplasmic (lane b) or the nuclear envelope fractions (lane e). In Fig. 5B the location of the ligation activity is shown. The religated product, 78 tRNA^{Tyr}, was produced by intact germinal vesicles (lane c), and nuclear contents (lane d), but not by the cytoplasm (lane b), or nuclear membranes (lane e). We conclude that both splicing activities are associated with the nuclear contents but are undetectable in the nuclear envelope (24).

DISCUSSION

It seems that for understanding of the intricate RNA processing pathway of an eukaryotic gene transcript, studying of single genes in great detail will be necessary. Using frog oocytes microinjected with a cloned yeast tDNA^{Tyr}, we were able to dissect a complex series of eukaryotic RNA processing events to some extent (14,16,21,24). The most striking observation was that base modifications are added in an invariant order which seems to be closely correlated with size alteration of the precursors. It is not immediately obvious why the processing of a small tRNA transcript should occur in such a complicated sequence of events. However, our finding that a single nucleotide change can stop the tRNA processing at some stage or alter the splicing efficiency significantly (21) may suggest one possible explanation. The intricate tRNA processing pathway itself may act as a selection mechanism in which only the normal and fully functional gene transcript can be produced and rendered to the cytoplasm where the gene product would be immediately utilized.

Although the processing of mRNA precursors is possibly very different from that of tRNA, it is worth keeping in mind that the presence of extra segments in mRNA precursors could also have considerable and unexpected effects in eukaryotic gene expression. Hopefully,it will be possible to extend the RNA processing studies presented here to protein-coding genes in the future.

ACKNOWLEDGMENTS

We thank Drs. J. Kurjan and B. D. Hall for a gift of mutant G37 and Dr. M. Olson for wild-type tRNA genes.

REFERENCES

1. Abelson, J., Ann. Rev. Biochem. 48, 1035 (1979).
2. Perry, R. P., J. Cell. Biol., in press.
3. De Robertis, E. M., and Gurdon, J. B., Scient. Amer. 241, 74 (1979).
4. Mertz, J. E., and Gurdon, J. B., Proc. Nat. Acad. Sci. U.S.A. 74, 1502 (1977).
5. De Robertis, E. M., and Mertz, J. E., Cell 12, 175 (1977).
6. Brown, D. D., and Gurdon, J. B., Proc. Nat. Acad. Sci. U.S.A. 74, 2064 (1977).
7. Gurdon, J. B., and Brown, D.D., Devl.Biol., 67, 346 (1978).
8. Kressman, A., Clarkson, S.G., Telford, J.L., and Birnstiel, M.L., Cold Spring Harb. Symp.Quant.Biol., 42, 1077 (1977).
9. Cortese, R., Melton, D.A., Tranquilla, T., and Smith, J.D., Nucleic Acids Res. 5, 4593 (1978).
10. Goodman, H.M., Olson, M.V., and Hall, B.D., Proc.Nat.Acad. Sci.U.S.A., 74, 5453 (1977).
11. Brownlee, G.G., in "Laboratory Techniques in Biochemistry and Molecular Biology" (J.S. Work and E. Work ed.), p.67, American Elsevier, New York, (1972).
12. Brownlee, G.G., and Sanger, F., Eur. J. Biochem. 11, 395 (1969).
13. De Robertis, E.M., and Olson, M.V., Nature 278, 137 (1979).
14. Melton, D.A., De Robertis, E.M., and Cortese, R., Nature 284, 143 (1980).
15. Sprinzl, M., Grueter, R., Spelzhaus, A., and Gauss, D.H., Nucleic Acids Res. 8, rl (1980).
16. Nishikura, K., and De Robertis, E.M., J. Mol. Biol. 145, 405 (1981).
17. Nishimura, S., Prog.Nucl.Acids Res.Mol.Biol. 12, 49 (1972).
18. Kurjan, J., Hall, B.D., Gillan, S., and Smith, M., Cell 20, 701 (1980).
19. Kang, H.S., Ogden, R.C., Knapp,G., Peebles, C.L., and Abelson, J., in "Eukaryotic Gene Regulation" (R. Axel, T. Maniatis, and C.F. Fox ed.), p.69, Academic Press, New York, (1979).
20. Müller, F., and Clarkson, S.G., Cell 19, 345 (1980).
21. Nishikura, K., Kurjan, J., Hall, B.D., and De Robertis, E.M., in preparation.
22. Peebles, C.L., Ogden, R.C., Knapp, G., and Abelson, J., Cell 18, 27 (1979).
23. Knapp, G., Ogden, R.C., Peebles, C.L., and Abelson, J., Cell 18, 37 (1979).
24. De Robertis, E.M., Black, P., and Nishikura, K., Cell 23, 89 (1981).
25 Callan, H.G., and Lloyd, L., Phil. Trans, Roy. Soc. B. 243, 135 (1960).

THE CONTROL OF ADENOVIRUS VA_I RNA TRANSCRIPTION

Roberto Weinmann[1]
Richard Guilfoyle[1]

The Wistar Institute
Philadelphia, Pennsylvania

ABSTRACT

Cloned fragments of adenovirus 2 DNA containing the genes for virus-associated (VA) RNA are faithfully transcribed in vitro by crude RNA polymerase III prepared as described by Wu (2). Using the precessive nuclease BAL 31 we have constructed a series of deleted VA_I genes, which have sequences of the viral DNA substituted by plasmid sequences from nucleotide -27 up to +70 counting from the first G of the VA_I RNA gene product. The deletions up to and including nucleotide +10 inside the VA_I gene support the synthesis of a product of a size similar to VA_I RNA. However, deletions that go up to +15, +20, +55 and +70 completely eliminate VA_I transcription, while VA_{II} transcription continues unabated. Deletion of sequences from the 3' end eliminates the VA_{II} RNA gene and the termination signal for VA_I RNA transcription, when nucleotides downstream of +76 are substituted with plasmid sequences. These variants show termination of transcription at sequences downstream, as detected by the different size RNA transcription products. Retention of the respective parental 5' or 3' end sequences of the RNA products was determined by fingerprint analysis. Therefore an internal control region for RNA polymerase transcription initiation for this particular gene is located between nucleotides +11 to +75

[1]Supported by NIH grants AI-13231, CA 21124 and CA-10815.

ISBN 0-12-137420-3

inside the VA_I gene, as has been previously reported for the 5S genes of *Xenopus* transcribed by RNA polymerase III (8, 9). Transcription competition analysis indicates that the primary binding site for a control factor involved in VA transcription is located between nucleotides +55 to +76. This region shows strong homology with the TψC loop of tRNAs (16). Competition experiments with the ala tRNA gene suggest that common factors might be required for transcription.

INTRODUCTION

The development of cell-free extracts in which prokaryotic (1) or eukaryotic (2-6) genes are faithfully transcribed has greatly enhanced our understanding of the molecular basis of control. Faithful eukaryotic *in vitro* transcription systems were first developed for genes transcribed by DNA dependent RNA polymerase III. The adenovirus coded low molecular weight RNA VA_I was first faithfully transcribed by Wu (2) and this system was later modified slightly (3). *Xenopus* 5S genes were first transcribed by microinjection into oocytes (4) and later a cell-free germinal vesicle extract was successfully used (5). These *in vitro* systems allow ready definition and purification of the proteins required for each step in transcription (6, 7) as well as definition of the nucleic acid requirements for transcription to occur (8-11). The templates used contain eukaryotic DNA sequences cloned in bacterial vectors. By recombinant DNA techniques it has been possible to manipulate these templates so as to define at the nucleotide level the minimum sequences required for faithful initiation and termination of *in vitro* transcription (8-12). Initially for the 5S genes (8, 9) and later for tRNAs and VA RNA (10-12) it was found that regions internal to the gene are required for initiation *in vitro* of transcription by RNA polymerase III (see also this volume). In the case of the 5S gene, a protein factor, which binds to this internal control region, is required for transcription (6). The regulation of expression of this gene might also be mediated by the factor, since the RNA gene product can also bind the protein (13, 14). We have constructed an ordered series of deletions into the VA_I RNA gene which allows us to identify two distinct domains in the internal control region of the gene.

MATERIALS AND METHODS

The experimental procedures used for the isolation and analysis of the deletion mutants have been described (11). The details of the reaction with the precessive nuclease BAL 31 from Pseudomonas aeruginosa (obtained from BRL, Gaithersburg, Va.) are indicated in the legend to Figure 1. Transcription reactions and analysis by polyacrylamide gel electrophoresis were as described previously (3, 11).

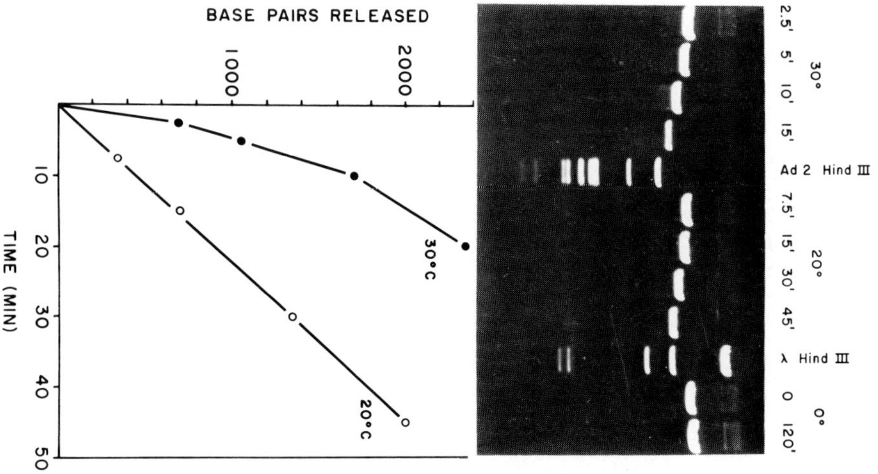

FIGURE 1. Calibration of BAL 31 digestion rate. A plasmid containing an Ad 2 SalI fragment from 26.5 to 45.9 map units inserted in the SalI site of pBR322 was cut with XbaI (20 μg of DNA) and digested for the times and temperatures indicated with 10 units of BAL 31 in a volume of 100 μl (23). Reactions, stopped by adding EDTA to 50 mM (24), were loaded onto a 0.7% agarose gel with appropriate markers. Lower curves in the graph were calculated from changes in molecular weight of the restriction fragments. Reaction rate at 20°C is 22 base pairs per minute per end.

RESULTS

A plasmid containing the VA_I RNA gene and single XbaI and BamHI sites was obtained by cloning a Sal-HindIII DNA fragment (coordinates 26.9-31.5 on the adenovirus genome) into the homologous sites of plasmid pBR322. Briefly (see Fig. 2), a plasmid containing the VA_I gene was linearized with XbaI for 5' end deletions, digested with BAL 31 for 1-4 minutes, and synthetic DNA linkers were added with T4 ligase. Appropriate restriction enzyme DNA fragments were electrophoretically purified, reinserted into pBR322 and amplified. The advantages of the use of BAL 31 are: a)

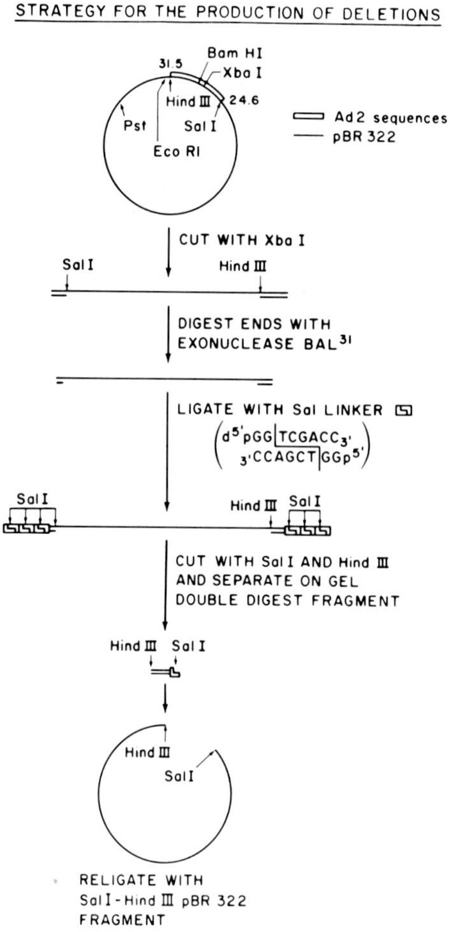

FIGURE 2. Preparation of cloned adenovirus DNA plasmids containing 5' end deletions. The protocol used is described in the text. The single XbaI site located at base pair -29 (all numbering is as described by Ohe and Weissman (21, 23), with the first G of VA_{IG} at position 1 and the A of the other species at position -3) was used to linearize these recombinant plasmids.

the size of the deletions can be regulated by the time and temperature of digestion (see Fig. 1), b) a gaussian distribution of deletion termini is generated for each incubation time point. These are closely spaced around the expected median and can often be obtained for each base pair around the region of interest, c) the sizing step allows preliminary selection of the deletions desired, reducing the screening effort, d) the addition of synthetic DNA linkers on the BAL 31 treated ends provides precise mileposts which simplify the analysis and greatly enhance the efficiency of religation. A diagram illustrating the deletion-substitutions obtained is shown in Fig. 3.

The DNAs from the deletions were either sequenced (wild type, dl 5' -20, dl 5' +1, dl 5' +10, dl 5' +15 and dl 5' +20) by the two methods described (25, 26) or their size estimated using as markers sequencing ladders and the sequenced deletions (dl 5' -27, dl 5' -21, dl 5' -15, dl 5' -14, dl 5' -12, dl 5' -5, dl 5' -3, dl 5' -2, dl 5' +55, dl 5' +70). The DNA from these plasmids was purified and used as a template for <u>in vitro</u> transcription. The RNAs were displayed in 8% polyacrylamide gels and autoradiograms are shown in Fig. 4. The VA_I RNA made <u>in vitro</u> not only

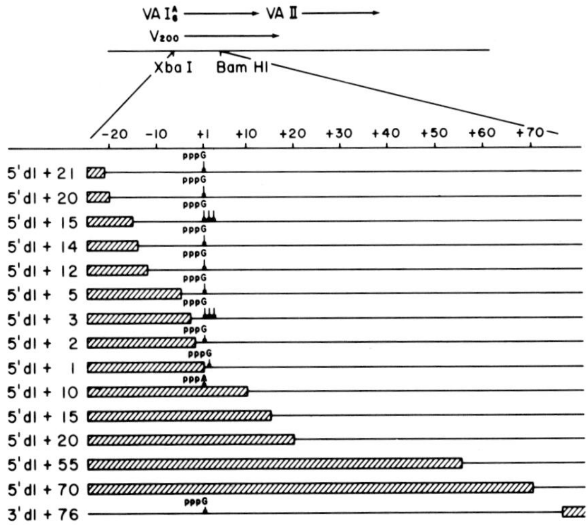

FIGURE 3. Deletion mutants at the VA_I gene. The schematic portion of the region of the viral genome with its restriction enzyme sites is indicated in the upper part. The symbols (⍖) indicate the initiation sites.

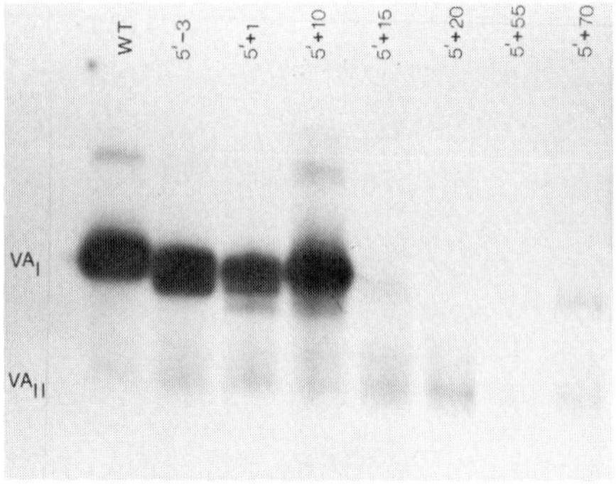

FIGURE 4. _In vitro_ transcription of deleted-substituted DNAs. RNA extracted from _in vitro_ transcription reactions performed in the presence of the different DNA deletion templates was analyzed on 8% polyacrylamide gels containing 8 M urea run at 800V.

comigrates with VA_I RNA made _in vivo_ (results not shown) but also has identical fingerprints (11). The results indicate that up to and including deletions of 10 base pairs into the gene sequence starting from the 5' end, and up to sequence +76 starting from the 3' end, no effect on the relative rate of initiation of transcription can be detected. We have shown that transcription of dl 5' +10 resulted in chimeric RNA, which contained the initial 10 nucleotides transcribed from the pBR322-_SalI_ linker foreign sequence (11). The deletion-substitution of sequences downstream of +76 results in RNA polymerase III transcripts of abnormal length due to removal of the termination sequences (Fig. 5, lane 1), but normal initiation levels and 5' oligonucleotides were found (11). The 5' triphosphate ends were determined by thin-layer analysis of RNAs made in the presence of all four radioactive ribonucleotide triphosphates (50 μM), VA RNA purification after electrophoresis and thin-layer analysis as described (15). As indicated in Fig. 3, initiations with GTP were detected in all cases, except for 5' dl +1 where an ATP initiation was found, and dl 5' -15 and dl 5' -3 where multiple GTP initiations were found. The control region therefore must span somewhere between nucleotides +11 to +76 on the VA_I RNA gene sequence. The deletions dl 5' +15, dl 5' +20, dl 5' +55, and dl 5' +70 are unable to support VA_I RNA gene transcription, although normal levels of VA_{II} RNA are detected.

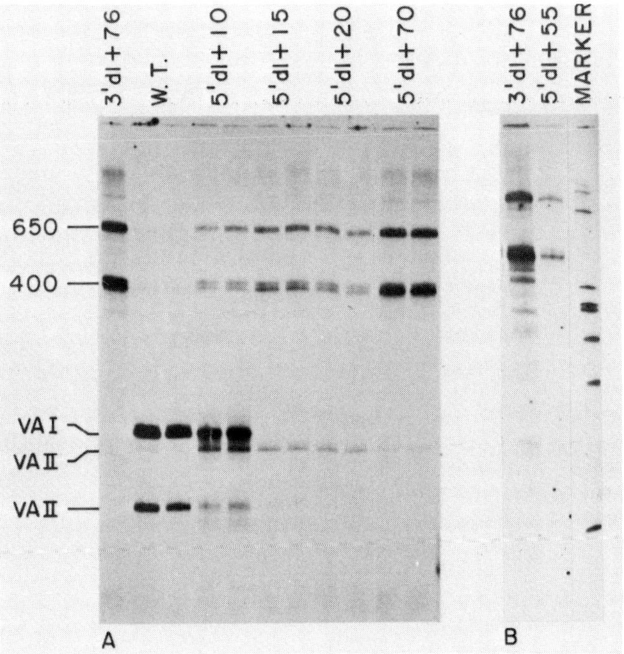

FIGURE 5. Competition of 5' VA, gene deletions for transcription of dl 3' +76 DNA. In dl 3' +76, sequences downstream of nucleotide +76 in VA_I, as well as VA_{II}, are replaced with new adenovirus sequences (11). Reaction mixtures contained 3 μg of dl 3' +76 DNA and four times the molar equivalent of the respective wild type or deletion DNA. Final concentrations of 40 μg/ml were attained by adding appropriate amounts of pBR322 DNA. Incubations in a volume of 50 μl contained 10 μCi of ^{32}P-GTP. RNAs were extracted and analyzed as described in Fig. 4. Gel slices for individual bands were counted by Cerenkov radiation. The level of competition for each competing DNA was expressed as the percentage of the value for the 400 and 650 nucleotide long transcripts obtained in the absence of competition. Values from three experiments (two shown in panel A and one in panel B, see ref. 11) were averaged for each competing plasmid tested: wild type, 16.4%; 5' dl +10, 30.8%; 5' dl +15, 39.9%; 5' dl +20, 28.5%; 5' dl +55, 30.9%; 5' dl +70, 80.9%. ^{32}P-labeled HaeIII-digested φX DNA was denatured and used as a marker.

However, to further define the control region we tested these deletions to see if they are still able to compete for transcriptional factors. To be able to use wild type VA_I RNA and distinguish between the competitor and wild type gene products, we used as a primary template dl 3' +76 (for construction see ref. 11), which gives larger RNA products of a distinguishable size (see Fig. 5, lane 1). Levels of templates were supplemented with pBR322 to maintain a constant concentration of DNA. As indicated in Fig. 5, when a level of competition was expected such that 80% of the transcripts should have been competed, we found

```
         10        20        30        40        50        60        70        80
UUUGUUUCUUCCGUGAUAGUUUAAUGGUCAGAAUGGGCGCUUGUCGCGUGCCAGAUCGGGGUUCAAUUCCCCGUCGCGGAG asp tRNA

GGGCACUCUUCCGUGGUCUGGUGGAUAAAUUCGCAAGGGUAUCAUGGCGGACGACCGGGGUUCGAACCCCGGAUCCGGCCGUCC VAI RNA

GGCUCGCUCCCUGUAGCCGGAGGGUUAUUUUCCAAGGGUUGAGUCGCAGGACCCCCGGUUCGAGUCUCGGGCCGGCCGGACU VAII RNA

AAAUACUCUCGGUAGCCAAGUUGGUUUAAGGCGCAAGACUUUAAAUCUUGAGAUCGGGCGUUCGACUCGCCCCGGGAGA sup4 tRNA
                                         |
                                  UUUAUCACUACGAA

GUUGGGGGCGUAGCUCAGAUGGUAGAGCGCUCGCUUAGCAUGCGAGAGGUACCGGGAUCGAUACCCGGCGCCUCCAAUAU ala tRNA
```

FIGURE 6. Comparison of VA_I and VA_{II} gene sequences to tRNA gene sequences. The first line shows the DNA sequence of yeast asp tRNA gene (from ref. 17). The second and third lines show the sequences of VA_I and VA_{II} RNA genes of Ad 2 (21). The fourth line shows the sequence of sup 4 tyr tRNA of yeast (12) with the sequence spliced out at the position of the arrow indicated on the fifth line. (This was done to include the point mutant described in ref. 12 and underlined here, within the control region of Ad 2 VA_I RNA.) The last line shows the sequence for <u>Bombyx mori</u> ala tRNA (22). There are several regions of strong homology between these genes. Asp tRNA and VA_I RNA are homologous between nucleotides 6-17 (partially outside the control region), between 16-27 for asp tRNA compared to 13-24 in VA_I RNA, from 28-41 in asp tRNA compared to 31-43 in VA_I RNA and 32-40 in sup 4 tyr tRNA. VA_I and ala tRNA are partially homologous between 19-25 for VA_I versus 20-26 for ala tRNA, 29-34 for both. All five sequences are very similar between nucleotides 54-73 for asp tRNA, 53-71 for VA_I RNA, 49-70 for VA_{II} RNA, 49-70 for tyr tRNA and 50-69 for ala tRNA. Outside of the control region of VA_I RNA the first 10 nucleotides also have strong homology with the first 10 of the sup 4 tyr tRNA primary transcript. The VA_I RNAs, VA_{II} RNAs and sup 4 tyr tRNA 5' ends of the primary transcripts are indicated by the first nucleotide. In the case of asp tRNA, the 5' end of the mature tRNA is nucleotide 9, but the primary transcription product in the only case determined is a tandem with an arg tRNA at the 5' end (17).

that indeed dl 5' +15, dl 5' +20 and dl 5' +55 were as efficient competitors as dl 5' +10, although they could not serve as VA_I RNA transcriptional templates. Since dl 5' +70 could not serve as competitor, it narrows down a region between +55 and +76 to which this factor present in limiting amounts binds. This DNA region has been shown to contain significant homologies to tRNA sequences (10, 11, 16, and see Fig. 6). We therefore tested the ability of tRNA genes to compete for VA_I RNA transcription and vice versa. The control in Fig. 7 shows that this heterologous system is unable to distinguish the 5' upstream control region

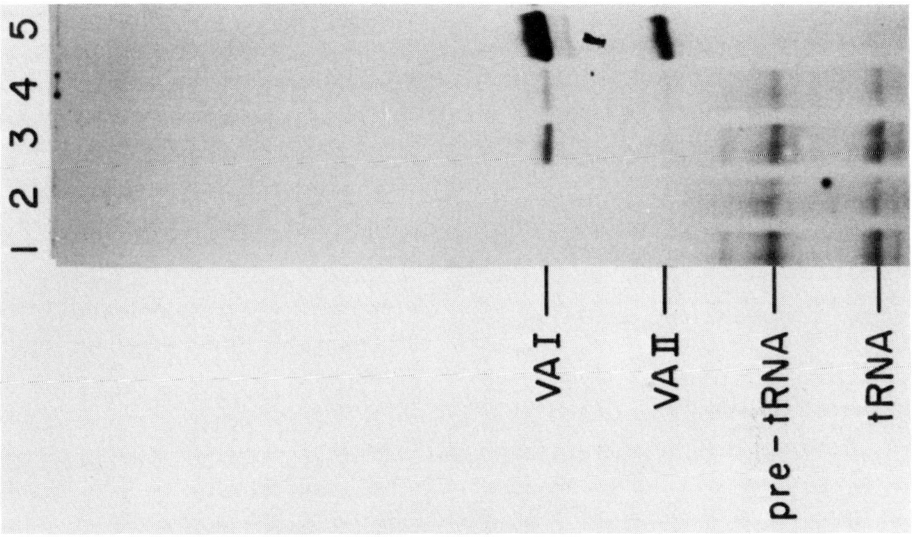

FIGURE 7. Competition between Bombyx mori ala tRNA gene and VA_I gene. The ala tRNA cloned DNAs (22), pBm11-L (intact) and pBm11Δ5' -11L (truncated) were transcribed in the system described above using 1.5 μg of each plasmid (lanes 1 and 2, respectively). Lanes 3 and 4 show identical reactions but with 0.3 μg of the wild-type VA_I plasmid added, representing a two- and four-fold molar excess of the intact and truncated tRNA genes, respectively. The control contained 0.3 μg of the VA_I plasmid alone. All reactions were adjusted to 40 μg/ml DNA with pBR322. Transcription products were analyzed as described in Fig. 5. Transcription of the intact and truncated cloned tRNA genes (about 2,500 cpm in each, lanes 1-4) reduced the synthesis of VA_I RNA to 17.5% (1,250 cpm) and 9.5% (680 cpm) of the control value (7,100 cpm).

previously reported (22). Furthermore, in this HeLa cell system, tRNA is processed to the mature forms.

The heterologous ala tRNA from _Bombyx mori_ used here is a very efficient competitor of VA transcription (Fig. 7) while 5S RNA gene from _Xenopus_ is not (results not shown). Conversely, the wild type VA_I gene can also compete for ala tRNA transcription although less efficiently (results not shown). It should be noted however that VA deletions dl 5' +10, dl 5' +55, and dl 5' +70 cannot compete as well as wild type VA_I RNA for ala tRNA transcription (results not shown). This suggests that for the competition to be efficient the primary binding site is not enough and other factors might be operating.

DISCUSSION

We have slightly modified methods for the _in vitro_ generation of deletions (8, 9) so that a larger multiplicity of deletions can be obtained and analyzed in a reduced amount of time.

The deletions were generated in the adenovirus coded VA_I RNA gene and allow us to give the limits of the DNA regions required for initiation of transcription. These have been located between nucleotides +11 to +76, starting from the major 5' end G of Ad 2 VA_I RNA. Similar results were recently obtained by Fowlkes and Shenk (10), with limits for the control region of +6 to +69 (using our numbering system). The transcripts obtained from dl 5' +10 template are the same size but have different 5' oligonucleotides than the parental DNA template (11). Therefore some measuring mechanism from the internal control region must reach out to the initiation site. Deletion of one of the initiation sites in dl 5' +1 results in initiation in the adjacent purine while substitution of the initial G purine (C on the template stand) by an A in dl 5' +10 results in A initiated VAs, without secondary initiations at any of the surrounding pyrimidines (Fig. 3). Contrary to the requirements of sequences upstream of the gene common to all prokaryotic promoters (1) or required for eukaryotic RNA polymerase II promoters (see this volume), the control region of genes transcribed by RNA polymerase III is internal. Some effect of 5' end sequences occurs on the alignment of the 5' end nucleotide for VA RNA (27), in the case of ala tRNA transcription (22) and here in the case of VA (Fig. 3). Termination sequences are located downstream as indicated by the longer transcripts of 3' dl +76 (Fig. 5).

Questions arise as to the mechanism which participates in the regulation of transcription of this viral gene. It has been shown that one of the factors required for 5S RNA transcription binds very strongly to the control region (b) as defined by genetic manipulation (8, 9). This 37 Kd protein also binds to the 5S RNA molecule (13, 14) and is at least partially responsible for the developmental control of 5S RNA transcription in *Xenopus*. This protein is distinct from the one required for VA and tRNA transcription, and can be separated by chromatography (6, 7). The transcription of VA RNA can be effected by extracts from uninfected cells and therefore it seems likely that the factors required are common to cellular ones, like those necessary for tRNA transcription. The sequence similarities between VA RNA and tRNA extend beyond the control region (see Fig. 6) It seems hard to determine which features of VA and tRNA gene sequence are due to common transcriptional requirements and which are due to some undetermined phylogenetic relationship between the two. The fact that strong competition between tRNA and VA genes occurs does not necessarily demonstrate that the homology is occurring at the initial binding site of the VA factor as defined genetically. The deletion mutants dl 5' +10, dl 5' +55 and dl 5' +70 were not able to compete with tRNA as efficiently as wild type VA_I does (results not shown). These results suggest that the relationship between VA and tRNA might be at the level of a second transcriptional specificity factor. A separation of a factor required for met tRNA versus one required for VA RNA transcription has been recently reported (7). Due to the phylogenetic relationship between VA RNA and several tRNAs previously suggested (10, 11, 16) the use of ala tRNA from *Bombyx* may not have been the best choice. The asp tRNA from yeast is more closely related to VA but a systematic search for an appropriate tRNA as well as use of partially purifed factors will help to establish which factors are distinct for ala tRNA and VA_I RNA and which are common. To illustrate the confusion generated by straight comparison of homologies, the asp tRNA shown in Fig. 6 is the second part of a dimeric transcript (17), where the control region of the first part is probably used for transcription.

There have been contradictory reports that viral DNA replication is required for VA RNA expression (18) or is not required (19). We have previously shown (20) that the increases in VA transcription after adenovirus infection are not due to changes in the level of RNA polymerase III. Regulation by binding of the initiation factor required for VA transcription to the homologous region on the VA RNA itself might explain some of the features of the regulation of transcription observed after adenovirus infection. Before DNA replication, few copies of viral DNA are present so that most of the transcription is from the cell tRNA genes. As viral DNA replicates, the increasing number of viral DNA molecules competes for the initiation factor and VA transcription ensues. As VA RNA is made in large amounts, it in turn can bind to the factor and partially repress VA RNA synthesis, as occurs at later times after infection (20). This model can and will be tested in vitro.

ACKNOWLEDGMENTS

We would like to thank Dr. K. Sprague for the generous gift of plasmids containing Bombyx mori ala tRNA genes.

REFERENCES

1. Chamberlin, M. J., in "RNA Polymerase" (R. Losick and M. J. Chamberlin, eds.), p. 17. Cold Spring Harbor Laboratory, New York, (1976).

2. Wu, G. J., Proc. Natl. Acad. Sci. USA 75, 2175 (1978).

3. Weil, P. A., Segal, J., Harris, B., Ng, S.-Y., and Roeder, R. G., J. Biol. Chem. 254, 6163 (1979).

4. Brown, D. D., and Gurdon, J. B., Proc. Natl. Acad. Sci. USA 74, 2064 (1977).

5. Birkenmeier, E. H., Brown, D. D., and Jordan, E., Cell 15, 1077 (1978).

6. Engelke, D. R., Ng, S.-Y., Shastry, B. S., and Roeder, R. G., Cell 19, 717 (1980).

7. Segall, J., Matsui, T., and Roeder, R. G., J. Biol. Chem. 255, 11,986 (1930).

8. Sakonju, S., Bogenhagen, D. F., and Brown, D. D., Cell 19, 13 (1980).

9. Bogenhagen, D. F., Sakonju, S., and Brown, D. D., Cell 19, 27 (1980).

10. Fowlkes, D., and Shenk, T., Cell 22, 405 (1980).

11. Guilfoyle, R., and Weinmann, R., Proc. Natl. Acad. Sci. USA, (1981) in press.

12. Koski, R. A., Clarkson, S. G., Kurjan, J., Hall, B. D., and Smith, M., Cell 22, 415 (1980).

13. Pelham, H. R. B., and Brown, D. D., Proc. Natl. Acad. Sci. USA 77, 4170 (1980).

14. Honda, B., and Roeder, R.G., Cell 22, 119 (1980).

15. Korn, L. J., Birkenmeier, E. H., and Brown, D. D., Nucleic Acids Res. 7, 947 (1979).

16. Ohe, K., and Weissman, S. W., Science 167, 879 (1970).

17. Abelson, J., and Soll, D., Nature 287, 750 (1980).

18. Mathews, M., and Pettersson, U., J. Mol. Biol. 119, 293 (1978).

19. Jones, N., and Shenk, T., Proc. Natl. Acad. Sci. USA 76, 3665 (1979).

20. Weinmann, R., Jaehning, J., Raskas, H. S., and Roeder, R. G., J. Virol. 17, 114 (1976).

21. Akusjarvi, G., Mathews, M. B., Andersson, P., Vennstrom, B., and Pettersson, U., Proc. Natl. Acad. Sci. USA 77, 2424 (1980).

22. Sprague, K. U., Larson, D., and Morton, D., Cell 22, 171 (1980).

23. Ohe, K., and Weissman, S. M., J. Biol. Chem. 246, 6991 (1971).

24. Legerski, R. J., Hodnett, J. L., and Gray, H. B., Nucleic Acid Res. 5, 1445 (1978).

25. Maxam, A. M., and Gilbert, W., in "Methods in Enzymology" (L. Grossman and K. Moldave, eds.), Vol. 60, p. 499. Academic Press, New York, (1980).

26. Seif, I., Khoury, G., and Dhar, R., Nucleic Acids Res. 9, 2225 (1980).

27. Thimmapaya, B., Jones, N., and Shenk, T., Cell 18, 947 (1979).

REGULATION OF ADENOVIRUS VA RNA GENE EXPRESSSION

Cary Weinberger, Bayar Thimmappaya[1], Dana M. Fowlkes[2], and Thomas Shenk

Department of Microbiology
Health Sciences Center
State University of New York
Stony Brook, New York 11794

ABSTRACT By constructing deletion mutations in a cloned adenovirus type 2 VAI RNA gene and measuring the ability of altered templates to direct transcription of VAI RNA in cell free extracts, we have located two VAI control regions. One region is located entirely within the VAI RNA coding sequences. This intragenic control region is essential for function of the VAI transcription unit *in vitro*. The second control region lies in the 5' flanking sequences of the VAI gene. This region serves to modulate the efficiency of *in vitro* transcription. Nucleotide sequence similarities are evident on comparison of the VAI intragenic control region to other genes transcribed by RNA polymerase III. We have also cloned the adenovirus VA RNA genes into the late region of the SV40 genome. VAI RNA is produced in cells infected by this recombinant virus, demonstrating that the VAI gene can be expressed *in vivo* in the absence of any additional adenovirus gene products. A model to account for the differential expression of VAI and VAII RNAs in infected cells is discussed.

INTRODUCTION

The adenovirus VA RNAs are small RNAs (about 160 nucleotides) of unknown function which are synthesized in large amounts at late times after infection. These RNAs are transcribed by RNA polymerase III (1,2). They are encoded by two different genes, designated VAI and VAII, which are located at about 30 map units on the adenovirus type 2 (Ad2).

[1]Present address: Department of Microbiology and Immunology, Northwestern University School of Medicine, Chicago, Illinois 60611
[2]Present address: Laboratory of Pathology, National Cancer Institute, Bethesda, Maryland 20205.

ISBN 0-12-137420-3

chromosome (3,4,5). The VA RNA genes of both Ad2 and Ad5 have been sequenced (6,7,8). The VAI RNA species is heterogeneous at both 5' and 3' ends (7,9,10,11). Two 5' ends have been identified: the longer species initiates with an A residue and the shorter with a G residue located three nucleotides downstream from the A start. Since the 5' ends of the VAI RNAs do not undergo a post-transcriptional cleavage (3,10,12,13), it is clear that transcription by RNA polymerase III initiates at the positions to which the 5' ends have been mapped.

Here we briefly review the VAI transcriptional control region, we demonstrate that VAI RNA can be expressed in animal cells in the absence of any other adenovirus genes, and we present a simple model to explain the regulation of VA RNA gene expression.

RESULTS AND DISCUSSION

<u>The VAI RNA gene contains two transcriptional control regions</u>. An essential VAI transcriptional control region

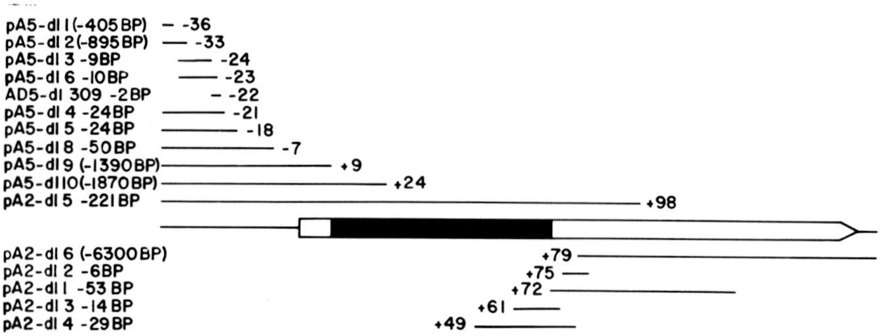

Figure 1. Diagrammatic representation of deletions within and surrounding the VAI RNA gene. The VAI gene coding region is designated by a thick arrow pointing in the 5' to 3' direction. The filled portion of the arrow represents the intragenic control region (+9 to +72 relative to the VAI[A] start). Deletions are represented by bars and their end points nearer the control region are indicated (relative to the VAI[A] initiation site which is +1). The size of each deletion is included. Numbers in parentheses are size estimates based on altered migration of restriction endonuclease-generated fragments. Even though the boundaries of these deletions have been determined at the nucleotide sequence level, the wild-type sequence between the two points has not been determined in its entirety.

was identified by constructing deletion mutations in a VAI gene cloned in pBR322 and measuring the ability of altered templates to direct transcription of VAI RNA in HeLa cell-free extracts. pA5-dl9, whose deletion extends through the 5' flanking region and into the VAI coding region to position +9 (Figure 1), still directed the synthesis of an RNA. This transcript initiated at a G residue within pBR322 sequences. Deletions extending to +24 or +98 (pA5-dl10 or pA2-dl5, Figure 1) prevented the synthesis of a VAI transcript. Thus the 5' boundary of the region essential for cell-free transcription of the VAI gene lies between nucleotide positions +9 and +24. Analysis of deletions extending from the other side of the control region fix its 3' boundary between +72 (pA2-dl1 is an active template, Figure 1) and +61 (pA2-dl3 is inactive). The segment essential for *in vitro* transcription of the VAI gene is located entirely within the VAI RNA coding region. The maximal extent of this intragenic control region is position +9 to +72, ignoring the possible effects of new sequences brought into the vicinity of the control region as a result of the deletions. The *Xenopus* 5S RNA gene (14,15, 16) and the yeast SUP4 $tRNA^{tyr}$ gene (17) have also been demonstrated to contain intragenic control regions.

A second control region for the VAI gene lies within its 5' flanking sequences. This region is not essential for cell-free transcription of VA RNA, but it modulates the efficiency of transcription. Mutant templates carrying deletions in their 5' flanking sequences (e.g. pA5-dl4 and pA5-dl7, Figure 1) are transcribed less efficiently *in vitro* than genes with a wild-type flanking sequence (demonstrated by competition experiments, ref. 18). Modulation by 5' flanking sequences *in vitro* has also been observed for *Drosophila* $tRNA^{lys}$ (19) and *Xenopus* $tRNA^{met}$ genes (20). In both cases, the 5' flanking sequences modulated transcription downward. When the 5' flanking sequences of a poorly transcribed *Drosophila* $tRNA^{lys}$ gene were replaced by pBR322 sequences, the modified template directed the synthesis of substantially increased amounts of product. Alterations in the 5' flanking regions of the VAI and tRNA genes produced opposite effects: VAI RNA transcription was reduced while tRNA transcription was increased. Apparently, the 5' flanking sequences can regulate expression by either optimizing (VAI gene) or reducing ($tRNA^{lys}$ gene) transcription.

The VAII gene, like the VAI gene, contains an intragenic control region. A small deletion within the VAII coding region (pA2-dl8, ref. 18) inactivates transcription of the gene. VAII RNA is synthesized in much smaller amounts than VAI RNA subsequent to infection. Consistent with the *in* vivo observation, the VAI gene is a strong competitive inhibitor of VAII RNA transcription in HeLa cell extracts (18). Presumably, the VAII 5' flanking region or intragenic control region (or both) is less efficient in directing transcription than the corresponding VAI region.

Conserved sequences in RNA polymerase III-transcribed genes. Two sequences within the adenovirus VAI intragenic control region appear to be conserved in a variety of different genes transcribed by polymerase III (Figure 2, and ref. 18). The first conserved sequence is near the 5' end

	+1 +10 +20 +30 +40 +50 +60 +70 +80 +90
AD-2 VAI(G)	GGGCACTCTTCCGTGGTCTGGTGGATAAATTCGCAAGGGTATCATGGCGGACGACCGGGGTTCGAACCCCGGATCCGGCCGTCCGCCGTGA...
AD-2 VAII	GGCTCGCTCCCTGTAGCCGGAGGGTTATTTTCCAAGGGTTGAGTCGCAGGACCCCCGGTTCGAGTCTCGGGCCGGCCGGACTGCGGCGAAC...
$tRNA^{TYR-SUP4}$	AAAUACTCTCGGTAGCCAAGTTGGTTTAAGGCGCAAGACTTTAATTTATCACTACGAAATCTTGAGATCGGGCGTTCGACTCGCCCCCGGG...
MOUSE 4.5S	GCCGGTAGTGGTGGCGCACGCCGGTAGGATTTGCTGAAGGAGGCAGAGGCAGAGGGATCACGAGTTCGAGGCCAGCCTGGGCTACACATTTTTT
HUMAN A36	AGGCTGGGAGTGGTGGCTCACGCCTGTAATCCCAGAATTTTGGGAGGCCAAGGCAGGCAGATCACCTGAGGTCAAGAGTTCAAGACCAACC...
XENOPUS 5S	GCCTACGGCCATACCACCCTGAAAGTGCCCGATATCGTCTGATCTCGGAAGCCAAGCAGGGTCGGGCCTGGTTAGTACTTGGATGGGAGAC...

Figure 2. Partial nucleotide sequence of several genes transcribed by RNA polymerase III. Sequences are the sense of the RNA encoded and +1 marks the 5'-end of the primary transcript. Arrowheads mark the intragenic control region boundaries of the VAI and Xenopus 5S genes. The conserved sequences within the control region are underlined.

of the control region and reads 5'-GTGGPyNNPuGTGG-3'. The second is near the 3' boundary of the control region and reads 5'-GGGTTCGAANCC-3'. Similar sequences are present in tRNA genes, mouse and hamster 4.5S RNA genes (21,22,23,24), human alu family polymerase III transcripts such as A36 (25) and the Xenopus 5S gene. The VAI sequences are not represented perfectly in each gene, but the substantial homology together with their relatively constant location within these genes indicates they are likely of critical importance. Indeed, if several base-pairs are removed from either sequence in the VAI RNA gene, the mutant template is no longer transcribed in vitro (18,26). Further, single base-pair changes within these homologous regions in the yeast SUP4 $tRNA^{tyr}$ gene can drastically affect transcription in vitro (17). A change at position +56 (numbered according to Sprinzl et al., ref. 27) prevented detectable initiation of transcription, while changes at +15 and +21 enhanced in vitro transcription. The change at +56 is within the 3' region and that at +15 within the 5' region of homology.

A recombinant SV40 virus carrying the Ad5 VA genes expresses VAI RNA subsequent to infection. Does the synthesis of VAI RNA within an infected cell require expression of other adenovirus-specific gene products? To answer this question the Ad5 VA RNA genes were cloned into the late transcriptional unit of SV40 (diagrammed in Figure 3). This defective recombinant was propagated by complementation

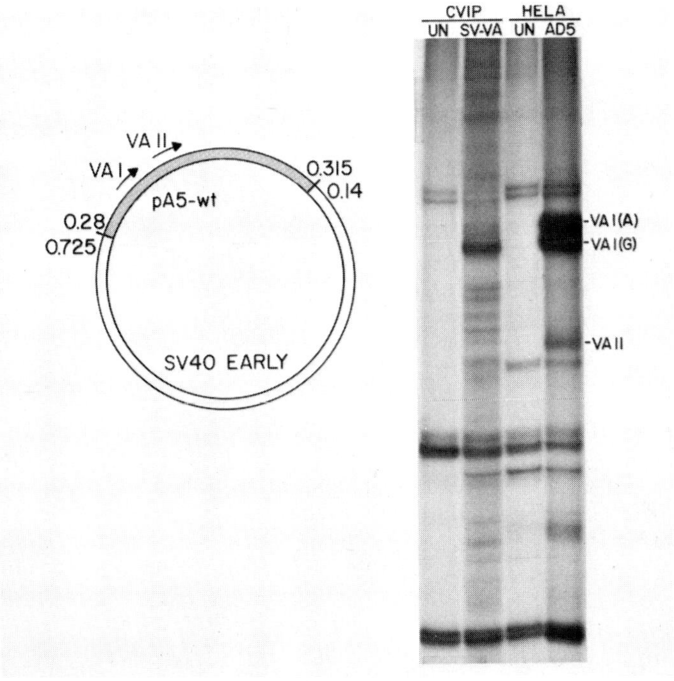

Figure 3. Autoradiogram of ^{32}P-labeled RNAs extracted from the cytoplasm of uninfected cells and cells infected with either Ad5 or an SV40 recombinant carrying the Ad5 VA RNA genes. HeLa cells were infected with Ad5 (multiplicity of infection = 5 pfu/cell) or CVIP cells were infected with a stock of the SV40/Ad5 recombinant depicted in the diagram plus its helper virus, tsA58 (approximate multiplicity of infection = 0.1 pfu recombinant virus/cell). Ad5-infected cells were labeled from 12-36 hr after infection and SV40-infected cells were labeled from 96-120 hr after infection. Cells were harvested immediately after labeling and total cytoplasmic RNA was prepared. Electrophoresis was at 20W for 16 hr in an 8% polyacrylamide gel (0.4 mm thick x 40 cm long) containing 100 mM Tris-OH, 100 mM boric acid, 2mM EDTA and 8M urea.

using a helper virus, tsA58. VAI RNA was synthesized in monkey kidney cells subsequent to infection with a mixed recombinant/helper virus stock (Figure 3). Similar observations have been made by Kaufman and Sharp (28). We conclude that VAI RNA can be expressed in the absence of any other adenovirus gene products.

Regulation of VA RNA expression within adenovirus-infected cells. Analysis of VA RNA synthesis within adenovirus-infected cells suggests that the two VA RNA species are differentially regulated (5). VAI and VAII RNAs are synthesized in similar amounts at early times after infection, but as DNA synthesis begins, the rate of VAII RNA synthesis levels off and VAI RNA becomes increasingly predominant. The differential expression of VAI and VAII RNAs probably simply reflects competition by the two genes for a limiting transcriptional factor. We have demonstrated in vitro that the VAI gene is a strong competitive inhibitor of VAII RNA synthesis (discussed above and ref. 18). It seems likely that at early times after infection with few VA gene templates present in a cell, there are adequate supplies of all factors required for optimal transcription of both VAI and VAII genes. As a result the two RNAs are synthesized in approximately equal amounts. However, as DNA synthesis begins the VA gene copy number dramatically increases within the cells. Very likely, one or more polymerase III transcription factors becomes limiting and the VAI gene product begins to predominate.

Thus, it is not necessary to postulate specific regulatory factors to explain VA RNA expression in adenovirus infected cells. The VA genes can be expressed independently of additional viral gene products (Figure 3). Further, the two VA genes are probably "on" and capable of functioning at all times after infection. Differential expression of the two RNAs very likely reflects the relative abilities of the two VA genes to compete for limiting transcription factors as the infection proceeds.

ACKNOWLEDGMENTS

This work was supported by a USPHS research grant from the National Cancer Institute (CA-28919). C.W. is the recipient of a National Cancer Institute predoctoral traineeship; D.M.F. was a postdoctoral fellow of the National Institute of Allergy and Infectious Disease and T.S. is an Established Investigator of the American Heart Association.

REFERENCES

1. Reich, P.R., Rose, J., Forget, B. and Weissman, S.M. (1966). J. Mol. Biol. 17, 428.
2. Weinmann, R., Raskas, H.J. and Roeder, R.G. (1974). Proc. Nat. Acad. Sci. USA 71, 3426.
3. Mathews, M.B. (1975). Cell 6, 223.
4. Pettersson, U. and Philipson, L. (1975). Cell 6, 1.
5. Soderlund, H., Pettersson, U., Venstrom, B., Philipson, L. and Mathews, M.B. (1976). Cell 7, 585.
6. Ohe, K. and Weissman, S.M.(1970). Science 167, 879.
7. Thimmappaya, B., Jones, N. and Shenk, T. (1979). Cell, 18, 947.
8. Akusjarvi, G., Mathews, M.B., Andersson, P., Vennstrom, B. and Pettersson, U. (1980). Proc. Nat. Acad. Sci. USA 77, 2424.
9. Celma, M.L., Pan, J. and Weissman, S.M. (1977). J. Biol. Chem. 252, 9032.
10. Celma, M.L., Pan, J. and Weissman, S.M. (1977). J. Biol. Chem. 252, 9043.
11. Vennstrom, B., Pettersson, U., and Philipson, L. (1978). Nucl. Acids Res. 5, 195.
12. Price, R. and Penman, S. (1972). J. Mol. Biol. 70, 435.
13. Vennstrom, B., Pettersson, U., and Philipson, L. (1978). Nucl. Acids Res. 5, 205.
14. Sakonju, S., Bogenhagen, D.F. and Brown, D.D. (1980). Cell 19, 3.
15. Bogenhagen, D.F., Sakonju, S. and Brown, D.D. (1980). Cell 19, 27.
16. Engelke, D.R., Ng, S.-Y., Shastry, B.S. and Roeder, R.G. (1980). Cell 19, 717.
17. Koski, R.A., Clarkson, S.G., Kurjan, J., Hall, B. and Smith, M. (1980). Cell 22, 415.
18. Fowlkes, D.M. and Shenk, T. (1980). Cell 22, 405.
19. DeFranco, D., Schmidt, O. and Soll, D. (1980). Proc. Nat. Sci. USA 77, 3365.
20. Clarkson, S.G., Koski, R.A., Corlet, J. and Hipskind, R.A. (1981). J. Sup. Struct. Cell. Biochem. Supp. 5, 425.
21. Peters, G.G., Harada, F., Dahlberg, J.E., Panet, A., Haseltine, W.A. and Baltimore, D. (1977). J. Virol. 21, 1031.
22. Jelinek, W. and Leinward, L. (1978). Cell 15, 205.
23. Harada, F. and Ikawa, Y. (1979). Nucl. Acids Res. 7, 895.
24. Harada, F., Kato, N. and Hoshino, H. (1979). Nucl. Acids Res. 7, 909.
25. Duncan, C.H., Jagadeeswaran, P., Wang, R.R.C., and Weissman, S.M. (1981). Gene, in press.
26. Guilfoyle, R. and Weinman, R. (1981). Proc. Nat. Acad. Sci. USA, in press.

27. Sprinzl, M., Grueter, F., Spelzhaus, A. and Gauss, D.H. (1980). Nucl. Acids Res. 8, rl.
28. Kaufman, R.J. and Sharp, P.A. (1981). J. Sup. Struct. Cell. Biochem. Supp. 5, 434.

RETROVIRUS ONCOGENES

J. Michael Bishop

Department of Microbiology and Immunology
University of California
San Francisco, California

Framing the Problem: Cancer and Normal Cellular Genes

The search for etiological agents has dominated much of the research and thinking on human cancer. But what are the targets for these agents within the cell, and by what means is the malignant phenotype sustained? In 1973, David Comings formulated a model that addressed these issues by drawing upon the mounting evidence for the existence of "cancer genes" - genetic loci within the human genome whose alteration or anomalous expression might underlie all forms of oncogenesis (1). The model envisioned a battery of genes (denoted by the generic term "Tr" for transforming) that direct cell division during the course of normal growth and development. Regulatory loci would dictate when Tr genes should and should not act. Any change (such as a mutation or chromosomal rearrangement) that relieved a Tr gene of the influence of its regulatory locus might trigger unremitting cell division and, hence, malignant growth. The model limited the effects of each Tr gene to the cells of one or another developmental lineage, in an effort to account for the apparent tissue-specificity of most of the identified "cancer genes" (2). With remarkable prescience, Comings further proposed that tumor viruses arose by the seizure (or "transduction", in formal terms) of Tr genes; once incorporated into a viral genome, a Tr gene might escape the inhibitory influences of the cell and thus become an "oncogene" - a genetic locus responsible for viral tumorigenesis.

DEVELOPMENTAL BIOLOGY USING PURIFIED GENES

ISBN 0-12-137420-3

If correct, Coming's proposal would have a two-fold significance for tumor virology. First, the actions of viral oncogenes might mirror the mechanisms of oncogenesis induced by other means - viral oncogenesis may be a general paradigm rather than an experimental curiosity. Second, the mechanisms of viral oncogenesis may be kindred to the mechanisms that regulate the normal course of growth and development: to study one set of mechanisms is to study the other.

Retrovirus Oncogenes and Differentiation

We can now argue that goodly portions of Comings' model may be correct. The evidence for this statement comes from studies of retroviruses, which have provided the most coherent and penetrating view of oncogenesis presently available to us. I will explain the evidence by reviewing recent work with a set of prototypic retroviruses isolated from tumors in chickens (Table 1). Notice that: i) these viruses can induce a wide variety of tumors; cells derived from each of the germ layers can be affected; ii) the viruses can be classified according to their pathogenicity; each class of virus reliably induces a specific type (or types) of tumor; iii) transformation by the viruses in cell culture duplicates the specificity displayed in the animal; iv) with one exception (the avian leukosis viruses), tumorigenesis by each class of virus is attributable to a specific viral oncogene, which we name according to the viruses in which they are found, eg., v-src of Rous sarcoma virus, v-myc of myelocytomatosis virus, v-erb of erythroblastosis virus, and v-myb of myeloblastosis virus.

The tissue specificity of transformation by retrovirus oncogenes evokes one of the major themes in the model proposed by Comings: cancer genes may be determinants of normal growth and development in another guise, and malignant transformation may therefore be viewed as an anomaly of differentiation. Retrovirus oncogenes unquestionably affect the differentiation of susceptible cells. For example, v-src blocks the conversion of myoblasts to myotubes and eradicates differentiated features of fibroblasts, chondroblasts and melanoblasts (3). These changes seem not to be central to the mechanism of malignant transformation by v-src, but they nevertheless speak of an interaction between v-src and cellular regulators of differentiation.

TABLE 1. Pathogenicity of Some Avian Retroviruses

Virus	Pathogenicity	Target in Culture	Oncogene	Proto-Oncogene
RSV	Sarcomas	Fibroblast	v-*src*	c-*src*
MCV	Carcinomas Myeloid tumors Sarcomas	Epithelial Myeloid Fibroblast	v-*myc*	c-*myc*
AEV	Erythroleukemia Sarcomas	BFU-E Fibroblast	v-*erb*	c-*erb*
AMV	Myelocytomatosis	Myeloid	v-*myb*	c-*myb*
ALV	B cell lymphoma	None	None	None

Leukemogenesis has elicited more explicit efforts to relate differentiation and oncogenesis: the pathogenesis of human leukemias has been attributed to the arrest of differentiation in specific developmental lineages (4); and erythroid cells transformed by avian erythroblastosis virus display an immature phenotype that can undergo further development if the activity of the viral oncogene is withdrawn (5). It would be premature, however, to conclude that arrest of differentiation accounts for tumorigenesis. For example, transformation of hematopoietic cells by either avian myeloblastosis virus or myelocytomatosis virus engenders what can best be described as "chaotic" phenotypes, comprising elements of not one but several different compartments in the myelomonocytic lineage (6): the oncogenes of these viruses do not arrest development at a defined point in the lineage; the "immature" phenotypes are likely to be manifestations rather than causes of malignant transformation.

The Products of Retrovirus Oncogenes: Mechanisms of Viral Oncogenesis

We can locate oncogenes as discrete loci within retrovirus genomes; we can isolate their nucleotide sequences in large quantities by the use of recombinant DNA; and we can identify the protein products by which the actions of the genes are mediated. We know the most about v-src and its product, a phosphoprotein ($pp60^{v\text{-}src}$) that apparently possesses the enzymatic activity of a protein kinase (7). Products of the other avian retrovirus oncogenes listed in Table 1 have also been identified, but nothing is known as yet about their functions (Table 2).

The protein kinase activity of $pp60^{v\text{-}src}$ surfaced with the discovery that the viral protein could phosphorylate antibodies to which it was bound (8,9). The protein has now been purified in several laboratories (10-12), and after rigorous scrutiny, it remains a protein kinase with the unexpected property of phosphorylating tyrosine (and only tyrosine) in protein substrates (7,11,12). The enzymatic activity of $pp60^{v\text{-}src}$ offers an attractive explanation for its biological effects, because phosphorylation is one of the central means by which the activity of many proteins is controlled (13). Thus, one enzyme, by phosphorylating several or even many proteins, can vastly alter the functioning of a cell.

TABLE 2. The Products of Avian Retrovirus Oncogenes*

	RSV	MCV	AEV		AMV
Oncogene	src	myc	erb		myb
Mr of product	60K	110K	74K	41K	?
Phosphorylated	+	+	+	?	?
Kinase	+	-	-	?	?
Subcellular location	PM	?	?	?	?
Cellular progenitor	+	+	+	+	+

*Symbols are as follows: +, detected; -, not detected; ?, uncertain; PM, plasma membrane.

Where in the cell does $pp60^{v\text{-}src}$ act? Cancer is a disturbance of cell division. We might therefore expect to find $pp60^{v\text{-}src}$ in the nucleus, where many of the events of cell division may originate. We would be wrong: the weight of the evidence now suggests that there is little if any $pp60^{v\text{-}src}$ in the nucleus of the cell; rather, the majority (perhaps all) of the protein is bound to the cytoplasmic aspect of the plasma membrane (14,15). We are forced to the conclusion that $pp60^{v\text{-}src}$ exerts its effects at the periphery of the cell, and that most if not all of the aberrations in the cell transformed by v-src must be traced to these peripheral effects. Moreover, $pp60^{v\text{-}src}$ is apparently not unique in its actions. At least five other products of retrovirus oncogenes are also protein kinases (7,16); all but one of these appear to phosphorylate tyrosine (7,16); and at least two are associated with the plasma membrane (16).

Identification of Proto-Oncogenes

The evolutionary origins of retrovirus oncogenes came into view with the discovery that the DNAs of normal birds and mammals contain a gene (c-src) that is closely related to the oncogene (v-src) of Rous sarcoma virus (17). These findings have since been generalized to all of the oncogenes described by Table 1: the DNA of every vertebrate tested contains a homologue for each of these oncogenes, and the composition of the homologues is extensively conserved throughout the vertebrate phyla (18). We designate the vertebrate homologues as "proto-oncogenes" because we believe that they gave rise to viral oncogenes during the course of evolution (19).

Make no mistake - proto-oncogenes are cellular genes, not viral genes in disguise (18,19): they are located at constant genetic loci in every member of a species, in striking contrast to the distribution and positioning of endogenous retrovirus genes in the same species; they behave as classical Mendelian loci in genetic crosses; they contain intervening sequences (or introns) - a hallmark of eukaryotic genes and, again, a telling contrast with the organization of retrovirus genes; they are not found within or even linked to proviruses of endogenous retroviruses; and their expression is not coordinated with that of endogenous retrovirus oncogenes. It seems unlikely that proto-oncogenes were introduced into vertebrate genomes by infection of ancestral species with retroviruses; instead, it appears that complex cellular genes have made their way, in part or entirely, into the genomes of pre-existent retroviruses, much as Comings envisioned (1).

Proto-Oncogenes are Expressed and Encode Proteins Similar to the Products of Viral Oncogenes

Each of the proto-oncogenes described in Table 1 gives rise to one or more distinctive RNA transcripts whose sizes are identical in different tissues and species (18). The constancy of these RNAs among widely diverged species testifies to the selective pressures that have apparently preserved the structure and function of proto-oncogenes across vast periods of evolutionary time. Where experimental reagents so far permit, we can also detect proteins encoded by proto-oncogenes. For example, all vertebrate cells examined to date produce a protein ($pp60^{c\text{-}src}$) that is virtually indistinguishable from the product of v-*src* ($pp60^{v\text{-}src}$): both cellular and viral proteins are tyrosine kinases (7), and both are bound to the plasma membrane (15).

Are Proto-Oncogenes the "Cancer Genes" of Normal Cells?

The principles enunciated above for the oncogenes of avian retroviruses are widely applicable to the oncogenes of other retroviruses: homologues of these genes (i.e., proto-oncogenes) can be found in vertebrate DNA, and many of these are expressed in phenotypically normal cells (16). The sole exception at present is the oncogene of the Spleen Focus Forming Virus of mice, which appears to be a recombinant form of the retrovirus env gene rather than the derivative of a cellular gene (20). Indeed, it appears that proto-oncogenes may comprise a family of genes whose interrelationships are akin to those found in the multigene families that encode immunoglobulins and histocompatability antigens. Even proto-oncogenes whose nucleotide sequences bear no detectable relationship to one another may nevertheless encode proteins of similar function and evolutionary origins (19).

The discovery and characterization of proto-oncogenes engendered a general theory of oncogenesis that embodies three brash predictions: i) that proto-oncogenes are none other than the Tr genes posited by Comings - genes active in normal growth and development and hence able to influence the course of cell division; ii) that proto-oncogene and viral oncogene are one and the same - viral oncogenesis is the consequence of presenting an unhealthy and sustained abundance of an otherwise normal gene product to the cell; and iii) that the identification of proto-oncogenes has revealed the common genetic effectors for many oncogenic agents. These immodest proposals have been abroad for some years; they have at last acquired some experimental substance.

Are proto-oncogenes either regulators or effectors of differentiation? We cannot yet say, but circumstantial evidence holds the possibility open. First, transcription from the proto-oncogenes for avian retroviruses is independently controlled from one tissue to another (unpublished results of T. Gonda, D. Sheiness and J.M.B.; see also ref. 19); thus, the proto-oncogenes are not coordinately expressed as a group, and the function of each gene may be required only in certain tissues. Second, primitive hematopoietic cells - but not cells of other origins - contain large amounts of the protein encoded by the proto-oncogene of a murine retrovirus that induces both sarcomas and erythroleukemia (21).

Third, the proto-oncogene for avian erythroblastosis virus is composed of two domains whose expression appears to be independently determined by the embryological lineage of tissues (unpublished results of B. Vennstrom and J.M.B.).

What can be said to sustain the suggestion that proto-oncogenes harbor the potential for transforming cells to malignancy? A proto-oncogene has been isolated from normal rodent cells by molecular cloning, linked to portions of retrovirus genomes that can facilitiate vigorous gene expression, and reinserted into cells, where it is assimilated into the cellular genome and expressed (22). The result: some of the recipient cells become malignant. The conclusion: achieve sufficient expression of a proto-oncogene and the events of viral oncogenesis are apparently duplicated.

Are proto-oncogenes the common mediators of oncogenesis, whatever its inciting cause - Tr genes which, when unleashed, set the cell on the path to malignancy? At the present, we have a single, immensely provocative clue derived from an unexpected source. Avian leukosis viruses have no oncogenes, yet infection of the bursa of young chickens by these viruses gives rise to a fatal B-cell lymphoma (Table 1). How do these tumors arise? Suppose that insertion of the leukosis virus provirus at certain sites in the genome of the bursal cell were mutagenic, and that this mutagenesis engendered the neoplastic phenotype? The supposition can be tested: viral DNA can be traced to its precise location in the cellular genome, and the adjacent cellular DNA can be identified and isolated. These tracking operations have now been done, and the results are startling; in all but a few of the tumors, viral DNA has been inserted in the immediate vicinity of a single proto-oncogene (c-myc, the progenitor for v-myc; see Table 1), and as a seeming consequence, expression of the proto-oncogene is greatly augmented (23). The full implications of these findings have yet to be explored. Why, for example, are the tumorigenic effects of the leukosis virus limited mainly to lymphoid tissues? And why aren't similar tumors induced by v-myc, the viral progeny of c-myc? But these nuances should not be permitted to obscure the likelihood that the discovery of proto-oncogenes has vindicated much of David Comings' prophesy. The study of retrovirus oncogenes and their evolutionary origins may have unveiled both a final common pathway for tumorigenesis and genetic effectors of normal growth and development.

ACKNOWLEDGMENTS

Work in the author's laboratory is supported by USPHS grants CA 12705, CA 19287, Training grant 1T32 CA 09043 and American Cancer Society grant MV48G.

REFERENCES

1. Comings, D.E. (1973). Proc. Nat. Acad. Sci. USA 70, 3324.
2. Knudson, A.G., Jr. In "Genes, Chromosomes, and Neoplastia" (33rd Annual Symposium on Fundamental Cancer Reserch) M.D. Anderson Hospital and Tumor Institute (F.E. Arrighi, P.N. Rao and E. Stubblefield, eds.), p. 453. Raven Press, New York.
3. Boettiger, D., and Durban, E.M. (1980). Cold Spring Harbor Symposia on Quant. Biol. 44, 1249.
4. Greaves, M., and Janossy, G. (1978). BBA (Reviews on Cancer) 516, 193.
5. Graf, T., Ade, N., and Beug, H. (1978). Nature 275, 496.
6. Durban, E.M., and Boettiger, D. (1981). Proc. Nat. Acad. Sci. USA, in press.
7. Hunter, T., and Sefton, B.M. (1981). In "Molecular Aspects of Cellular Regulation" (P. Cohen and S. Van Heznigen, eds.) Elsevier/North Holand.
8. Collett, M.S., and Erikson, R.L. (1978). Proc. Nat. Acad. Sci. USA 75, 2021.
9. Levinson, A.D., Oppermann, H., Levintow, L., Varmus, H.E., and Bishop, J.M. (1978). Cell 15, 561.
10. Maness, P.F., Engeser, H., Greenberg, M.E., O'Farrell, M., Gall, W.E., and Edelman, G.M. (1979). Proc. Nat. Acad. Sci. USA 76, 5028.
11. Erikson, R.L., Collett, M.S., Erikson, E., and Purchio, A.F. (1979). Proc. Nat. Acad. Sci USA 76, 6260.
12. Levinson, A.D., Oppermann, H., Varmus, H.E., and Bishop, J.M. (1980). J. Biol. Chem. 255, 11973.
13. Rubin, C.S., and Rosen, O.M. (1975). Ann. Rev. Biochem. 44, 831.
14. Willingham, M.C., Jay, G., and Pastan, I. (1979). Cell 18, 125.

15. Courtneidge, S.A., Levinson, D., and Bishop, J.M. (1980). Proc. Nat. Acad. Sci. USA 77, 3783.
16. Bishop, J.M. (1981). In "International Symposium on Aging and Cancer", in press.
17. Spector, D., Varmus, H.E., and Bishop, J.M. (1978). Proc. Nat. Acad. Sci. USA 75, 4102.
18. Bishop, J.M., Gonda, T., Hughes, S.H., Sheiness, D.K., Stubblefield, E., Vennstrom, B., and Varmus, H.E. (1980). In "Mobilization and Reassembly of Genetic Information", p. 261.
19. Bishop, J.M. (1981). Cell 23, 5.
20. Linemeyer, D.L., Ruscetti, S.K., Scolnick, E.M., Evans, L.H., and Duesberg, P.H. (1981). Proc. Nat. Acad. Sci. USA 78, 1401.
21. Scolnick, E.M., Weeks, M.O., Shih, T.Y., Ruscetti, S.K., and Dexter, T.M. (1981). Mole. and Cellular Biol. 1, 66.
22. Oskarsson, M., McClements, W.L., Blair, D.G., Maizel, J.V., and Vande Woude, G.F. (1980). Science 207, 1222.
23. Hayward, W.S., Neel, B.G., and Astrin, S.M. (1981). Nature 290, 475.

ISOLATION OF A NEW NONDEFECTIVE ADENOVIRUS-SV40 HYBRID VIRUS FROM *IN VITRO* CONSTRUCTED DEFECTIVE VIRUSES[1]

Suzanne L. Mansour
Carl S. Thummel
Robert Tjian

Department of Biochemistry
University of California
Berkeley, California 94720

Terri Grodzicker

Cold Spring Harbor Laboratory
P.O. Box 100
Cold Spring Harbor, New York 11724

ABSTRACT In the course of constructing and characterizing defective adenovirus-SV40 hybrid viruses, we have isolated a novel nondefective recombinant virus. Six isolates were characterized in detail and found to synthesize a 26,000 dalton protein that shares antigenic determinants with the carboxy-terminus of SV40 large T antigen. As expected, these viruses plaque with one-hit kinetics on monkey cells and contain an insertion of SV40 DNA from map position 0.30 to 0.15. At the position of the SV40 insertion is a deletion of approximately 2500 bp of adenoviral DNA that maps between 79 and 86 map units. The SV40 sequences are oriented such that transcription can proceed from the adenoviral E3 early promoter.

I. INTRODUCTION

A variety of naturally occurring nondefective human

[1]Supported by grants from the NIH and ACS.

ISBN 0-12-137420-3

adenovirus 2-SV40 hybrid viruses have been described (1). All of these recombinants were isolated from the same original lysate and encode the carboxy-terminal portion of SV40 large T antigen. This region of T antigen has been found to provide a helper function that allows adenovirus to propagate on otherwise nonpermissive monkey cells (2,3). These nondefective hybrid viruses therefore grow efficiently on both human and monkey cells with one-hit kinetics. All of the nondefective viruses share a common 3' SV40-adenovirus junction, but vary in the length of the SV40 sequences incorporated and in the amount of adenoviral sequences that are deleted. In this paper, we describe a new nondefective adenovirus-SV40 hybrid virus that arose from our *in vitro* constructions of defective hybrid recombinants.

The construction and propagation of defective adenovirus-SV40 hybrid viruses that express full-length SV40 T antigen has been described (4). In these experiments, two *in vivo* interserotypic recombinants of adenovirus type 2 and adenovirus type 5, designated 1x51i and 4x225b, were used as vectors (5,6). Each of these recombinants contains two Bam sites located at map positions 29 and 59.5 in 1x51i and at 42 and 59.5 in 4x225b. DNA isolated from these viruses was digested with Bam in order to excise the internal region of the adenoviral genome. A Bam fragment containing the SV40 A gene was then ligated to the vector fragments and the ligation mix was transfected into human 293 cells in the presence of parental helper DNA. We then exploited the helper function of SV40 large T antigen as a strong biological selection for hybrid viruses. Only those adenovirus recombinants that express at least the carboxy-terminal portion of the SV40 A gene will propagate efficiently on monkey cells (2,3). Thus by passaging and plaque-purifying the constructed recombinants on monkey cells, we can effectively isolate hybrid viruses that are producing SV40 T antigen. Because we are removing essential adenovirus sequences during our construction of adenovirus-SV40 hybrid genomes, the resultant virus is defective and requires the presence of a wild type helper virus in order for it to propagate. Therefore, as expected, all of the hybrid viruses that express full-length T antigen show two-hit kinetics in a plaque assay on monkey cells. In some cases, however, we have identified recombinant viruses that plaque on monkey cells with one-hit kinetics indicative of a nondefective virus. This infrequent generation of nondefective viruses only occurs when 1x51i is used as a vector and is not seen once a plaque-purified stock of a defective hybrid virus has been established. Each isolate has an identical stable genome structure and synthesizes a 26,000 dalton protein encoded by the carboxy-terminal portion of the SV40 A gene. These viruses are probably generated by rare unequal crossing-over events that allow the wild type helper genome to rescue SV40 sequences from the constructed recombinant DNA.

II. RESULTS AND DISCUSSION

1. Analysis of SV40-encoded Proteins Produced by the Nondefective Viruses

By studying the naturally occurring nondefective adenovirus-SV40 hybrid viruses, it was discovered that only the carboxy-terminal portion of SV40 large T antigen was needed to allow adenovirus to propagate on monkey cells (2,3). Since all of our constructed hybrid viruses were selected by their ability to grow efficiently on monkey cells, they should encode at least this part of the T antigen molecule.

In order to determine whether the nondefective hybrid viruses express any SV40 T antigen, infected monkey cells were pulse labeled with ^{35}S-methionine and the extracts were subjected to immunoprecipitation with a serum directed against the T antigen-related protein, D2 (7,8). Figure 1A shows that a protein of 26,000 daltons is selectively precipitated by the anti-D2 serum but not by a control serum. Furthermore, of ten separate isolates that we have studied, all encode a T antigen-related protein of this size (data not shown). This result indicates that only a portion of the SV40 A gene is being expressed by the nondefective hybrid viruses.

In order to identify which region of the T antigen molecule is represented in the 26,000 dalton protein, we have exploited the specificity of four monoclonal antibodies directed against separate regions of the large T polypeptide. The recognition sites of these monoclonal antibodies, DL-3C4, DL-3C5, L7 and L19 are depicted in Figure 1. Extracts prepared from infected cells and labeled with ^{35}S-methionine were subjected to immunoprecipitation with each of the four monoclonal antibodies and the precipitated proteins were displayed by gel electrophoresis. As expected, L7, which is directed against the carboxy-terminus of T antigen, is able to precipitate the 26,000 dalton protein,

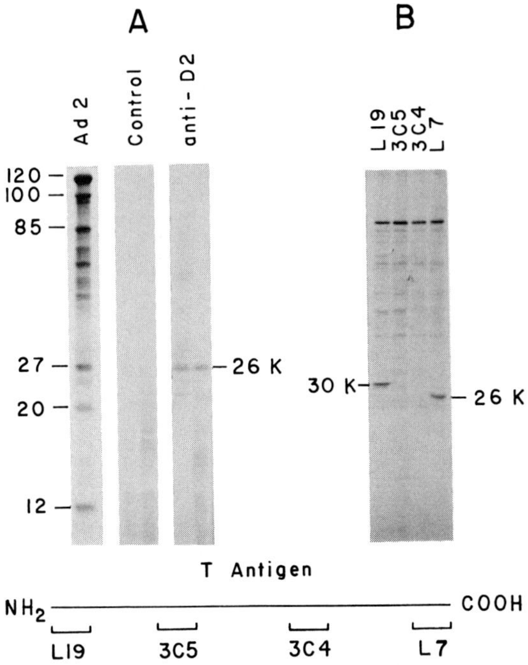

Figure 1. Immunoprecipitation of the truncated T antigen encoded by 1x51iND

A) Subconfluent monolayers of monkey cells were infected with two separate isolates of plaque-purified 1x51iND virus. After 48 hours at 37°C the cells were pulse-labeled with ^{35}S-methionine, lysed, and the extracts were subjected to immunoprecipitation with either rabbit preimmune control serum or rabbit serum directed against the D2 protein (8). The immunocomplexes were collected on protein A-bearing *Staphylococcus aureus* cells and then fractionated on a 7 to 15% gradient of polyacrylamide containing SDS, as described (4). Molecular weight standards from adenovirus type 2 infected cells are shown in the first lane.

B) Four monoclonal antibodies: DL-3C4, DL-3C5 (gifts from David Lane) L7, and L19 (gifts from Ed Harlow) were used to immunoprecipitate the 26,000 dalton truncated T antigen from extracts of cells infected with 1x51iND. The approximate regions of the T antigen molecule recognized by these monoclonal antibodies are depicted in the lower part of the figure (Clark *et al.*, manuscript submitted; E. Harlow, personal communication). The 30,000 dalton protein immunoprecipitated by L19 has been identified as a host cell protein that cross-reacts with the antibody (E. Harlow *et al.*, manuscript submitted).

while the other monoclonal antibodies do not (Figure 1B). This is consistent with the fact that it is the carboxy-terminal region of large T antigen that provides the helper function allowing the growth of adenovirus on monkey cells (2,3).

2. Viral Genome Structure

The genome structures of several recombinant viruses were examined by restriction endonuclease mapping and by Southern blot hybridization using nick-translated SV40 DNA as a probe. The map derived from this analysis is shown in Figure 2. Because all six of the viral genomes analyzed in this manner have an identical structure, we have designated this nondefective hybrid virus 1x51iND.

The restriction fragments of 1x51iND DNA are identical to those of the parental vector DNA except for fragments in the region between map positions 77.9 and 91.9 where there has been a deletion of approximately 2500 bp of adenoviral sequences. Since the fiber protein, whose initiation codon maps near position 86.1 (9), is essential for viability, the deletion must end before this gene. In place of the deleted adenovirus sequences there is an insertion of SV40 DNA that is flanked by a Bam and a Hind III site. The orientation of the SV40 sequences was determined by analyzing the Bcl I and Bam digestion patterns (see Figure 2). This orientation and the appearance of the 26,000 dalton protein early in infection (data not shown) suggests that the SV40 DNA may be transcribed from the adenovirus E3 promoter. Finally, 1x51iND contains two new Hind III sites and one new Bcl I site adjacent to the SV40 sequences. Since these new sites do not correspond to any of the possible junctions from the original ligation mix, some rearrangement of adenovirus sequences must have occurred in addition to a simple insertion/deletion event.

The extent of the SV40 sequences in the 1x51iND genome was determined by restricting full-length SV40 DNA with several enzymes, separating the fragments by gel electrophoresis, transferring the fragments to nitrocellulose, and hybridizing with nick-translated 1x51iND DNA (see Figure 3). As expected, no SV40 sequences extending beyond the Bam site into the SV40 late region hybridize to the 1x51iND probe. The left end of the inserted SV40 sequences is within the HaeIII A fragment but does not extend as far as map position

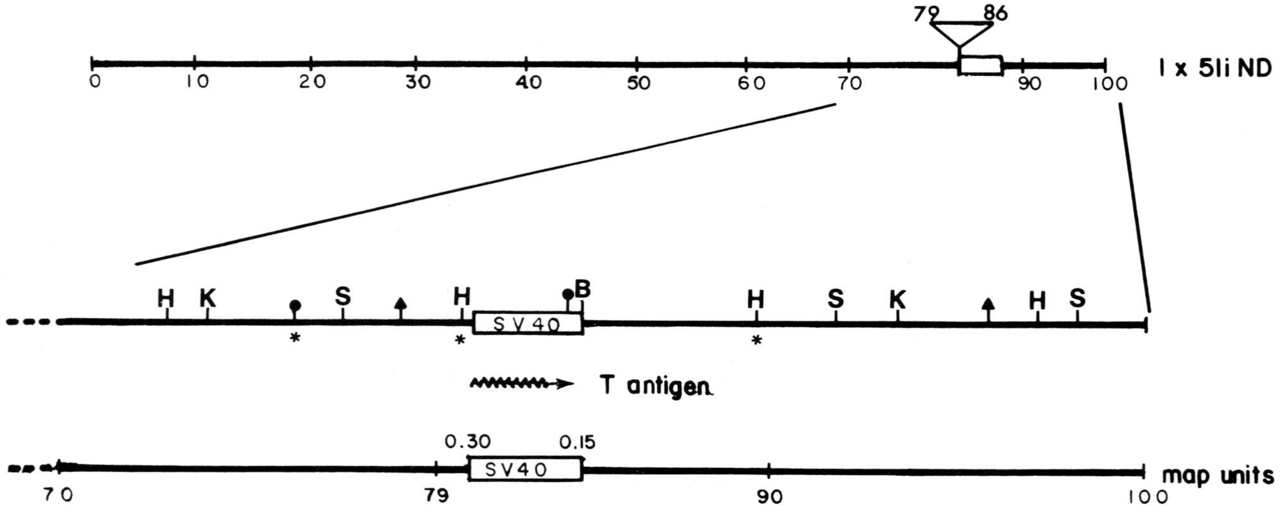

Figure 2. The Genome Structure of 1x51iND

DNA extracted from monkey cells infected with 1x51iND was digested with either Bam (B), Hind III (H), Kpn I (K), Bcl I (●), Sma I (S), or Bgl II (▲) and the fragments were separated by agarose gel electrophoresis. The DNA was then blotted onto nitrocellulose by the Southern procedure (10) and hybridized to nick-translated SV40 DNA as described (4). This figure summarizes our results in the form of a restriction map. The rearranged region is shown in detail with the unusual restriction sites marked by asterisks.

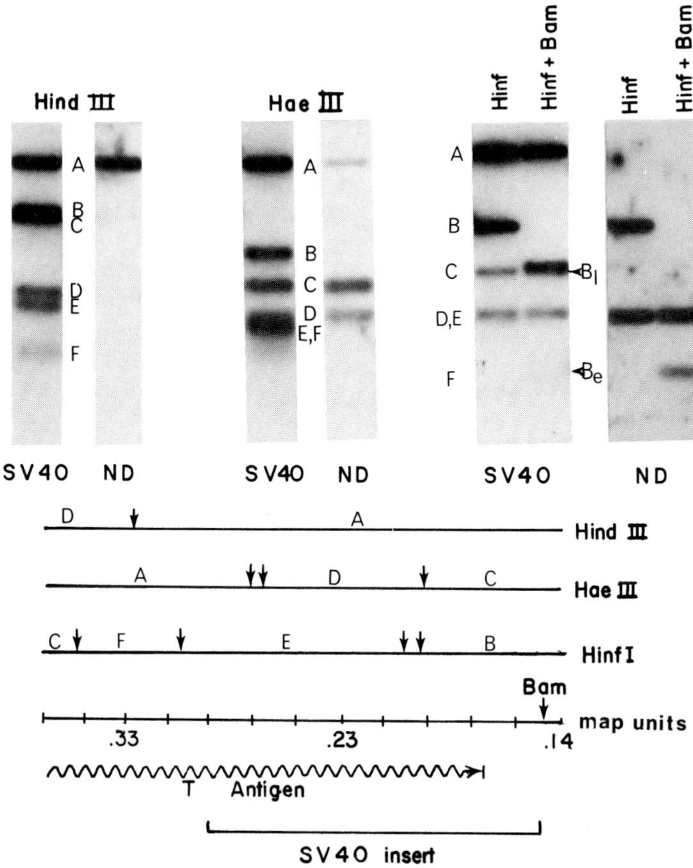

Figure 3. Determination of the Extent of SV40 DNA Carried by 1x51iND

SV40 DNA was restricted with either Hind III, Hae III, or Hinf I and the fragments were separated by 1.4% agarose gel electrophoresis. The DNA was then transferred to nitrocellulose and the filters were probed with nick-translated 1x51iND DNA. Only the SV40 Hind III A fragment, Hinf I B and E fragments and the Hae III C and D fragments hybridize strongly with the recombinant probe. In order to determine whether any late SV40 sequences located beyond the Bam site are contained in the 1x51iND genome, the Hinf I B fragment was subcut with Bam forming a 795 bp late fragment (B_l) and a 290 bp early fragment (B_e). Only the 290 bp fragment was found to hybridize with the 1x51iND probe suggesting that the Bam site marks the right-hand end of the SV40 insert. The left-hand end appears to lie near 0.30 map units. A restriction map of part of the SV40 early region from 0.37 to 0.14 map units is depicted below the blots.

0.31 because the Hinf I F fragment does not hybridize to the 1x51iND DNA. The right end of the SV40 sequences most likely terminates at the Bam site because the probe does not hybridize to the SV40 Hinf I-Bam fragments from the SV40 late region. Thus, this nondefective virus has a right-hand adenovirus-SV40 junction located at about map position 0.15 on the SV40 map, rather than at position 0.11 where the junction occurs in the $Ad2^{+}ND$ series (1). The total amount of SV40 DNA in the insertion is approximately 800 base pairs or enough DNA to code for a protein of 26,000 daltons.

It is possible that unequal crossing-over between 1x51i helper DNA and a constructed defective genome produced the nondefective 1x51iND virus. During the ligation of Bam cut adenovirus DNA to the SV40 DNA, a multitude of different products are presumably generated. Upon transfection of this mixture into human cells, the recombinant molecules can recombine both with each other and with the helper DNA. A nonhomologous recombination event could allow the 1x51i helper DNA to rescue SV40 sequences from a defective recombinant genome thereby forming a nondefective hybrid virus, such as 1x51iND. Strong selective pressure is then exerted by passaging the virus population in monkey cells resulting in the amplification of both defective and nondefective adenovirus-SV40 hybrid viruses. In the presence of low levels of helper virus, nondefective hybrid viruses have a distinct advantage over the defective recombinants and can quickly dominate the population.

We have previously described defective hybrid viruses derived from 1x51i (Ad-SVR15) and 4x225b (Ad-SVR5) that contain the SV40 A gene at the original sites of insertion and that carry no rearrangements (4). In addition, we also isolated a number of defective hybrid viruses derived from 1x51i (Ad-SVR1, R6) that had undergone rearrangements. In all these cases, the SV40 A gene was in the opposite orientation to that in Ad-SVR15. Apparently, the original R1 and R6 recombinants required a rearrangement in the adenovirus sequences in order to form a defective genome that expresses T antigen. If T antigen cannot be expressed in some original 1x51i recombinants, then subsequent rearrangement events may lead to the production of both defective and nondefective hybrid viruses that express T antigen helper function.

Unlike the 1x51i recombinants, the hybrid viruses constructed with the 4x225b vector can apparently express T antigen without further rearrangement. Therefore it is less likely that rearranged 4x225b nondefective viruses generated by subsequent rare recombination events will overgrow the population and be isolated. Thus, both the choice of vectors as well as the method of selection can determine the number and type of recombinant genomes that can be isolated.

ACKNOWLEDGMENTS

We thank Taffy Mullenbach for help with the tissue culture work and Karen Erdley for preparing the manuscript.

REFERENCES

1. Grodzicker, T. (1980) in "Molecular Biology of Tumor Viruses; pt. 2" (J. Tooze, ed.), pp. 577-614. Cold Spring Harbor Laboratory, Cold Spring Harbor, New York.
2. Rabson, A.S., O'Conor, G.T., Berezesky, K., and Paul, G.T. (1964) Proc. Soc. Exp. Biol. Med. 116, 187.
3. Grodzicker, T., Lewis, J.B., and Anderson, C.W. (1976) J. Virol. 19, 559.
4. Thummel, C., Tjian, R., and Grodzicker, T. (1981) Cell 23, 825.
5. Sambrook, J., Williams, J., Sharp, P.A., and Grodzicker, T. (1975) J. Mol. Biol. 97, 369.
6. Grodzicker, T., Anderson, C., Sambrook, J., and Mathews, M.B. (1977) Virol. 80, 111.
7. Hassell, J.A., Lukanidin, E., Fey, G., and Sambrook, J. (1978) J. Mol. Biol. 120, 209.
8. Tjian, R., Robbins, A., and Lane, D. (1979) in "Eukaryotic Gene Regulation" (R. Axel, T. Maniatis, and C.F. Fox, eds.), p. 637. Academic Press, New York.
9. Zain, B.S., and Roberts, R.J. (1979) J. Mol. Biol. 131, 341.
10. Southern, E.M. (1975) J. Mol. Biol. 98, 503.

CONSTRUCTION AND TRANSFER OF RECOMBINANT RETROVIRUS CLONES CARRYING THE HSV-1 THYMIDINE KINASE GENE

Alexandra Joyner
Yusei Yamamoto
Alan Bernstein

The Ontario Cancer Institute, and
Department of Medical Biophysics
University of Toronto
Toronto, Canada

I. ABSTRACT

To determine whether retroviruses can be used as vectors for the transfer of cloned genes into animal cells, we have constructed 2 types of recombinant clones containing DNA segments from Friend spleen focus-forming virus and the thymidine kinase (*TK*) gene of Herpes simplex virus-1. The first group of clones contained the *TK* gene, including 5' sequences involved in the initiation of *TK* transcription, flanked by the long terminal repeats (LTR) of Friend virus. The second group of clones contained the *TK* coding sequences, with or without 5' sequences necessary for the initiation of transcription, inserted either 200 bp or 1.2 kb downstream from the Friend 5' LTR. From gene transfer experiments and Southern gel analysis of recipient cell DNA we conclude: (1) TK sequences are stably maintained as part of the retrovirus genome; (2) the LTRs do not interfere with the expression of functionally active DNA fragments inserted within a retrovirus genome; (3) the LTRs can activate the expression of heterologous coding sequences deleted of their homologous promoter region. These observations suggest that these viruses may be useful vectors for the introduction of genomic or cDNA fragments into animal cells.

ISBN 0-12-137420-3

II. INTRODUCTION

The introduction of specific DNA segments into animal cells in culture can lead to heritable changes in cellular phenotype. Thus, DNA-mediated gene transfer provides a biological assay system for identifying those molecular features important to gene expression in higher eucaryotes. Many phenomena in developmental biology and cellular differentiation undoubtedly have their biochemical basis in changes in the arrangement or regulation of certain genes; analysis of the mechanisms controlling the expression of cloned genes transferred into various differentiated cell types may provide insight into these important biological problems. In microbes, the analysis of gene expression has been facilitated by the development of gene transfer vectors, such as sex factors and transducing phages, that mediate the high frequency transfer of cloned genes into bacterial cells. In mammalian systems, the DNA tumor virus, SV40, has been used as a vector for the introduction of cloned DNA segments into cultured animals cells (1,2).

In this paper, we will summarize experiments designed to examine whether RNA tumor viruses, or retroviruses, can also be used as vectors for the transfer and activation of heterologous genes in animal cells. Retroviruses may be particularly attractive as transducing vectors for various reasons. Recent studies on the integrated DNA form of retrovirus genomes have revealed a number of similarities between these tumor viruses and transposable genetic elements found in procaryotes and lower eucaryotes (3-5). The termini of the integrated retrovirus, or provirus, contains long terminal repeats (LTR) at the junction between the host and viral genomes (Fig. 1). These LTRs, which are 500-600 bp in length, are formed during reverse transcription and DNA synthesis of the provirus (6-8), and are composed of sequences derived from both the 5' and 3' ends of the genomic viral RNA. Nucleotide sequence analysis of the termini of integrated retroviruses has revealed a number of similarities with sequences of the insertion (IS) elements that form the corresponding termini of bacterial transposons. These include a short (3-6 bp) direct repeat of host DNA at the junction between the provirus and host DNA (4, 9, 10) the presence of a short (11 bp) inverted repeat at the termini of each LTR (3-5, 9), and the presence of transcriptional control signals, including regions implicated in the initiation and termination of transcription of eucaryotic genes (11, 12). As well as these structural similarities, both retroviruses and transposable elements have the ability to integrate into a very large number of possible target sites in cellular DNA. Thus, LTRs appear to be involved in a number of essential steps in the replication cycle of retroviruses, including the reverse transcription and integration of the proviral genome and the transcription of genomic and sub-genomic RNAs.

The genomes of rapidly transforming retroviruses also include sequences responsible for initiating and maintaining oncogenic transformation (13). These viral oncogenes appear to have been acquired, by recombination, from host genetic information because the DNA of normal cells also contains single-copy sequences that are highly related to these viral oncogenes. Thus, retroviruses appear to act as natural cloning and transfer vectors for at least some cellular genes.

In this study, we have investigated whether retroviruses can also serve as gene transfer vectors for other eucaryotic genes. For this purpose, the thymidine kinase (TK) gene of Herpes simplex virus-1 (HSV-1) was chosen because growth conditions exist which select against those cells not expressing a functional TK gene. Recombinant clones were constructed between the DNA of Friend spleen focus forming virus ($SFFV_P$), a murine leukemia virus, and the TK gene, with or without its own promoter region. Gene transfer with these clones indicates that heterologous genes are stably maintained as part of a retrovirus genome when transfected into mouse cells. In addition, evidence is presented indicating that the TK gene coding sequences in these clones can be expressed in mouse cells under control by its own promoter region or through activation by presumptive promoter regions located within retrovirus LTRs.

III. MATERIALS AND METHODS

1. Cell Culture. Two TK^- recipient cell lines were used in the eucaryotic transformation experiments. Murine LTA cells, an $APRT^-$ derivative of Ltk^- cells (14) derived by R. Hughes and P. Plagemann, were obtained from C. P. Stanners. Murine IT22 cells (15), a TK^- derivative of Swiss mouse 3T3 cells derived by C. Croce, were obtained from R. Godbout. Both cell lines were maintained in α-MEM (16) supplemented with 5% calf serum and 5% fetal calf serum. TK^+ cells were selected in the same medium supplemented with HAT (0.1 mM hypoxanthine, 1.0 μM aminopterin, 40 μM thymidine).

2. Eucaryotic Cell Transformation. 7×10^5 LTA cells or 5×10^5 IT22 cells seeded in 100 mm petri dishes were transfected with cloned DNA as described (17) with modifications (18). The pLTR-TK clones were mixed with carrier DNA derived from LTA cells and λSFFV-TK clones were mixed with carrier DNA extracted from TK^+ mouse cells.

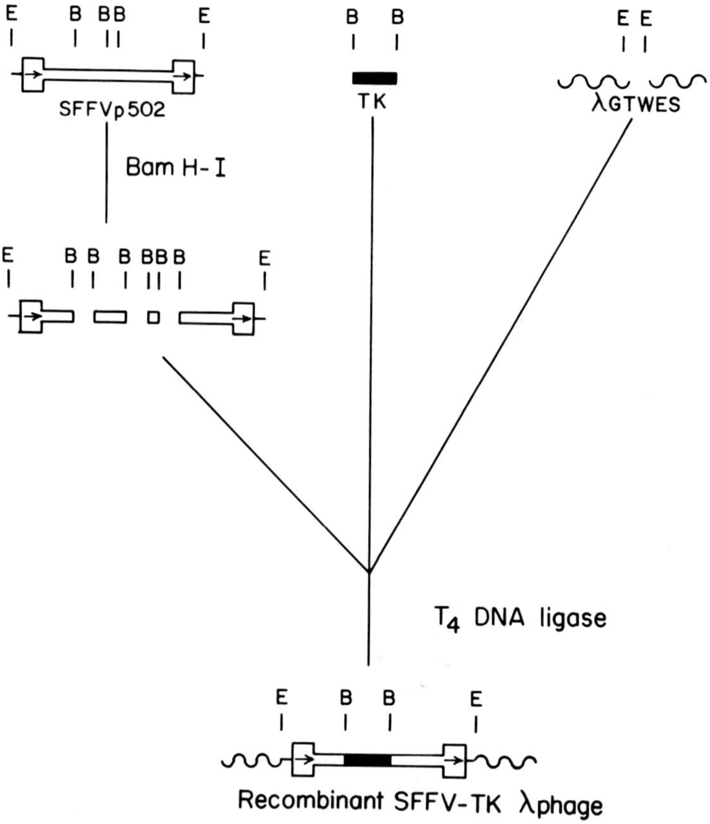

FIGURE 1. Strategy for the construction of λSFFV-TK recombinant retroviruses. $SFFV_P502$ DNA is indicated by open area, TK DNA by solid area, λgtWES.λB by ∿ lines and rat sequences flanking $SFFV_P502$ by straight lines. The restriction enzymes EcoR1 and BamH1 are indicated by E and B respectively.

IV. RESULTS

A. Construction of Recombinant λ Phages Containing Two $SFFV_P502$ LTRs and HSV-1 TK

To determine whether cloned retrovirus DNA molecules could serve as transfer vectors for heterologous genes, we first constructed, in phage λ, recombinants consisting of two LTRs, derived from a molecular clone of Friend $SFFV_P$ (19),

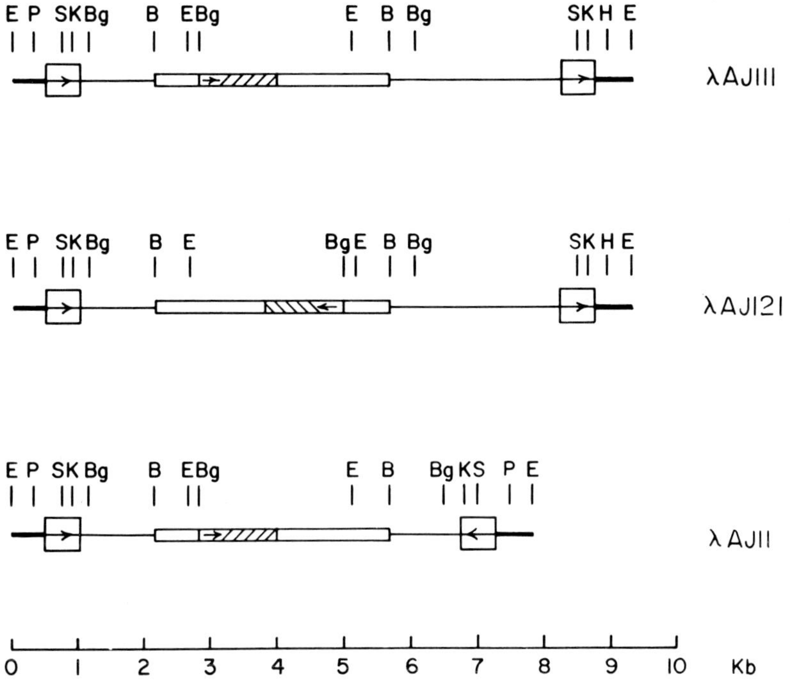

FIGURE 2. Physical maps of λSFFV-TK recombinant clones. The simplified maps indicate the $SFFV_P$502 long terminal repeats ( ), $SFFV_P$502 sequences ( ), rat host flanking sequences ( ), TK coding sequences ( ), HSV-1 TK BamH1 clone ( ) and direction of transcription ( ). Relevant restriction endonuclease sites are shown: B, BamH1; Bg, BglII; E, EcoR1; H, HindIII; K, Kpn1; P, Pst1; S, Sst1. All clones were made in λgtWES.λB.

surrounding the HSV-1 BamH1 fragment containing the TK gene (λSFFV-TK). The cloning strategy used is shown diagramatically in Figure 1. The infectious integrated clone of Friend SFFVp, λSFFVp502, includes the entire SFFVp genome plus 5' and 3' rat flanking sequences cloned as a 7.4 kb EcoR1 fragment into λgtWES·λB. The restriction endonuclease BamHI cuts the SFFVp genome 3 times, but leaves the LTRs intact. Recombinant λ phages containing the TK DNA segment surrounded by two SFFVp502 LTRs were constructed by digesting purified EcoR1 7.4 kb SFFVp502 DNA with BamHI, and then ligating this DNA with the BamHI TK fragment isolated from the plasmid pxl (20) and left and right arms of the EcoR1 cloning vector, λgtWES·λB. The cloning procedure was such that any λ clone containing TK must be flanked by two end fragments (containing LTRs) of $SFFV_p502$. 2,000 plaques were screened and twenty-four of the sixty positive clones containing TK sequences were characterized by restriction enzyme and Southern analysis (21). 17 of the clones contained TK flanked by two 5' ends of $SFFV_p502$ (λAJ11) and 7 contained TK flanked by a 5' and 3' end of $SFFV_p502$. Of these 7, 3 contained TK in the same transcriptional orientation as the viral LTRs (λAJ111), and 4 contained TK in the opposite orientation (λAJ121). Figure 2 shows these three types of clones schematically. AJ111 and AJ121 can be described as $SFFV_p502$ recombinant retroviruses containing intact 5' and 3' LTRs with the TK fragment inserted between the 5' and 3' most BamH1 site of $SFFV_p502$. The clone λAJ11 does not have two direct LTRs but has two inverted 5' LTRs surrounding the TK fragment.

B. Transforming Activity of λSFFV-TK Clones

The transforming activity of the three λSFFV-TK clones (λAJ11, λAJ111, λAJ121) was tested by transferring these recombinant phages into TK^- mouse cells (LTA and IT22) by the calcium phosphate technique and selecting for TK^+ transformants in HAT medium. Table 1 lists the combined results of three DNA transfer experiments with either intact or BamH1-digested λSFFV-TK clones. Between 30 to 230 or 15 to 80 TK^+ colonies/plate were obtained with LTA and IT22 recipient cells respectively, with the quantities of cloned DNA used in these experiments. Carrier DNA alone gave 0-2 colonies/plate. The results presented in Table 1 indicate that both undigested and BamH1 digested λSFFV-TK DNA from all three λ clones was able to transform TK^- mouse cells with high efficiency.

To characterize the organization of the TK sequences in the transformed cells, high molecular weight DNA was prepared from IT22 cells transformed to a TK^+ phenotype with either λAJ111 or λAJ121 and analyzed by the technique of Southern (21) with a TK specific probe. The cellular DNA of these clones was digested with either Sst1 which cleaves SFFV-TK DNA within the LTRs of $SFFV_p502$, Kpn1 which cleaves once within the LTRs and once at the 5' end of the TK gene, or BamH1 which cleaves at the junction between $SFFV_p502$ and TK sequences. As shown in Figure 3, identical restriction enzyme patterns were obtained with both the cleaved λSFFV-TK DNA and cellular DNA from the TK^+ transformant, indicating that this 1 cell line contained a SFFV-TK genome which was co-linear with the intact clone AJ111 over a region extending between the 2 LTRs. Four additional transformants have been analyzed in this manner with identical results.

The restriction enzyme Sst1 cleaves within the LTRs of

TABLE I. TK Transforming Activity of Recombinant SFFV-TK λ Phages

Recombinant Phage	Restriction Enzyme	Recipient Cell	Transformation Efficiency (no. colonies/pmole phage)
λAJ111	BamH1	LTA	9,400 ± 1,700 (3)
λAJ111	none		7,600 ± 800 (3)
λAJ121	BamH1		18,000 ± 3,800 (3)
λAJ121	none		14,000 ± 8,300 (4)
λAJ11	BamH1		6,700 ± 1,400 (3)
λAJ11	none		2,500 ± 1,500 (3)
λAJ111	BamH1	IT22	1,300 ± 150 (2)
λAJ111	none		600 ± 190 (2)
λAJ121	BamH1		720 ± 700 (4)
λAJ121	none		560 ± 360 (5)
λAJ11	BamH1		272 (1)
λAJ11	none		240 ± 96 (3)

Schematic presentation of each recombinant λ phage is shown in Figure 2. 3 μg and 600 ng of each recombinant phage DNA was used to transform IT22 cells and LTA cells, respectively. The transformation efficiency is given as the mean ± S.D. of _n_ determinations (in parenthesis).

λAJ111 and λAJ121 yielding a 7.4 kb fragment consisting of $SFFV_P$ sequences and the _TK_ fragment. To determine whether the 7.4 kb fragment detected in cellular DNA with a _TK_ probe also included viral sequences, the DNA of a number of TK^+ transformants was analyzed by the Southern technique (21) for the presence of a _Sst1_ digestion product which would hybridize to a ^{32}P-labelled cDNA probe to Friend leukemia virus complex. The results of one such experiment, shown in Figure 3b, show that a 7.4 kb _Sst1_ fragment of cellular DNA contains viral, as well as _TK_, sequences.

C. Transforming Activity of Plasmids Containing the $SFFV_P$502 5' LTR and _TK_ Gene Coding Sequence

The results above indicate that the LTRs of $SFFV_P$ did not interfere with the expression of a DNA segment containing the _TK_ coding sequence as well as its homologous transcriptional promoter sequences. It was of interest to determine whether a retrovirus LTR could also activate the expression of _TK_ coding sequences, deleted of 5' transcriptional controlling

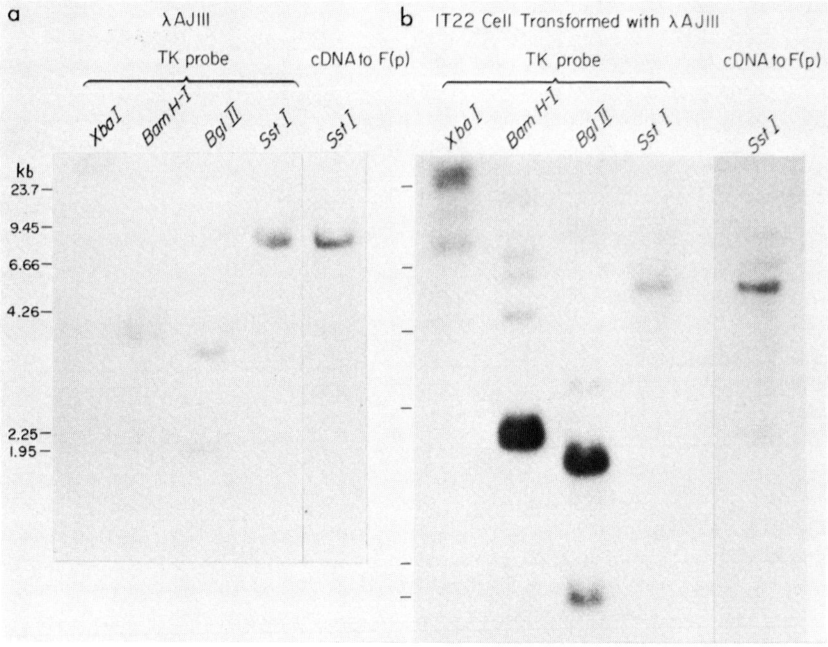

FIGURE 3. Analysis of the TK and SFFVp sequences in λAJ111 (a) and in an IT22 cell clone transformed by this λ recombinant (b). The DNA was cleaved by the restriction enzymes indicated and analyzed by the technique of Southern (21) with either a ^{32}P-labelled BamH1 TK fragment probe or ^{32}P-labelled cDNA probe to Friend leukemia virus complex (F(p))cDNA).

sequences. We therefore constructed clones containing the 5' LTR of SFFVp and fragments of the HSV-1 TK gene coding sequence (22).

The 5' end fragment of $SFFV_p$502, extending from the EcoR1 site in the 5' rat flanking sequences to the BamH1 site 1.2 kb 3' of the LTR, was subcloned into pBR322 (pYY508) (22). This fragment has one BglII site approximately 200 nucleotides 3' of the LTR. DNA sequencing data and mRNA mapping of the HSV-1 TK gene indicate that there is a BglII site in the leader mRNA sequence of this gene.(23). Thus digestion of the plasmid px1 with BglII and BamH1 produces a 2.8 kb fragment containing the intact TK coding sequence but lacking 5' sequences required for expression of the TK gene in mammalian cells. This 2.8 kb BglII/BamH1 TK fragment was also subcloned into pBR322 (pAJ2) (22). Eight plasmid clones consisting of either the complete BamHI TK or the BamHI/BglII TK fragment inserted into the BamHI or BglII sites of pYY508, in either orientation, were

TABLE 2. TK transforming activity of recombinant pLTR-TKs.

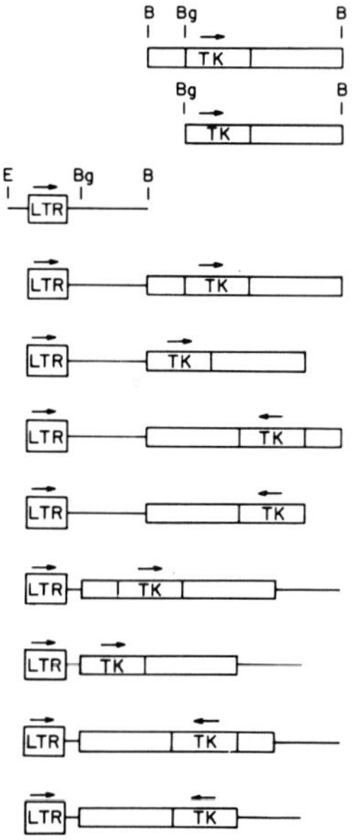

Plasmid	Schematic of plasmid	Transformation efficiency (no.TK$^+$ colonies/p mole plasmid)	
px1		13,000	(6)
pAJ2		0	(5)
pYY508		0	(3)
pAJ11		20,000	(3)
pAJ21		0	(5)
pAJ12		12,000	(5)
pAJ22		0	(6)
pAJ13		13,000	(6)
pAJ23		14,000	(5)
pAJ14		22,000	(5)
pAJ24		0	(6)

Approximately 100 ng of each plasmid was used to transform LTA cells. The numbers in parentheses are the number of plates used to calculate the transformation efficiency. The schematic diagrams are drawn approximately to scale: (LTR) long terminal repeat of SFFV$_p$502; (TK) HSV-1 thymidine kinase coding sequence; (——) SFFV$_p$502 sequences; (TK) HSV-1 sequences surrounding the TK gene; (→) direction of transcription; (B) BamHI; (Bg) BglII; (E) EcoRI.

constructed (pLTR-TK) (22). These clones are shown schematically in Table 2.

The transforming activity of the 8 pLTR-TK recombinant plasmids, as well as the parental clones used to construct these recombinants, was tested by gene transfer into LTA cells. Table 2 shows the combined results of two transfer experiments with the 11 clones. The results demonstrate that the transforming efficiency of the intact BamHI TK fragment was similar whether TK was inserted into pBR322 alone (px1) or inserted in either orientation into the BglII or BamHI sites of pYY508. As expected, the pBR322 plasmid containing the BglII/BamHI TK fragment (pAJ2) did not transform LTA cells to a TK^+ phenotype. Similar observations have been described using BglII-digested pxl (20). Of the four recombinant pLTR-TK clones containing the BglII/BamHI TK fragment, only one clone containing the TK coding sequence inserted at the BglII site in the same polarity as the LTR transformed LTA cells to a TK^+ phenotype. The transformation efficiency of this clone was similar to that of the clones containing the complete BamHI TK fragment. The remaining three clones, containing the TK segment inserted at either the viral BglII site in the opposite orientation as the LTR, or at the downstream BamHI site in either orientation, had no detectable transforming activity on LTA cells (Table 2).

V. DISCUSSION

In these experiments, we sought to determine whether a murine retrovirus could serve as a cloning and transfer vector for a eucaryotic gene. To test this possibility, two kinds of recombinant DNA clones were constructed and transfected into TK^- mouse cells. The first group of clones included the Herpes TK coding sequence as well as 5'sequences known to be involved in the transcriptional activation of this gene (23), inserted between the LTRs of the retrovirus Friend $SFFV_P$. Southern analysis demonstrated that this 3.4 kb TK insert was stably maintained as part of the retrovirus genome when transfected into mammalian cells. Molecular clones containing the 3.4 kb BamHI TK DNA segment in either orientation relative to the 3' and 5' LTRs were all active on gene transfer into TK^- mouse cells. Thus, we conclude that this TK insert does not depend on the LTRs for its expression, and more importantly, the presence of the viral LTRs does not interfere with transcription initiated within the homologous TK promoter region.

The second group of recombinant clones constructed were designed to determine whether retrovirus LTRs, which include a presumptive promoter region involved in the initiation of transcription (11, 12), could activate the expression of a DNA segment containing TK coding sequences but deleted of homologous Herpes sequences involved in the initiation of transcription. DNA transfer experiments with these clones demonstrated that the Friend 5' LTR can activate the expression of the TK coding sequence only when the TK fragment is inserted 200 bp downstream from, and in the same orientation as, the LTR. The lack of activation of TK sequences by the LTR in clones where TK was inserted an extra 1 kb downstream from the LTR suggests that the 1 Kb of $SFFV_P$ sequences may interfere with the transcription or translation of TK sequences. Alternatively, a functionally inactive $SFFV_P$/TK polyprotein may be synthesized.

The observation that a retrovirus LTR, when placed upstream from an inactive cellular coding sequence, can activate a silent gene extends observations by others that cellular DNA sequences homologous to the viral oncogenes present on certain retrovirus genomes can be activated when located near a retrovirus LTR (24). Our results are also consistent with recent observations on the mechanisms of viral carcinogenesis by the slowly transforming avian leukosis virus (25,26). These studies have suggested that the induction of bursal tumors by ALV requires integration of the ALV provirus adjacent to specific cellular sequences. These cellular sequences, which appear to be related to transforming sequences on the genome of the acute leukemia virus, MC-29, are then activated by the adjacent ALV LTR. Thus, there are at least three mechanisms by which potential cellular oncogenes can become closely linked to, and activated by, a retrovirus LTR: (a) an *in vivo* recombination event which results in the stable transduction of these cellular sequences as part of a retrovirus genome (13); (b) *in vitro* ligation using recombinant DNA techniques similar to those described here (24); or (c) a rare integration event *in vivo* resulting in the random integration of the provirus genome adjacent to a potential cellular oncogene (25, 26).

The finding that heterologous genes are stably maintained and functionally expressed *in vivo* when inserted within a retrovirus genome satisfies major prerequisites for using these viruses as transfer vectors into mammalian cells. Such clones should be replication-defective because they are deleted of viral genes that encode viral proteins involved in virion assembly and DNA synthesis. Experiments are currently in progress to determine whether a recombinant retrovirus-TK virus can be rescued from TK^+ transformants by superinfection of these cells with a replication-competent helper virus. Such phenotypically mixed particles should mediate high efficiency transfer of the *TK* gene into a wide variety of animal cells both in culture and *in vivo*.

ACKNOWLEDGMENTS

We thank G. Cheong and V. Prideaux for excellent technical assistance. Alexandra Joyner is the recipient of a studentship from MRC of Canada. This work was supported by grants from the NCI and MRC of Canada.

REFERENCES

1. Hamer, D. H., and Leder, P., *Nature 281*, 35 (1979).
2. Mulligan, R. C., Howard, B. H., and Berg, P., *Nature 277*, 108 (1979).
3. Ju, G., and Skalka, A. M., *Cell 22*, 379 (1980).
4. Shinotohno, K., Mizutani, S. and Temin, H. H., *Nature 285*, 550 (1980).
5. Sutcliffe, J. G., Shinnick, T. M., Verma, I. M., and Lerner, R. A., *Proc. Natl. Acad. Sci. 77*, 3302 (1980).
6. Dina, D., and Benz, E. W., *J. Virol. 33*, 377 (1980).
7. Gilboa, E., Goff, S., Shields, A., Yoshimura, F., Mitra, S., and Baltimore, D., *Cell 16*, 863 (1979).
8. Hughes, S., Shank, P., Spector, D., Kung, H., Bishop, J., Varmus, H., Vogt, P., and Breitman, L., *Cell 15*, 1397(1978).
9. Dhar, R., McClements, W., Enquist, L., and Vande Woude, G., *Proc. Natl. Acad. Sci. 77*, 3937 (1980).
10. Majors, J. E., and Varmus, H. E., *Nature 289*, 253 (1981).
11. Benz, E. W., Wydro, R. M., Nadal-Ginard, B., and Dina, D., *Nature 288*, 665 (1980).
12. Yamamoto, T., de Crombrugghe, B., and Pastan, I., *Cell 22*, 787 (1981).
13. *Cold Spring Harbor Symp. on Quant. Biol. 44*, (1980).
14. Kit, S., Dubbs, D., Piekarski, L., and Hsu, T., *Exp. Cell Res. 31*, 291 (1963).
15. Croce, C. M., *Proc. Natl. Acad. Sci. 73*, 3248 (1976).
16. Stanners, C. P., Eliceiri, G. L., and Green, H., *Nature New Biol. 230*, 52 (1971).
17. Graham, F., and Van der Eb, A., *Virology 52*, 456 (1973).
18. Wigler, M., Pellicer, A., Silverstein, S., Axel, R., Urlaub, G., and Chasin, L., *Proc. Natl. Acad. Sci. 76*, 1373 (1979).
19. Yamamoto, Y., Gamble, C., Clark, S., Joyner, A., Bernstein, A., and Mak, T. *Submitted for publication.*
20. Enquist, L., Vande Woude, G., Wagner, M., Smiley, J., and Summers, W., *Gene 7*, 335 (1979).
21. Southern, E. M., *J. Mol. Biol. 98*, 503 (1975).
22. Joyner, A., Yamamoto, Y., and Bernstein, A., *In prep.*
23. McKnight, S., *Nucleic Acids Res. 8*, 5949 (1980).
24. Oskarsson, M., McClements, W., Blair, D., Maizel, J., and Vande Woude, G., *Science 207*, 1222 (1980).
25. Neel, B., Hayward, W., Robinson, H., Fang, J., and Astrin, S., *Cell 23*, 323 (1981).
26. Payne, G., Courtneidge, S., Crittenden, L., Fadly, A., Bishop, J., and Varmus, H., *Cell 23*, 311 (1981).

RAT INSULIN GENE COVALENTLY LINKED TO BOVINE PAPILLOMAVIRUS DNA IS EXPRESSED IN TRANSFORMED MOUSE CELLS

Nava Sarver[1]
Peter Gruss[2]
Ming-Fan Law[1]
George Khoury[2]
Peter M. Howley[1]

[1]Laboratory of Pathology
[2]Laboratory of Molecular Virology

National Cancer Institute
Bethesda, Maryland

ABSTRACT

A novel eukaryotic vector derived from the transforming region of bovine papillomavirus (BPV) was established and demonstrated to be highly effective for introducing foreign genes into animal cells. Using the rat preproinsulin gene as a model we demonstrate that the foreign DNA covalently linked to the vector replicates as an episome, is actively transcribed, and the transcripts are translated into an authentic gene product.

I. INTRODUCTION

We have been studying the biology of the bovine papillomaviruses and in particular their ability to transform mouse cells. In the course of these studies, several observations were made which suggested the potential usefulness of BPV DNA as a vector for introducing foreign genes into cells: 1. The molecularly cloned BPV-1 DNA as well as a cloned 69% subgenomic fragment of the BPV-1 genome is very efficient in inducing transformed foci in susceptible mouse cells (8,16). 2. BPV-transformed cells contain multiple copies (10-120 per cell) of the viral DNA. These copies exist exclusively as

ISBN 0-12-137420-3

unintegrated extrachromosomal molecules (14), thus offering a natural means of amplifying foreign DNA sequences which are covalently linked to the BPV transforming segment. 3. Since integration of the viral genome does not occur (14), the physical contiguity of the "passenger" DNA segment should be preserved. 4. The transformed phenotype provides a marker for selecting those cells that have incorporated the foreign DNA segment. Thus, any cell line susceptible to BPV transformation is a potential recipient. 5. BPV transformed cells grow faster than their non-transformed counterparts. This should facilitate the large scale production of cells containing the exogenous gene and possibly, therefore, the gene product.

In this communication we report the construction of a recombinant BPV DNA containing the rat preproinsulin gene I and show that the foreign DNA is faithfully expressed in transformed cells.

A. Experimental Procedures

1. Cells and DNA Transformation. DNA transformation was performed using the calcium precipitation method followed by enhancement with 25% dimethyl sulfoxide (DMSO) (5,20). Portions representing 0.8 μg of the recombinant DNA were added to mouse C127I cells (17) in 60 mm Petri dishes containing 4 ml of fresh medium. DNA adsorption was at 37° for 4 hrs. DMSO enhancement was at room temperature for 4 min. Following DNA transfection, cultures were incubated at 37°C and refed with fresh medium every 3 days.

2. Protein Analysis. Cells in 100 mm plates were washed 3 hr before labeling with Earle's balanced salt containing 5% normal medium and 2% dialyzed fetal bovine serum (GIBCO) and labeled in the same medium with 200 uCi/ml of ^{35}S L-cysteine (855.6 Ci/mmol, New England Nuclear) for 4 hr at 37°C.

Lysis of cells was performed in 1 ml of Tris buffered saline (pH 7.6) containing 1% NP40, 1 mM DTT, 2 mM phenyl methylsulfonyl-fluoride and 2 mM N-tosylphenylalanine chloromethyl ketone. The proteins were then immunoprecipitated (10) with anti-bovine insulin serum (Miles Yeda, Israel). For competitive binding studies the antiserum was first neutralized with bovine insulin (2 μg bovine insulin per 6 μl of antiserum, 30 min at 4°C) after which the mixture was added to the samples.

Immunoprecipitated proteins were analyzed on a 10-17.5% linear gradient SDS-polyacrylamide gels (12) followed by fluorography (13).

II. RESULTS

A. Construction of $pBPV_{69T}$-rI Recombinant

The rat preproinsulin gene (RI_1) employed in this study was originally isolated as a λ clone (15) from rat chromosomal DNA library (19). A 5.3 kb fragment containing the rI_1 gene was cloned into pBR322 and a 1.62 kb segment containing the coding sequences, the intervening sequences and the regulatory signals of the rI_1 gene was then generated from the cloned 5.3 kb fragment by Bam HI/Hinc II digestion (Fig. 1). Synthetic Hind III linkers (Collaborative Research) were joined to the Hinc II site (18). The modified 1.62 kb segment was cloned into pBR322 and amplified in E. coli (9,22, 4). Separation of this fragment from the plasmid was accomplished by Bam HI/Hind III digestion followed by fractionation through an agarose gel.

A recombinant plasmid, $pBPV_{69T}$, consisting of the 69% transforming region of BPV-1 DNA (equivalent to 5.52 kb) has been previously described (8,16). The purified 5.52 kb viral fragment (BPV_{69T}) was ligated to the 1.62 kb insulin segment (rI_1), the resulting products digested with Hind III and then cloned at the Hind III site of pBR322. A physical map

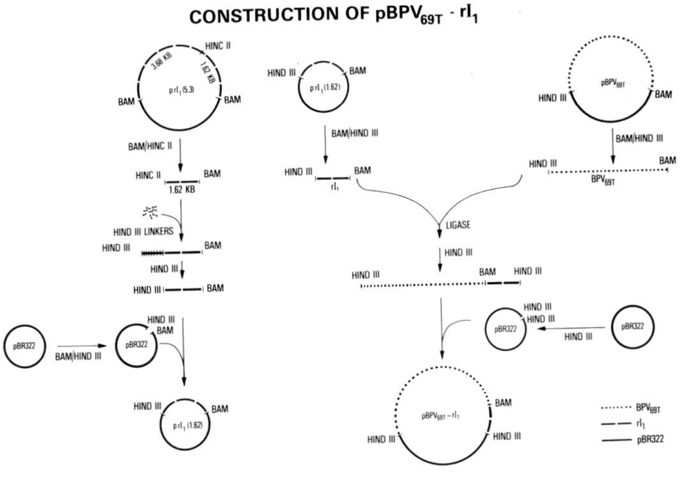

FIGURE 1. Construction of BPV recombinant containing the rat preproinsulin gene I: dotted lines - BPV_{69T} sequences; broken lines - rI_1 sequences, solid lines - pBR322 sequences, for details see text. Published with permission.

of the recombinant thus constructed is presented elsewhere (Sarver et al., submitted).

B. Transformation of Cells

BPV_{69T}-rI_1 DNA was separated from plasmid DNA by Hind III digestion and the resulting products used to transform mouse cells as described in Materials and Methods. Cells were then incubated at 37°C and observed daily for formation of transformed foci.

Foci of transformed cells were first observed seven days after transfection and by day eleven were of sufficient size to be isolated. This was true for cells transformed either by BPV DNA alone or by the recombinant DNA. Moreover, the linearized recombinant DNA (BPV_{69T}-rI_1) transformed cells with high efficiency (approximately 200 transformed foci per μg DNA) indicating that linking the exogenous rat insulin DNA to BPV sequences does not interfere with its ability to transform. Transformed colonies were isolated and established as cell lines. It should be noted that BPV_{69T}-rI_1 transformed cells referred to in this study were each propagated from a single transformed focus rather than from singly-cloned cells.

As an initial screening for the presence of rI_1 sequences in BPV_{69T}-rI_1 transformed cells, the cells were denatured *in situ* (21) and hybridized with ^{32}P-labeled rI_1 DNA. Using this approach all of the transformed colonies isolated (48/48) hybridized with the insulin DNA probe whereas untransformed cells did not (not shown). A detailed analysis of the state of the DNA in BPV_{69T}-rI_1 transformants has shown that these cells contain multiple copies of rI_1 sequences and that these sequences exist predominantly, if not exclusively, as free non-integrated episomes (Sarver *et al*., submitted).

C. RNA Analysis

To ascertain whether insulin specific RNA transcripts are produced in BPV_{69T}-rI_1 transformed cells, the endonuclease S1-exonuclease VII mapping method of Berk and Sharp (2) was adopted. The DNA fragments resulting after S1 nuclease analysis and separation in alkaline agarose gel are depicted in Fig. 2 right panel. When authentic rat preproinsulin mRNA, derived from rat insulinoma cells (3) was hybridized with ^{32}P-labeled SVL_1-rI_1 DNA probe, a single band of 402 nucleotides was detected. This band represents the coding sequences at the 3' end of the insulin gene. The 5' end of this fragment delineates the 3' terminus of the excised intervening sequence. The expected 42 nucleotide leader sequence is not resolved in the gel system used. RNA from BPV_{69T}-rI_1 trans-

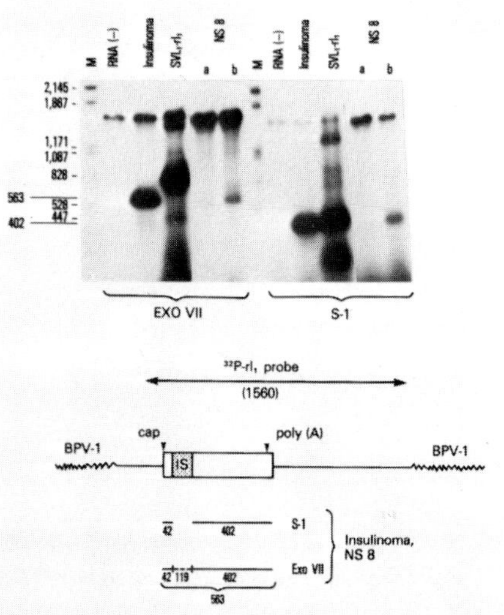

FIGURE 2. Poly(A) selected RNA (11,1) from 2 x 10^6 and 5 x 10^6 cells (lanes a and b respectively) were mixed with 10,000 cpm of ^{32}P-labeled rat preproinsulin DNA (2 x 10^6 cpm/μg) purified from recombinant SVL1-rI_1 virus by Hae II - Bam HI digestion (6). The mixture was precipitated with ethanol, resuspended in 20 μl of formamide buffer and hybridized for 3 hr at 50°C. RNA-DNA duplexes were treated with endonuclease S1 or exonuclease VII (2) and the digests analyzed by electrophoresis through a 1.4% alkaline agarose gel. The gel was exposed for 72 hr at -70°C. Numbers to the left of the gel indicate the size in base pairs of SV40 DNA segments. The 1560 base species represents the self-annealed ^{32}P-labeled DNA probe. DNA protected by authentic preproinsulin mRNA is 402 nucleotides in the case of endonuclease S1 analysis and 563 nucleotides for exonuclease VII analysis. RNA(-)-probe contained no added RNA; insulinoma - RNA from rat insulinoma cells; SVL1-rI_1 - RNA from AGMK cells infected with SV40-insulin recombinant DNA; NS8 - RNA from BPV_{69T}-rI_1 transformed cells. The diagram depicts the classes of DNA fragments expected if only insulin regulatory signals are involved in transcription.

formed cell line (NS8) also protected a single 402 base DNA fragment. This band was seen in every BPV_{69T}-rI_1 transformed cell line tested but not in cells transformed by BPV alone (not shown). Since the size of the DNA fragments obtained in these two cases is identical and since the 5' end is fixed by the splice junction, we conclude that the entire coding region of the preproinsulin gene is represented in mRNA produced in BPV_{69T}-rI_1 transformed cells and that the polyadenylation signal at the 3' end of the gene is faithfully recognized.

The exonuclease VII analysis is shown in Fig. 2 left panel. Exonuclease VII digests only single-stranded termini but not internal single-stranded loops. The size of the fragment should therefore be increased by 161 nucleotides, corresponding to the size of the intron plus the 5' terminal portion. As expected, analysis of insulinoma RNA resulted in a band of 563 bases. RNA from BPV_{69T}-rI_1 transformed cells also gave rise to a single 563 nucleotide fragment, indicating that the 5' ends of the RNA's in both cases are similar if not identical.

The presence of only one preproinsulin transcript in BPV_{69T}-rI_1 transformed cells suggests that viral regulatory signals at either the 3' or 5' end are not involved in the transcription of the gene. As such, these results differ from those obtained using SV40 as a vector in which viral regulatory sequences are effectively used in the initiation and termination of the preproinsulin mRNA (6,7; Fig. 2).

D. Protein Analysis

Having identified insulin specific RNA in BPV_{69T}-rI_1 transformed cells the efficacy of the transcripts in directing translation and processing of polypeptides was investigated. Cells were labeled with ^{35}S-cysteine, an extract was prepared and immuno-precipitated with hamster anti-bovine insulin serum. The precipitated proteins were then analyzed on SDS polyacrylamide gels. As a reference, the pattern obtained with lysate of African Green monkey kidney cells infected with SV40-rat preproinsulin recombinant virus (SVL1-rI_1) (6) is presented (Fig. 3). In this case a proinsulin polypeptide is immuno-precipitated which shows a migration pattern similar to that of authentic bovine proinsulin marker (bovine proinsulin migrates slightly ahead of rat proinsulin due to its smaller size). Analysis of BPV_{69T}-rI_1 transformed cells [NS6(-), NS8(-), NS24(-)] demonstrated the presence of a prominent band identical in its electrophoretic mobility to proinsulin from SVL_1-rI_1 infected cells (Fig. 3). This band was absent from the lysates prepared from cells transformed by BPV alone (ID14). In the competitive binding study, samples were incubated with antiserum previously neutralized with 2 μg of bovine insulin. Under these conditions, no proinsulin-like protein was immunoprecipitated [NS6(+), NS8(+), NS24(+)].

Analysis on culture media of BPV_{69T}-rI_1 transformed cells shows that the rat proinsulin is secreted into the medium (not shown) as it is in SVL_1-rI_1 infected monkey cells (6).

The levels of insulin-like proteins produced in transformed cells were estimated by a quantitative radioimmunoassay. As shown in Table 1 media from BPV_{69T}-rI_1 transformed cells contained from 10 μU/ml to more than 400 μU/ml (1 U = 48 ng) of material immunoreactive with anti-insulin serum. This represents a 2 to 80-fold increase over the 5 μU/ml present in medium from untransformed C127 cells. When the medium of cells transformed by BPV_{69T} alone was analyzed, only background levels (<5 μU/ml) were detected. This indicates that secretion of insulin or insulin-like material is not a property of BPV-transformed cells in general but rather a function of the exogenous DNA used in transformation.

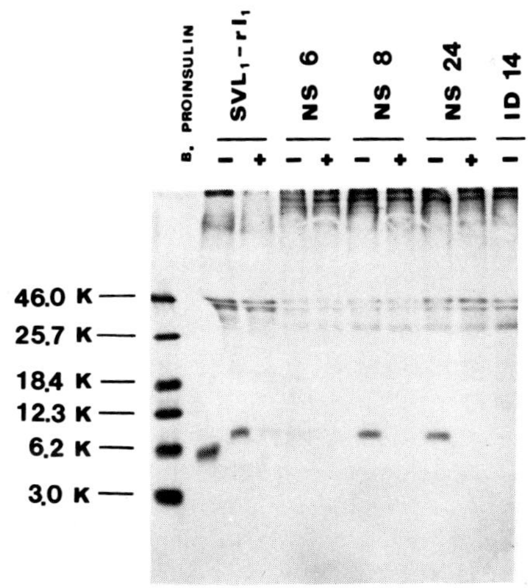

FIGURE 3. SDS-polyacrylamide gel electrophoresis of BPV_{69T}-rI_1 transformed cells. NS6, NS8 and NS24 are cells transformed by BPV. (-) indicates no competition; (+) indicates with competition. Published with permission.

TABLE I.

Mouse cell lines	Insulin secreted in 24 hr by 10^6 cells (uU/ml)	No. of lines/ total	[%]
Transformed by BPV_{69T}-rI_1	>400	24/48	50%
	200-400	11/48	23%
	40-200	10/48	21%
	10-40	3/48	6%
Transformed by BPV_{69T}	<5	6/6	100%
Untransformed C127	<5	6/6	100%

Rate of insulin production for a representative BPV_{69T}-rI_1 transformed cell line (NS8) = 400 μU/12 hr/10^6 cells. Assay was performed by Hazelton Laboratories, Vienna, Va.

III. DISCUSSION

Previous studies have established that mouse cells transformed by BPV-1 contain viral DNA sequences which exist exclusively in a free extrachromosomal state (14). This is true for cells transformed by the intact virus, by cloned linearized viral DNA or by a cloned 69% subgenomic DNA fragment. The unique ability of the papillomaviruses to transform cells in the absence of integration prompted us to assess the potential use of BPV-1 DNA as a eukaryotic cloning vector.

Our experimental design consisted of the construction of a recombinant DNA molecule containing the 69% transforming region of BPV-1 DNA and the rat preproinsulin I gene (rI_1) which contains all the regulatory signals (the putative promoter, polyadenylation site and intervening sequences) necessary for faithful transcription.

Mouse cells transformed by the recombinant molecules were isolated and tested for the expression of the exogenous gene. Using several criteria we have demonstrated that (i) the exogenous gene is transcribed into mRNA similar if not identical to authentic preproinsulin mRNA; (ii) these transcripts direct the synthesis of proinsulin protein, and (iii) the gene product is secreted by the cells into the tissue culture medium in large amounts.

It is noteworthy that 100% of the transformed cell lines tested (48/48) contained both BPV and rI_1 DNA sequences. Since these cell lines were initially selected solely on the basis of their ability to grow as foci in an untransformed cell monolayer it seems that transformation *per se* is a sufficient criterion for the isolation of cells that have stably incorporated the exogenous DNA.

A combined endonuclease S1-exonuclease VII analysis of the RNA's produced in transformed cells revealed the presence of a single insulin specific transcript representing the entire coding region of the gene. All the detectable insulin specific RNA's initiate from a preproinsulin promoter, are correctly spliced and are accurately terminated. Since the BPV promoter does not initiate detectable levels of transcripts of the foreign DNA, this system should be ideal for localizing regulatory elements of various genes and for assessing the effects of induced and naturally occurring mutation on promoter function.

We have also demonstrated that substantial levels of proinsulin protein are synthesized in transformed cells and secreted into the medium. Insulin, normally produced in the β cells of the pancreas, is first synthesized as preproinsulin. Two post-translational processing events, the removal of the leader sequence followed by removal of the internal C peptide convert the preproinsulin to proinsulin and insulin, respectively. The presence of proinsulin in BPV_{69T}-rI_1 transformed cells thus indicates that the first processing event occurs in these cells.

REFERENCES

1. Aviv, H., and Leder, P. (1972). Proc. Natl. Acad. Sci. USA 69, 1408.
2. Berk, A. J., and Sharp, P. A. (1978). Proc. Natl. Acad. Sci. USA 75, 1274-1278.
3. Chick, W. L., Warren, S., Chute, R. N., Like, A. A., Lauris, V., and Kitchen, K. C. (1977). Proc. Natl. Acad. Sci. USA 74, 628.

4. Clewell, D. B. (1972). J. Bact. 110, 667.
5. Graham, F. L., and van der Eb, A. J. (1973). Virology 52, 456.
6. Gruss, P., and Khoury, G. (1981). Proc. Natl. Acad. Sci. USA 78, 133-137.
7. Hamer, D. H., and Leder, P. (1979). Nature 281, 35.
8. Howley, P. M., Law, M.-F., Heilman, C., Engel, L., Alonso, M. C., Lancaster, W. D., Israel, M. A., and Lowy, D. R. (1980). In "Viruses in Naturally Occurring Cancers" (M. Essex et al., eds.), Cold Spring Harbor Laboratory, N. Y., in press.
9. Hutchinson, K. W., and Halvorson, H. O. (1980). Gene 8, 267.
10. Kessler, S. W. (1975). J. Immunol. 115, 1617.
11. Khoury, G., Howley, P., Nathans, D., and Martin, M.A. (1975). J. Virol. 15, 433.
12. Laemmli, U. K. (1970). Nature (London) 227, 680.
13. Laskey, R. A., and Mills, A. D. (1975). Eur. J. Biochem. 56, 335-341.
14. Law, M.-F., Lowy, D. R., Dvoretzky, I., and Howley, P. M. (1981). Proc. Natl. Acad. Sci. USA 78, in press.
15. Lomedico, P., Rosenthal, N., Efstratiadis, A., Gilbert, W., Kolodner, R., and Tizard, R. (1979). Cell 18, 545.
16. Lowy, D. R., Dvoretzky, I., Shober, R., Law, M.-F., Engel, L., and Howley, P. M. (1980). Nature 287, 72.
17. Lowy, D. R., Rand, E., and Scolnick, E. M. (1978). J. Virol. 26, 291.
18. Maniatis, T., Hardison, R. C., Lacy, E., Lauer, J., O'Connel, C., Quon, D., Sim, G. K., and Efstratiadis, A. (1978). Cell 15, 687.
19. Sargent, I.O., Uw, J., Sala-Trepat, J. M., Wallace, R. B., Reyes, A. A., and Benner, J. (1979). Proc. Natl. Acad. Sci. USA 76, 3256.
20. Stowe, N. D., and Wilkie, N. M. (1976). J. Gen. Virol. 33, 447.
21. Villareal, L. P., and Berg, P. (1977). Science 196, 183-186.
22. Wesnik, P. C., Finnegan, D. J., Donelson, J. E., and Hogness, D. S. (1974). Cell 3, 315.

THE DEVELOPMENT OF HOST VECTORS FOR DIRECTED GENE TRANSFERS IN PLANTS

Jeff Schell[1]
Marc Van Montagu
Marcelle Holsters

Laboratory of Genetics, Rijksuniversiteit Gent, Belgium

Jean-Pierre Hernalsteens
Jan Leemans
Henri De Greve

Laboratory GEVI, Vrije Universiteit Brussel, Belgium

Lothar Willmitzer
Leon Otten
Jo Schröder
Charles Shaw

Max-Planck-Institut für Züchtungsforschung, Köln, FRG

I. ABSTRACT

For the study of plant developmental biology nature has provided us with an unexpected system with the double advantage of being an efficient gene-vector that already contains a set of purified genes directly involved in the control of developmental processes in plants. This system is the crown gall tumor inducing Ti plasmid of *Agrobacterium tumefaciens*.

[1] Present address: Max-Planck-Institut für Züchtungsforschung, Köln, FRG.

ISBN 0-12-137420-3

II. INTRODUCTION

A description of the mechanism that underlies the formation of crown galls on plants as a result of an interaction with the soil bacterium *A. tumefaciens*, is justified in this symposium for two reasons.

(1) The formation of the plant cell tumors is the direct consequence of the transfer, integration and expression of "oncogenes" in the plant nucleus. These are genes the products of which directly or indirectly control cellular growth and differentiation of plant cells. The identification of these oncogenes and of their structure and function may therefore provide a unique system to study the developmental biology of plant cells.

(2) This transformation system is brought about by a natural gene vector for plants evolved by these soil bacteria in order to genetically engineer the plant cells for the benefit of the engineering organism.

We shall describe how this natural gene vector can be adapted to serve as a general vector to introduce genes in plants and will therefore allow us to experimentally probe the structure-function relations of plant genes. Before giving a description of the Ti plasmid and its properties, I want to briefly mention other gene-vector systems for plants that are currently being studied.

A. DNA viruses (cfr SV40 system)

Only very few plant DNA viruses are known. The best studied virus is CaMV, which infects some crucifer plants. Its genome consists of a small (± 8 Kb) double-stranded circular DNA. The entire genome of CaMV was cloned in an *E. coli* vector and this cloned genome was shown to be infective on turnips (1). The whole of the nucleotide sequence has been worked out by Franck *et al.* (2) and work is in progress to try to develop this virus genome as a gene vector.

B. Selectable marker-genes (cfr TK system)

Several groups are currently attempting to develop such marker genes. One promising model is the nitrate reductase (NR) gene. Nitrate reductase-deficient cell lines have been isolated (e.g. in tobacco) and selective conditions have been worked out to select for cells in which NR activity has been restored. Such a cloned NR gene has, however, not yet been isolated. Recently, one has started to look for drug resistance markers as selectable marker genes. The prominent candidates that are being studied here are bacterial genes coding for phosphorylation or acetylation enzymes that inactivate such antibiotics as kanamycin, neomycin, G418 and gentamycin. The fact that some of these genes have recently been found to be expressed in yeast cells and some others in mammalian cells, lends support to these efforts to use these or similar marker genes for plant cells. No success has, however, been reported yet and we do not know whether plant cells will readily take up DNA of isolated genes without significant degradation and, if so, whether the bacterial genes will be readily expressed. The use of liposomes to introduce DNA in plants has recently been reported and will be discussed in the next talk.

III. THE Ti PLASMID AS A NATURAL VECTOR

Ti plasmids are harbored by a group of soil bacteria (*Agrobacteria*) and are responsible for the capacity of these bacteria to induce so-called "crown gall" tumors on most dicotyledonous plants. (For recent reviews, see 3-9) Crown gall tumors proliferate autonomously in tissue culture on simple media devoid of growth hormones. The tumorous state of crown gall tissue results from the transfer of Ti plasmid DNA from *Agrobacterium tumefaciens* to the plant cells. This system represents a natural gene vector for plant cells, evolved by and for the benefit of the bacteria that harbor Ti plasmids (6). Crown gall cells, as a direct result of genetic transformation by the Ti plasmid, produce various substances called opines. Free living *Agrobacteria* utilize these opines as sources of carbon and nitrogen. The genetic information both for the synthesis of opines in transformed plant cells and for their catabolism by free living *Agrobacteria* is carried by Ti plasmids. Because different Ti plasmids induce and catabolize different opines, the Ti plasmids can be classified according to the type of opine they determine (10).

Bacteria are known to be able to conquer an ecological niche by acquiring the capacity to catabolize certain organic compounds, not readily degradable by most other bacterial species. In several cases, the genes determining this degradative capacity have been found to be part of extrachromosomal plasmids. Several groups of soil bacteria, especially those living in and around the rhizosphere of plants, are able to decompose organic compounds released by plants. Clearly, with the advent of Ti plasmids, Agrobacteria have carried this capacity one step further, by genetically forcing plant cells - via a gene transfer mechanism - to produce specific compounds (opines) which they are uniquely equipped to catabolize. This novel type of parasitism has therefore been called "genetic colonization" (6).

By comparing the properties of Agrobacterium strains with and without a given Ti plasmid, it has been possible to define Ti plasmid encoded functions. Ti plasmids are responsible for the following properties of Agrobacteria : (i) crown gall tumor induction; (ii) specificity of opine synthesis in transformed plant cells; (iii) catabolism of specific opines; (iv) agrocin sensitivity; (v) conjugative transfer of Ti plasmids; and (vi) catabolism of arginine and ornithine (11-24, 10).

In order to further establish whether these properties are determined by plasmid genes, mutant plasmids were isolated by transposon-insertion mutagenesis (7, 25-29) and by deletion (26, 30). The different mutations were localized on the physical maps established for nopaline (31) and octopine (32) Ti plasmids, thus allowing the establishment of functional genetic maps for these plasmids (26).

A number of general conclusions can be derived from this genetic analysis of Ti plasmids. Several regions of Ti plasmids must be directly or indirectly involved in determining the different steps leading to neoplastic transformation. Indeed, mutants affecting so-called Onc (oncogenicity) functions are distributed over about half of the Ti plasmid map; the other half does not appear to carry any genes involved in transformation.

One class of mutations localized within the so-called T region of Ti plasmids is of particular interest, since this region is transferred and maintained in transformed plant cells (see below). The mutations thus far localized in this region are of four general types.

(1) Mutant plasmids inducing crown gall tumors in which no opine synthesis occurs. These mutations demonstrate that opine synthesis and tumor induction and maintenance are coded for by different genes. Furthermore, since these opine synthesis deficient Ti plasmid mutants still allow Agrobacteria harboring them to catabolize opines, it must be concluded that genes involved in opine synthesis (after transfer to the plant cells) are not in any way involved in opine catabolism.

(2) Mutants that totally or incompletely abolish tumor formation.

(3) Mutants that strikingly influence the organogenic acti-

vity of transformed plant cells. Three types of such mutants have been isolated thus far, each localized in a well-defined segment of the T region : (i) shoot formers, (ii) root formers and (iii) shoot and root formers. From this last class of mutants normal plants can be regenerated which synthesize opines in all their tissues (stems, leaves and roots).

4) Mutants that do not visibly affect any of the known phenotypes of crown galls.

On the basis of the localization of these various mutants, a functional map of the T region can be drawn (Fig. 1). The most important conclusion from this genetic analysis is that the T region must contain genes that play a direct, active and specific role after transfer into plant cells. Some of these genes specify opine systhesis, whereas other genes directly or indirectly control plant cellular growth and differentiation.

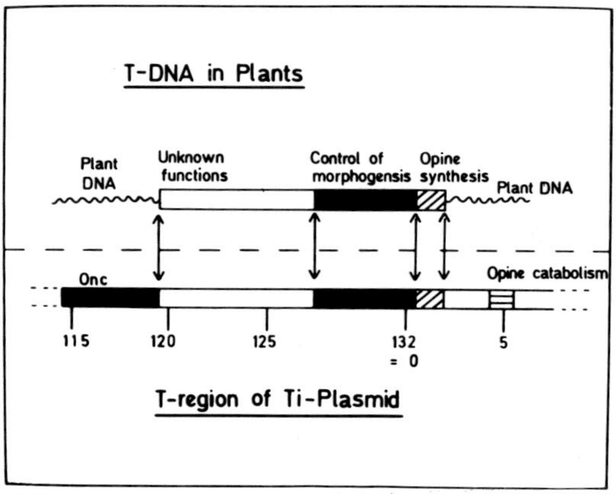

Figure 1. Functional map of the T region

IV. PROPERTIES OF THE GENE-TRANSFER MECHANISMS THAT ARE OF IMPORTANCE FOR THE USE OF THE Ti PLASMIDS AS A GENERAL GENE VECTOR FOR PLANTS

A. Host range

Most dicotyledonous plants that have been studied were found to be sensitive to one or more *Agrobacterium tumefa-*

ciens strains (33). Some Agrobacterium strains have a broad host range, whereas others have been shown to have a limited host range. It was recently shown that these differences in host range were determined by the type of Ti plasmids carried by these strains (34).

Unfortunately, no strains have thus far been found to transform cereals. It has been suggested (35) that monocotyledonous plants lack Agrobacterium adherence sites essential for the initial stages of transformation. If this would be the case, one could hope that the introduction of Ti plasmid DNA via liposomes or micro-injection, would allow transformation of cereals.

B. Transfer and stable maintenance of large DNA sequences via the T region of the Ti plasmid

The transfer and stable maintenance of Ti plasmid DNA in plant cells results from its integration in nuclear DNA. This was studied first by Southern blotting analysis and more recently by reisolation of junction fragments linking T DNA to plant DNA (6, 36-41).

The results show that all crown gall tumors contain a DNA segment, called the T DNA, which is homologous to DNA sequences in the Ti plasmid used to induce the tumor line. In all cases, this T DNA corresponds to - and is colinear with - a continuous stretch of Ti plasmid DNA, which has therefore been called the T region (see above).

The T DNA segment present in different tumors induced on tobacco by the nopaline strains C58 and T37 was found to be identical and had in each case a size of about 23 Kb. The T DNA in octopine tobacco crown gall lines was found to be smaller (about 14-15 Kb) and somewhat variable in size, although always derived from the T region.

By preparing DNA from purified nuclei, chloroplasts and mitochondria, isolated from crown gall tissues, it was shown that the T DNA is located in the nucleus and not in chloroplast or mitochondria (42, 43). When left- or right-hand border fragments of the T-region were used as probes in the Southern blot hybridization analysis, a small number of bands (usually 1 to 3) became apparent. The exact number and the sizes of these composite fragments, presumably consisting of T DNA covalently linked to plant DNA, varied from tumor line to line (and sometimes within a given line) as a function of the restriction enzyme used. These observations indicated that more than one copy of T DNA is integrated at more than one site in plant DNA.

In contrast, when internal fragments of the T region

were used as hybridization probes, only single bands were revealed, corresponding in each case to the full length of the T region restriction fragment used as probe. The bands hybridizing to different internal fragments of the T region demonstrated the same relative intensity.

Taken together, these data suggested that tumor cells contain a limited number of complete copies of T DNA associated with different plant DNA sequences. Further demonstration of the integration of T DNA in plant DNA and an understanding of the structure of T DNA was obtained when a complete EcoRI digest of T37 tobacco crown gall DNA, was cloned in a phage λ vector, thus allowing the isolation and detailed study of T DNA border sequences (40). Two clones were studied in detail. One clone contained the left and right borders of the T region linked together and the other clone contained the right end of the T DNA linked to repetitive plant DNA sequences. These data therefore suggested that the T DNA in nopaline tumors is organized as tandem repeats which can be inserted into repetitive sequences of plant DNA. As described in the abstract of poster 1199 presented at this meeting, Zambryski et al. recently isolated and studied several independent border fragments from T37 induced tumors on tobacco W38. Thus is was found that T-DNA can be covalently linked not only to highly repeated plant DNA sequences but also to moderately repeated or to unique sequences. It was also found by these authors that the right border of the T DNA was exactly (to the base pair) the same in two independent nopaline tumor lines.

Thomashow et al. (37,38) and De Beuckeleer et al. (in preparation) studied octopine tumor lines. A comparison between a number of octopine and nopaline T DNAs demonstrated that in all cases the T DNA consisted of a continuous stretch of DNA homologous to the T region of the inducing Ti plasmid. T DNAs in nopaline crown galls are generally significantly longer than those from octopine crown galls. Nopaline tumors carry from one to at least four different T DNA copies, whereas all investigated octopine tumor lines were found to contain a single T DNA copy.

The sum of the evidence derived from comparative studies involving different tumor lines induced with the same Ti plasmid; tumor lines induced with Ti plasmid mutants carrying transposon-insertions in the T region (39); and also from a detailed study of the nucleotide sequences bordering the T DNA segment in cloned T DNA fragments (40) suggest that the "ends" of the T region are involved in the integration of the T DNA. Either the T region ends recombine before integration into the plant DNA or the entire T region is excised and subsequently inserted (Fig. 2)

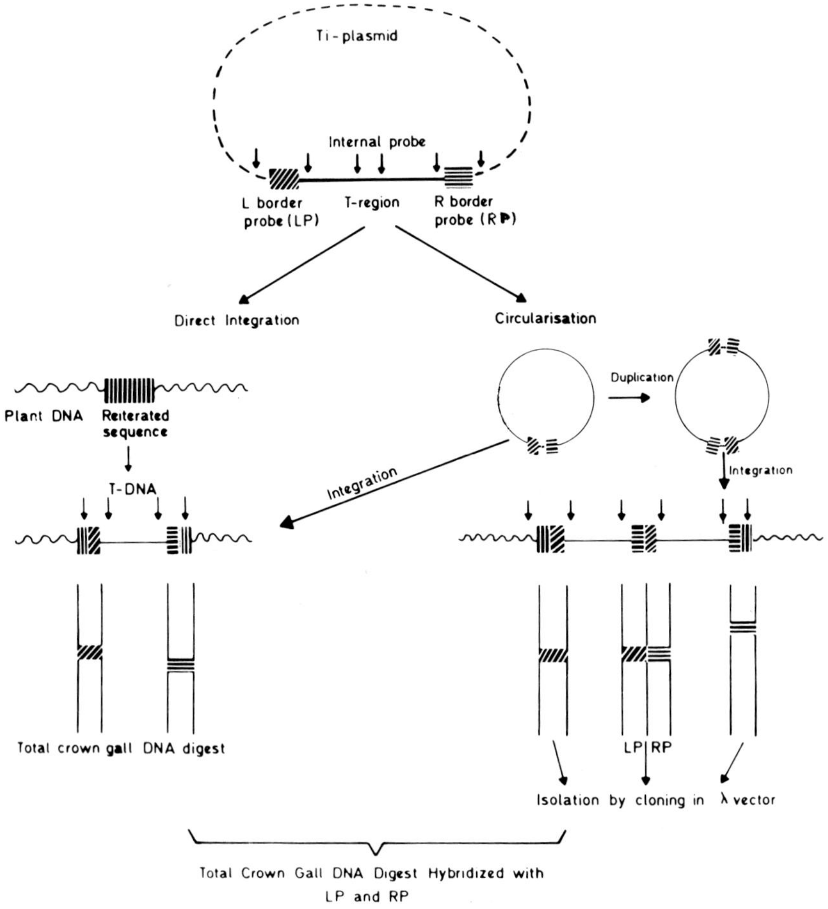

Figure 2. Transfer of T region to plant nucleus

The T DNA therefore has one central property in common with bacterial transposons since it appears to be a discrete unit of DNA with the capacity to integrate into non-homologous plant DNA.

The fact that the integration of the T DNA fragment is apparently directed by specific sequences bordering the T region, is of particular interest for its use as a gene-vector. Indeed, one could expect that any DNA sequence, integrated in between these recognition sequences will be cotransferred to the plant DNA. This was shown to be the case by inserting a bacterial transposon (Tn7 - 9.2 Md) in the T region of a nopaline Ti plasmid and demonstrating that the plant tumors induced by this plasmid contained a complete and intact Tn7 sequence inserted in the T DNA sequence (44). That the Tn7 inserted in the plant DNA was not subjected to sequence rearrangements was demonstrated by recloning the Tn7 from the plant DNA. The recombinant cosmids containing Tn7 DNA covalently linked to plant DNA were identified by directly selecting for the expression of Tn7 (streptomycin, spectinomycin, trimethoprim resistance)(Holsters et al., unpublished results).

C. The gene-vector should allow or possibly promote the expression of the transferred genes

The first positive evidence that the T·DNA is transcribed within the plant cell was reported by Drummond et al. (45), who demonstrated that total RNA isolated from crown gall tumors hybridized to a specific fragment of the Ti plasmid assumed to be present within the plant cell. A more extensive study was performed by Gurley et al. (46), who studied transcription in three different octopine tumors and found that the right part of the T DNA was transcribed most actively, whereas little transcription was detected from the left part. A serious draw-back in both studies, however, was that the extent of the T DNA in the tumor lines studied was not known at the time. Therefore, these studies did not allow the construction of a detailed picture of the transcribed parts of the T DNA. Whereas the transcription studies by Drummond et al. (45) and Gurley et al. (46) were performed with total RNA, we decided to study in detail the transcription of the T DNA in two octopine tumors with special emphasis on a comparison between nuclear and polysomal RNA. For these studies, tumor lines with accurately mapped T DNAs were used.

The following approach was used : (i) nuclear RNA was isolated from crown gall cells and labelled by polynucleotide kinase. Furthermore radioactive cDNA was made from nuclear RNA using oligo dT as primer. (ii) Polysomal RNA was isolated from crown gall cells. After separation into polyA$^+$ and polyA$^-$ fractions by oligo dT-cellulose chromatography, the RNAs were labelled by polynucleotide kinase. (iii) Highly purified and physiologically active nuclei were isolated from crown gall cells and incubated under conditions allowing RNA synthesis by endogenous RNA polymerase. Synthesized RNA was labelled by incorporation of α-^{32}P-UTP.

To map transcribed regions, the labelled RNA was hybridized to Southern blots containing cloned parts of the T DNA digested by various restriction enzymes. To find out which RNA polymerase is responsible for the transcription of the T DNA, transcription was studied both in the presence and in the absence of various concentrations of α-amanitin.

These studies have indicated that the T DNA is transcribed over its entire length. However, well defined regions of the T DNA are reproducibly actively transcribed, whereas other parts are only weakly transcribed. Surprisingly, no extensive differences were found between nuclear transcripts and polysomal transcripts. This might indicate that extensive processing steps do not occur in the formation of polysomal RNA. T DNA derived transcripts were found both in the polyA$^+$ and in the polyA$^-$ fraction of the polysomal RNA. Both fractions gave the same relative hybridization pattern of T DNA fragments. By comparing the transcriptional pattern of the T DNA with the functional organization of the T region (Fig. 1), it was observed that some of the areas that are most actively transcribed correspond to those that are genetically correlated with opine synthesis (right end of T region)

Finally, the transcription of the T DNA in isolated nuclei was shown to be inhibited by low concentrations of α-amanitin (0.7 μg/ml). This concentration is known to inhibit specifically RNA polymerase II from plants. Therefore the transcription of the prokaryotic derived T DNA seems to be provided by the host RNA polymerase II.

Recently we (Willmitzer et al., unpublished results) were able to obtain Northern gels of T DNA specific transcripts allowing a more precise mapping of these transcripts. The results confirmed those obtained by the previous Southern gel analysis and demonstrated that the A6 octopine tumor line contains 7 - 8 different T DNA specific transcripts coming from distinct regions of the T DNA. Most, if not all, of these transcripts start and terminate within the T DNA.

To demonstrate that the transcription of T DNA results in the production of functional mRNAs, T DNA specific transcripts were isolated, translated in vitro, and the radioactive protein products identified (47).

A distinct protein with the molecular weight of lysopine dehydrogenase (responsible for the synthesis of octopine) was formed only with RNA from transformed cells annealed to the right part of the T region (see Fig. 1). Immunoprecipitations with antiserum raised against highly purified lysopine dehydrogenase showed that this distinct protein was specifically recognized by the antibodies, thus indicating that the structural gene for lysopine dehydrogenase is part of the T DNA.

The DNA from the left part of the T region selected for a mRNA coding for one or more small proteins (14 Kb). The significance of these proteins is still unknown.

To summarize these studies it can be said that the T DNA contains the necessary signals to start and terminate functional transcripts in plants. Some of these transcription signals (promotors) could therefore be used to express inserted genes in plants. That this is very likely to be the case was indicated by observations involving tobacco crown gall lines containing the Tn7 transposon.

The expression of the Tn7 insertion in the nopaline synthase locus (see above) was assayed as a model system for other genes inserted in this area. Tn7 codes for a dihydrofolate reductase which is resistant to methotrexate (48). Suspension cultures of both untransformed and of crown gall tobacco tissue were found to be completely inhibited by 0.5 μg/ml of methotrexate. In contrast, cultures established with Tn7 containing crown gall cells grew well on liquid media containing up to 2 μg/ml of methotrexate. In addition, nuclei isolated from these methotrexate tobacco lines were shown to synthesize Tn7 transcripts and a polyA mRNA fraction derived from purified polysomes was also shown to contain Tn7 transcripts. Further experiments are in progress to determine whether transcription of the inserted Tn7 is initiated by a T DNA promotor or by a Tn7 promotor. These observations are very important because they support the likelihood that other bacterial genes can be brought to expression in plant cells, e.g. other selectable markers such as the genes coding for enzymes inactivating toxic substances as herbicides and antibiotics, like G418, kanamycin, neomycin and gentamycin.

D. How can one introduce isolated genes in the T region via in vitro recombination ?

Because of the size of Ti plasmids (about 130 Md) and because a large number of Ti plasmid genes are involved in the transformation mechanism, it did not appear to be feasible to develop a "mini-Ti" cloning vector with unique cloning sites at appropriate locations within the T region and with all functions essential for T DNA transfer and stable maintenance. An alternative way was therefore developed to introduce genes at specified sites in the T region of a functional Ti plasmid. The principle was to make an "intermediate vector" consisting of a common *E. coli* cloning vehicle, for example the pBR322 plasmid, into which an appropriate fragment of the T region of a Ti plasmid is inserted. Single restriction sites in this T region fragment can then be used to insert the chosen DNA sequence. This "intermediate vector" is subsequently introduced, via transformation or mobilization, into an *A. tumefaciens* strain carrying a Ti plasmid which has been made constitutive for transfer (26) and which already carries antibiotic resistance markers (e.g. streptomycin, sulfonamide) cloned into the same T region restriction site as that chosen to insert the DNA to be transferred. These resistance markers were introduced into this site by a procedure essentially identical to the one described here. Recombination *in vivo* will transfer the DNA of interest into the appropriate site of the Ti plasmid. We have thus created a situation analogous to the one described above and therefore expected that the Ti plasmid, thus engineered to contain the desired DNA, would transfer this DNA into plant cells. This was recently shown to be the case with a model system in which a DNA fragment derived from the plasmid R702 and carrying streptomycin and sulfonamide resistance genes, was successfully introduced into tobacco crown gall cells. It should be noted that the introduction of inserts and subtitutions in all parts of the T region by these methods should allow us to probe further for the different T DNA functions, this approach being equivalent to "site specific" mutagenesis.

E. Can normal plants be regenerated from T-DNA containing plant cells ?

The ultimate aim of many gene transfer attempts in plants is to produce fertile cultivars harbouring and transmitting new genetic properties. It was, therefore, essential to determine whether T DNA transfer could be dissociated from neoplastic transformation and whether normal, fertile plants could be derived from T DNA transformed plant cells. In order to answer this question, a large set of insertion mutants were obtained with an octopine TiB6S3 plasmid (29). One of the mutant plasmids (pGV2100) was clearly less oncogenic on tobacco and sunflower hypocotyls when compared to the wild-type plasmid. Tumors appeared only after prolonged incubation time. Furthermore, shoots proliferated from the greenish tumors, in contrast to the undifferentiating white tumors induced by strains harboring the wild-type TiB6S3 plasmid. The Tn7 insertion in pGV2100 was mapped and found to be located in the left arm of the common DNA of the T region (Fig. 1).

As a test for transformation, tumor tissue and the shoots on tobacco were assayed for the presence of lysopine dehydrogenase (49). The tumor tissue was found to be positive. Most of the shoots were negative but some of the proliferating shoots were positive. One such shoot was grown further on growth-hormone-free media and found to develop roots and later to grow into a fully normal, flowering plant. Each part of this plant, leaves, stem and roots was found to contain lysopine dehydrogenase activity and polysomal RNA was

TABLE 1. MENDELIAN TRANSMISSION OF LpDH in E40-2 TOBACCO

CROSSES	N° OF PROGENY TESTED		LpDH positive		LpDH negative
E40.2 ♂ x E40.2 ♀	145		110 (76%)		35 (24%)
		semi-quant test	++	+	-
	200		42 (21%)	95 (48%)	63 (31%)
E40.2 ♂ x Wildtype ♀	248		124 (50%)		124 (50%)
Wildtype ♂ x E40.2 ♀	187		81 (43%)		106 (57%)
PLANTLETS DERIVED from Anther cultures of E40-2 Tobacco	102		47 (46%)		55 (54%)

Conclusions:

♂ \ ♀	LpDH	-
LpDH	++	+ -
-	-+	- -

- LpDH is transmitted through meiosis as a single dominant factor
- LpDH$^+$ eggs and pollen are fertile
- LpDH gene behaves as a single locus on a single chromosome
- E40-2 is HEMYZYGOTE LpDH/ -

found to contain T DNA transcripts homologous to the opine synthesis locus. No transcripts of the conserved segment of the T region were observed.

These observations therefore demonstrate that normal plants can be obtained from plant cells transformed with Ti plasmids genetically altered in specific segments of the T region. Seeds obtained by self-fertilization of these plants produced new plants with active T DNA linked genes, thus demonstrating that genes introduced in plant nuclei, via the Ti plasmid, can be sexually inherited. A series of sexual crosses were therefore designed to study the transmission pattern of the T DNA specified genes. The results of these crosses are presented in Table 1 and demonstrate very convincingly that the T DNA-linked genes (LpDH) are transmitted as a single Mendelian factor both through the pollen and through the eggs of the originally transformed plant. These crosses also showed that the original transformed plant was a hemizygote containing T DNA on one locus only of a pair homologous chromosomes. By these crosses tobacco plants homozygotes for the altered T DNA were obtained.

V. THE USE OF PLANT CELLS TRANSFORMED BY MODIFIED Ti PLASMIDS FOR THE STUDY OF THE STUDY OF THE DEVELOPMENTAL BIOLOGY OF PLANT CELLS

At the end of the previous section we described the origin of plants derived from crown gall cells induced with mutant Ti plasmids. Leaf cells from these plants contained T DNA specific transcripts derived from the opine synthase genes but did not contain measurable amounts of transcripts from the oncogenes. On media containing added growth hormones (cytokinins), however, these plant cells grew as partially organized teratomas. In these teratomas transcripts of the T DNA oncogenes were again observed.

These observations lend strong support to the notion that the T DNA does harbor oncogenes, the expression of which is directly or indirectly responsible for the abnormal growth pattern of the transformed cells. It is interesting to note that the expression of these oncogenes can, under certain circumstances, be controlled by known plant growth hormones and or by the developmental growth pattern of these cells.

The phenomenon of crown gall formation on plants is therefore of fundamental importance to the study of the molecular biology of plant development. The abnormal growth pattern of crown gall plant cells has been shown to be the direct consequence of the presence and expression of specific "tumor" genes. These genes have been transferred into the plant nuclei by Gram-negative bacteria, which contain natural gene vectors in the form of large Ti plasmids. These Ti plasmids carry plant tumor-inducing genes and can transfer them to the plant nucleus. Because these tumor genes and a series of mutations have been isolated which directly affect the growth and morphogenetic properties of the transformed plant cells, the crown gall system represents a model system uniquely suited to the study of the genes and gene products involved in the control of cellular growth and differentiation.

A comparison of the crown gall system with tumors induced by animal oncogenic viruses yields the following similarities : oncogenes are transferred and integrated in the chromosomes of transformed cells in both systems.

In the case of animal oncogenic viruses, such as the retroviruses, the oncogene (src) is derived from the genome of the uninfected mammal cells. The viruses apparently acquired these cellular genes by recombination, in other words, the assimilation of oncogenes by these viruses may be a by-product (accidental ?) of the integration of the normal viral genome in the host genome. In sharp contrast to this situation, we did not observe any homology between the oncogenes of the T DNA and the genomes of the uninfected plants (Tobacco, Arabidopsis, Petunia). The oncogenes of the T DNA do not correspond to normal plant genes, and one could argue that the crown gall oncogenes have specifically evolved as part of the transferred T DNA in order to produce autonomous proliferation of the initially transformed cells. The products of the T DNA oncogenes appear to produce a tumorous growth pattern by blocking differentiation, since inactivation of the oncogenes by mutation results in the formation, by transformed cells, of normally organized tissues. The actual blocking of differentiation may well depend on the relative dosage of the oncogenes. It was indeed observed (G. Wullems, personal communication) that somatic hybrids - made by fusion of crown gall tobacco cells with normal untransformed tobacco cells - could differentiate to normally organized plants.

Finally, it should be stressed that, although the T DNA is part of a prokaryotic plasmid, it is functionally an eukaryotic DNA sequence. Any speculation on its origin is bound to yield some unexpected and unorthodox concepts with relation to the exchange of DNA between prokaryotes and eukaryotes and a fortiori between different types of eukaryotes. It might also be worth speculating that crown galls are not a unique instance of natural gene transfers involved in oncogenesis. In fact, it might well be a more general phenomenon playing an important role in the directed genetic modification of host cells by some of their symbiotic or parasitic partners.

ACKNOWLEDGMENTS

The investigations reported here were supported by grants from the "Kankerfonds van de A.S.L.K", from the "Instituut tot aanmoediging van het Wetenschappelijk Onderzoek in Nijverheid and Landbouw" (248/A), from the "Fonds voor Wetenschappelijk Geneeskundig Onderzoek" (3.0052.78), and the "Onderling Overlegde Akties" (12052179). J.P.H. is a Research Associate of the Belgian National Fund for Scientific Research (N.F.W.O.).

REFERENCES

1. Howell, S.H., Walker, L.L., and Dudley, R.K. (1980) Science 208, 1265.
2. Franck, A., Guilley, H., Jonard, G., Richards, K., and Hirth, L. (1980). Cell 21, 285.
3. Gordon, M.P. (1979). In "Proteins and Nucleic Acids, Vol. 6, (A. Marcus, ed.), in press, Academic Press, New York.
4. Schilperoort, R.A., Hooykaas, P.J.J., Klapwijk, P.M., Koekman, B.P., Nuti, M.P., Ooms, G., and Prakash, R.K. (1979). In "Plasmids of medical, environmental and commercial importance" (K. Timmis and A. Pühler, eds), p. 339. Elsevier, Amsterdam.
5. Schilperoort, R.A., Klapwijk, P.M., Ooms, G., and Wullems, G.J. (1980). In "Genetic origins of tumour cells" (F.J. Cleton and J.W. Simons, eds), p. 87. Martinus Nijhoff, The Hague.
6. Schell, J., Van Montagu, M., De Beuckeleer, M., De Block, M., Depicker, A., De Wilde, M., Engler, G., Genetello, C., Hernalsteens, J.P., Holsters, M., Seurinck, J., Silva, B., Van Vliet, F., and Villarroel, R. (1979). Proc. R. Soc. Lond. B 204, 251.
7. Van Montagu, M., and Schell, J. (1979). In "Plasmids of medical, environmental and commercial importance" (K. Timmis and A. Pühler, eds), p. 71. Elsevier, Amsterdam.
8. Schell, J., and Van Montagu, M. (1980). In "Genome Organization and Expression in Plants" (C.J. Leaver, ed.), p. 453. Plenum Press, New York.
9. Van Montagu, M., Holsters, M., Zambryski, P., Hernalsteens, J.P., Depicker, A. De Beuckeleer, M.,

Engler, G., Lemmers, M., Willmitzer, M., and Schell, J. (1980). Proc. R. Soc. B 210, 351.
10. Guyon, P., Chilton, M.-D., Petit, A., and Tempé, J. (1980). Proc. Natl. Acad. Sci. USA 77, 2693.
11. Zaenen, I., Van Larebeke, N., Teuchy, H., Van Montagu, M., and Schell, J. (1974). J. Mol. Biol. 86, 109.
12. Van Larebeke, N., Engler, G., Holsters, M., Van den Elsacker, S., Zaenen, I., Schilperoort, R.A., and Schell, J. (1974). Nature 252, 169.
13. Van Larebeke, N., Genetello, C., Schell, J., Schilperoort, R.A., Hermans, A.K., Hernalsteens, J.P., and Van Montagu, M. (1975). Nature 255, 742.
14. Schell, J. (1975). In "Genetic manipulations with plant materials" (L. Ledoux, ed.), p. 163. Plenum Press, New York.
15. Watson, B., Currier, T.C., Gordon, M.P., Chilton, M.-D., and Nester, E.W. (1975). J. Bacteriol. 123, 255.
16. Engler, G., Holsters, M., Van Montagu, M., Schell, J., Hernalsteens, J.P., and Schilperoort, R.A. (1975). Molec. Gen. Genet. 138, 345.
17. Bomhoff, G., Klapwijk, P.M., Kester, H.C.M., Schilperoort, R.A., Hernalsteens, J.P., and Schell, J. (1976). Molec. Gen. Genet. 145, 177.
18. Genetello, Ch., Van Larebeke, N., Holsters, M., Depicker, A., Van Montagu, M., and Schell, J. (1977). Nature 265, 561.
19. Kerr, A., Manigault, P., and Tempé, J. (1977). Nature 265, 560.
20. Petit, A., Dessaux, Y., and Tempé, J. (1978a). Proc. IVth Int. Conf. Plant Path. Bact.-Angers, 143.
21. Petit, A., Tempé, J., Kerr, A., Holsters, M., Van Montagu, M. and Schell, J. (1978b). Nature 271, 570.
22. Firmin, J.L., and Fenwick, G.R. (1978). Nature 276, 842-844.
23. Klapwijk, P.M., Scheuldermon, T., and Schilperoort, R.A. (1978). J. Bacteriol. 136, 775.
24. Ellis, J., Kerr, A., Tempé, J., and Petit, A. (1979). Molec. Gen. Genet. 173, 263.
25. Hernalsteens, J.P., De Greve, H., Van Montagu, M., and Schell, J. (1978). Plasmid 1, 218.
26. Holsters, M., Silva, B., Van Vliet, F., Genetello, C., De Block, M., Dhaese, P., Depicker, A., Inzé, D., Engler, G., Villarroel, R., Van Montagu, M., and Schell, J. (1980). Plasmid 3, 212.
27. Ooms, G., Klapwijk, P.M., Poulis, J.A., and Schilperoort, R.A. (1980). J. Bacteriol. 144, 82.
28. Garfinkel, D.J., and Nester, E.W. (1980) J. Bacteriol. 144, 732.

29. De Greve, H., Decraemer, H., Seurinck, J., Van Montagu, M., and Schell, J. (1981). Plasmid, in press.
30. Koekman, B.T., Ooms, G., Klapwijk, P.M., and Schilperoort, R.A. (1979). Plasmid 2, 347.
31. Depicker, A., De Wilde, M., De Vos, G., De Vos, R., Van Montagu, M., and Schell, J. (1980). Plasmid 3, 193.
32. De Vos, G., De Beuckeleer, M., Van Montagu, M., and Schell, J. (1981). Plasmid, in press.
33. De Cleene M., and De Ley, J. (1976). Botan. Rev. 42, 389-466.
34. Thomashow, M.F., Panagopoulos, C.G., Gordon, M.P., and Nester, E.W. (1980a). Nature 283, 794.
35. Lippincott, J.A., and Lippincott, B.B. (1978). Science 199, 1075.
36. De Beuckeleer, M., De Block, M., De Greve, H., Depicker, A., De Vos, R., De Vos, G., De Wilde, M., Dhaese, P., Dobbelaere, M.R., Engler, G., Genetello, C., Hernalsteens, J.P., Holsters, M., Jacobs, A., Schell, J., Seurinck, J., Silva, B., Van Haute, E., Van Montagu, M., Van Vliet, F., Villarroel, R., and Zaenen, I. (1978). Proc. IVth Int. Conf. Plant Path. Bact.-Angers, 115.
37. Thomashow, M.F., Nutter, R., Montoya, A.L., Gordon, M.P., and Nester, E.W. (1980b). Cell 19, 729.
38. Thomashow, M.F., Nutter, R., Postle, K., Chilton, M.-D., Blattner, F.R., Powell, A., Gordon, M.P., and Nester, E.W. (1980). Proc. Natl. Acad. Sci. USA 77, 6448.
39. Lemmers, M., De Beuckeleer, M., Holsters, M., Zambryski, P., Depicker, A., Hernalsteens, J.P., Van Montagu, M. and Schell, J. (1980). J. Mol Biol. 144, 355.
40. Zambryski, P., Holsters, M., Kruger, K., Depicker, A., Schell, J., Van Montagu, M., and Goodman, H.M. (1980). Science 209, 1385.
41. Yadav, N.S., Postle, K., Saiki, R.K., Thomashow, M.F., and Chilton, M.-D. (1980). Nature 287, 458.
42. Willmitzer, L., De Beuckeleer, M., Lemmers, M., Van Montagu, M., and Schell, J. (1980). Nature 287, 359.
43. Chilton, M.-D., Saiki, R.K., Yadav, N., Gordon, M.P., and Quetier, F. (1980). Proc. Natl. Acad. Sci. USA 77, 4060.
44. Hernalsteens, J.P., Van Vliet, F., De Beuckeleer, M., Depicker, A., Engler, G., Lemmers, M., Holsters, M., Van Montagu, M., and Schell, J. (1980). Nature 287, 654.
45. Drummond, M.H., Gordon, M.P., Nester, E.W., and Chilton, M.-D. (1977). Nature 269, 535.
46. Gurley, W.B., Kemp, J.D., Albert, M.J., Sutton, D.W., and Callis, J. (1979). Proc. Natl. Acad. Sci. USA 76, 2828.
47. Schröder, J., Schröder, G., Huisman, H., Schilperoort, R.A., and Schell, J. (1981). Nature, in press.

48. Tennhammer-Ekman, B., and Sköld, O. (1979). Plasmid 2, 334.
49. Otten, L.A. and Schilperoort, R.A. (1978). Biochim. Biophys. Acta 527, 497.

THE TI PLASMID AS A VECTOR FOR THE GENETIC ENGINEERING OF PLANTS

Robert B. Simpson[1,2] Department of Microbiology
and Immunology
University of Washington
Seattle, Washington

ABSTRACT

Agrobacterium tumefaciens causes tumors on plants and during the process transfers a defined portion of the bacterial Ti plasmid to the plant cells. The fate of this transferred DNA, called T DNA, suggests the use of the plasmid as a vector for the introduction, stable maintenance and expression of DNA in plant cells. Characteristics of the interaction pertinent to such a goal are discussed as well as two relevant lines of recent experiments. The first identifies sites and techniques for the introduction of DNA into the Ti plasmid for subsequent transfer to the plant. The second describes the reversion of tumors to phenotypically normal plant cells and the regeneration of fertile plants from these cells. The revertant plants retain a portion of the T DNA even through meiosis.

I. INTRODUCTION

An understanding of how genes are controlled in development will probably result from a variety of approaches. A biochemical approach, which has recently become available due to advances in molecular cloning and characterization of DNA sequences, correlates chromosomal and RNA structures with expression. Structural changes which have been observ-

---

1 Current address: ARCO Plant Cell Research Institute, 6905 Sierra Ct., Dublin, CA 94566.

2 Supported by NIH and ACS grants to E.W. Nester and M.P. Gordon and a Damon Runyon-Walter Winchell Cancer Fund Postdoctoral Fellowship.

ISBN 0-12-137420-3

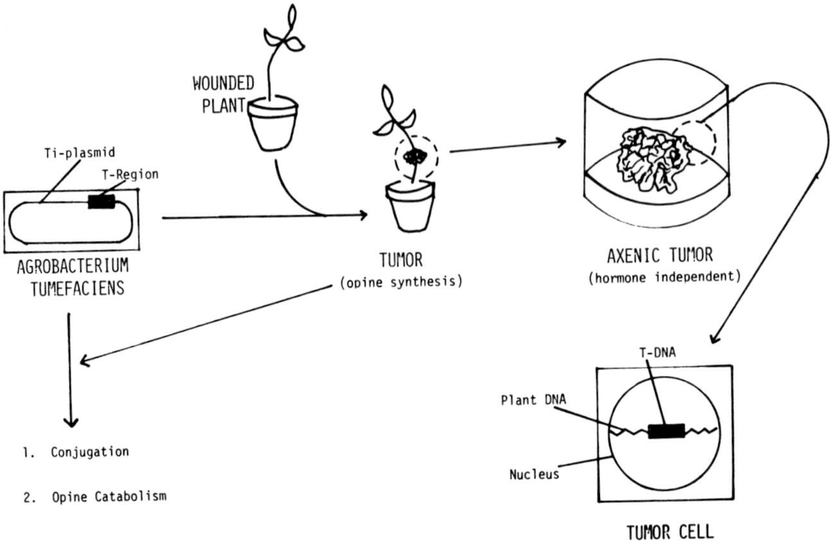

FIGURE 1. Crown gall tumors and the ecology of the Ti plasmid.

ed include DNA modification, gene amplification, gene loss and DNA rearrangement as well as transcriptional, posttranscriptional and translational control (Reviewed in 1). A traditional genetic approach, which involves isolation and characterization of mutants with altered phenotypes, provides a basic framework for models of development. The "new genetic" approach, where DNA is systematically mutagenized *in vitro* and then returned to the organism to test for functional effects, provides a new avenue to dissect developmental processes. Characteristics of the *Agrobacterium tumefaciens*/plant interaction suggest this interaction may be exploited in this "new genetics" approach to developmental questions.

As schematically illustrated in Figure 1 and detailed in recent reviews (2-4), the infection of a plant wound with *Agrobacterium tumefaciens* containing the Ti plasmid results in a tumor (crown gall) with several important characteristics. First, the tumor produces one or more members of a class of unusual molecules called "opines" (e.g. octopine, nopaline or agropine) which are undetectable in either uninfected plants or free-living bacteria. Second, the tumor tissue, freed of bacteria, can grow *in vitro* on a simple defined growth media in the absence of plant hormones. In contrast, untransformed plant tissue requires the addition of two hormones, an auxin and a cytokinin. Third, tumor cells but not untransformed plant cells stably maintain in their nuclei DNA called "T DNA" which was transferred from the bacteria, attached to plant DNA and homologous to a specific region of the Ti plasmid called the "T Region".

The opines produced by the tumor are thought to leak into the surrounding environment and induce in bacteria two functions specified by the Ti plasmid, opine catabolism and conjugation. Since to a first approximation bacteria containing the Ti plasmid are the only organisms in the soil which can use opines as a sole carbon and nitrogen source, the bacteria with the plasmid have created their own ecological niche through tumor formation. Also, the opines induce Ti plasmid transfer to other bacteria and consequently result in the spread of the Ti plasmid.

Studies of the organization of the T DNA in individual tumor lines (5,6) have revealed several features which are common and some which are not. In each case only a limited amount of Ti plasmid DNA is maintained in the tumor and this T DNA is for the most part colinear with the corresponding T Region of the Ti plasmid. Also particular sites on the plasmid seem to serve as preferred sites of attachment to the plant DNA. However, only tumors producing nopaline contain T DNA segments of identical size arranged as tandem repeats (6). Octopine-type tumors contain T DNA of somewhat variable extent and frequently in only a single copy (5). Since the initial integration events have not been studied directly but only inferred from the DNA organization after many cell divisions and possible DNA rearrangements, the possibility still exists that a discrete unit of DNA is initially integrated.

The T DNA is transcribed in the tumor and this RNA is 5'-capped, 3'-polyadenylated, transported to the polysomes, and, at least _in vitro_, translationally active (7 and references therein). Indirect evidence suggests that one protein product may be the enzyme synthesizing opines in the tumor.

II. TI PLASMID AS VECTOR

The Ti plasmid is a natural vector whose function is to transfer a specific piece of DNA into the nucleus of a plant where it is stably maintained and expressed. One might reasonably expect to use the Ti plasmid as a vehicle to transport a DNA segment of choice into a plant cell and ultimately observe expression of this DNA in a fertile plant. Figure 2 diagrams one possible pathway to achieve such a goal.

First (Step I), the DNA of choice would be inserted into the Ti plasmid at a location that guarantees its transfer to the plant cell and stable maintenance in the result-

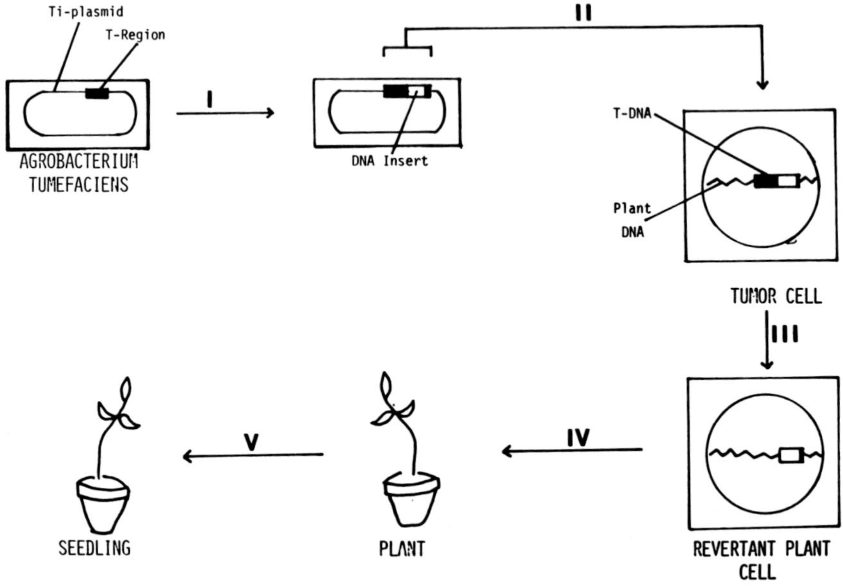

FIGURE 2. Model for the use of the Ti plasmid to isolate fertile plants containing inserted DNA. See text for details.

ing tumor. The boundaries of T DNA that always appears to be transferred are known. The next section describes experiments which determine sites within this region which do not inactivate T DNA functions, and suggests methods to accomplish the insertion. Next (Step II), the bacteria containing the chimeric Ti plasmid would be used to incite a tumor. Since the T DNA appears to be colinear with the T Region, the inserted DNA should also be transferred with fidelity to the plant cell. Such appears to be the case for a Tn7 insertion in the T DNA (8). Further since the T DNA probably must be expressed for tumor maintenance, the inserted DNA will be located in a transcriptionally active region of the plant genome.

Since in many cases, the study of the inserted DNA in a tumor background is not desirable, revertant plant cells would be isolated which perhaps have lost most of the T DNA but retain the inserted DNA (Step III). The unique regeneration ability of plant cells would then be exploited to obtain an entire plant from the revertant cell (Step IV). Finally, for some purposes, it would be desirable to transmit the inserted DNA through meiosis (Step V). Unfortunately, in one case of reversion, post-meiotic revertant tissues lost all detectable traces of T DNA (6,9). However, the fourth section of this paper describes the isolation of tumor revertants whose properties suggest that steps III-V of Figure 2 are in fact feasible.

III. SITE-SPECIFIC MUTAGENESIS OF THE T REGION

Aside from answering basic questions about the neoplastic process, a fine-structure genetic map of the T Region is necessary for the "new genetic" approach. Such a map should identify sites where DNA can be inserted and not hamper transfer of DNA from the bacteria to the plant nor interfere with tumorigenesis. Although tumor formation may not be desirable, the resulting hormone independence *in vitro* is currently the only selectable marker to identify those few plant cells that received the T DNA and stably maintained it.

Several insertions of transposable elements in the T Region have been reported but the techniques are laborious because of the size of the Ti plasmid, about 50 times the size of pBR322. For example Garfinkel and Nester (10) isolated 8,900 independent Tn5 insertions, 25 of which resided on the Ti plasmid with only 3 in the T Region.

To saturate the T Region with insertions, we (11) have used a streamlined version of the "marker exchange" technique described by Ruvkun and Ausubel (12). In *Escherichia coli*, a plasmid vehicle (pRK290) containing a cloned fragment from the T Region is mutagenized with the transposon Tn5 and the products characterized. Plasmid isolates containing desired insertions in the T Region are then introduced into *Agrobacterium tumefaciens* containing the wild type Ti plasmid. Subsequently and rarely, DNA on each side of the Tn5 in the clone undergoes homologous recombination with the corresponding T Region DNA on the Ti plasmid. Such a double recombination event produces a Ti plasmid with an insert at the identical location within the T Region to that in the original mutant clone. To select for this rare occurrence, we introduce a third plasmid which is incompatible with the cloning vehicle and contains a drug resistance marker distinct from the drug resistance of the transposon. Among the cells which are resistant to both drugs are some which have transferred the transposon to the Ti plasmid by the homologous recombination events described above and then

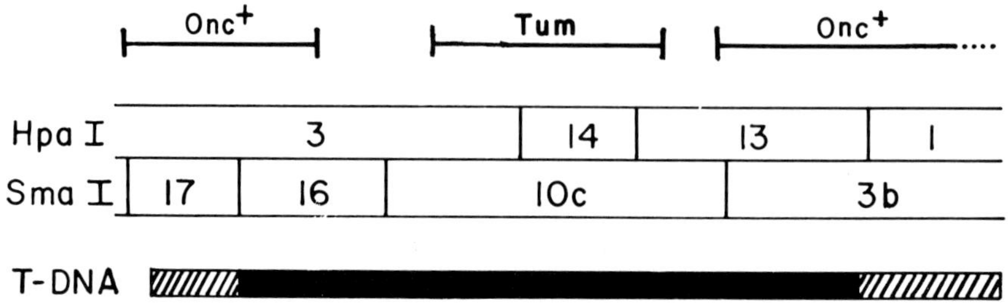

FIGURE 3. Functional map of the T Region of the A6 Ti plasmid. "T DNA" is the DNA found in the A6S/2 tumor (5).

lost the cloning vehicle (which now contains a wild-type fragment of the T Region). Biochemical analysis of the plasmid structures verifies the presumed structure of the Ti plasmid. Thus with relatively little biochemical manipulation, a large number of specific insertions within the T Region are isolated.

As a preliminary functional map of the T Region of the A6 octopine-type Ti plasmid demonstrates (Figure 3), insertions with similar phenotypes cluster on a physical map of the DNA. Mutant phenotypes include those with altered tumor morphology (Tum) such as growth of shoots or the growth of roots under conditions where the tumors induced by wild-type Ti plasmids only appear as disorganized callus. More important for this discussion are the mutants (onc^+) which possess DNA inserts yet have no discernable difference in virulence. These mutants define suitable targets for insertion of other DNA sequences using a process similar to the "marker exchange" technique described above.

IV. TUMOR REVERTANTS

The ultimate goal of many plant genetic engineers is to produce fertile plants containing new genetic information. The Ti plasmid has desirable traits as a vehicle in this process since DNA is naturally transferred to plant cells and identification of plant cells that received, stably maintain and express the T DNA is possible by selecting those cells able to grow in culture without phytohormones.

However, these cells are neoplasms and it is desirable to isolate revertants from these tumors which are phenotypically normal yet retain the new genetic information.

Treatment of a cloned nopaline-type tumor (BT37) with a cytokinin results in the formation of relatively normal-appearing shoots (13). We have induced such shoots to form roots and set viable seed. In contrast to the parental tumor tissue, the derived tissues are not phytohormone independent, do not produce nopaline, and have lost most of the T DNA. The revertant tissues have lost the central portion of the T DNA which contains the "common DNA" sequences, a highly conserved region of Ti plasmids that is incorporated in all tumors studied. Thus, these sequences appear necessary for oncogenicity and tumor maintenance, and their loss is probably directly related to tumor reversal. However, in contrast to the previously reported example (6,9), the reverted plants as well as the plants obtained from seed do retain sequences homologous to the ends of the T DNA present in the parental tumor. The persistance of foreign sequences during the process of meiosis and seed formation suggests the feasibility of the scheme outlined in Figure 2 for the directed genetic engineering of plants.

V. CONCLUSION

The Ti plasmid/plant interaction should be useful in the study of plant development in two distinct ways. First, the molecular basis for the uncontrolled growth of crown galls, especially the relationships between hormone production, tumor phenotype, and T Region mutants should provide clues to normal hormonal regulation in plants. Second, this system potentially provides a powerful tool for the study of many plant developmental systems since *in vitro* - altered DNA could be introduced into fertile plants and its expression studied subsequently in developmental states such as embryogenesis.

One limitation to the use of the Ti plasmid as a vector is the observation that only dicotyledonous plants are susceptible to infection by virulent *Agrobacterium* (14). This is a large host range but plants outside this category include corn, wheat, rye and barley. However, Ti plasmid DNA encapsulated in phospholipid vesicles can transform tobacco (a dicot) protoplasts producing hormone independent tissue which produces octopine and contains T DNA (15). This direct use of Ti plasmid DNA without the requirement of bacterial functions such as attachment to the plant and transfer of plasmid DNA may extend the range of plants susceptible to transformation. Another exciting prospect is that only a small portion of the T DNA may be required for transformation of protoplasts.

ACKNOWLEDGMENTS

I thank D. J. Garfinkel, A. M. Montoya and P. J. O'Hara for criticisms of the manuscript and K. J. Spangler for typing the manuscript.

REFERENCES

1. Brown, D.D. (1981) Science 211, 667.
2. Nester, E.W. and Kosuge, T. (1981) Ann. Rev. Microbiol., in the press.
3. Schilperoort, R.A., Klapwijk, P.M., Ooms, G. and Wullems, G.J. (1980) in Genetic Origins of Tumour Cells (F.J. Cleton and J.W. Simons, eds.), p. 87, Martinus Nijhoff, The Hague.
4. Van Montagu, M., Holsters, M., Zambryski, P., Hernalsteens, J.P., Depicker, A., DeBeuckeleer, M., Engler, G., Lemmers, M., Willmitzer, M., and Schell, J. (1980) Proc. R. Soc. B 210, 351.
5. Thomashow, M.F., Nutter, R., Montoya, A.L., Gordon, M.P. and Nester, E.W. (1980) Cell 19, 729.
6. Lemmers, M., DeBeuckeleer, M., Holsters, M., Zambryski, P., Depicker, A., Hernalsteens, J.P., Van Montagu, M. and Schell, J. (1980) J. Mol. Biol. 144, 353.
7. Gelvin, S., Gordon, M.P., Nester, E.W. and Aronson, A. (1981) Plasmid, in the press.
8. Hernalsteens, J.P., Van Vliet, F., DeBeuckeleer, M., Depicker, A., Engler, G., Lemmers, M., Holsters, M., Van Montagu, M. and Schell, J. (1980) Nature 287, 654.
9. Yang, F.-M., Montoya, A.L., Merlo, D.J., Drummond, M.H., Chilton, M.-D., Nester, E.W. and Gordon, M.P. (1980) Mol. Gen. Genet. 179, 707.
10. Garfinkel, D.J. and Nester, E.W. (1980) J. Bacteriol. 144, 732.
11. Garfinkel, D.J., Simpson, R.B., Ream, L.W., Gordon, M.P. and Nester, E.W. (1981) in preparation.
12. Ruvkun, G.B. and Ausubel, F.M. (1980) Nature 289, 85.
13. Yang, F.-M. and Simpson, R.B. (1981) Proc. Nat. Acad. Sci. USA, in the press.

14. DeCleene, M. and DeLey, J. (1976) Bot. Rev. 42, 389.
15. Dellaporta, S., Giles, K., Fraley, R.T., Papahadjopoulos, D., Powell, A.T., Thomashow, M., Nester, E.W. and Gordon, M.P. (1981), submitted for publication.

USE OF CAULIFLOWER MOSAIC VIRUS DNA AS A MOLECULAR VEHICLE IN PLANTS

Stephen H. Howell
Joan T. Odell
Richard M. Walden
R. Keith Dudley
Linda L. Walker

Department of Biology, C-016
University of California, San Diego
La Jolla, CA 92093

I. INTRODUCTION

Cauliflower mosaic virus (CaMV) is frequently cited as a potential vehicle for introducing foreign DNA into plant cells. CaMV is one of the few plant viruses to contain a double-strand DNA genome (see Shepherd, 1979). CaMV is attractive as a plant vehicle because DNA extracted from the virus can be introduced directly into plants by rubbing the DNA onto leaves. CaMV DNA is not known to integrate into the host plant genome nor to replicate in episomal form, and so as a vehicle, CaMV DNA would be propagated in the plant as virions.

The CaMV genome is a double-strand DNA circle of about 8 kb. The genome has unusual secondary structure in that it is "tangled" or "twisted" (Hull and Shepherd, 1977). The genome obtained from virions is not covalently closed due to the presence of 2-3 site-specific single strand interruptions (Hull and Howell, 1978). The genome of CaMV isolate Cabb-S has recently been entirely sequenced (Franck _et al._, 1980). The sequence data reveal six closely spaced potential coding regions and an intergenic region of about 1 kb.

For the purposes of using CaMV DNA as a molecular vehicle, it was of concern whether the unusual secondary structure of the genome was required for infectivity. As shown in Fig. 1, linearized CaMV DNA, which has lost its "tangled" form, infects plants. Furthermore, when the site-specific

DEVELOPMENTAL BIOLOGY USING PURIFIED GENES

ISBN 0-12-137420-3

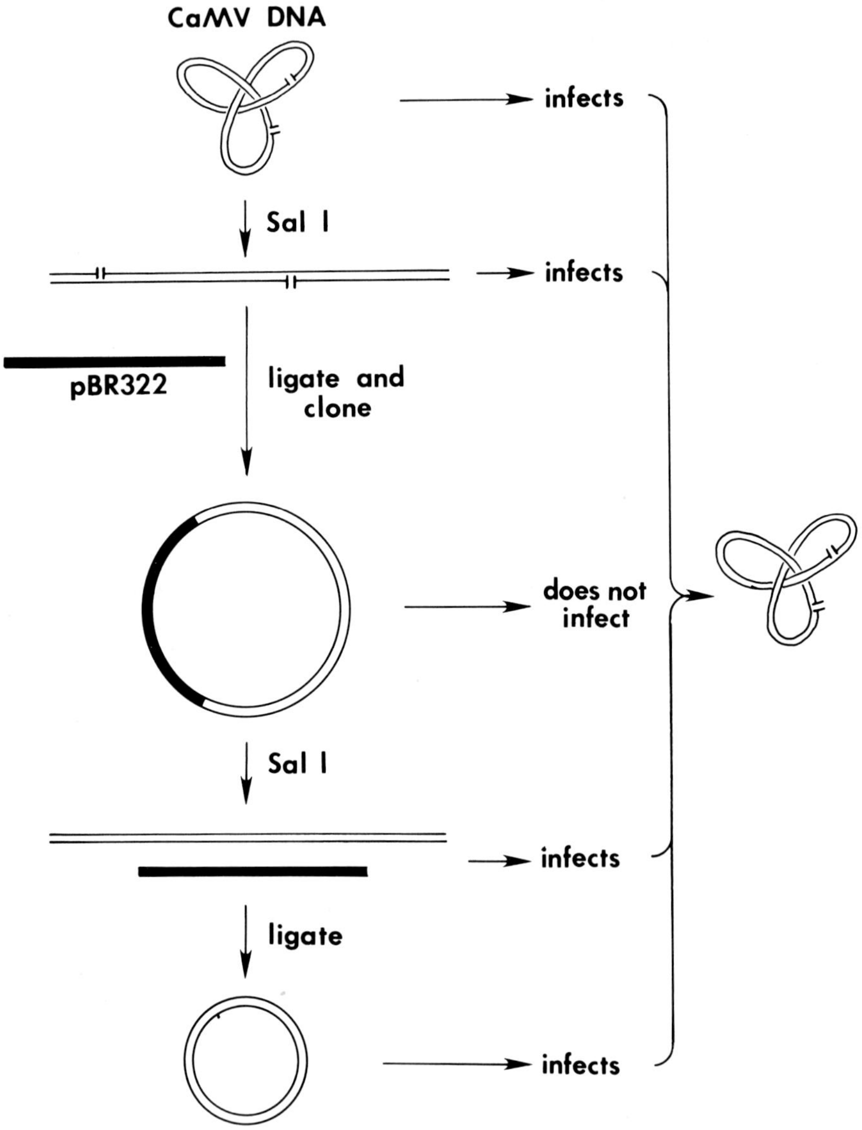

Fig. 1. Infectivity of various forms of CaMV DNA. Each infective form produces virus particles containing DNA with typical secondary structure.

single-strand breaks are sealed during the cloning of CaMV DNA in a bacterial plasmid, the cloned viral DNA is fully capable of infecting plants provided it is excised from its parent plasmid (Howell et al., 1980). It is interesting that infection with cloned CaMV DNA, possessing neither single strand breaks nor "tangled" form, produces virus particles containing DNA that has reacquired these properties (Fig. 1).

It has been the goal of recent experiments to find suitable sites in the CaMV genome into which foreign DNA can be inserted. Insertions of foreign DNA should neither destroy the infectivity of the viral DNA nor interfere with the movement of the virus in the plant. (The virus moves throughout the plant in establishing a systemic infection.) As a first step to test the ability of viral DNA to tolerate such modifications, we inserted Eco RI linkers -- oligodeoxynucleotides of defined sequence (8 bp) containing an Eco RI site -- at various Alu I sites in the cloned CaMV genome (Fig. 2). Deletions were also generated by cleaving the CaMV genome at more than one Alu I site, then rejoining the DNA. An array of linker-insertion and deletion mutants modified in different regions of the viral genome (Fig. 3) were tested for infectivity. Most insertions and deletions were virus-lethal in that neither virus nor infection symptoms were produced following inoculation. An exception was pLW414-R212 (indicated in Fig. 3 and Table 1 simply as 212), which will be discussed later.

Since most of the mutants described above did not infect plants when inoculated on their own, we attempted to infect plants with pairwise combinations of the mutants. In most, but not all cases, infection was successful when pairs of plasmids with mutations at nonoverlapping sites were used to inoculate plants, as in the case of 189 and 153 (Table 1). In no case, when pairs of mutants with overlapping mutations were used, such as with 189 and 213, did productive infection occur. In general, it appeared that almost any defective genome can rescue any other defective genome as long as the mutant sites were either non-overlapping or not too close to each other.

Such a rescue process might occur by complementation or by intermolecular recombination. Successful coinfection by the pair 173 and 49 suggested that mutant genomes were not rescued by complementation. The plasmid 173 has a large deletion, which leaves no intact potential coding regions in the remaining viral DNA. (Only portions of coding regions IV and V remain.) Without any intact coding regions, 173 probably could not supply any normal gene products to complement the corresponding defective product encoded by

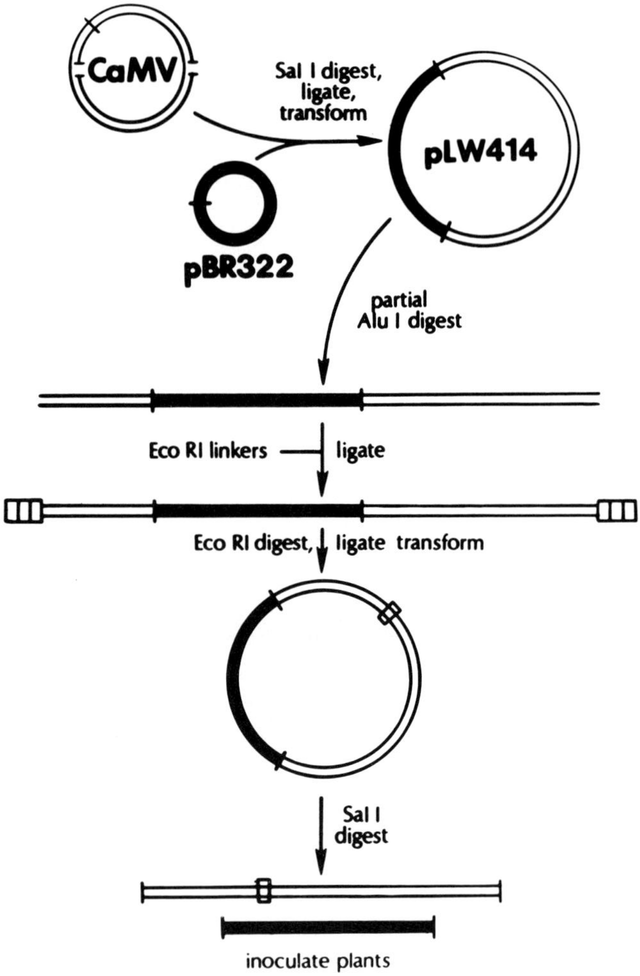

Fig. 2. Procedure modified from Heffron et al. (1978) for the generation of linker-insertion or deletion mutants of cloned CaMV DNA (pLW414). Recombinant plasmid containing cloned CaMV DNA was linearized by cutting at about 1 site per molecule with Alu I. Eco RI linkers d(GGAATTCC) were ligated onto the ends of linear molecules. Linear molecules were recircularized and used to transform E. coli C600. Clones containing modified plasmids were selected after examining Eco RI digests of small-scale plasmid preparations. To inoculate plants, isolated plasmid DNA was digested with Sal I and rubbed, with an abrasive, onto the leaves of plants. From Howell, Walker and Walden, submitted for publication.

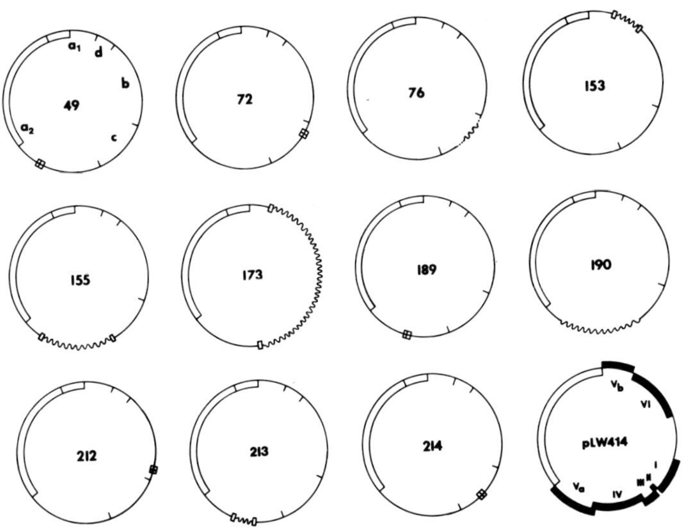

Fig. 3. Maps of modified CaMV DNA containing plasmids with linker insertions (-⊞-) or deletions (∿∿). Major Eco RI fragments a_1-d are indicated on the map of plasmid 49 in the upper left corner. Potential coding regions I-VI from the sequencing data of Frank et al. (1980) are redrawn onto pLW414 map in lower right hand corner. Open bars represent pBR322 DNA, lines represent CaMV DNA. From Howell, Walker and Walden, submitted for publication.

TABLE I. Infectivity of Modified Plasmids[a]

Infects in pairwise combinations			Infects singly
49 x 72	49 x 76	49 x 153	212
49 x 173	49 x 214	72 x 190	
76 x 155	76 x 189	76 x 190	
153 x 189	153 x 190	153 x 213	
155 x 214	189 x 214	213 x 214	

[a]See maps of plasmids in Fig. 3

49. Rescue, in such case, would have to occur by recombination.

To determine more directly how rescue of defective genomes occur, we examined the DNA from virus particles resulting from successful coinfection. We found in all cases, that the viral DNA was restored to a normal form. This indicated that defective genomes were rescued not by complementation but by intermolecular recombination events that produce normal virus DNA and expel insertion or deletion modifications (Fig. 4).

For the purposes of using the viral DNA as a molecular vehicle, our findings at this point tend to rule out the use of a helper virus system for CaMV similar to that used for SV40 in animal cells (see, for example, Mulligan et al., 1979).

Instead our attention has turned toward the use of CaMV as a vehicle by inserting foreign DNA into the genome at sites that do not destroy infectivity. Two sites which tolerate various DNA modifications have been found. One mentioned earlier is in the mutant plasmid 212. This plasmid has a linker insertion in the intergenic region (Fig. 3) and can infect plants when inoculated on its own. The modification is stable in the plant in that DNA resulting from infection by 212 retains the linker insertion (Fig. 4). Gronenborn et al. (1980) have found a modification in a genic region, coding region II, that can be tolerated by the virus. A comparison of various virus isolates shows that this coding region is extensively modified in nature (Hull, 1980). The viral isolate we have studied has about a 400 bp deletion in this 500 bp coding region (Hull and Howell, 1978). The maximum size of a piece of foreign DNA which one can insert in these sites is under investigation, but constraints in packaging viral DNA might limit the size of an insert.

It would seem ideal to insert foreign DNA into an active viral gene in such a way so that the inserted element would be under viral control. One of the most active genes in the CaMV genome encodes the synthesis of a non-virion protein called P66 (Odell and Howell, 1980). P66 is apparently the matrix protein of the cytoplasmic viral inclusion body (Covey and Hull, in press). P66 is encoded by coding region VI (Franck et al., 1980, see Fig. 3) and by an abundant 19S RNA (Odell et al., in press). To date, we have not been able to modify or insert foreign DNA into the P66 gene.

We have described here the insertion of foreign DNA (a linker, in this case) into the intergenic region near coding region I. The intergenic region does not code for any viral protein, but is actively transcribed. CaMV DNA codes for a large, full genome-length transcript which is initiated in the intergenic region (Dudley, Odell and Howell, in preparation). Hence, the linker inserted in the mutant plasmid 212 lies at the 5' end of the large transcript. The coding function of the large transcript, however, is not known.

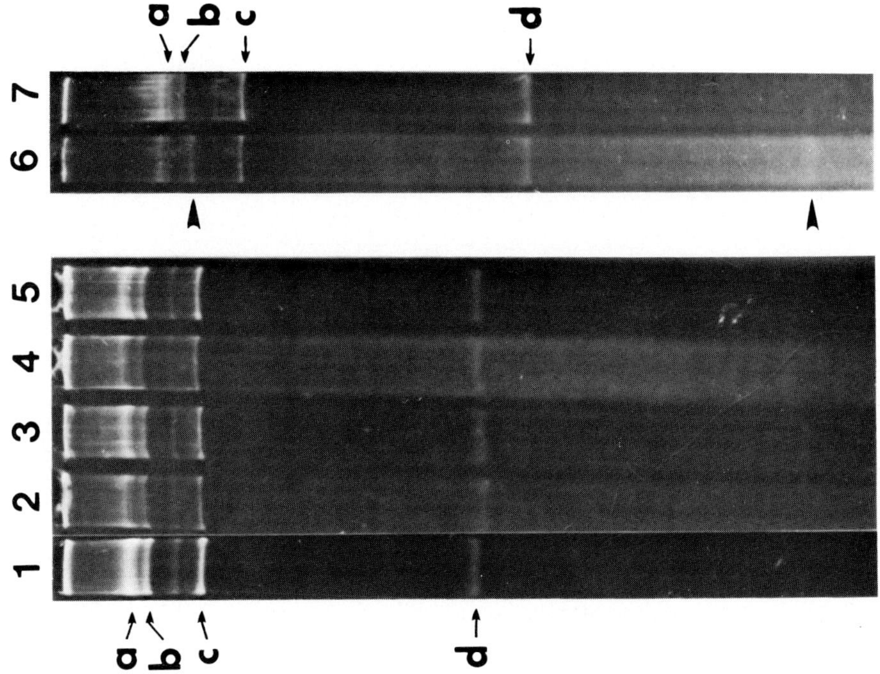

Fig. 4. Eco RI digest pattern of CaMV DNA obtained from small-scale viral DNA preparations of infected leaves. Viral DNA from leaves infected with (1) pLW414, the parent plasmid, (2) 49 x 72, (3) 49 x 76, (4) 49 x 173, (5) 76 x 155, (6) 212 and (7) pLW414. Eco RI fragments a-d as indicated are shown on the map of 49 in Fig. 3. From Howell, Walker and Walden, submitted for publication.

ACKNOWLEDGMENTS

This work was supported by the National Science Foundation under grant PCM7913707 and the Science and Education Administration of the U.S. Department of Agriculture under grant 5901-04010-8-0178-0. We thank R. Hull for preprinted material.

REFERENCES

Covey, S.N. and Hull, R. Virology (in press).
Franck, A., Guilley, H., Jonard, G., Richards, K., and Hirth, L. (1980). Cell 21, 285.
Gronenborn, B., Gardner, R., and Shepherd, R.J. 6th EMBO Annual Symposium: Molecular Biologists Look at Green Plants, abs. (1980).
Heffron, F., So, M., and McCarthy, B. (1978). Proc. Natl. Acad. Sci. USA 75, 6012.
Howell, S.H., Walker, L.L., and Dudley, R.K. (1980). Science 208, 1265.
Hull, R. (1980). Virology 100, 76.
Hull, R., and Howell, S.H. (1978). Virology 86, 482.
Hull, R., and Shepherd, R.J. (1977). Virology 79, 216.
Mulligan, R.C., Howard, B.H., and Berg, R. (1979). Nature 277, 108.
Odell, J.T., Dudley, R.K., and Howell, S.H. Virology (in press).
Odell, J.T., and Howell, S.H. (1980). Virology 102, 349.
Shepherd, R.J. (1979). Ann. Rev. Plant Physiol. 30, 405.

EVIDENCE FOR THE TRANSFORMATION OF *DICTYOSTELIUM DISCOIDEUM* WITH HOMOLOGOUS DNA

David I. Ratner[1*], Thomas E. Ward[2**], and Allan Jacobson[2**]

*Department of Cellular Biology
Scripps Clinic and Research Foundation
La Jolla, CA 92037

**Department of Molecular Genetics and Microbiology
University of Massachusetts Medical School
Worcester, MA 01605

ABSTRACT A transformation system for *Dictyostelium discoideum* should facilitate the isolation and *in vivo* study of genes important for slime mold differentiation. We are developing a system in which mutant slime mold amoebae unable to utilize *Bacillus subtilis* as a food source are converted to *Bacillus* "prototrophy" by treatment with wild type slime mold DNA. The *bsgA5* mutation, preventing growth of *Dictyostelium* amoebae on *B. subtilis*, is recessive and non-leaky, maps at a single locus, and reverts infrequently, all factors favorable for transformation. Mutant amoebae, after DNA treatment and brief permissive growth, are plated selectively upon lawns of *B. subtilis*. High molecular weight DNA preparations from a Bsg^+ slime mold strain generate Bsg^+ colonies at an average frequency of 0.24×10^{-7}, roughly one log above that of reversion of parallel cultures treated with calcium phosphate alone. We have ligated restricted wild type slime mold DNA to DNA from the pBR322-yeast-slime mold plasmid ARS22. We are now analyzing Bsg^+ colonies obtained after treatment of Bsg cells with this recombinant DNA. One transformant contained two DNA fragments that hybridize to pBR322; the untreated Bsg strain contains no such fragment. These data constitute molecular evidence for transformation and offer the possibility of recovery of the *bsgA* gene. Isolation of this gene should allow high frequency transformation of *Dictyostelium*.

[1]Supported by NIH Grant GM26309.

[2]Supported by NIH Grant GM27757. T.E.W. Was the recipient of a Post-doctoral Fellowship from the Muscular Dystrophy Association. A.J. has received a Faculty Research Award from the American Cancer Society.

ISBN 0-12-137420-3

INTRODUCTION

We are interested in gene expression during the growth and development of the cellular slime mold *Dictyostelium discoideum*. Because the transformation of cells with purified DNA has become a powerful tool for the analysis of eukaryotic genomes, we wish to use this approach with *Dictyostelium*. The development of a transformation system requires a suitable selective marker. In other systems, cloned genes conferring prototrophy or drug resistance have been used to great advantage (see other papers in this volume). In *Dictyostelium*, however, the known mutations conferring drug resistance are almost all recessive, while the few auxotrophic selections possible in the slime mold pose technical difficulties. If a wide spectrum dominant drug resistance gene isolated from some other species is to be used (1), one can never know, *a priori*, if it will function in a given foreign host. Thus, we decided to explore the use of conditional growth mutations used previously in slime mold genetic analyses.

A *Dictyostelium* mutation that prevents utilization of *Bacillus subtilis* as a food source (*bsgA*; bacillus sensitive growth) has been very useful in the selection of diploids via parasexual methods (2,3). Although the physiological basis for the phenotype is unknown, the mutation has several genetic properties that make it potentially useful as a selective marker for transformation. The *bsgA5* allele is phenotypically tight and its reversion frequency low (2;4; see below). Because the mutation is recessive, transformants should be selectable even in the heterozygous condition. The *bsgA* locus has been assigned to genetic linkage group III (5), and the mutation appears to be the result of a single lesion. Consequently, we have attempted to transform mutant Bsg amoebae to *Bacillus* "prototrophy" using wild-type Bsg^+ slime mold DNA.

METHODS

Strains. Recipient cells used for transformation are of strain XP55 (ref. 5; genotype *cycA5*, *bsgA5*). The unlinked *cycA5* mutation results in cycloheximide resistance. Cells of the axenic strain AX3 (6) were used as a source of Bsg^+ *Dictyostelium* DNA for transformation. XP55 amoebae are grown permissively on *Klebsiella aerogenes* strain OXF1 and screened for transformation using the nonsporulating *B. subtilis* strain 36.1 (2). The pBR322-yeast-slime mold plasmid ARS22 was constructed in the laboratory of R. Davis (7).

Preparation of DNA. Whole cell Dictyostelium DNA to be used for transformation was prepared as described by Firtel et al. (8). DNA preparations from axenic cells or from cells grown on K. aerogenes were of high molecular weight, in excess of 50 kilobases. DNA for hybridization analysis from cells grown on B. subtilis was initially prepared by the same method, but this occasionally failed to give a high molecular weight product. For these cells, a better procedure involved suspension of the cells in 2 mM Tris·HCl pH 7.5, 1 mM EDTA, followed by rapid sequential addition of (final concentrations) 50 mM EDTA pH 8.0, 100 mM Tris base, 0.5% SDS, and 1% diethylpyrocarbonate (9). Only a single round of cesium chloride/ethidium bromide centrifugation was used.

Stinchcomb et al. (7) constructed the ARS22 plasmid into which we have shotgunned wild type Dictyostelium DNA. Slime mold nuclear DNA (18) was digested to completion with either BamHl or Sall, and ligated (roughly 2 to 1 weight ratio of slime mold to vector DNA) to the similarly restricted vector ARS22. After ethanol precipitation, the entire ligation mixture was used for transformation of amoebae.

Transformation and Bsg^+ Selection. DNA was co-precipitated with calcium phosphate (10) and then added to ten volumes of XP55 amoebae suspended in water at 10^7/ml. The final buffer was thus of relatively low ionic strength, well suited to Dictyostelium. Cells and DNA were incubated at 22°C with aeration. In a typical experiment, 3 X 10^7 washed vegetative cells were incubated with 3-30 μg of DNA for 1-4 hours, followed by 4-15 hours of permissive growth prior to plating on B. subtilis. Permissive growth occurred by resuspension of the amoebae, centrifuged free of the transformation buffer, at 2 X 10^6/ml in 10 mM $Na_xK_yH_2PO_4$ pH 6.0, 10 mM NaCl, 10 mM KCl, 2 mM $CaCl_2$, 100 μg/ml dihydrostreptomycin sulfate and 10^{10}/ml K. aerogenes. Under these conditions, with aeration at 22°C, amoebae show a lag of several hours, double with a generation time of roughly 4 hours, and plateau at a titer of approximately 2 X 10^7/ml. In some experiments, amoebae were plated directly upon B. subtilis without any permissive growth, while in another variation the permissively grown amoebae were induced to sporulate on buffered Millipore filters (11) before B. subtilis selection.

Selection of Bsg^+ colonies followed the methods used in parasexual analysis (2,3), with the following minor modifications. Selective plates contained 12 μg/ml dihydrostreptomycin sulfate, so that no further antibiotic was added at the time of plating (giving a more uniform B. subtilis lawn). Plates were incubated overnight at 27°C (aiding bacterial growth), and then shifted to the standard temperature of 22°C.

10^6 XP55 amoebae or 10^7 spores were screened per 10 cm petri plate. Although more spores than amoebae can be screened per plate, the variable recovery of spores coupled with their occasionally low viability detracted from this screening method. A typical screen examined 5 X 10^7-10^8 cells, so that if extensive permissive growth was allowed to occur, only a portion of the cells was tested.

Presumptive Bsg^+ colonies appearing on the selective plates after 5 to 7 days were tested for their Bsg and Cyc phenotypes using standard parasexual methods (5). All Bsg^+ isolates retained the cycloheximide resistance marker present in XP55, ensuring that they were not the result of contamination with wild type amoebae. Clones that consistently grew on *B. subtilis* lawns when introduced by toothpicking vegetative amoebae or spores were deemed Bsg^+. Bsg^+ isolates were purified clonally on either *K. aerogenes* or *B. subtilis* plates. The permissive plates were needed in a few cases in which the amoebae grew too slowly upon *B. subtilis* to plaque before the death of the bacterial lawn. Bsg^+ cells could also be grown in *B. subtilis* suspension by innoculating amoebae and bacteria into the standard *Dictyostelium* axenic medium (ref. 6; but with only 10 μg/ml streptomycin).

Filter Hybridization. DNA from XP55 and its Bsg^+ derivatives was digested with an excess of *EcoRl* (a gift of J. Gottesfeld), electrophoresed in 0.8% agarose gels, and transferred to nitrocellulose filters following conventional procedures (12). As a hybridization probe, we have used primarily pBR322 (a gift of T. Gilroy); we have also used ARS22 as well as the pBR322 derivatives pCM3 and p65M, which contain *Dictyostelium* inserts (gifts of J. Brandiss). Nick translation, following standard methods (13), gave probes with specific activities in excess of 10^7 cpm/μg. Hybridization was in 6 X SSC, 0.1% bovine serum albumin, 0.1% polyvinylpyrrolidone, 0.1% Ficoll, 0.1% $Na_4P_2O_7$, 20 μg/ml salmon sperm DNA, 65°C for 60 hours. Filters, after 65°C washes ultimately in 2 X SSC, were autoradiographed using preflashed Kodak XR-5 film with an intensifying screen.

RESULTS

Statistical Evidence for Transformation. Pilot experiments indicated the possible utility of the Bsg phenotype as the basis for a transformation selection. 10^6 Bsg^- XP55 amoebae, or even 5 X 10^6 (viable) XP55 spores, can be plated upon a *B. subtilis* lawn without any apparent growth due to reversion or phenotypic leakiness. If, on the other hand, 10^6 Bsg^- cells are mixed with a small number (10^2) of Bsg^+

NC4 amoebae, these prototrophs will clone with unimpaired efficiency (data not shown). Thus, rare Bsg^+ transformants should be able to grow and be identified under these selective conditions.

Not knowing the molecular basis for the Bsg phenotype, we can use only wild-type Bsg^+ *Dictyostelium* as a certain source of transforming DNA. We purified high molecular weight DNA from strain AX3 and administered it, using the calcium phosphate method (10), to Bsg^- cells (see Methods). As a control, equal numbers of amoebae were treated in parallel with calcium phosphate alone. Table 1 summarizes the results of 10 experiments conducted over several months and using 4 different DNA preparations. The frequency of spontaneous

TABLE 1. Transformation with high molecular weight DNA

Experiment	-DNA: Bsg^+ colonies / cells screened[a]	+DNA: Bsg^+ colonies / cells screened[a]
1	Not tested	1/4 X 10^7
2	0/6 X 10^7	b
3	0/8 "	b
4	0/3 "	5/2 X 10^7 c
5	0/1 "	1/1 "
6	1/15 "	4/13 " d
7	1/2 "	0/4 "
8	0/2 "	0/9 "
9	0/5 "	1/10 "
10	0/2 "	0/6 "
Total	2/44 X 10^7 Freq=4.5 X 10^{-9}	12/49 X 10^7 Freq=2.4 X 10^{-8}

a) The data in each sample have been corrected for viability. Differing spore recoveries are largely responsible for the difference in the number of cells screened ± DNA in several experiments.

b) The DNA preparation used in these experiments subsequently was found to be badly degraded. No Bsg^+ colonies were observed from 6 X 10^7 and 4 X 10^7 cells in experiments 2 and 3.

c) The Bsg^+ colonies observed are presumed to be independent since no growth of the cells occurred prior to selection.

d) These colonies may not be independent since several generations of permissive growth preceded selection.

Bsg^+ revertants, estimated from these data, is 4.5×10^{-9} (2 colonies/44×10^7 cells tested without DNA). Other experiments, in part using alternative means of DNA treatment (see below), in which absolutely no Bsg^+ colonies were observed in either the DNA-treated or control groups, are consistent with this low value (a total in excess of 30×10^7 additional cells examined). Treatment with slime mold DNA yielded a higher frequency of Bsg^+ colonies (12 colonies per 49×10^7 cells screened; frequency equals 2.4×10^{-8}). Although the number of Bsg^+ colonies arising after DNA treatment is still low, the increased frequency is statistically very significant. Arguing the null hypothesis that reversion is the source of all colonies, the likelihood that, of 14 revertants, 12 or more would occur by chance in the DNA-treated samples is less than 1%. (Analogous to 12 or more heads in 14 tosses of a coin.) Alternatively, if the "true" reversion frequency is as low as was calculated above, the likelihood that the sampling of DNA-treated cells contains 12 revertants is vanishingly small. (Given an expected value of 2 "hits", the Poisson probability of 12 or more is of the order 10^{-6}.) From these statistical arguments, the DNA treated Bsg^+ colonies are most unlikely to be revertants. We believe them to be the result of DNA-mediated transformation.

Characterization of Transformants. The Bsg^+ transformants grow well permissively but vary considerably in their ability to grow under selective pressure. Growth was usually assessed by the rate of expansion of clones or picks (a deposit of approximately 10^5 cells) on *Bacillus* plates. Most transformants could be recloned on selective plates, although the small (1-2 mM) plaques required 6 days to appear, whereas wild type NC4 clones are detectable by the 4th day. The most slowly growing transformants we recovered would not clone under selective conditions; rather, the *Bacillus* lawn lysed (day 8) before any plaques appeared. When grown upon *Bacillus* in suspension, the generation times of one typical and one slow transformant were 12 and 15 hours, respectively, as opposed to 8-10 hours for NC4.

Although the reduced rate of growth under selective conditions might suggest an instability of the transforming DNA, as occurs in other species, we have not been able to document the appearance of Bsg^- segregants. That is, in the cases examined, Bsg^+ cells passaged permissively for 1 or 2 weeks (many generations) on *K. aerogenes* plates continued to show a clonal plating efficiency on *B. subtilis* as high as that of cells kept under selective pressure throughout. It may be that unstable transformants, growing even more poorly, could not appear in our original selection; or that processes, such as integration, occurring after the initial transformation event generate stable isolates.

We have tried to increase the transformation frequency by varying the physiological state of the recipient cells (vegetative versus starved for 15 hours in suspension), varying the amount of DNA per cell (over the range 1 to 10 μg/10^7 cells), omitting the permissive growth period after DNA treatment, and screening spores rather than amoebae. Starved cells gave no transformants in several attempts. Variations in the other conditions gave inconsistent results.

Researchers introduce DNA into cells of other species using polyethylene glycol (PEG; 14) or liposomes (15) instead of calcium phosphate. Our preliminary tests of these methods were unsuccessful. PEG treatment of Dictyostelium caused extensive death; most troublesome was the irreproducibility of the extent of killing. Although slime mold amoebae can be efficiently labelled by liposomes containing a fluorescent dye (D.R., unpublished observations; large unilamellar vesicles were generously prepared by R. Fraley), the incorporation into these vesicles of very large Dictyostelium DNA fragments was inefficient, and no Bsg^+ clones were obtained from 8 X 10^7 cells.

Transformation Using an ARS Vector. Stinchcomb et al. (7) have described the isolation of DNA fragments from several eukaryotes, including Dictyostelium, that stimulate yeast transformation by allowing extrachromosomal replication of the covalently linked selectable marker. These automonously replicating sequences (ARS) may, assuming they are replicons in their own species, aid the transformation of such cells. The pBR322 sequences in the ARS plasmids make possible the analysis of transformants by means of DNA filter hybridization. We used ARS22 (7) as a vector into which we hoped to introduce the bsg^+ gene. ARS22 consists of pBR322, a 1.1 kb yeast ura3 gene and a 1.4 kb EcoRl fragment of Dictyostelium DNA. The design of this work is given in Figure 1. BamHl and Sall endonucleases were chosen to restrict wild-type Dictyostelium DNA because these enzymes generate, on average, very large fragments from the GC-poor slime mold genome (16), and we hoped such fragments would include the intact bsg^+ gene. From 3 transformation experiments using, alternately, the Bam- and Sal- shotgunned material, 4 Bsg^+ clones were obtained (4 clones / 18 X 10^7 cells screened/34 μg of ligated DNA). The transformants were grown upon B. subtilis in liquid, their DNA purified, restricted with EcoRl, and analyzed by the method of Southern (12) for the presence of fragments containing sequences homologous to pBR322. Three of the 4 transformants contained no EcoRl fragments that hybridized to pBR322.

FIGURE 1.

CONSTRUCTION OF A *Dictyostelium* VECTOR

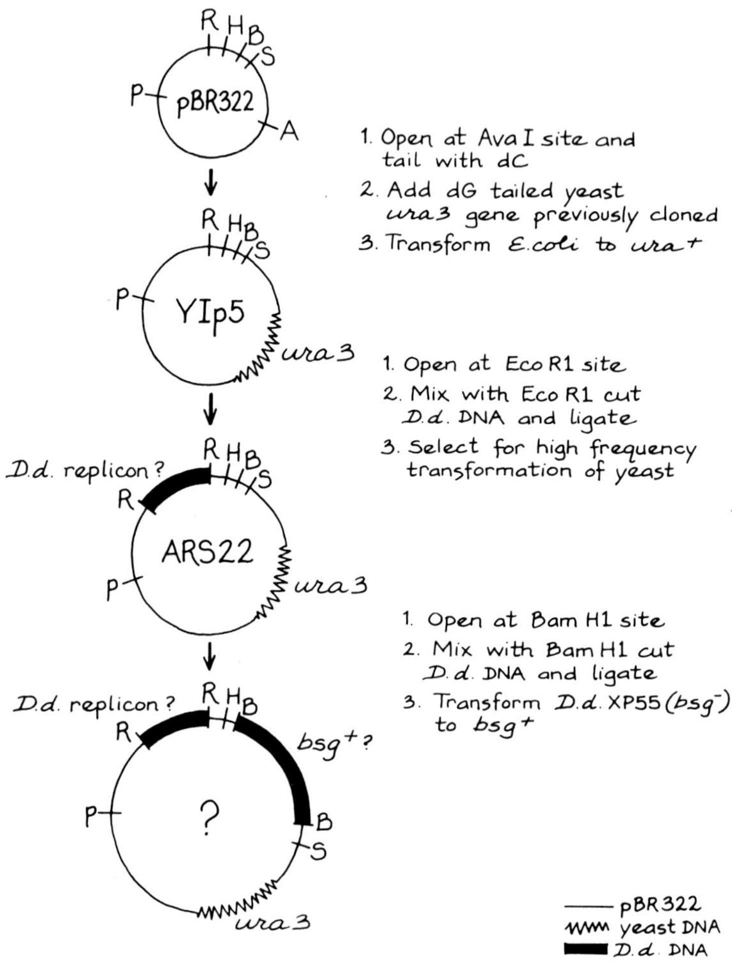

Figure 2 reveals that the other isolate, HDR12, contained one major and one minor band with homology to pBR322. These bands were detected in two separate hybridizations (lanes b and d), and were absent from the DNA of the Bsg^- recipient XP55 (lanes a and c). The estimated sizes of the major and minor bands are 10 and 7 kilobases, respectively. Their presence indicates the existence of plasmid sequences in the Bsg^+ strain and is further evidence for the transformed nature of

FIGURE 2.

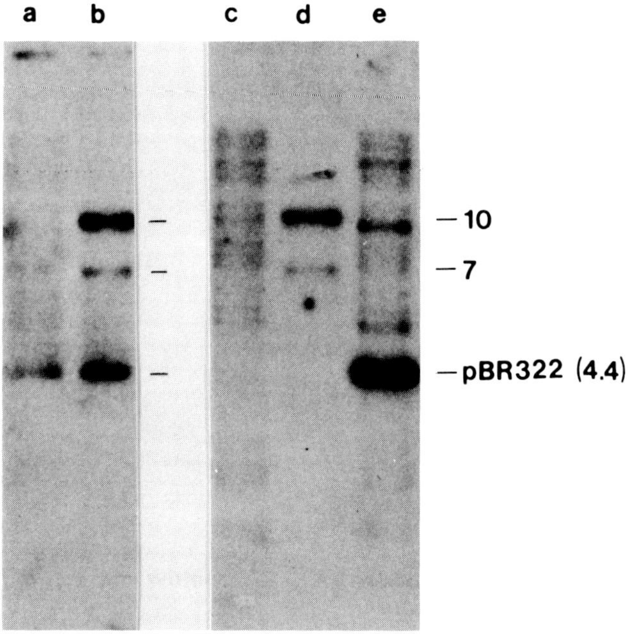

FIGURE 2. Filter hybridization of XP55 and HDR12 DNA to pBR322. EcoR1 restricted DNA of the Bsg recipient XP55 and of the Bsg^+ transformant HDR12 was electrophoresed and transferred to nitrocellulose. Lanes a and b were probed with nick translated pBR322, lanes c-e, with a pBR322 derivative pCM3 from which, in this particular plasmid preparation, the single copy Dictyostelium insert had been inadvertantly lost. (The greater background of lanes c-e may be due to the retention in the pCM3 probe of some repeated slime mold sequences.) Lanes a and c: 1 μg of XP55 DNA. Lanes b and d: 0.3 μg of HDR12 DNA. Lane e: 1 μg of XP55 DNA plus 10^{-4} μg of pBR322 DNA (roughly the equivalent of 1 plasmid copy per Dictyostelium genome). Fragment sizes were determined relative to the migration of EcoR1 fragments of phage lambda run on the same gel. The presence of a band at 4.4 kb in lanes a and b is due to spillover of some pBR322 sample added in huge excess (0.5 μg) in an adjacent lane (excised prior to this exposure); this band is absent as expected in the corresponding lanes c and d. The band visible at 9 kb in lane e is also due to pBR322, resulting from a trace amount of higher molecular weight plasmid. The bands specific for the transformant HDR12 are indicated at 10 and 7 kb.

these cells. However, no conclusions can be drawn as yet concerning the physical relationship between the hybridizing sequences and the selected *bsg*$^+$ gene.

From the relative intensities of hybridization of the probes to HDR12 DNA (lanes b and d) and to pBR322 DNA added externally (lane e), we suspect that the pBR322 sequences are present at less than 1 copy per transformed cell. In fact, we have been unable to demonstrate pBR322 homology in several later DNA preparations from strain HDR12. Two interpretations come to mind. The first is that *Dictyostelium* can, as is true of yeast (14) and especially of *Neurospora* (17), incorporate the homologous, selected gene into its genome with the concomitant loss of foreign DNA. (HDR12 does remain stably Bsg$^+$.) A second possibility is that the initial preparation of DNA from HDR12 was contaminated with some material homologous to, but bigger than, pBR322. Our technical precautions and the absence at the time of other plasmid work in the laboratory (of D.R.) make this explanation seem unlikely.

SUMMARY

This is the first report of the DNA-mediated transformation of *Dictyostelium*. Given the small number of Bsg$^+$ isolates, and the irreproducibility of the hybridization analysis, our evidence is not conclusive. Nevertheless, this work offers encouragement to those trying to develop an efficient slime mold transformation system. It is our hope that, by use of the ARS vector, we shall be able to recover the intact *bsg*$^+$ gene, at which point the poor transformation efficiency reported here should be dramatically improved.

Acknowledgment. We wish to thank R. Davis for providing us with plasmid ARS22.

REFERENCES

1. Jimenez, A. and Davies, J. (1980). *Nature*, *287*, 869.
2. Newell, P.C., Henderson, R.F., Mosses, D. and Ratner, D.I. (1977), *J. Gen. Microb.*, *100*, 207.
3. Newell, P.C., Ratner, D.I. and Wright, M.D. (1977), in *Development and Differentiation in the Cellular Slime Moulds* (P. Cappuccinelli and J. Ashworth, eds.), p. 51, Elsevier, Amsterdam.
4. Ross, F.M. and Newell, P.C. (1979). *J. Gen. Microb.*, *115*, 289.
5. Ratner, D.I. and Newell, P.C. (1978). *J. Gen. Microb.*, *109*, 225.

6. Loomis, W.F., Jr. (1971). Exp. Cell Res., 64, 484.
7. Stinchcomb, D.T., Thomas, M., Kelly, J., Selker, E. and Davis, R.W. (1980). Proc. Nat. Acad. Sci., USA, 77, 4559.
8. Firtel, R.A., Cockburn, A., Frankel, G. and Hershfield, V. (1976). J. Mol. Biol., 102, 831.
9. Cameron, J.R., Philippsen, P. and Davis, R.W. (1977). Nucleic Acids Res., 4, 1429.
10. Wigler, M., Pellicer, A., Silverstein, S., Axel, R., Urlaub, G. and Chasin, L. (1979). Proc. Nat. Acad. Sci., USA, 76, 1373.
11. Sussman, M., (1966). Methods in Cell Phys., 2, 397.
12. Southern, E.M. (1975). J. Mol. Biol., 98, 503.
13. Rigby, P., Dieckmann, M., Rhodes, C. and Berg, P. (1977). J. Mol. Biol., 113, 237.
14. Hinnen, A., Hicks, J.B. and Fink, G.R. (1978). Proc. Nat. Acad. Sci., USA, 75, 1929.
15. Fraley, R., Subramani, S., Berg, P. and Papahadjopoulos, D. (1980). J. Biol. Chem., 255, 10431.
16. Firtel, R.A. and Bonner, J. (1972). J. Mol. Biol., 66, 339.
17. Case, M.E., Schweizer, M., Kushner, S.R. and Giles, N.H. (1979). Proc. Nat. Acad. Sci., USA, 76, 5259.
18. Jacobson, A. (1975). Methods in Molecular Biology, 8, 161.

NUCLEAR AND GENE TRANSPLANTATION IN THE MOUSE

Karl Illmensee
Kurt Bürki

Department of Animal Biology
University of Geneva
Geneva, Switzerland

Peter C. Hoppe

Jackson Laboratory
Bar Harbor, Maine

Axel Ullrich

Genentech
South San Francisco, California

ABSTRACT

In *nuclear transplantations*, the pronuclei of fertilized mouse eggs have been replaced by nuclei of cells from various embryonic stages in order to reveal their developmental capacities and to determine whether irreversible alterations of the nuclear potential are correlated with a particular cell type. At the preimplantation stage during blastocyst formation, nuclei of trophectoderm cells when transplanted into enucleated eggs participate only in early preimplantation development, bringing about abnormal cleavage embryos. On the other hand, transplanted nuclei of the inner cell mass are capable of promoting normal differentiation to the adult stage. The production of fertile mice originating from these embryonic nuclei reveal their full developmental potential, identical to the zygotic genome. Following the preimplantation period, nuclear transplantations have been extended to the

ISBN 0-12-137420-3

postimplantation embryo at day 7. While nuclei from the ectoplacental cone, extraembryonic ectoderm and distal endoderm do not support development of recipient eggs beyond early preimplantation, nuclei from embryonic ectoderm and proximal endoderm give rise to adult mice and, therefore, still retain a pluripotent genome. These results are discussed in connection with cell lineage segregation and cellular diversification during mouse embryogenesis.

In *gene transplantations*, recombinant DNA molecules composed of human DNA including the entire insulin gene and the bacterial plasmid pBR322 have been microinjected into fertilized mouse eggs. After *in vitro* culture to the blastocyst stage, the injected embryos were transferred into the uteri of pseudopregnant females to allow implantation and late fetal development to occur. Fetuses with their placentas were isolated from the uteri at days 16-19 of pregnancy and their DNA extracted and analyzed using various restriction enzymes. In two normally developed fetuses and their corresponding placentas, the human insulin gene with the flanking regions and bacterial plasmid sequences could be demonstrated in Southern blot hybridization experiments. We present first evidence for the presence of a cloned human gene in developing mouse fetuses and its possible integration into the host genome.

NUCLEAR TRANSPLANTATION

Nuclear transplantations carried out in mammals have met so far with rather limited success. Initial attempts to introduce somatic nuclei into mouse eggs via inactivated Sendai virus-mediated cell fusion resulted at best, in a few abnormal cleavage divisions of the treated eggs (Graham, 1969; Baranska and Koprowski, 1970; Lin et al., 1973). Failure to develop beyond the early cleavage stage could probably be attributed to several technical and biological factors such as damaging effects during virus-induced fusion, inadequate culture conditions, or introduction of large amounts of donor-cell cytoplasm, particularly during fusion of eggs with embryonic blastomeres. In recent studies, the possible damage due to the viral fusion procedure has been prevented by microsurgically injecting early embryonic nuclei into unfertilized rabbit eggs (Bromhall, 1975). The transfer of radioactively labeled nuclei from cells of morulae into unfertilized and parthenogenetically activated mature oocytes resulted in the development of preimplantation embryos in which weakly labeled nuclei were identified in about one-half of their cells. Because of the lack of chromosomal analysis and donor-specific markers, it was not possible to determine whether these embryos were derived exclusively from the implanted nuclei or from the residing egg nucleus together with the injected somatic nucleus which had participated in early preimplantation development. In order to identify clearly the transplanted nucleus from the egg nucleus both biologically and genetically, Modlinski (1978) transferred chromosomally marked nuclei from cells of the CBA/H-T6 mouse morula into fertilized mouse eggs of a different strain. Some of the developing morulae and blastocysts revealed a tetraploid karyotype including the T6 translocation chromosomes of the donor nuclei, thus indicating functional participation of the transferred nuclei in preimplantation development. However, the injection of a somatic nucleus into a fertilized egg does not allow one to trace the donor nucleus through normal embryogenesis, which is obscured by the developmental contributions of the residing zygote nucleus of the recipient egg, thereby giving rise to a tetraploid embryo.

We attempted to circumvent these complications by injecting a genetically marked nucleus into a fertilized but enucleated mouse egg in order to examine the developmental potential of the transplanted nucleus in the absence of the nuclear genome of the recipient egg (Fig. 1).

Preimplantation Donor

At the late preimplantation stage, about four days after fertilization, the mouse embryo develops into a blastocyst composed of the trophectoderm (TE) and the inner cell mass (ICM). During normal development, ICM cells will give rise to the embryo proper (reviewed by Gardner, 1978), whereas TE cells do not contribute to the embryo but to an extraembryonic cell lineage (reviewed by Gardner and Rossant, 1976), cease to divide and become polyploid during implantation (Barlow and Sherman, 1972), synthesize particular proteins (van Blerkom et al., 1976) and contain intermediate filaments similar to differentiated epithelial cells (Jackson et al., 1980). The question of whether irreversible alterations in gene function occur during differential segregation into TE and ICM cells has been answered in nuclear transplantations.

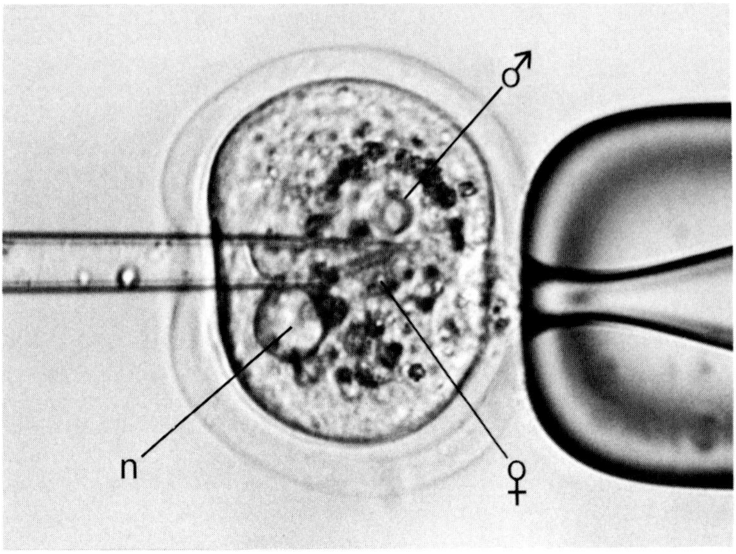

FIGURE 1. Nuclear transplantation into fertilized mouse egg.

The C57BL/6 egg (approx. 70 μm in diameter) is attached to a blunt holding pipette (right) and the injection pipette (left) is pushed through the zona pellucida into the fertilized egg which still contains its two pronuclei separated at this developmental stage. The nucleus (n) previously taken from an embryonic cell is injected and the two pronuclei are subsequently removed from the egg, the female pronucleus (♀) being already sucked into the pipette and followed by the male pronucleus (♂). After this three-step operation, the egg reaching again its original size is cultured *in vitro* for about four days in order to allow preimplantation development to proceed. (For further details, see Illmensee and Hoppe, 1981a).

After injection of TE nuclei into fertilized but enucleated eggs, the resulting early embryos showed abnormalities and their development was usually arrested during the early preimplantation period. On the contrary, nuclei from ICM cells initiated development of the recipient eggs and gave rise to normally appearing blastocysts. Biochemical and karyotypic analyses were carried out on the nuclear-transplant embryos in order to determine whether the transplanted nuclei had actually supported development of the injected eggs. Allelic variants of the enzyme glucosephosphate isomerase

(GPI) and chromosomal markers of the T6 translocation enabled us to distinguish between the genome of the transplanted nucleus and the recipient egg (Illmensee and Hoppe, 1981a).

Preimplantation embryos derived from TE nuclei exhibited predominantly the egg-specific GPI pattern and only occasionally, in addition, enzyme activity originating from the nuclear donor. Apparently, nuclei from TE cells failed to promote normal development and appeared already limited in their differentiation capacity. On the other hand, enzyme tests carried out on ICM nuclear-transplant embryos showed that a considerable proportion of them expressed exclusively enzyme activity characteristic of the donor genome. In addition, chromosomal analysis of the ICM-derived embryos revealed a diploid karyotype with the two short T6 translocation chromosomes of the donor nucleus.

Are these nuclear-transplant blastocysts able to continue in development and, if so, reach the adult stage ? For this reason, we transferred some of them together with nonoperated control embryos into the uteri of pseudopregnant females to allow development to term. Three live-born mice were derived genetically from the transplanted ICM nuclei as judged by their coat color, karyotype and GPI enzyme pattern (Illmensee and Hoppe, 1981a). In additional breeding tests, two of the nuclear-transplant mice, one female and one male, proved to be fertile and transmitted the ICM genome to their progeny, all of which exhibited the nuclear-donor phenotype. Obviously, the nuclear genome of ICM cells can still express all the genes necessary to code for the entire animal.

Postimplantation Donor

More recently, we extended our ontogenetic analysis into the postimplantation period with an attempt to determine the developmental potential of nuclei derived from different cell types of the day-7 embryo (Illmensee and Hoppe, 1981b). During this particular postimplantation stage, the mouse embryo comprises of an outer monocellular layer of distal endoderm, an inner layer of proximal endoderm and, more internally, a multicellular layer of embryonic and extraembryonic ectoderm as well as the ectoplacental cone (reviewed by Snell and Stevens, 1968). These regions of the embryo may conveniently be separated from one another by mechanical means and digestive enzyme treatment (Fig. 2) and be further dissociated into single cells so that each cell type can then be tested in nuclear transplantation experiments. Do these different cell

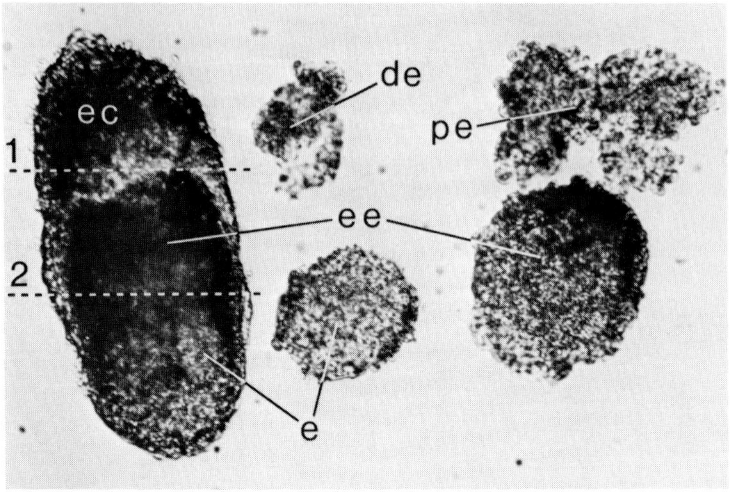

FIGURE 2. Dissection of day-7 mouse embryo and separation of different cell types.

After isolation of the entire embryo from the desidual region of the uterus, the ectoplacental cone (ec) is first removed mechanically (line 1). Subsequently, with digestive enzyme treatment, the peripherally located distal endoderm (de) is isolated, then followed by the more internally located proximal endoderm (pe). The remaining ectoderm is divided (line 2) into the embryonic (e) and extraembryonic (ee) portion. Each of the different cell clusters is further dissociated enzymatically into single cells to be used in nuclear transplantations (Illmensee and Hoppe, 1981b).

types also exhibit different nuclear potencies or, alternatively, are their similarities in developmental capacities between the various nuclei irrespective of their cellular origin and fate ?

Although these extensive series of nuclear transplantations have not yet been completed, some of the results obtained may be already presented and discussed concerning the nuclear potential of the different cell types. Nuclei from distal endoderm cells did not promote development of the enucleated eggs beyond early cleavage divisions. Similarly, nuclei taken from cells of the extraembryonic ectoderm and ectoplacental cone were not able to bring about normal development of recipient eggs which usually cleave but then invariably arrest after a few divisions. Contrary to these restrictions in nuclear potential, when nuclei from cells of the embryonic ectoderm and proximal endoderm were transplanted into enucleated eggs, they gave rise to well advanced preimplantation embryos some of which expressed exclusively the GPI enzyme pattern of the nuclear donor type or showed a diploid karyotype containing the T6 translocation chromosomes of the donor strain. But are these nuclear-transplant embryos which have progressed quite normally through the preimplantation stages also capable of developing further to term ?

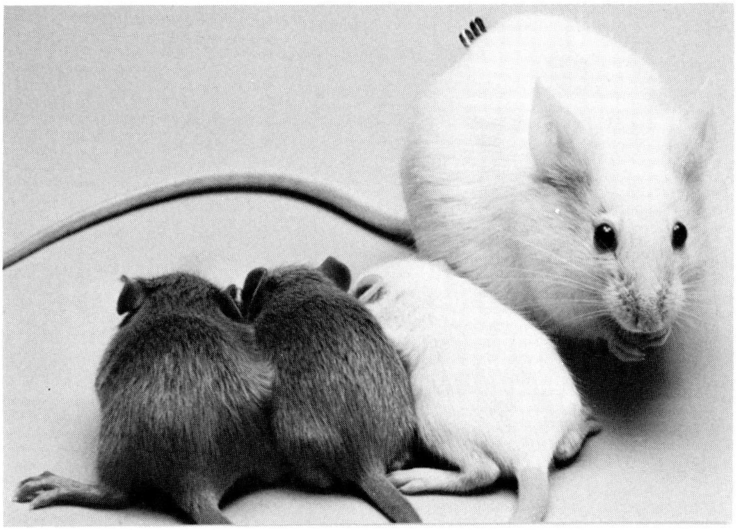

FIGURE 3. Nuclear-transplant doublets derived from proximal endoderm nuclei of day-7 mouse embryo.

The ICR/Swiss foster female (white) with her litter shows one of the nonoperated control mice (white) together with the nuclear-transplant offspring (colored). The "twin" females originating from transplantations of proximal endoderm nuclei into fertilized C57BL/6 eggs are genetically identical with the DBA/J donor embryo and reveal the coat color and various enzyme markers characteristic for the nuclear donor strain.

When embryos originating from nuclei of proximal endoderm or embryonic ectoderm were transferred into the uteri of pseudopregnant foster females, some of them developed into young mice bearing exclusively the genetic markers of the transplanted nuclei. In one instance, female doublets have been clonally derived from proximal endoderm cell nuclei of a single donor embryo (Fig. 3). Our data suggest that nuclei from these two different cell types are still able to express all genes necessary for adult development. Nuclear transplantation in the mouse may therefore provide a vigorous bioassay to functionally determine the genomic capacity of the entire nucleus, to uncover the progressive loss of nuclear potential in differentiating cells, and to search for the mechanisms of gene inactivation during cell diversification. In this respect, it will be important to extend nuclear transplantations to more advanced embryonic and fetal stages in order to gain more insight into the biological and genetic consequences of nuclear changes during mammalian cell differentiation.

GENE TRANSPLANTATION

Although DNA-mediated transformation of cultured cells or living embryos has been demonstrated in a variety of eukaryotic organisms (reviewed in Celis et al., 1980), only recently, by using cloned viral or eukaryotic genes in conjunction with restriction enzyme analysis it became feasible to trace these DNA sequences to the recipient cell genome and eventually determine their chromosomal integration sites (Wigler et al., 1979). Some of the problems related to the low frequency of transformation and the selection procedures required for gene transfer into cultured mammalian cells have been overcome by microinjecting the DNA molecules directly into the recipient cells (Graessmann and Graessmann, 1976; Capecchi, 1980; Anderson et al., 1980). In general, however, it has proven rather difficult to use established cell lines for the study of cell lineage-specific activity and control of the introduced genes (Mantei et al., 1979; Wold et al., 1979; Pellicer et al., 1980).

Alternatively, the expression and regulation of cloned genes may be analyzed in the developing organism, which would require the injection of those genes into eggs or early embryos. In mammals, preimplantation embryos of the mouse seem to be suitable recipients for gene transplantation because of their well investigated genetic background and the relative ease with which they can be collected from pregnant females, manipulated and cultured *in vitro*, and then retransferred into foster mothers for further development. In this way, the integration and possible expression of the foreign genes can be most effectively studied at various stages of embryonic, fetal and adult life and, eventually, transmitted to the next generation for genetic mapping of their location in the host genome. This *in vivo* approach was first used by injecting purified simian virus 40 DNA into the cavity of mouse blastocysts and by demonstrating the presence of injected DNA copies in some organs of surviving mice (Jaenisch and Mintz, 1974). More recently, recombinant DNA plasmids composed of segments of Herpes simplex virus, simian virus 40 and the bacterial vector pBR322 were injected directly in

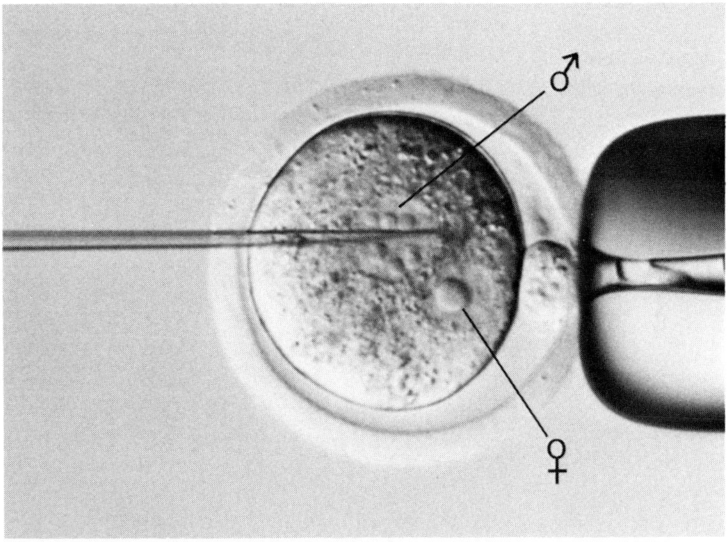

FIGURE 4. Gene transplantation into fertilized mouse egg.

After having attached a fertilized C57BL/6 egg to the holding pipette (right) in the appropriate position, the injection pipette (left) is introduced preferably into one of the two pronuclei. A volume of 10 pl of DNA solution injected into each egg contains about 30 000 copies of a recombinant plasmid composed of a 12.5 kb fragment of human DNA including the entire insulin gene and the 4.3 kb bacterial vector pBR322. (For further details, see Bürki et al., 1981).

fertilized mouse eggs in order to increase the probability of integration of donor DNA into all cells of the developing embryos. Two of the newborn mice derived from plasmid-injected eggs were found to contain donor DNA-specific sequences, though in abnormally rearranged form (Gordon et al., 1980).

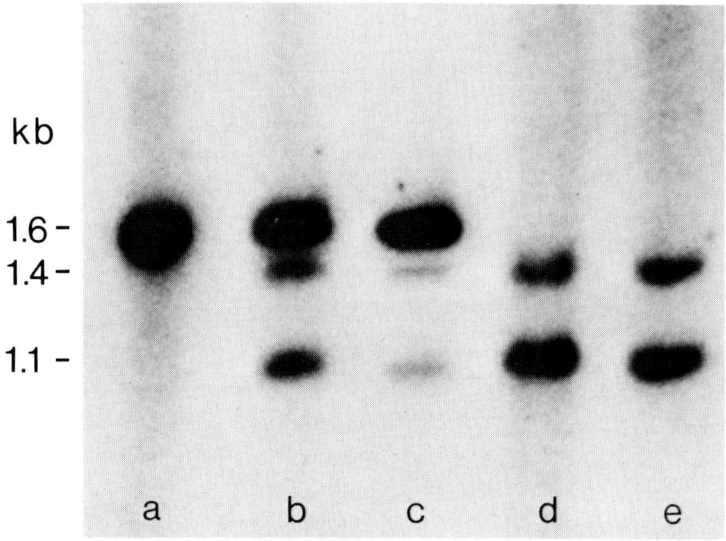

FIGURE 5. Detection of human insulin gene sequences in plasmid-injected mouse fetus.

High molecular weight DNA from plasmid-injected fetuses and their placentas is extracted and digested with Pvu II endonuclease. A 310 bp fragment, which contains most of the human preproinsulin gene (Ullrich et al., 1980), is used as radioactively labeled hybridization probe. In controls, total human DNA (a) digested with Pvu II yields a 1.6 kb band detectable by hybridization with the 310 bp probe. One fetus (b) and its placenta (c) show this 1.6 kb hybridizing fragment, whereas another fetus (d) and its placenta (e) remain negative. Both fetuses and their placentas, however, show two bands of 1.1 and 1.4 kb in size corresponding to the two endogenous mouse insulin genes which are detected by cross-hybridization with the 310 bp probe. (For details, see Bürki et al., 1981).

We attempted to introduce recombinant DNA molecules containing a defined eukaryotic gene, the human insulin gene, into developing mouse eggs via microinjection (Fig. 4). The cloned plasmid we employed is comprised of a 12.5 kb region of the human genome including the entire insulin gene (Ullrich et al., 1980) and the 4.3 kb bacterial vector pBR322 (Bolivar et al., 1977). Following DNA injection and *in vitro* culture to the blastocyst stage, the developing embryos were then transferred into the uteri of pseudopregnant foster females. Fetuses and their corresponding placentas were isolated from the uteri at late pregnancy and their DNA was extracted and screened for the presence of the injected DNA sequences by using a radioactively labeled 310 bp probe that contains most of the human preproinsulin (Ullrich et al., 1980). With the aid of restriction endonuclease digestion in conjunction with Southern blot hybridization we found that in two normally developed fetuses at day 18 of pregnancy, both fetal and placental tissues contained the human insulin gene including the flanking regions and bacterial plasmid sequences (Bürki et al., 1981). The Pvu II endonuclease digestion revealed a hybridizing fragment that corresponded in size to the human insulin gene (Ullrich et al., 1980), indicating that the size of the injected gene was maintained during *in situ* development (Fig. 5).

Although we have demonstrated the presence of the human insulin gene within the fetal mouse genome, it remains to be shown whether this particular gene is functionally expressed during cellular differentiation, organ-specifically controlled at the adult stage, and eventually transmitted through the germline to the progeny.

ACKNOWLEDGMENTS

We thank M.F. Blanc, F. Bourquin, R. Illmensee, E. Markert, and E.Y. Wong for excelllent technical assistance. Inbred mice were kindly donated by the Füllinsdorf Institute of Biomedical Research, Switzerland. During his sabbatical year at the University of Geneva, P.C. Hoppe was a recipient of an American Cancer Society - Eleanore Roosevelt - International Fellowship awarded by the International Union against Cancer. K. Illmensee should like to acknowledge support from the Swiss Science Foundation (FN3.442.0.79), the March of Dimes Birth Defects Foundation (1-727), the National Institutes of Health (CA27713-02) as well as appropriations from the Fonds Marc Birkigt and Schmidheiny.

REFERENCES

Anderson, W.F., Killos, L., Sanders-Haigh, L., Kretschmer, P.J. and Diacumakos, E.G. (1980). Proc. Natl. Acad. Sci. USA 77, 5399-5403.

Baranska, W. and Koprowski, H. (1970). J. Exp. Zool. 174, 1-14.

Barlow, P.W., Sherman, M.I. (1972). J. Embryol. Exp. Morphol. 27, 447-465.

Bolivar, F., Rodriguez, R.L., Greene, P.J., Betlach, M.C., Heyneker, H.L., and Boyer, H.W. (1977). Gene 2, 95-113.

Bromhall, J.D. (1975). Nature 258, 719-721.

Bürki, K., Wong, E.Y., and Ullrich, A. (1981). In preparation.

Capecchi, M.R. (1980). Cell 22, 479-488

Celis, J.E., Graessmann, A., and Loyter, A. (1980). "Transfer of Cell Constituents into Eukaryotic Cells". Vol. 31, NATO Advanced Study Institutes Series. Plenum Press, New York.

Gardner, R.L. (1978). In "Results and Problems in Cell Differentiation", 9 (W. Gehring, ed.), p. 205-241. Springer-Verlag, Heidelberg.

Gardner, R.L., and Rossant, J. (1976). In "Embryogenesis in Mammals". Ciba Foundation Symposium 40 (new series), p. 5-18. Elsevier, Excerpta Medica, Amsterdam.

Gordon, J.W., Scangos, G.A., Plotkin, D.J., Barbosa, J.A., and Ruddle, F.H. (1980). Proc. Natl. Acad. Sci. USA 77, 7380-7384.

Graessmann, M., and Graessmann, A. (1976). Proc. Natl. Acad. Sci. USA 73, 366-370.

Graham, C.F. (1969). In "Heterospecific Genome Interaction" (V. Defendi, ed.), p. 13-35. Symposium Monograph Vol. IX. Wistar Institute Press, Philadelphia.

Illmensee, K., and Hoppe, P.C. (1981a). Cell 23, 9-18.

Illmensee, K. and Hoppe, P.C. (1981b). In "Progress in Developmental Biology" (H.W. Sauer, ed.), Gustav Fischer Verlag, Stuttgart, in press.

Jackson, B.W., Grund, C., Schmid, E., Bürki, K., Franke, W.W., and Illmensee, K. (1980). Differentiation, 17, 161-179.

Jaenisch, R., and Mintz, B. (1974). Proc. Natl. Acad. Sci. USA 71, 1250-1254.

Lin, T.P., Florence, J., and Oh, J.O. (1973). Nature 242, 47-49.

Mantei, N., Boll, W., and Weissmann, C. (1979). Nature 281, 40-46.

Modlinski, J.A. (1978). Nature 273, 466-467.

Pellicer, A., Wagner, E.F., El Kareh, A., Dewey, M.J., Reuser, A.J., Silverstein, S., Axel, R. and Mintz, B. (1980). Proc. Natl. Acad. Sci. USA 77, 2098-2102.

Snell, G.D., and Stevens, L.C. (1968). In "Biology of the Laboratory Mouse" (E.L. Green, ed.), p. 205-245. Dover Publications, New York.

Ullrich, A., Dull, D.J., Gray, A., Brosius, J., and Sures, I. (1980). Science 209, 612-614.

Van Blerkom, J., Barton, S.C., and Johnson, M.H. (1976). Nature 259, 319-321.

Wigler, M., Sweet, R., Sim, G.K., Wold, B., Pellicer, A., Lacy, E., Maniatis, T., Silverstein, S., and Axel, R. (1979). Cell 16, 777-785.

Wold, B., Wigler, M., Lacy, E., Maniatis, T., Silverstein, S., and Axel, R. (1979). Proc. Natl. Acad. Sci. USA 76, 5684-5688.

RESTRICTION ENDONUCLEASES, DNA SEQUENCING, AND COMPUTERS[1]

Richard J. Roberts

Cold Spring Harbor Laboratory
P.O. Box 100
Cold Spring Harbor, New York 11724

INTRODUCTION

Among the 250 Type II restriction endonucleases now characterized, there are more than 70 different specificities (1) and yet there is no indication that the range of specificities is exhausted. Indeed, there is good reason to believe that hundreds, if not thousands, of different specificities would be found if a diligent search were carried out. One reason for this speculation is illustrated in Table 1, which shows the range of sequence patterns with which different Type II restriction endonucleases interact. Among the simple symmetric hexanucleotide sequences designated here as Class A, almost half of the possible sequence patterns are already represented by well-characterized enzymes. There is no reason to believe that a similar number of enzymes will not be found for the other patterns in Classes B through F. Similarly, it seems likely that enzymes recognizing degenerate patterns, like HgiAI and AccI, are not the sole representatives of the class. Within the last year alone, five new classes (C, D, F, N, and O) were added to this list.

[1] This work is supported by grants from the National Cancer Institute (CA13106 and CA27275) and the National Science Foundation (PCM 79-19882 and PCM 80-11709).

ISBN 0-12-137420-3

Table 1

Hexanucleotide Recognition

General Pattern		Specific Example		Known	Possible
A.	XYZZ'Y'X'	NarI	GG↓CGCC	30	64
B.	XYZNZ'Y'X'	SauI	CC↓TNAGG	2	64
C.	XYZNNNZ'Y'X'	Tth111I	GACN↓NNGTC	1	64
D.	XYZNNNNZ'Y'X'	XmnI	GAANNNNTTC	1	64
E.	XYZNNNNNZ'Y'X'	BglI	GCCNNNN↓NGGC	1	64
F.	XYZNNNNNNZ'Y'X'	HgiEII	ACCNNNNNNGGT	1	64
G.	XYPyPuY'X'	HgiCI	G↓GPyPuCC	2	16
H.	XPyYY'PuX'	AcyI	GPu↓CGPyC	1	16
I.	XPuYY'PyX'	AvaI	C↓PyCGPuG	1	16
J.	PuXYY'X'Py	XhoII	Pu↓GATCPy	2	16
K.	$\binom{A}{T}$XYY'X'$\binom{A}{T}$	HaeI	$\binom{A}{T}$GG↓CC$\binom{A}{T}$	1	16
L.	X$\binom{A}{T}$YY'$\binom{A}{T}$X'	HgiAI	G$\binom{A}{T}$GC$\binom{A}{T}$↓C	1	16
M.	XY$\binom{A}{C}\binom{G}{T}$Y'X'	AccI	GT↓$\binom{A}{C}\binom{G}{T}$AC	1	16
N.	PyXXXXX	GdiII	Py↓GGCCG	1	?
O.	XXXPuXX	Tth111I	CAAPuCA	1	?

Other enzymes in Class A are: ApaI (GGGCC↓C), AsuII (TT↓CGAA), AvaIII (ATGCAT), AvrII (CCTAGG), BalI (TGG↓CCA), BamHI (G↓GTACC), BclI (T↓GATCA), BglII (A↓GATCT), ClaI (AT↓CGAT), EcoRI (G↓AATTC), HindIII (A↓AGCTT), HpaI (GTT↓AAC), KpnI (GGTAC↓C), MstI (TGC↓GCA), NruI (TCGCGA), PstI (CTGCA↓G), PvuI (CGAT↓CG), PvuII (CAG↓CTG), RruI (AGT↓ACT), SacI (GAGCT↓C), SacII (CCGC↓GG), SalI (G↓TCGAC), SmaI (CCC↓GGG), SnaI (GTATAC), SphI (GCATG↓C), StuI (AGG↓CCT), XbaI (T↓CTAGA), XhoI (C↓TCGAG), XmaIII (C↓GGCCG). In Class B: BstEII (G↓GTNACC). In Class G: HindII (GTPy↓PuAC). In Class K: HaeII (PuGCGC↓Py). Full references may be found in (1), except for XmnI (2), ApaI (M. van Montagu, personal communication), and NarI, NruI (I. Schildkraut, personal communication).

If we turn now to enzymes recognizing pentanucleotide sequences (Table 2), we see that only two basic patterns occur--those in which a unique pentanucleotide sequence is recognized, and those in which the outer four bases possess a dyad axis of symmetry. The site of cleavage differs dramatically between these two classes. In the former class, the point of cleavage lies within (or immediately adjacent to) the recognition sequence, whereas in the latter class, the site of cleavage is well removed from the recognition sequence and, furthermore, differs from one enzyme to another.

Table 2

Pentanucleotide Recognition

	General Pattern	Specific Example		Known	Possible
A.	XY($^{C}_{G}$)Y'X'	Cau II	CC($^{G}_{C}$)GG	1	16
B.	XY($^{A}_{T}$)Y'Z'	Bbv I	GC($^{A}_{T}$)GC	3	16
C.	XXXXX	Sfa NI	GCATC	4	480

Other examples in Class B are Ava II (G↓G(A/T)CC), EcoR II (↓CC(A/T)GG). In Class C, Hga I (GACGC - 10/5), Hph I (GGTGA - 9/8), Mbo II (GAAGA - 9/8). The numbers following the recognition sequence indicate the sites of cleavage; e.g., for Hga I, a full representation would be

5' GACGCNNNNNNNNNN↓ 3'
3' CTGCGNNNNN↑

The final group of enzymes, which recognize tetranucleotide sequences, are shown in Table 3. As with the pentanucleotide-recognizing enzymes, there is a gross difference in the position of the cleavage site when the recognition sequence does not contain a dyad axis of symmetry.

Table 3

Tetranucleotide Recognition

	General Pattern	Specific Example		Known	Possible
A.	XYY'X'	RsaI	GT↓AC	8	16
B.	XYNY'X'	Fnu4HI	GC↓NGC	4	16
C.	XXXX	MnlI	CCTC	1	240

Other examples in Class A are: AluI (AG↓CT), FnuDII (CG↓CG). HaeIII (GG↓CC), HhaI (GCG↓C), HpaII (C↓CGG), MboI (↓GATC), TaqI (T↓CGA). In Class B: AsuI (G↓GNCC), DdeI (C↓TNAG), HinfI (G↓ANTC). MnlI cleaves about 8 nucleotides 3' to the recognition sequence (I. Schildkraut, personal communication).

Examination of the specific sequences recognized by these enzymes reveals a preponderance of GC-rich sequences. But more startling is the complete absence of sequences containing only AT base pairs. Given the large number of enzymes now characterized, it seems unlikely that this is some statistical anomaly and it rather looks as though there may be some biological reason why such sequences are not recognized. A number of explanations have been proposed for this, but my own current favorite is that the problem arises not with the restriction enzyme recognizing only AT-base pairs but with the methylase that would be required to protect the host DNA against its action. If such AT-rich sequences are involved in interaction with polymerases, repressors, or other regulatory molecules, it may be that by methylation this interaction is destroyed. This could serve to prevent the evolution of methylases recognizing only AT-rich sequences. Even with this restraint, it is still clear that the present number of known specificities is a rather small percentage of the likely total. One might reasonably ask if there is any need for more diversity in the restriction endonucleases that are available to the molecular biologist. I believe the answer to this question is a whole-hearted "Yes!"

More than 500kb of DNA sequence have now been determined and new sequence is accumulating at an exponential rate. The availability of a given piece of DNA sequence means that experiments requiring the manipulation of that sequence can often be planned in a rather precise manner. However, the experimental possibilities are often limited by the availability of a restriction endonuclease which cuts at a convenient site. For example, the Southern blotting technique, which allows one to probe the inner depths of the eucaryotic genome with precision, demands that suitable restriction enzyme sites be present. Experiments by Kan et al. (3) show the feasibility of genetic diagnosis by Southern blotting and, in principle, this technique could be extended to many genetic disorders if appropriate restriction enzymes are available. Another example, in which even more detailed changes are examined, comes from the work of Flavell et al. (4). In this case, two restriction endonucleases (HpaII and MspI) which recognize the same nucleotide sequence but show differential sensitivity to methylation patterns within that sequence, are used to examine the state of methylation at individual bases within the human genome. It would clearly be desirable if other such pairs of enzymes were available.

DNA Sequence Analysis

Among the many technological advances which have been made possible by the discovery of the Type II restriction endonucleases has been the rapid progress in the determination of DNA sequences. Two main techniques have been used--the chemical method devised by Maxam and Gilbert (5) and the primed synthesis method developed by F. Sanger and his colleagues (6, 7). The principle of the primed synthesis method is outlined in Figure 1.

In the chemical method, similar sets of oligonucleotides are produced by means of base-specific cleavage. In both cases, the resulting four sets of oligonucleotides are fractionated by polyacrylamide gel electrophoresis under conditions that resolve oligonucleotide chains differing by a single base. The sequence is then read directly from the gel and, under appropriate conditions, sequences in excess of 300 nucleotides can be deduced from a single reaction. Until recently, the chemical method of sequencing has been preferred by most investigators. In part, this was due to

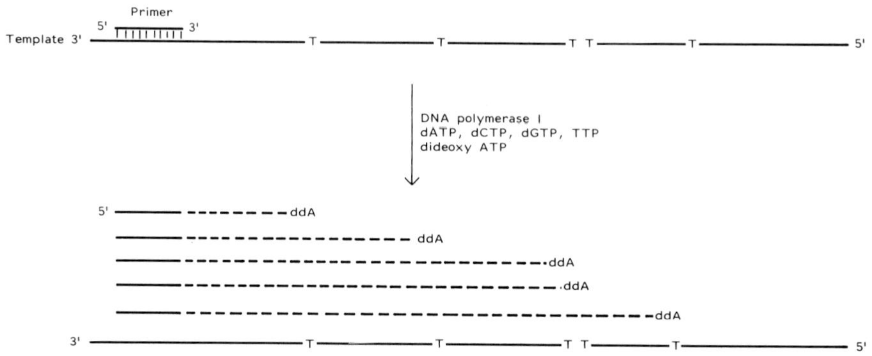

Figure 1: Schematic illustration of the primed synthesis method using chain terminators. Following annealing of a template and primer, the primer is extended, using the Klenow fragment of DNA-polymerase I, in the presence of a mixture of all four dNTPs plus one dideoxynucleoside triphosphate (ddNTP). Incorporation of the ddNTP takes place by chance and leads to chain termination because this analogue lacks a 3'-hydroxyl group. The products are a mixture of oligonucleotides with a common 5'-terminus and 3'-termini which, in this example, correspond to the position of T-residues in the template. Four separate reactions are carried out using each of the four ddNTPs.

the technical difficulties of preparing suitable single-stranded DNAs for use as templates in the primed synthesis method. However, it also reflects the carefully produced protocols and extensive trouble-shooting guide prepared by Alan Maxam. Within the last two years, there have been substantial advances in the methodology for primed synthesis; first, by the introduction of the chain termination procedure (7) which replaced the earlier plus-minus method (6), and then by the use of exonucleases, such as exonuclease III and T7 exonuclease, as a means of preparing single-stranded templates (8). More recently, the M13 cloning-sequencing system has been developed (9) and seems likely to be the method of choice for the determination of large DNA sequences. The key features of this method are illustrated in Figure 2.

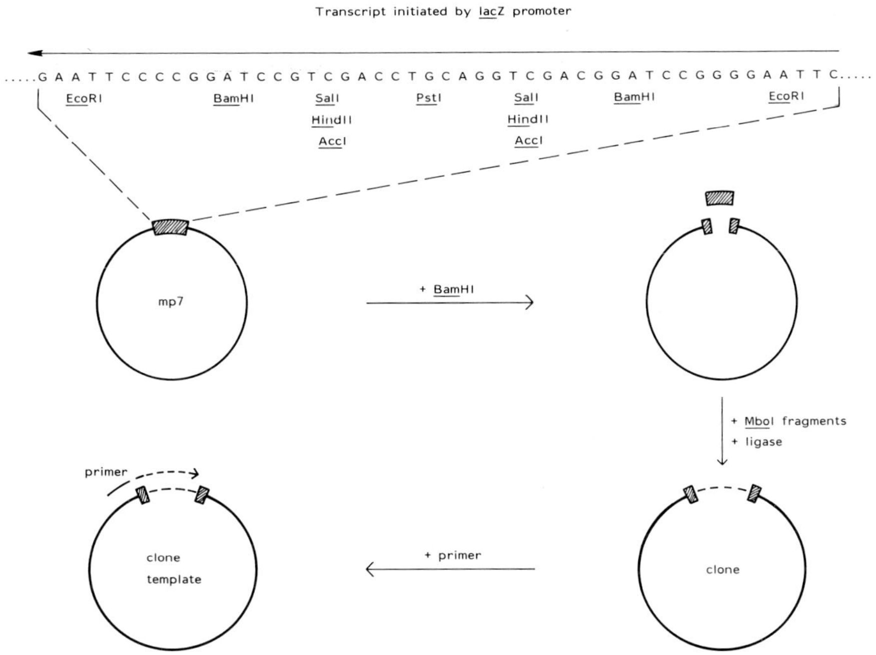

Figure 2: The M13 cloning-sequencing system. The vector mp7 contains a synthetic sequence of 42 nucleotides inserted at an EcoRI site in the lacZ gene. It is shown at the top of the figure and indicated elsewhere by hatched lines. It has multiple restriction sites into which fragments can be cloned and was constructed by J. Messing and his colleagues (12). Insertion of a fragment at any of the sites within this sequence usually interrupts expression of the N-terminal sequences of lacZ and prevents α-complementation. Recombinants are then easily identified on an appropriate indicator strain (see 12).

Briefly, the DNA to be sequenced is cleaved into small fragments by means of appropriate restriction enzymes and these fragments are cloned into a derivative of the filamentous phage M13, called mp7 (12), which contains a variety of restriction sites (see Figure 2). A convenient genetic test exists by which recombinants may be identified and a suitable number can then be selected for further study. At this point, it becomes relevant to consider the biology of M13. Upon infection of *E. coli*, this phage replicates via a double-stranded replicative form intermediate and produces single-stranded DNA circles which are packaged into the virion. The virions are then extruded from the cell and accumulate in the culture medium. The infected cells do not lyse and so following prolonged growth, extremely high phage titers can be found in the culture fluid. The infected cells are removed by centrifugation and the phage can be precipitated from the supernatant with polyethylene glycol. Single-stranded phage DNA is then easily prepared by phenol treatment. From 1 ml of a good supernatant, enough single-stranded DNA can be isolated to serve as template in about 50 sequencing reactions. The clone bank is conveniently analyzed by first running single-channel reactions with each clone to avoid repetitive sequencing of multiple isolates of the same clone, followed by four-channel sequencing reactions of a representative set. For each of these sequencing reactions, the same primer is used because, as illustrated in Figure 2, the primer is actually complementary to a segment of the vector DNA lying immediately adjacent to the site of insertion. Short, synthetic oligonucleotides, which make excellent primers, are available from several commercial sources. Among the clones that are isolated, both strands of each fragment should be present, although it is relatively easy to turn around an insert, if that should be desired. This is because of the symmetrical nature of the restriction sites in mp7 (see Figure 2). By preparing several clone banks using different fragment sets, it is usually straightforward to obtain overlapping sequences and, hence, to reconstruct the complete sequence of the original fragment. Such reconstruction is most easily accomplished by using a computer.

Computer Methods

During the last few years, many individuals have taken advantage of computer methods, both to analyze and to assemble DNA sequences (11). Numerous programs have been written for their analysis and sequence features, such as dyad symmetries, codon usage, base composition, etc. are easily obtained. One ingenious program will locate tRNA-like structures (12), while others (13, 14) assist in the determination of restriction endonuclease recognition sites. The computer has also played an important role during the assembly of DNA sequences and it is this area that is most pertinent to the topic of this symposium.

Two programs have been described (15, 16) which can be useful at the very start of a DNA sequencing project. These programs are concerned with the generation of restriction enzyme maps. For the construction of a map defining the cleavage sites of, say, three restriction enzymes, six digests are carried out--the three single digests, the three double digests, and the triple digest. The fragment lengths observed in each case serve as the primary data upon which the algorithms work. In one approach, all possible maps are generated, and those which do not conform to the data are eliminated. Depending upon the accuracy with which the fragment lengths are known, this can sometimes lead to a unique map, but more often, to a set of a few possible maps which can be distinguished by further experimentation. The second approach which attempts to overcome the problems of length determination, uses a least-squares method to improve the accuracy with which the size of each restriction fragment is known. An iterative procedure is used until the final map position shows the best fit with the data. The major limitation with both of these programs lies not with the algorithms, but rather with our technical ability to determine accurately the length of restriction fragments.

The primary data from which a long sequence is deduced consists of a number of short sequence stretches that may be up to 300 nucleotides in length. These must then be linked by means of overlapping stretches of common sequence to reconstruct the complete structure of the original molecule. Two sets of programs have been written to aid in this process (17, 18). Both programs provide three essential functions. The first of these is to form a master archive which contains all of the primary data. A second function allows the identification of blocks of sequence that contain homologous or complementary stretches--the overlaps. A key feature here lies in distinguishing between sequences that overlap by chance and sequences that are truly contiguous. This is particularly difficult if the two areas of overlap lie at the ends of the primary sequence data, when perhaps the last few nucleotides have been squeezed from the gel and, hence, are more prone to errors. At this point in the analysis, it is often very helpful if some restriction enzyme mapping data is available or other, auxiliary information which can help decide between chance and genuine overlaps. Once the overlaps have been identified, the third function is that of melding two strings of nucleotides into a single continuous sequence. If discrepancies occur at certain positions in the two strings, then the original data must be consulted so that some decision can be made between the two possibilities.

One problem that frequently arises during the alignment of the primary data stems from trivial human errors introduced during the manual recording of the data. This often leads to a base being skipped or channels being mistaken. We have been attempting to overcome errors at this stage by using a digitizing tablet and signal pen to transfer the data directly from the radioautograph into the computer. The tablet operates by sending to the computer the location of any point on the surface of the pad, once this point has been touched by the signal pen. The location is represented in the form of a digitized set of X and Y coordinates. Several readings of the same gel may be taken and the computer will then match the two readings, will point out discrepancies, and will allow that region of the gel to be re-checked before the final sequence is entered. A typical set-up is illustrated in Figure 3. While the gel is being read, the data can be displayed on a CRT screen, although this is rather inconvenient since it means constantly transferring attention from the gel to the screen. Recently, we have improved this feature by incorporating a "voice box" into the setup. As each base is read from the sequence, the box vocalizes the base, thus confirming that the computer received the intended data. Our present system is based upon the "Speak and Spell" chip produced by Texas Instruments and has the drawback that it speaks rather slowly. We are looking into ways to improve the speed, as the advantages of this approach are significant.

Let me conclude by discussing a few simple auxiliary programs which we have found useful while sequencing Adenovirus-2 DNA (about 36 kb in length). A great deal of information is available about the polypeptides expressed by this genome and many of the individual transcripts have been positioned by electron microscopy. Good restriction enzyme maps are available, and these sites provide convenient points of reference along the genome. As the sequence data accumulates, it has often been possible to correlate the sequence with these previously-identified features. However, keeping track of these correlates can become complicated. For instance, even the simple task of annotating the sequence is difficult when absolute coordi-

Figure 3: A digitizing tablet and CRT terminal used for reading sequencing gels.

nates with reference to the termini of the genome, cannot be given because of gaps in the sequence. We have overcome this difficulty by using not position coordinates, but sequence strings as a means of locating and identifying interesting regions. This is achieved by creating, as a companion to the main sequence file, an information file. This file contains a string of nucleotides (say, 10 or 12 long), together with an appropriate comment that annotates the occurrence of those nucleotides--for instance, a transcriptional start point. A program, called "COMBINE", will then condense the sequence file with the information file to produce an annotated version of the sequence file, as illustrated in Figure 4. In this way, the annotated version of the sequence file can be automatically updated as the sequence is updated.

```
        3960          3970          3980          3990          4000
TCTCAGCAGC    TGTTGGATCT    GCGCCAGCAG    GTTTCTGCCC    TGAAGGCTTC
AGAGTCGTCG    ACAACCTAGA    CGCGGTCGTC    CAAAGACGGG    ACTTCCGAAG


   10.98% Terminator UAA for polypeptide IX
          11.01% AAUAAA for early Ib and polypeptide IX mRNAs
        4010             *             *          4040          4050
CTCCCCTCCC    AATGCGGTTT    AAAACATAAA    TAAAAACCAG    ACTCTGTTTG
GAGGGGAGGG    TTACGCCAAA    TTTTGTATTT    ATTTTTGGTC    TGAGACAAAC
                                                              *
                                  Poly-A addition site for IVa2 mRNA  11.07%


11.10% Poly-A addition site for polypeptide IX mRNA
11.08% Bcl I
    *   4060  *       4070          4080          4090          4100
GATTTTGATC    AAGCAAGTGT    CTTGCTGTCT    TTATTTAGGG    GTTTTGCGCG
CTAAAACTAG    TTCGTTCACA    GAACGACAGA    AATAAATCCC    CAAAACGCGC
                                             * *
                              AAUAAA for IVa2 mRNA  11.16%
                        Terminator UAA for IVa2 polypeptide  11.17%


  11.23% Xma I
        *             4120          4130          4140          4150
CGCGGTAGGC    CCGGGACCAG    CGGTCTCGGT    CGTTGCAGGG    TCCTGTGTAT
GCGCCATCCG    GGCCCTGGTC    GCCAGAGCCA    GCAACGTCCC    AGGACACATA


        4160          4170          4180          4190          4200
TTTTTCCAGG    ACGTGGTAAA    GGTGACTCTG    GATGTTCAGA    TACATGGGCA
AAAAAGGTCC    TGCACCATTT    CCACTGAGAC    CTACAAGTCT    ATGTACCCGT
```

Figure 4: An example of output from the program COMBINE. This sequence is from the left end of the Adenovirus-2 genome (T.R. Gingeras, R.E. Gelinas, and R.J. Roberts, unpublished results).

Finally, we are beginning to face the problem of comprehending the information that is available in a sequence of this length. To gain an overall view of that information content, some diagrammatic representation is necessary. Usually, this is achieved by an artist manually converting the information to a diagram. However, this is slow and tedious and requires constant updating unless the initial data, in this case the sequence, is itself complete. We have turned our attention to the use of the computer for this purpose, aiming to produce on the CRT terminal a graphic display which will summarize certain kinds of information. Our initial attempts in this direction are illustrated in Figure 5, which is the product of a program called "DSPLAY" which will take a sequence and translate it in all three reading frames from both strands. A condensed graphic display is

```
         RESULTS OF TRANSLATING THE FILE :
RIGHT END OF AD-2 (2/23/81)

     LENGTH OF SHORTEST POLYPEPTIDE DISPLAYED IS   50 AMINO ACIDS



       -------->                      ----------------->

                        -------->                         ----> ------>

              ---------->     -------->

*******|*******|*******|*******|*******|*******|*******|*******|*******|*******|
     383    766     1149     1532    1915    2298     2681    3064    3447    3830
  98.95%  97.90%  96.85%  95.80%  94.75%  93.70%  92.65%  91.61%  90.56%  89.51%
*******|*******|*******|*******|*******|*******|*******|*******|*******|*******|

               <---        <----------

                      <--------                                     <------------

                                     <------
```

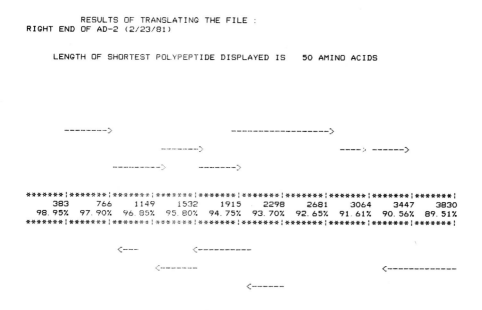

Figure 5: An example of output from the program DISPLAY. The input sequence contains 4,000 nucleotides from the right end of the Adenovirus-2 genome (D. Sciaky, T.R. Gingeras, and R.J. Roberts, unpublished results).

then provided within the 80 character field of a standard CRT terminal. Although the resolution obtained by such a display is not terribly good, the principal features can be seen easily. Two features of the program increase its usefulness. First, it is possible to select the minimum length of open reading frame is to be displayed and, secondly, the program has a zoom feature which enables individual regions to be expanded. The zoom feature can, in fact, go to the nucleotide level, if desired. We have found this program extremely useful in assessing the effects of slight changes in the sequence as new data becomes available. The program is highly interactive and a complete display for a sequence that is 15 kb long can be generated using a PDP 11/60 computer in about 2 seconds. Similar types of programs which would present restriction enzyme maps or other sequence features in a graphic mode on a standard terminal would be extremely useful and of great practical value.

REFERENCES

1. Roberts, R.J. (1981) Nucleic Acids Res. 9: r75.
2. Lin, B-C., Chien, M-C. and Lou, S-Y. (1980) Nucleic Acids Res. 8: 6189.
3. Kan, Y.W. and Dozy, A.M. (1978) Proc. Natl. Acad. USA 75: 5631.
4. van der Ploeg, L.H.T. and Flavell, R.A. (1980) Cell 19: 947.
5. Maxam, A.M. and Gilbert, W. (1977) Proc. Natl. Acad. Sci. USA 74: 560.
6. Sanger, F. and Coulson, A.R. (1975) J. Mol. Biol. 94: 441.
7. Sanger, F., Nicklen, S., and Coulson, A.R. (1977) Proc. Natl. Acad. Sci. USA 74: 5463.
8. Smith, A.J.H. (1979) Nucleic Acids Res. 6: 831.
9. Sanger, F., Coulson, A.R., Barrell, B.G., Smith, A.J.H., and Rowe, B.A. (1980) J. Mol. Biol. 143: 161.
10. Messing, J., Crea, R., and Seeburg, P.H. (1981) Nucleic Acids Res. 9: 309 and references therein.
11. Gingeras, T.R., Roberts, R.J. (1980) Science 209: 1322.
12. Staden, R. (1980) Nucleic Acids Res. 8: 817.
13. Fuchs, C., Rosenvold, E.C., Honigman, A., and Szybalski, W. (1978) Gene 4: 1.
14. Gingeras, T.R., Milazzo, J.P., and Roberts, R.J. (1978) Nucleic Acids Res. 5: 4105.
15. Stefik, M. (1978) Artif. Intell. 11: 85.
16. Schroeder, J.L. and Blattner, F. (1978) Gene 4: 167.
17. Staden, R. (1979) Nucleic Acids Res. 6: 2601.
18. Gingeras, T.R., Milazzo, J.P., Sciaky, D., and Roberts, R.J. (1979) Nucleic Acids Res. 7: 529.

HIGH RESOLUTION TWO-DIMENSIONAL RESTRICTION ANALYSIS OF METHYLATION IN COMPLEX GENOMES

Steven S. Smith, J. Garrett Reilly and C.A. Thomas, Jr.

Department of Cellular Biology
Scripps Clinic and Research Foundation
10666 N. Torrey Pines Road
La Jolla, Ca. 92037

ABSTRACT We have used two-dimensional gel electrophoresis to study methylation at the CCGG site in DNA from adult Drosophila and two tissues of the mouse. In this recently perfected technique (20), BamHI digested DNA is end-labelled using DNA polymerase I and fractionated by electroelution. Individual fractions are then split into two aliquots. One aliquot is digested to completion with HpaII and the other with its isoschizomer MspI. The resulting digests are separated by electrophoresis in adjacent lanes of long slab gels to form an interdigitated array. Autoradiographs of the dried gels give information on the relative copy number for multiply represented sequences, information on the degree of methylation at the internal cytosine in the CCGG sequence, and information which may be used to identify interspersed sequences.

In this report, we present evidence for the absence of methylation at this site in Drosophila and evidence for different patterns of modification at this site in the repeated DNA from tissues of the mouse.

INTRODUCTION

The function of the postsynthetic modification of DNA is not clear. The methylated base 5-methyl-cytosine (MeC) has been found in a variety of organisms, and is the only modified base thus far detected in the vertebrates (1,2). Hypothetical roles for this type of base modification in eukaryotic

ISBN 0-12-137420-3

gene regulation have been suggested by several groups (3-6); and evidence has been accumulating in favor of such roles (7-9). The evidence today suggests that methylation at cytosine is somehow involved in the "selective silencing" (6) of eukaryotic genes. Thus, heavily methylated genes generally are also inactive. For example, germ line transmitted proviral sequences are highly modified (10) as is the integrated (inactive) form of adenovirus type 12 DNA (11). Recent evidence also supports the suggestion that DNA methylation plays a role in silencing the HPRT (hypoxanthine phosphoribosyl transferase) locus when it resides on the inactive X chromosome (12,13). On the other hand, it has been suggested that the extent of cytosine methylation seems to be a species-specific characteristic rather than a phenomenon universally related to development (14). This view is supported by the observation that the level of MeC varies greatly between species (15), with a trend toward higher levels of methylcytosine as one ascends the phylogenetic tree (16).

In this report we summarize recent work on the application of high resolution two-dimensional displays of end-labelled restriction fragments (17-20) to the study of this phenomenon.

METHODS

The data described below were obtained using the methods described in reference 20. The electroelution device used is available from J.M. Specialty Parts, 5605 Sandburg Ave., San Diego, Ca. 92122.

RESULTS

MeC has not been detected in DNA from adult *Drosophila* (21,22); however, the DNA of the mouse contains about 3% MeC (23,24). This contrast is clearly shown in the experiment of Fig. 1. In this experiment, DNA from whole adult *Drosophila melanogaster* and from the spleens of six adult Balb/c mice was digested to completion with *Bam*HI and end-labelled as previously described (20). A portion of each of the end-labelled digests was digested to completion with either *Hpa*II or *Msp*I. The DNA fragments were then separated by electrophoresis in 1.6% agarose slab gels. Autoradiographs produced from the dried gels form the composite shown in Fig. 1. *Hpa*II and *Msp*I recognize the sequence CCGG (25). *Hpa*II will not cleave this sequence if the second cytosine is methylated. *Msp*I on the other hand will cleave this sequence whether or not the second cytosine is modified (see ref. 26). Thus, diminished cleavage by *Hpa*II relative to *Msp*I is evidence of methylation at the internal cytosine in this sequence. Lane 1 of the figure shows DNA of Drosophila after digestion with BamHI and end-labelling. Lane 2 shows this same end-labelled DNA after secondary digestion with HpaII and lane 3 shows the DNA after secondary digestion with MspI. Careful inspection of the autoradiograph shows that HpaII and MspI cleaved Drosophila DNA in a completely equivalent fashion, producing the same characteristic "signature" of multiply represented fragments (detected as intense bands in the digest) and the same number-average fragment size for each double digest. Lanes 4, 5 and 6 show the same experiment performed on mouse DNA. Here the results are quite different. Again, a characteristic signature of multiply represented DNA fragments was observed in the end-labelled BamHI digest of mouse DNA (lane 4). After secondary digestion with HpaII, a small but significant amount of digestion occurred as evidenced by the lower number-average molecular weight of the digest, but none of the multicopy bands produced by BamHI appears to have been cleaved. MspI, on the other hand, cleaved all but one of these bands extensively, and produced a double digest of significantly lower molecular weight.

Lack of Methylation at CCGG in Drosophila. The apparently equivalent behavior of HpaII and MspI on Drosophila DNA was studied further with the two-dimensional display technique. An end-labelled BamHI digest of adult Drosophila DNA was separated by electroelution through agarose gel, and the resulting fractions were concentrated by alcohol precipitation. Identical aliquots of pooled fractions were then subjected to digestion with HpaII or MspI, loaded into adjacent lanes of long 1.5% agarose slab gels and separated by electrophoresis as previously described (20). Autoradiographs of the dried gels were used to form the composite shown in Fig. 2. Three kinds of information can be inferred from this type of experiment.

Information on sequence multiplicity derives from the recent observations of Gilroy (unpublished) that nearly all members of a sampling of clones from very dense bands in these gels hybridize with one another. Thus, it seems that most dense bands in these gels do not arise by the superposition of unrelated sequences of identical length, but rather are superpositions of very closely related or identical sequences. Examples are marked A and B in Fig. 2. Those marked B are derived from the 4.8 and 5.0 kb BamHI repeating units in the histone gene cluster. The BamHI-HpaII segments of 2.05 and 1.03 kb are expected to be produced from the 5.0 kb repeat and the 1.85 and 1.03 from the 4.8 kb repeat (27). Indeed, we observed these relationships in experiments (see 20) in which

Fig. 1

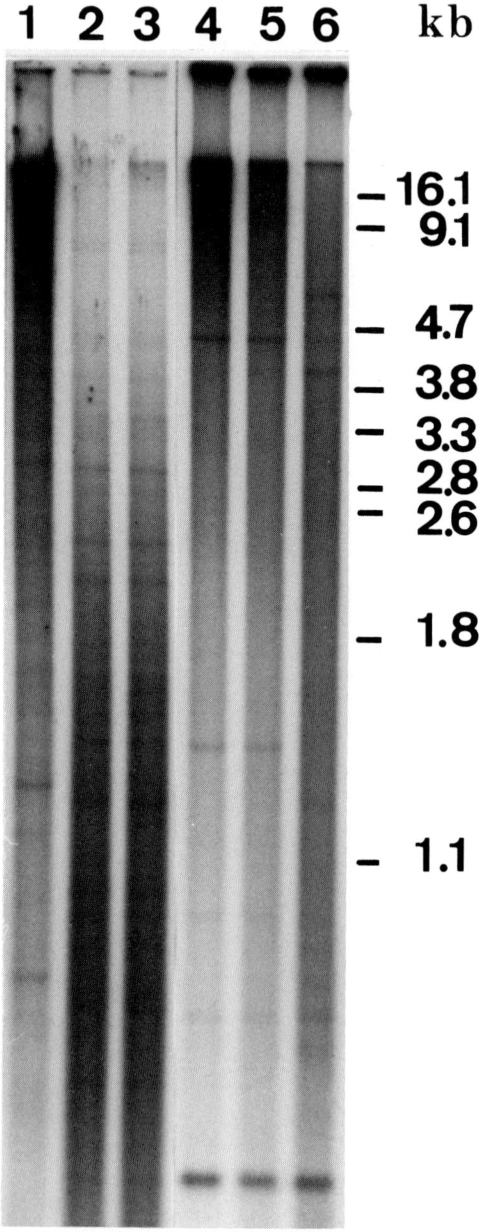

the 4.8 and 5.0 kb BamHI fragments separated during the primary fractionation. Further, the observed band densities were found to reflect the relative copy number determined by Goldberg (27), for both types of histone gene repeat.

Information on sequence interspersion may be inferred from the observation of "lines". Lines are secondary fragments of defined length that are produced by the digestion of primary fragments of variable length. These bands appear at the same position across the display. A representative "line" is marked with the letter L in Fig. 2. We think that the simplest explanation of these "lines" is that they are derived from multiple copies of sequences that are integrated into diverse locations in the chromosomal DNA.

Information on the state of methylation can be inferred by the use of HpaII and MspI to cleave aliquots of the same mixture of end-labelled primary fragments. Given the likelihood that an intense band arises by the superposition of fragments of nearly identical sequence, and given the specificities of HpaII and MspI, the ratio of the intensity of a band in the HpaII lane to that in the MspI lane is a measure of the degree of methylation at the internal cytosine in the CCGG cleavage site.

The experiment in Fig. 2 shows that HpaII and MspI produce equivalent displays of segments in every fraction tested. Thus, modification at the CCGG sequence was not detected by

---

Fig. 1. Methylation at the CCGG Sequence in the DNA of Drosophila and Mouse.

DNA from Drosophila and mouse was digested to completion with BamHI and labelled by terminal repair with DNA polymerase I using α^{32}PdGTP. Aliquots of each end-labelled digest were then digested to completion with either HpaII or MspI. The products of the digests were separated on 1.6% agarose slab gels, and autoradiographs were produced to form the composite gel. Lane 1) BamHI digested Drosophila DNA. Lane 2) BamHI digested Drosophila DNA after secondary digested with HpaII. Lane 3) BamHI digested Drosophila DNA after secondary digestion with MspI. Lane 4) BamHI digested mouse DNA. Lane 5) BamHI digested mouse DNA after secondary digestion with HpaII. Lane 6) BamHI digested mouse DNA after secondary digestion with MspI. As molecular length markers, the positions of λDNA fragments produced by double digestion with BamHI and EcoRI are shown in ink to the right of the figure. These bands correspond to 16.1, 9.1, 4.7, 3.8, 3.3, 2.8, 2.6, 1.8 and 1.1 kb.

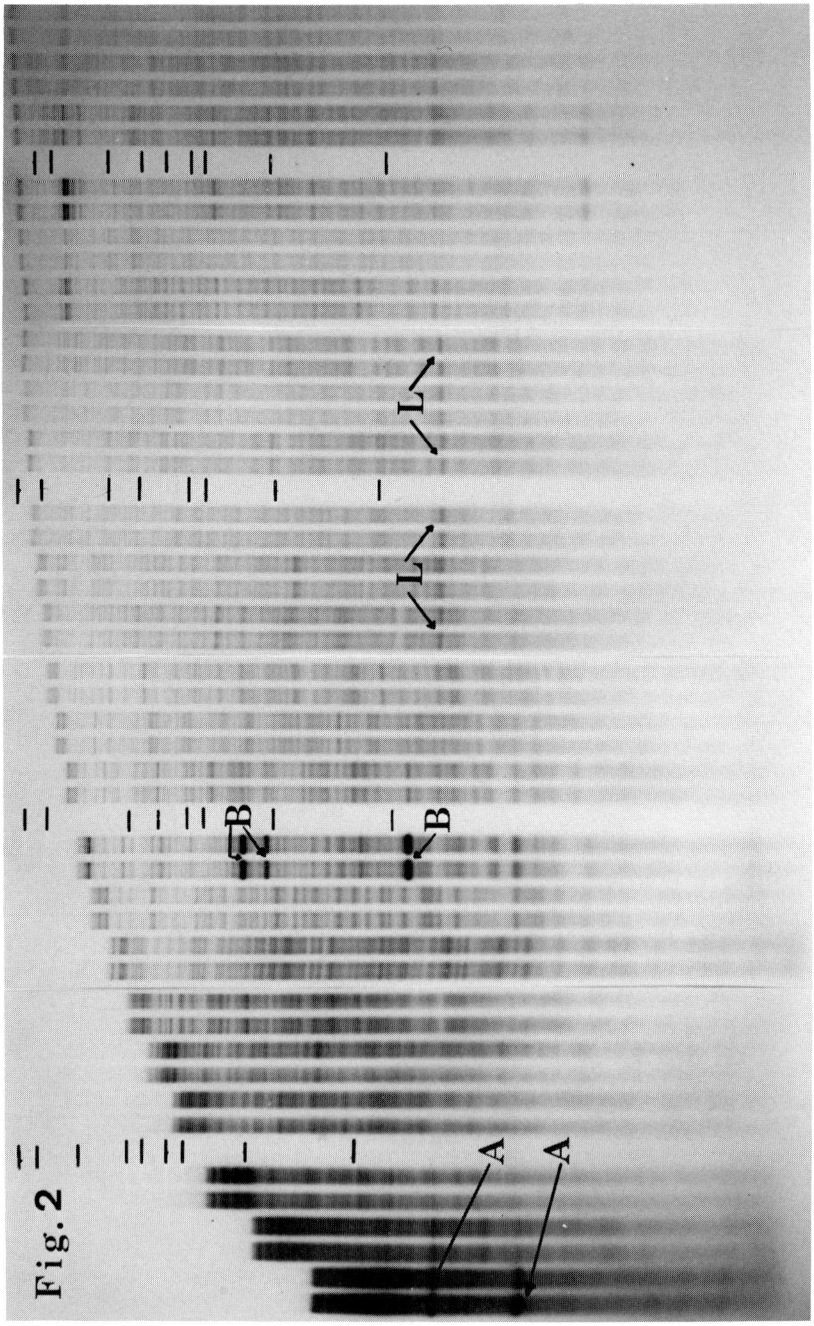

Fig. 2

this method in adult *Drosophila*. However, when the same experiment was performed with mouse tissues, rather different results were obtained.

A Pattern of Methylation in the Repeated DNA of the Mouse. Fig. 3 shows the pattern obtained from mouse spleen DNA. A complex array of bands is observed. Since the genome of mouse is about thirty-fold larger than that of *Drosophila*, it is expected that the bands we see represent multicopy sequences above a background of single copy sequences which are not resolved by the technique. Assuming that individual bands in the pattern are superpositions of closely related or identical sequences, many are intensely methylated, examples are marked with an *I* in the figure. Others appear to be weakly methylated at the *Hpa*II site, and examples are marked *W* in the figure. In addition, bands which are disposed on "lines" and thus likely to be interspersed, are generally found in a single methylation state. Taken together these results suggest that a pattern of methylation exists in the repeated sequences of the mouse.

We considered the possibility that simple satellite DNA might contribute to the pattern, but our own experiments and the recently published sequence of the mouse satellite DNA (28) show that this DNA lacks cleavage sites for *Bam*HI. Thus it is not likely to contribute to the patterns shown in Fig. 3 and Fig. 4.

Since the data suggest that there is a pattern of methylation within the repeated sequences of the mouse, we tested the possibility that there might be tissue specific differences in this pattern. Fig. 4 shows the pattern obtained with DNA isolated from the testes of the same mice used in the preparation of Fig. 3. The general pattern observed in the testis DNA is quite similar to that of spleen DNA, but differences were observed. Homologous bands marked *D* in

---

Fig. 2. A Comparison of *Hpa*II and *Msp*I Digestion Products of *Bam*HI cleaved *Drosophila* DNA.

Drosophila DNA was digested to completion with *Bam*HI, end-labelled and separated by electroelution. Fractions were then pooled and identical samples of each pool were digested to completion with *Hpa*II or *Msp*I. The digests were loaded into adjacent lanes of four separate slab gels with *Hpa*II digests on the left and *Msp*I digests on the right in each pair. Autoradiographs of the dried gels were used to produce the composite shown. Markers in the central lane of each panel are the same as shown in Fig. 1. Lettering is explained in the text.

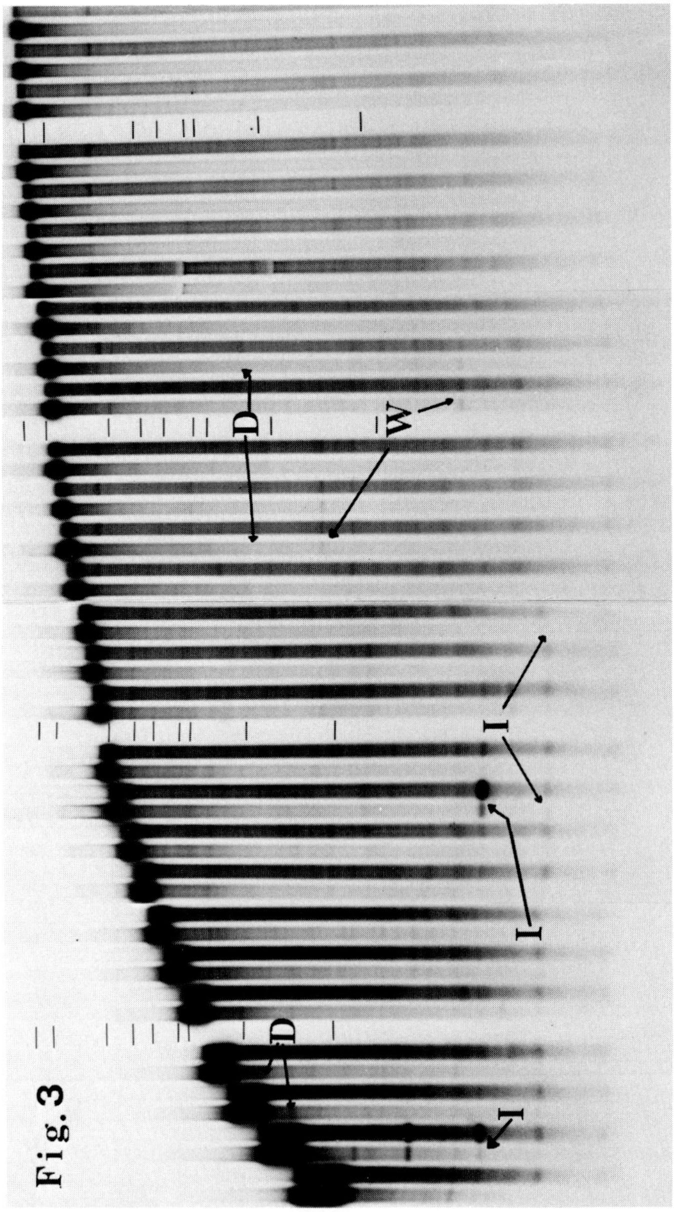
Fig. 3
D
W
I
I
D
I

Fig. 3 and Fig. 4 appear to show different levels of methylation at the CCGG site in the two tissues. We have repeated this experiment with DNA isolated from another litter of Balb/c mice. In this experiment, the patterns from testis DNA were quite similar; however, there were differences between spleen DNA patterns (data not shown). Thus, both tissue specific and individual differences appear to have been detected in the repeated DNA of the mouse.

DISCUSSION

The data shown above illustrate the utility of the two-dimensional display technique in the study of DNA methylation. The strength of the method is best shown by the *Drosophila* pattern in Fig. 2. About 1800 bands are present in the pattern. Since some bands spread over more than one primary fraction, there are perhaps only 1000 different sequences represented in the pattern. Many of these are multiply represented, like those of the histone gene cluster. If we take the average copy number for the bands we visualize in the array to be between 1 and 5, then we have probed 1000 to 5000 discrete CCGG sequences for methylation, and have found none. This greatly strengthens similar conclusions, using these enzymes, that were based on diffuse distributions of DNA segments in one-dimensional gels (21,22,29). Nevertheless, studies with antibodies to MeC indicate some antibody binding to polytene chromosomes from *Drosophila* larvae (29).

The results we have presented suggest that there is a pattern of methylation present in the repeated (non-satellite) DNA of this organism. Moreover, this pattern appears to vary somewhat between tissues and between individuals, although further experimentation will be required in order to determine the source of these variations.

The observations presented here are consistent with the existence of a stable, clonally heritable, pattern of MeC in the repeated sequences of the mouse. The apparent absence of methylation in *Drosophila*, on the other hand, argues against the universality of models (4,5) in which changes in cell commitment are explained in terms of methylation at cytosine.

Fig. 3. A Comparison of *HpaII* and *MspI* Digestion Products of *BamHI*-Cleaved Mouse Spleen DNA.

A two-dimensional array was constructed, essentially as described in the legend to Fig. 2, from the pooled spleen DNA of six Balb/c mouse littermates. Lettering is explained in the text.

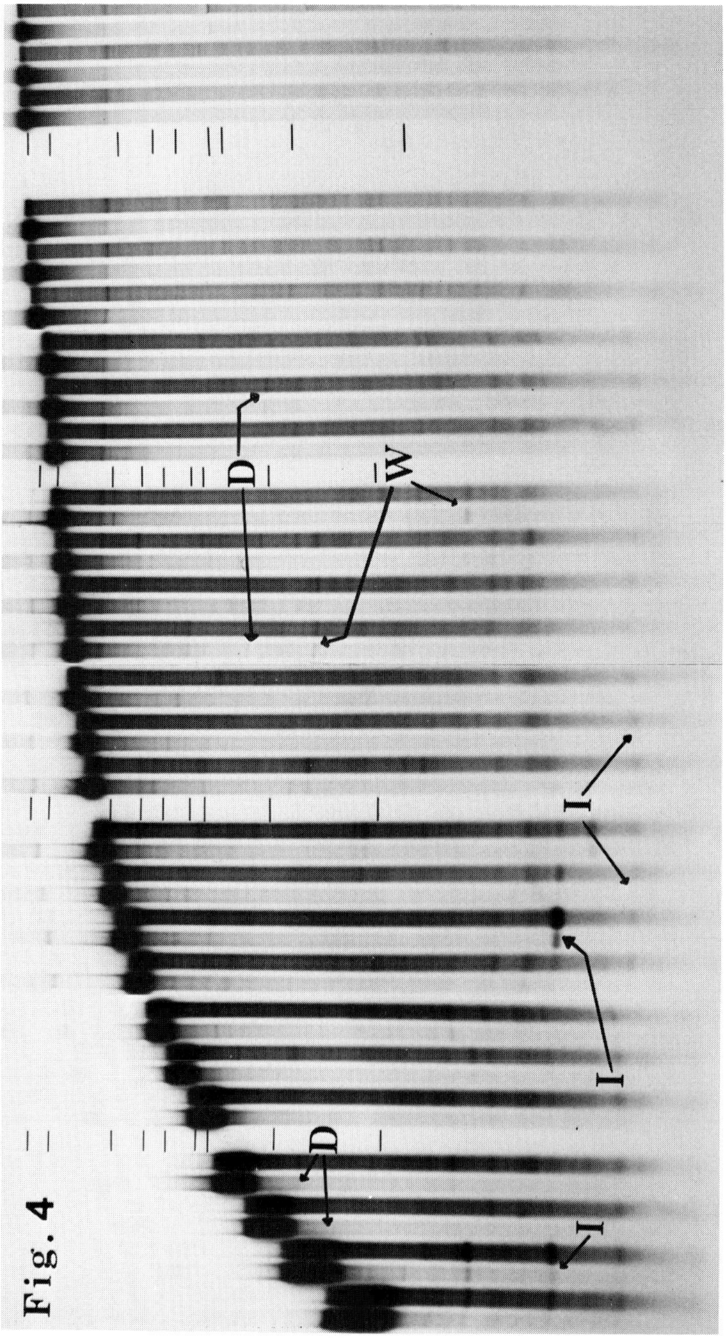
Fig. 4
D
W
I
I
D
I

This, however, does not preclude a role for DNA methylation in the hierarchy of control mechanisms involved in vertebrate gene regulation (30). Finally, it seems that the two-dimensional display procedure described above may prove useful in the future.

ACKNOWLEDGMENTS

This research was supported by grants from the NIH (GM 25531) and the NSF (PCM 78-16282). S.S.S. was the recipient of an NIH postdoctoral fellowship (GM 07065), and J.G.R. of an ACS postdoctoral fellowship (J-457-01). Special thanks to Tom Gilroy for communicating unpublished results, and to Dave Ratner for restricting and modifying the manuscript.

REFERENCES

1. Vanyushin, B.F., Tkacheva, S.D., and Belozersky, A.N. (1970) Nature 225, 948-950.
2. Taylor, J.H. (1979) In: Molecular Genetics, Part III, Chromosome Structure, J.H. Taylor Ed., Academic Press, New York, pp. 89-115.
3. Scarano, E. (1971) Adv. Cytopharmacol. 1, 13-24.
4. Riggs, A.D. (1975) Cytogenet. Cell. Genet. 14, 9-25.
5. Holliday, R. and Pugh, J.E. (1975) Science 187, 226-232.
6. Sager, R. and Kitchin, R. (1975) Science 189, 426-433.
7. McGhee, J.D. and Ginder, G.D. (1979) Nature 280, 419-420.
8. Van der Ploeg, L.H.T. and Flavell, R.A. (1980) Cell 19, 947-958.
9. Jones, P.A. and Taylor, S.M. (1980) Cell 20, 85-93.
10. Cohen, J.C. (1980) Cell 19, 653-662.
11. Sutter, D. and Doerfler, W. (1980) Proc. Nat. Acad. Sci. (USA) 77, 253-256.
12. Mohandas, T., Sparkes, R.S., and Shapiro, L.J. (1981) Science 211, 393-396.
13. Liskay, R.M. and Evans, R.J. (1980) Proc. Nat. Acad. Sci. (USA) 77, 4895-4898.
14. Brown, D.D. (1981) Science 211, 667-674.

Fig. 4. A Comparison of HpaII and MspI Digestion Products of BamHI-Cleaved Mouse Testis DNA.

The two-dimensional array was constructed from the pooled testis DNA of the same Balb/c littermates used to construct Fig. 3. Lettering is explained in the text.

15. Bird, A. and Taggart, M.H. (1980) Nucleic Acids. Res. 8, 1485-1497.
16. Bird, A.P. (1980) Nucleic Acids Res. 8, 1499-1504.
17. Potter, S.S. and Newbold, J.E. (1976) Analyt. Biochem. 71, 452-458.
18. Potter, S.S., Bott, K. and Newbold, J.E. (1977) J. Bact. 129, 492-500.
19. Potter, S.S. and Thomas, Jr., C.A. (1977) Cold Spring Harbor Symp. Quant. Biol. 42, 1023-1031.
20. Smith, S.S. and Thomas, Jr., C.A. (1981) Gene, in press.
21. Rae, P.M.M. and Steele, R.E. (1979) Nucleic Acids Res. 6, 2987-2995.
22. Cedar, H. et al. (1979) Nucleic Acids Res. 6, 2125-2132.
23. Singer, J., Roberts-Ems, J. and Riggs, A.D. (1979) Science 203, 1019-1021.
24. Singer, J. et al. (1979) Nucleic Acids Res. 7, 2369-2385.
25. Mann, M.B. and Smith, H.O. (1977) Nucleic Acids Res. 4, 4211-4221.
26. Sneider, T.W. (1980) Nucleic Acids Res. 8, 3829-3840.
27. Goldberg, M.L. (1979) Sequence Analysis of Drosophila Histone Genes. Ph.D. Dissertation submitted to Stanford University.
28. Hörz, W. and Altenburger, W. (1981) Nucleic Acids Res. 9, 683-696.
29. Eastman, E.M., Goodman, R.M., Erlanger, B.F. and Miller, O.J. (1980) Chromosoma (Berl.) 79, 225-239.
30. Razin, A. and Riggs, A.D. (1980) Science 210, 604-610.

A RAPID AND SENSITIVE IMMUNOLOGICAL METHOD FOR *IN SITU* GENE MAPPING[1]

Pennina R. Langer and David C. Ward

Department of Human Genetics
Yale University
New Haven, Connecticut

ABSTRACT A method for *in situ* localization of specific DNA sequences has been developed which exploits the interaction between modified nucleotides and antibodies directed against the modification. Cloned sequences of *Drosophila melanogaster* DNA are nick translated *in vitro* in the presence of *E. coli* DNA polymerase I and an analogue of dUTP which contains a biotin molecule covalently linked to the C-5 position of the pyrimidine ring. The nick translated probe, with approximately 1-5% of its nucleotides containing biotin, is hybridized *in situ*, according to standard protocols, to *Drosophila* salivary gland chromosomes. After hybridization, the slides are incubated with monospecific rabbit anti-biotin immunoglobulins followed by FITC-goat anti-rabbit IgG. After counterstaining with Evans Blue, fluorescent yellow-green bands, corresponding to the map location of the cloned DNA are seen against a red fluorescent background of the salivary gland chromosomes. Methods for refining the system for use in localization of unique sequences on mammalian metaphase chromosomes are also discussed.

INTRODUCTION

The mapping of genes or their transcripts to specific loci on chromosomes has proven to be a tedious and time consuming occupation, mainly involving techniques of cell-fusion and somatic cell genetics. Although *in situ* hybridization has been employed successfully for mapping single-copy gene

[1]This work was supported by United States Public Health Service Grants GM-20124 and CA-16038.

ISBN 0-12-137420-3

sequences in species that undergo chromosome polytenization, such as *Drosophila*, detection of unique sequence genes in most higher eukaryotic chromosomes has been extremely difficult (1). The necessity for polynucleotide probes of very high specific radioactivity to facilitate autoradiographic localization of the hybridization site results in rapid radiodecomposition of the probe and a concomitant increase in the background noise of silver grain deposition. The use of hybridization probes with low to moderate specific radioactivities requires exposure times which are often impractical. The objective of this study was to test the feasibility of a novel approach to *in situ* hybridization which circumvents the radiochemical and temporal limitations of the normal hybridization methods, and which offers the potential of rapid and specific analysis of single-copy sequences. This approach exploits the interaction between a modified nucleotide and antibodies directed against this modification.

Biotin has many features which make it an ideal modification probe (2). Davidson and his colleagues (3-5) took advantage of the biotin-avidin interaction by chemically crosslinking biotin to RNA via cytochrome C or polyamine bridges. Sites of hybridization were then detected in the electron microscope through the binding of avidin-ferritin or avidin-methacrylate spheres. Although successful in the specialized cases examined, this method has not proven to be of general utility. It is likely that this is due to the nonspecific interaction of avidin with DNA or chromatin which we and others (2,6,7) have observed. To circumvent this limitation we have developed immunological reagents for biotin detection which do not exhibit nonspecific binding to chromosomes and have used these in conjunction with biotin labeled nucleotides which can be enzymatically incorporated into a hybridization probe.

METHODS

Synthesis of Biotinyl-Deoxyuridine 5'-Triphosphate. An analogue of dUTP that contains a biotin molecule covalently bound to the C-5 position of the pyrimidine ring through an allylamine linker arm was synthesized according to the scheme illustrated in Figure 1. Allylamine-dUTP was prepared from 5-mercuri-dUTP (8) by reaction with a 10-fold molar excess of allylamine and one nucleotide equivalent of potassium tetrachloropalladium (K_2PdCl_4) in 0.25 M sodium acetate buffer, pH 5.0 for 18-24 hours. Similar palladium-catalyzed alkylation reactions have been used previously for the synthesis of a variety of C-5 substituted pyrimidine nucleoside compounds (9,10). The allylamine-dUTP product was purified by ion

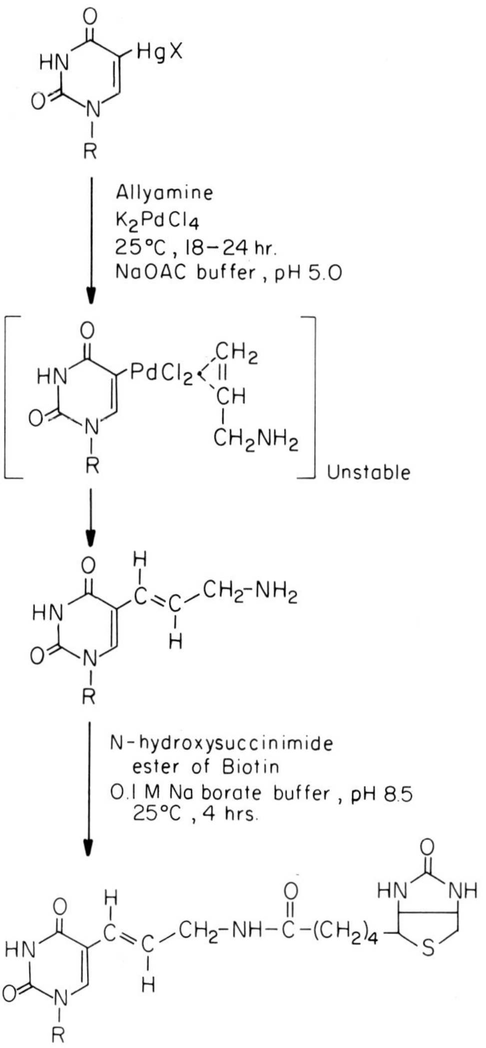

FIGURE 1. Reaction scheme used for synthesis of the biotinylallylamine-derivative of dUTP (Bio-dUTP).

exchange chromatography on DEAE-cellulose and by HPLC reverse phase chromatography on a Partisil ODS-2 support. The biotinylallylamine-derivative of dUTP (Bio-dUTP) was prepared by reaction with two equivalents of biotinyl-N-hydroxysuccimide ester in 0.1 M sodium borate buffer, pH 8.5, for 4 hours at room temperature and purified by DEAE-cellulose chromatography. Details on the synthesis, purification and characterization of Bio-dUTP will be presented elsewhere (11).

Preparation of Rabbit Anti-biotin Antibodies. Biotin-BSA, prepared as described (12), was used to raise antibody in rabbits. The immunization schedule was modified from that used by Berger (13) and is detailed elsewhere (manuscript in preparation). The animals were immunized initially with 1.2 ml of antigen (2 mg/ml biotin-BSA in 0.9% NaCl diluted with an equal volume of complete Freunds adjuvant) and boosted every 14 days with 1 ml of 0.9% NaCl containing 2 mg biotin-BSA. Within 6 weeks after immunization, the animals were producing anti-biotin antibodies as determined by ouchterlony immunodiffusion. Beginning at this time, 30-40 cc of blood was collected 10-11 days after each boost. The blood was incubated at 37°C for 1-2 hours, and then refrigerated at 4°C for a minimum of 4 hours. Serum was separated from clotted material by centrifugation and stored at -20°C.

In Situ Hybridization. Cloned *Drosophila* DNA, nick translated in the presence of ^{3}H dATP and either TTP or Bio-dUTP, was hybridized to *Drosophila* polytene chromosomes as described (14). The acetylation step described by Hayashi *et al.* (15) was included before denaturation of the chromosomes (personal communication, M. Pardue). After hybridization, non-specifically bound material was removed by washing in 2X SSC at 60°C and then at room temperature. The slides were then rinsed in PBS and incubated with rabbit anti-biotin (2.5 μg/ml in PBS + 10 mg/ml BSA) at 37°C. The slides were again rinsed in PBS and the chromosome spreads were incubated with FITC-goat anti-rabbit IgG (Miles, diluted 1:100 in PBS + 10 mg/ml BSA). After rinsing in PBS, the chromosomes were counterstained for 2 minutes with Evans Blue (0.5% w/v in PBS + 1% fetal calf serum), rinsed in PBS, mounted in 0.1 M Tris pH 8.0, 90% glycerol, and viewed with an epi-illuminated phase-fluorescent microscope.

RESULTS

Preparation of Biotin-substituted Hybridization Probes. Bio-dUTP can replace TTP as a substrate for a variety of DNA polymerases of both prokaryotic and eukaryotic origin

(11). Similarily, the ribonucleotide analogue, Bio-UTP, can substitute for UTP in reactions catalyzed by the RNA polymerases of *E. coli* and bacteriophage T7 (11). Thus, biotinyl-DNA and biotinyl-cRNA hybridization probes can be synthesized conveniently and rapidly by straightforward enzymatic reactions. The substrate properties of Bio-dUTP with *E. coli* DNA polymerase I, using either the nick-translation protocol of Rigby et al. (16) or the gap-filling reaction of Bourguignon et al. (17), are illustrated in Figure 2. Although the

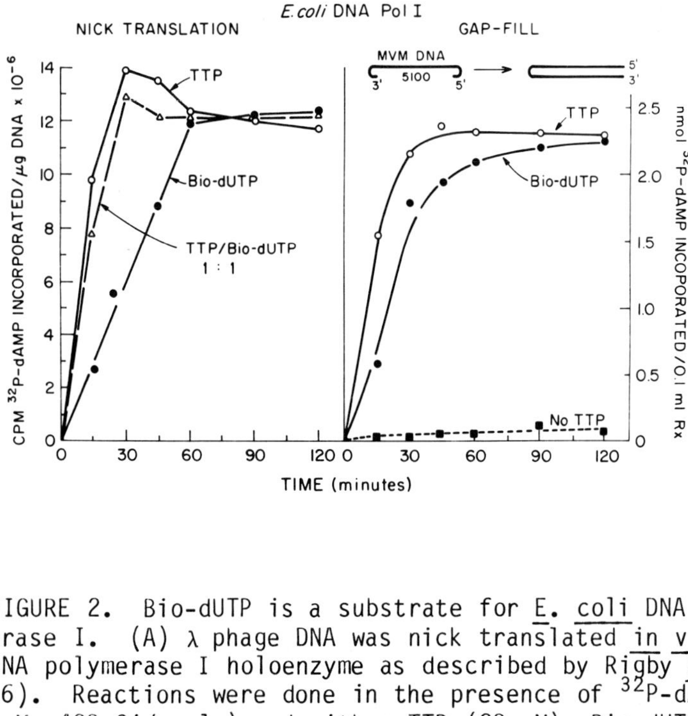

FIGURE 2. Bio-dUTP is a substrate for *E. coli* DNA polypolymerase I. (A) λ phage DNA was nick translated *in vitro* with DNA polymerase I holoenzyme as described by Rigby *et al.* (16). Reactions were done in the presence of ^{32}P-dATP (1.35 µM, 400 Ci/mmole) and either TTP (20 µM), Bio-dUTP (20 µM), or an equal concentration of TTP and Bio-dUTP (10 µM each). (B) Minute Virus of Mice (MVM) DNA, a 5 Kb single-stranded DNA molecule with terminal hairpin duplexes (20), was converted to double-stranded (RF) DNA by reaction with DNA polymerase I (Klenow fragment) as described by Bourguignon *et al.* (17). The three nucleotide reaction contained only 0.1 mM of dCTP, dGTP and ^{32}P-dATP (50 µCi/µmole). TTP or Bio-dUTP reactions were supplemented with the appropriate triphosphate at a final concentration of 0.1 mM.

analogue is incorporated at initial rates which are only 30-40% that of the control (TTP-containing) reactions, the final specific activities (and the extent of polymerization) that can be achieved are essentially the same.

The T_m of biotinated polynucleotide duplexes decreases as the biotin-content of the polymer increases (11). However, DNA in which every TMP residue in one strand is replaced by a Bio-dUMP residue (e.g., Bio-MVM RF DNA in Figure 2), has a T_m value that is only 5°C less than that of unsubstituted control DNA. Furthermore, nick translated DNA probes that have between 0.2 and 2.0% of their total nucleotides biotinized exhibit reassociation rates that are essentially the same as those observed with biotin-free DNA (11). Since a substantial number of biotin molecules can be introduced into a polynucleotide without significantly altering its hybridization characteristics, the same in situ hybridization conditions have been used for both the control and biotin-substituted DNA probes employed in the experiments described below.

Purification of Anti-biotin Antibodies by Affinity Chromatography. Biotin specific antibodies from immune serum (see Methods) were purified by affinity chromatography on ovalbumin sepharose and biotin-ovalbumin sepharose. The resins were prepared by coupling ovalbumin or biotin-ovalbumin (12) to cyanogen bromide activated sepharose 4B (18). Serum was loaded onto ovalbumin sepharose to remove any ovalbumin binding component. The flow through from.this column was loaded directly onto biotin-ovalbumin sepharose. This resin was washed extensively with phosphate-buffered saline (PBS) until the flow through contained no detectable protein. Protein specifically bound to the resin was then eluted with 3 M KSCN in PBS, concentrated in an amicon filter unit, and desalted on a Sephadex G-25 column equilibrated in PBS. The antibody recovered in the void volume was adjusted to a concentration of ≈ 0.5 mg/ml in PBS with 5-10 mg/ml BSA, and stored at -20°C.

This antibody was assayed for specificity against biotin-labeled DNA by immunoprecipitation with Staph aureus (19). DNA, nick translated with α^{32}P-dATP and either TTP or Bio-dUTP, was used in these assays. When unmodified DNA was incubated in the absence of serum, with nonimmune serum, or with rabbit anti-biotin the ^{32}P counts remained in the supernatant. Bio-DNA, either in the absence of serum or in the presence of nonimmune rabbit serum also remained in the supernatant. However, Bio-DNA was immunoprecipitated with affinity purified rabbit anti-biotin in the presence of Staph A protein. The percentage of the total Bio-DNA radioactivity found in the immune precipitate was dependent upon the antibody concentration and the time of incubation with the antigen. Under optimum conditions, greater than 90% of the Bio-DNA is immunoprecipitable.

In Situ Hybridization. A variety of cloned *Drosophila* sequences were used to test the specificity and sensitivity of the immunological method of hybrid detection. Results obtained with two specific clones, designated pPW 539 and pAC 104, will serve to illustrate the potential advantages of this procedure over conventional autoradiographic detection. Clone pPW 539, obtained from Otto Schmidt (Yale University), contains a 22 Kb fragment encoding a methionine tRNA gene inserted into a pMB9 plasmid (S. Sharp *et al.*, manuscript in preparation). This sequence is known to map to band 61D on chromosome 3L. Clone pAC 104, obtained from V. Pirotta (EMBO), contains 6 Kb of a transposition element inserted into the plasmid pAcyc 184 (V. Pirotta, personal communication). This cloned fragment maps to numerous loci throughout the Drosophila genome. The probes were nick translated *in vitro* with ^{3}H dATP and either TTP or Bio-dUTP. Kinetics of Bio-dUMP and TMP incorporation were similar in both cases to those illustrated in Figure 2. These probes were then hybridized to chromosomes *in situ* for 10-12 hours at 65°C (14,15). After removal of nonspecifically bound material, the slides were washed and incubated with antibodies as described in the methods section. A reasonable fluorescent signal was obtained with 1 hour incubations of both antibodies, but the optimum time for a strong signal was determined to be 4 hours for the first antibody and 2 hours for the FITC-goat anti-rabbit IgG. Figures 3 and 4 are sets of phase, immunofluorescent, and autoradiographic exposures of the same chromosome spreads.

Figure 3, showing hybridization of a unique *Drosophila* sequence (clone pPW 539) establishes the clarity and specificity of the signal obtained by immunofluorescence. Hybridizations using ^{3}H labeled -- but Biotin-free -- DNA followed by antibody staining, showed no detectable fluorescent signal even though autoradiographic exposures indicated that accurate and efficient hybridization had occurred.

Figure 4, illustrating hybridization of the clone pAC 104, shows the tremendous increase in resolving power offered by this technique. This is easily seen by a comparison of regions indicated by the arrows in the fluorescent and autoradiographic exposures. The number of bands at these sites are difficult to determine from the autoradiogram, however, they are easily counted in the immunofluorescent picture.

Overall, the immunological method of hybrid localization offers four immediate advantages over autoradiography. These include a decrease in the time required for localization,

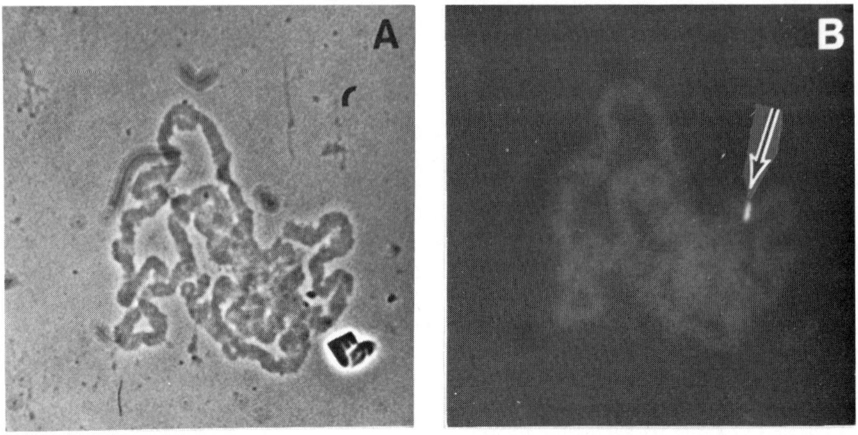

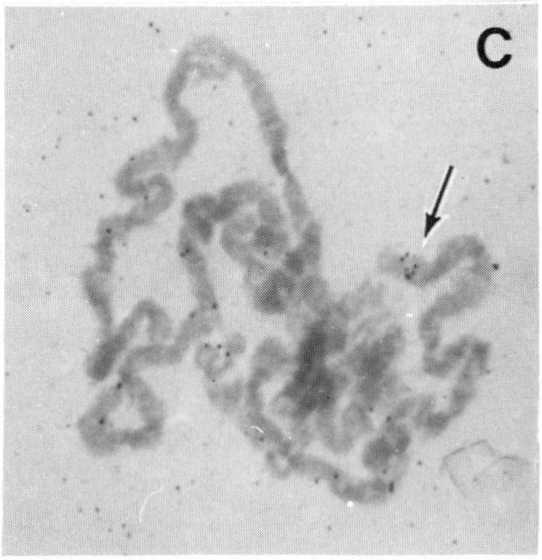

FIGURE 3. Results of *in situ* hybridization using clone pPW 539 DNA that contained both ^{3}H-dAMP (8.8×10^{5} cpm/μg) and Bio-dUMP (7% of TMP residues substituted). (A) phase picture, (B) immunofluorescence picture, (C) autoradiographic exposure (9 days).

FIGURE 4. Results of *in situ* hybridization using clone pAC 104 DNA that contained both ^{3}H-dAMP (5×10^{6} cpm/μg) and Bio-dUMP (17% of TMP residues substituted). (A) phase picture, (B) Immunofluorescence picture, (C) autoradiographic exposure (11 days).

improved resolution, a lower background noise and the ability to prepare stable hybridization probes which are not subject to the decomposition problems found with radiolabeled probes.

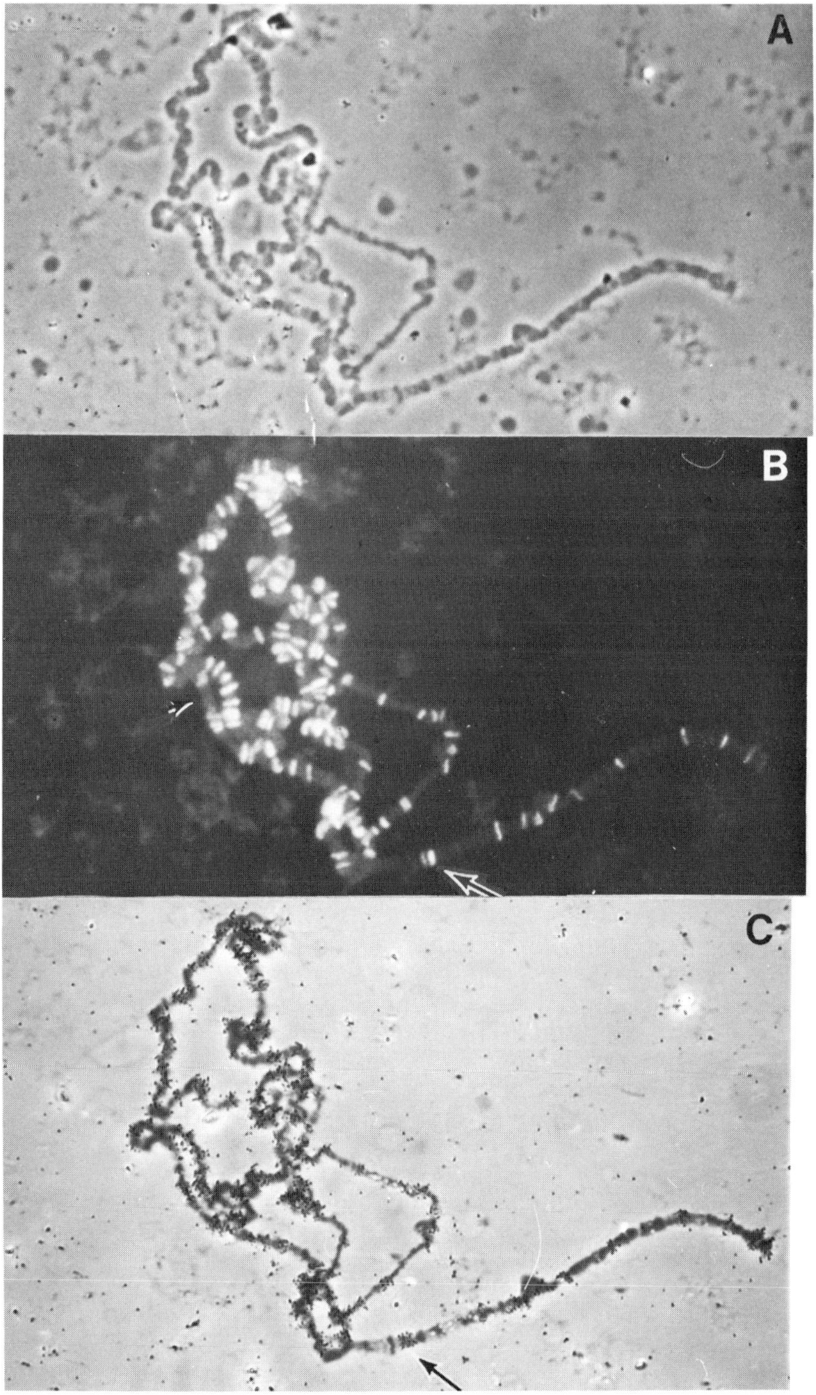
A
B
C

DISCUSSION

The results outlined above demonstrate that the interaction between biotinyl-nucleotides and anti-biotin antibodies can be used effectively for the _in situ_ localization of DNA sequences within the _Drosophila polytene_ chromosome. The fluorescent bands observed in Figures 3 and 4 represent the signal produced by about 100-200 copies of biotinized DNA probes that can hybridize to 6-22 Kb of _Drosophila_ DNA. This estimate comes from the polytenization number of 500 and a hybridization efficiency to alkali-denatured chromosomes of 20-40% (21). Additional studies (Langer and Ward, manuscript in preparation) have shown that bands with fluorescent signals only 5-10% as strong as those illustrated here can be visualized and photographically recorded. The latter observation suggests that the simple "antibody-sandwich" method described here is sensitive enough to detect a chromosomal locus which hybridizes 10-20 copies of a sequence a few kilobases in size or 100-200 copies of a sequence a few hundred nucleotides long.

This immunological method of hybrid detection is not restricted to _Drosophila_ polytene chromosomes since we have used it successfully to determine the chromosomal loci of reitterated sequences, e.g., satellite DNA's, on mammalian metaphase chromosomes (Langer and Ward, manuscript in preparation). In addition, Bio-cRNA probes have been used in conjunction with anti-biotin antibodies to localize the histone gene clusters on the lampbrush chromosome of the Newt, Notophthalmus (J. Gall, personal communication). Nevertheless, further refinements of the basic method will be required before single-copy sequences can be visualized on mammalian metaphase chromosomes. Fortunately, there are a variety of approaches that can be used to either increase the fluorescent signal generated by the probe or to enhance the sensitivity of signal detection. For example, by using a "haptene-antibody sandwich" technique (22) the fluorescent signal can be enhanced by up to 50-fold over that achieved by the standard double-antibody method (23-24). Another approach to signal amplification is to treat the chromosomes alternatively with streptavidin and a biotinated, FITC-labeled "carrier" protein. Streptavidin is a tetrameric protein produced by _Streptomyces avidinii_ that binds four biotin molecules, each with a K_{dis} of 10^{-15} (25). Since this protein, unlike avidin, does not exhibit non-specific binding to chromosomes (A. Chi, P. Langer and D. Ward, unpublished), one can hopefully exploit the high affinity of the streptavidin-biotin interaction to deliver many layers of the fluorescent "carrier" protein rapidly and specifically.

One of the practical problems with weak fluorescent signals is that they often bleach rapidly, thus making it difficult to record the signal photographically. This limitation can be circumvented, however, by using low-light intensity illumination in conjunction with electronic image intensification. Alternatively, one can employ histochemical detection by coupling the primary or secondary antibody to enzymes, such as peroxidase, alkaline phosphatase, and β-galactosidase (26). The enzymatic conversion of soluble substrates to insoluble colored precipitates at the site of hybridization would also permit visualization by standard light microscopy. We are currently exploring a variety of such procedures to increase the sensitivity of probe detection with the objective of developing a protocol whereby single-copy sequences in mammalian chromosomes can be mapped with speed and precision.

REFERENCES

1. Steffensen, D.M. (1977). In "Molecular Structures of Human Chromosomes" (J.J. Yunis, ed.), p. 59. Academic Press.
2. Bayer, E.A., and Wilchek, M. (1980). Methods of Biochem. Anal. 26, 1.
3. Manning, J.E., Hershey, N.D., Broker, T.R., Pellegrini, M., Mitchell, H.K., and Davidson, N. (1975). Chromosoma (Berlin) 53, 107.
4. Broker, T.R., Angerer, L.M., Yen, P.H., Hershey, N.D., and Davidson, N. (1978). Nucleic Acid Res. 5, 363.
5. Sodja, A., and Davidson, N. (1978). Nucleic Acid Res. 5, 385.
6. Heggeness, M.H. (1977). Stain Technol. 52, 165.
7. Heggeness, M.H., and Ash, J.F. (1977). J. Cell Biol. 73, 783.
8. Dale, R.M.K., Martin, E., Livingston, D.C., and Ward, D.C. (1975). Biochemistry 14, 2447.
9. Bergstrom, D.E., and Ogawa, M.K. (1978). J. Am. Chem. Soc. 100, 8106.
10. Bigge, C.F., Kalaritis, P., Deck, J.R., and Mertes, M.P. (1980). J. Am. Chem. Soc. 102, 2033.
11. Langer, P.R., Waldrop, A.A., and Ward, D.C. (1981). Proc. Natl. Acad. Sci. U.S.A., in press.
12. Heitzmann, H., and Richards, F.M. (1974). Proc. Natl. Acad. Sci. 71, 3537.
13. Berger, M. (1979). Methods in Enzymology 62, 319.
14. Pardue, M.L., and Gall, J. (1975). Methods in Cell Biol. 10, 1.
15. Hayashi, S., Gillam, I.C., Delaney, A.D., and Tener, G.M. (1978). J. Histochem. Cytochem. 26, 677.

16. Ribgy, P.W.J., Dieckmann, M., Rhodes, C., and Berg, P. (1977). J. Mol. Biol. 113, 237.
17. Bourguignon, G.J., Tattersall, P.J., and Ward, D.C. (1976). J. Virol. 20, 290.
18. Porath, J., Axen, R., and Ernback, S. (1967). Nature 215, 1491.
19. Kessler, S.W. (1975). J. Immunol. 115, 1617.
20. Chow, M.B., and Ward, D.C. (1978). In "Replication of Mammalian Parvoviruses" (D.C. Ward and P. Tattersall, eds.), p. 205. Cold Spring Harbor Laboratory.
21. Szabo, P., Elder, R., Steffensen, D.M., and Uhlenbeck, O.C. (1977). J. Mol. Biol. 115, 539.
22. Lamm, M.E., Koo, G.C., Stackpole, C.W., and Hämmerling, U. (1972). Proc. Natl. Acad. Sci. U.S.A. 69, 3732.
23. Cammisuli, S., and Wofsy, L. (1976). J. Immunol. 117, 1695.
24. Wallace, E.F., and Wofsy, L. (1979). J. Immunol. Method. 25, 283.
25. Green, N.M. (1975). Adv. Protein Chem. 29, 85.
26. Nakane, P.K. (1975). Ann. N.Y. Acad. Sci. 254, 203 (and references therein).

A STRATEGY FOR HIGH-SPEED DNA SEQUENCING

Joachim Messing

Department of Biochemistry
University of Minnesota
St. Paul, Minnesota

Peter H. Seeburg

Division of Molecular Biology
Genentech, Inc.
South San Francisco, California

I. HIGH-SPEED DNA SEQUENCING

Two DNA sequencing methods have been developed recently, one which involves chemical modification and cleavage of DNA (Maxam & Gilbert, 1977), and another which involves *in vitro* DNA synthesis using chain terminators (Sanger *et al.*, 1977). Although 400 nucleotides can be read off a sequencing gel from a single reaction (Sanger & Coulson, 1978), these methods alone do not allow the rapid sequencing of long sequences. The time consuming steps in DNA sequencing are not the sequencing reactions but the prior steps: the isolation of single DNA fragments in small 400 nucleotide long pieces.

To obtain these fragments for the chemical method (Maxam & Gilbert, 1977) the DNA is cut by restriction endonucleases into smaller fragments, labeled at their ends with radioactive ^{32}P, cut with a second restriction endonuclease and the differently sized fragments purified by gel electrophoresis. Two problems arise. The number of DNA fragments generated by restriction endonucleases increases with the length of the DNA. Consequently the purification of DNA fragments by gel electrophoresis becomes impossible. In addition, for proofreading both strands have to be sequenced. If the two strands

ISBN 0-12-137420-3

are separated by gel electrophoresis (Maxam & Gilbert, 1977) the purification of end labeled DNA can only be accomplished with a lcss complex set of fragments. Alternatively one could pursue a sequential cutting pattern and reduce the complexity of fragments stepwise. This requires more time, a larger quantity of material, and more restriction endonuclease.

The dideoxy method (Sanger et al., 1977) follows a similar scheme. Purified fragments, however, do not have to be labeled and are used as a primer or as template. Both are needed, but either the primer or the template has to be purified single-stranded DNA. Using these previous procedures the chemical method is more convenient and, therefore, was more widely used.

To understand eukaryotic genome organization and eventually function, the knowledge of the primary structure of long stretches of DNA is essential. With cloning systems like the lambda phage vectors and the lambda packaging system (Sternberg et al., 1977) eukaryotic DNA has been dissembled into libraries of 10^6 to 10^7 phages per species (Maniatis et al., 1978). One recombinant phage, however, still contains up to 18 kb of foreign DNA. Depending on how much of the sequence of this DNA needs to be determined up to 18 kb have to be further broken down into small pieces of DNA of around 400 bp. The resulting large number of DNA fragments are difficult to handle in the approach so far described.

II. A CLONING SYSTEM FOR DNA SEQUENCING

Our approach to the problem of sequencing long stretches of DNA involves a shotgun scheme. The DNA is cleaved into a set of overlapping fragments which are cloned into M13mp7 replicative form (RF) and recombinants are sequenced randomly. Such a shotgun cloning approach is similar to the generation of genomic libraries. The genomic DNA is cleaved by partial digestion with a restriction endonuclease into fragments of a length of 18 kb and shotgun cloned into a lambda vector. The lambda library contains overlapping fragments of the total genomic DNA and the physical location of an individual fragment is determined by common sequences of overlapping fragments.

Although cloning can help to sort out the large number of restriction fragments, the separation of the DNA strands remain a problem. Fortunately both can be combined since it has been shown that single-stranded DNA phages can be used to clone DNA (Messing et al., 1977). The replicative double-stranded form (RF) of the phage is linearized with a restriction endonuclease. A DNA fragment to be cloned is joined to linearized RF molecules with DNA ligase, and the product is used to transform suitable host cells. Transformed cells amplify the double-stranded recombinant DNA and late in cell growth the gene V product selectively binds to the (+) strand of the phage DNA before it is packaged and extruded from the cells without lysis. The mode of replication (Staudenbauer et al., 1978) and the selective binding of gene V product results in the separation of any DNA into its complementary strands. Since a given fragment may be incorporated into the RF in two possible orientations, both strands can be easily obtained for sequencing. Furthermore only the viral single-stranded DNA (SS) is extruded from host cells and all other nucleic acids are easily removed with the intact host cells. Therefore this cloning procedure allows rapid DNA purification. In summary, fragment families of a larger piece of eukaryotic DNA can be stored in a single-stranded DNA phage vector system and individiual templates of a fragment family can be readily prepared by a simple purification procedure.

III. CLONING AS AN INTEGRATIVE PART IN DNA SEQUENCING

The dideoxy method allows either to vary the primer and keep the template invariant or vice versa. Since in the single-stranded phage vector system the template is readily purified, it is of advantage to keep the primer constant. This has led to the construction of a master primer: first a 96 bp Eco RI fragment (Heidecker et al., 1980) and later a synthetic 15 mer (Messing et al., 1981) which is now commercially available. Thus all the components of the sequencing reaction are invariant. This includes the radioactive label, the deoxynucleotidetriphosphates, the buffers, the Klenow fragment of DNA polymerase, the primer, the dideoxynucleotidetriphosphates. The template variation is created by shotgun cloning of the fragment families into the phage vector. Since this step is the major time consuming part in the procedure, the phage vector system has been tailored for these needs.

One useful character of a cloning system is to have a rapid detection test of the successful integration of a piece of DNA into the vector. Therefore a part of the *E. coli lac* operon had been inserted into M13RF DNA and an Eco RI cloning site created in the region encoding the N-terminus of the β-galactosidase (Messing et al., 1977; Gronenborn & Messing, 1978). The described master primer has been constructed for this Eco RI cloning site. Using Eco RI linkers any DNA fragment can be shotgun cloned into the Eco RI site of M13mp2 and sequenced with the primer (Heidecker et al., 1980). The linker step, however would interfere with a rapid cloning component in an integrative sequencing strategy. Therefore a DNA sequence encoding a new N-terminus of β-galactosidase has been engineered with synthetic DNA and the aid of DNA repair to give rise of a new structural gene encoding a modified N-terminus of β-galactosidase. The additional codons contain new cloning sites within the Eco RI site, but do not destroy gene function. The new phage M13mp7 is a vector allowing the direct cloning of different DNA fragment families without having to use the Eco RI linkers. Since all the cloning sites are contained within the Eco RI site the same master primer can be used for all the various recombinant phages (Messing et al., 1981).

IV. THE STRATEGY

The cloning sites in M13mp7 important for the shotgun cloning of large fragment families are those which either allow the blunt end cloning of DNA fragments obtained by different means than the use of restriction endonucleases or the direct cloning of restriction fragments. Out of the 16 possibilities for a 4-nucleotide long sequence, eight are available for direct cloning using the M13mp7 system.

As shown in Table I Hinc II cleaved M13mp7 can be used to clone Alu I, Rsa I, Fnu DII, and Hae III fragments. The Acc I ends of M13mp7 are "sticky" to DNA fragments with Taq I, Msp I and Sci NI ends. Sau 3A fragments fit into the Bam HI cloning site of M13mp7. Although some of the cloning sites cannot be recut after insertion of the described fragments, Sau 3A clones can be cut out by flanking Eco RI sites, and all other clones can have the insert excised either by Eco RI or Bam HI (Messing et al., 1981). This offers the potential of turning an insert around to obtain the sequence of the complementary strand. This is not possible for Eco RI' fragments. The specificity of Eco RI' activity as well as its potential to include a large fragment family is described elsewhere (Gardner, Howarth, Hahn, Shepherd & Messing, manuscript in preparation). DNA cleaved by mechanical forces or

M13mp7 Cloning Site	Shotgun Fragment Family	Possible Sites within Fragment Family
Hind II GTC↓GAC	Alu I AG↓CT	Hind III A↓AGCTT Pvu II C AG↓CTT Sac I G AGCT↓C
	Rsa I GA↓TC	Rru I A GT↓ACT Kpn I G GTAC↓C
	Fnu DII CG↓CG	Sac II C CGC↓GG
	Hae III GG↓CC	Stu I A GG↓CCT Xma III C↓GGCCG Bal I A GG↓CCA Hae I (4) A GG↓CCA T GG↓CCT Stu I, Bal I
Acc I GT↓CGAC	Taq I T↓CGA	Cla I A T↓CGAT Xho I C↓TCGAG Sal I G↓TCGAC Asu II T T↓CGAA
	Msp I C↓CGG	Sma I C CC↓GGG Xma I C↓CCGGG Cau II (2) C C↓CGG C C↓GGG
	Sci NI G↓CGC	Hae II (4) A GCGC↓C A GCGC↓T G GCGC↓C G GCGC↓T Mst I T GCGC A
Bam HI G↓GATCC	Sau 3 A ↓GATC	Bgl II A↓GATCT Pvu I C GAT↓CG Bam HI G↓GATCC Bcl I T↓GATCC Xho II (4) G↓GATCT A↓GATCC Bam HI Bgl II
Eco RI G↓AATTC	Eco RI′ N↓AATTC G↓AATTN G↓ANTTC G↓AANTC G↓AATNC	— —

Table I. Restriction fragment families of the M13mp7 cloning system.

other nucleases can be treated with the enzyme Bal 31 in order to be blunt end cloned into the Hinc II site of M13mp7 (Messing et al., 1981).

Thus cloning of DNA fragments produced by the high number of todays known restriction endonucleases into a corresponding cloning site of a vector has been solved by a relatively small modification of the lac region in M13mp7.

Whatever means is used to produce the particular fragment family the flow of the shotgun DNA sequencing is as follows (Fig. 1). M13mp7 RF is cleaved with a restriction endonuclease to produce sticky or blunt ends as needed to clone the particular fragment family (1). The fragment family is added to the RF and the ligation initiated (2). Competent host cells (JM 103) are prepared, the ligation mix added and the cells heat shocked (3). Cells are plated in the presence of IPTG, Xgal, and exponentially growing cells. They are then incubated overnight at 37°C (4). Colorless plaques are picked, and the infected cells are cultured individually in 1.5 ml of growth medium (5). The phages are isolated and the single-stranded DNA (SS) is extracted (6). The templates are annealed with the master primer (7). For each template/primer mix 4 DNA synthesis reactions are done, each with one of the 4 chain terminators, ddG, ddA, ddT or ddC (8). The reactions are analysed by polyacrylamide gel electrophoresis under denaturing conditions. Reactions are loaded in the order GATC. After the run the gel is exposed to an x-ray film (9). The DNA sequence is read off of the autoradiogram and the data processed by a computer assisted program (Staden, 1980) (10).

V. RANDOM CLEAVAGE OF DNA

Following this flow of steps fragment families from long stretches of DNA can be rapidly sequenced in M13mp7. To reconstruct the original DNA sequence a program is needed to align single sequences for complementarities and overlaps. Therefore the most randomly generated template bank involves the minimal sequencing load for reconstructing the original DNA sequence. The same principle applies for shotgun cloning to create genomic libraries. Two avenues can be used to enhance random cloning.

Mechanical shearing or the use of "non-specific" endonucleases can be used to reduce large DNA into small fragments. DNA fragments of less than 500 bp in length should

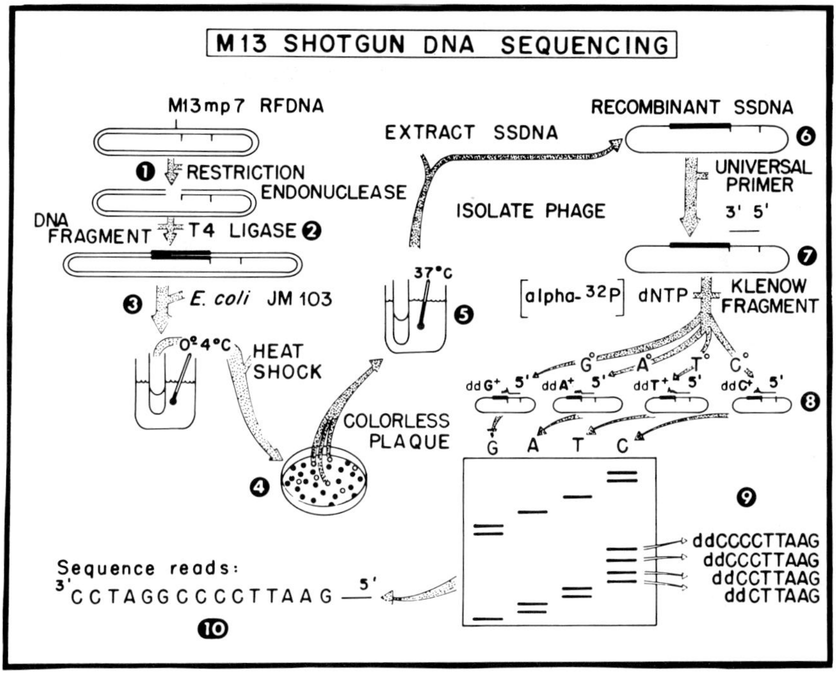

Fig. 1. The scheme of shotgun DNA sequencing.

not be included in the ligation to the M13 vector and are conveniently separated from bigger and smaller fragments by acrylamide gel electrophoresis. To avoid ligation of two fragments to the M13 vector the 5' phosphate termini are removed from the DNA fragments by alkaline phosphatase (Seeburg et al., 1977). Thus a maximum of sequence information per template is ensured and complications are prevented arising from recombinant phage carrying two or more DNA fragments from different regions in the original DNA. The flow step is as follows:

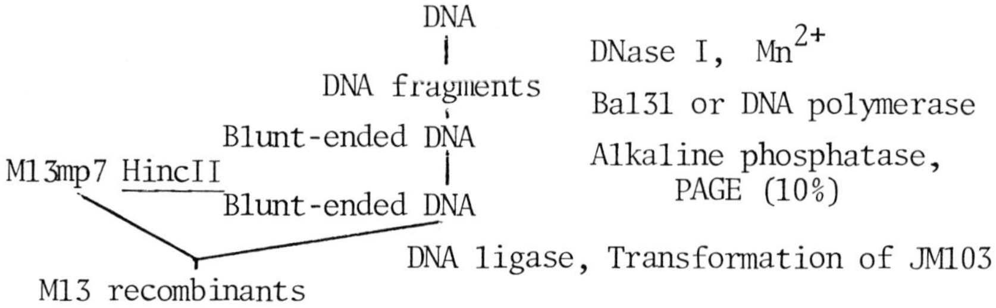

The DNA, probably in most cases a double-stranded circular molecule consisting of a plasmid vector and the insert is treated with DNase I in the presence of Mn^{2+} (Ehrlich *et al.*, 1973). After removal of the nuclease by phenol extraction the DNA fragments are run through a 10% polyacrylamide gel. By comparison to a size marker fragments ranging from 500-1000 bp are excised and eluted from the gel. The ends of the purified fragments are blunt-ended by treatment with DNA polymerase from *E. coli* or with nuclease *Bal 31* (Legerski *et al.*, 1978). Consecutive steps are the same as in Fig. 1, steps 5-10.

Fig. 2 shows the application of such a procedure for a pBR 322 recombinant containing about 8 kb of foreign DNA. If colorless plaques are picked randomly and gridded on two nitrocellulase filters they are screened readily by using a radioactive probe made from 100 ng of pBR322 or insert DNA. As expected about 1/3 of the recombinants light up with the pBR322 probe (Fig. 2a). The sequence of such a clone is determined by using a synthetic primer (Messing *et al.*, 1981) (Fig. 2c).

The second avenue is based on using a number of different restriction endonucleases to fragment the DNA to be sequenced. Many enzymes are available with different cleavage characteristics. Therefore the combinatorial strategy of 9 different enzymes (Table I) may have a potential of high randomness. Using the restriction endonuclease approach, fragments categorized into their different families can be directly cloned into their appropriate sites in M13mp7. To prevent the cloning of many small restriction fragments DNA is cleaved only partially. Using the appropriate cloning conditions which avoid the insertion of two fragments into one vector molecule, partial cleavage gives the additional advantage that linkage groups within one fragment family may be established independent from overlaps with a second fragment family. A problem, of course, may be the large size distribution generated by a partial digestion, especially with DNA

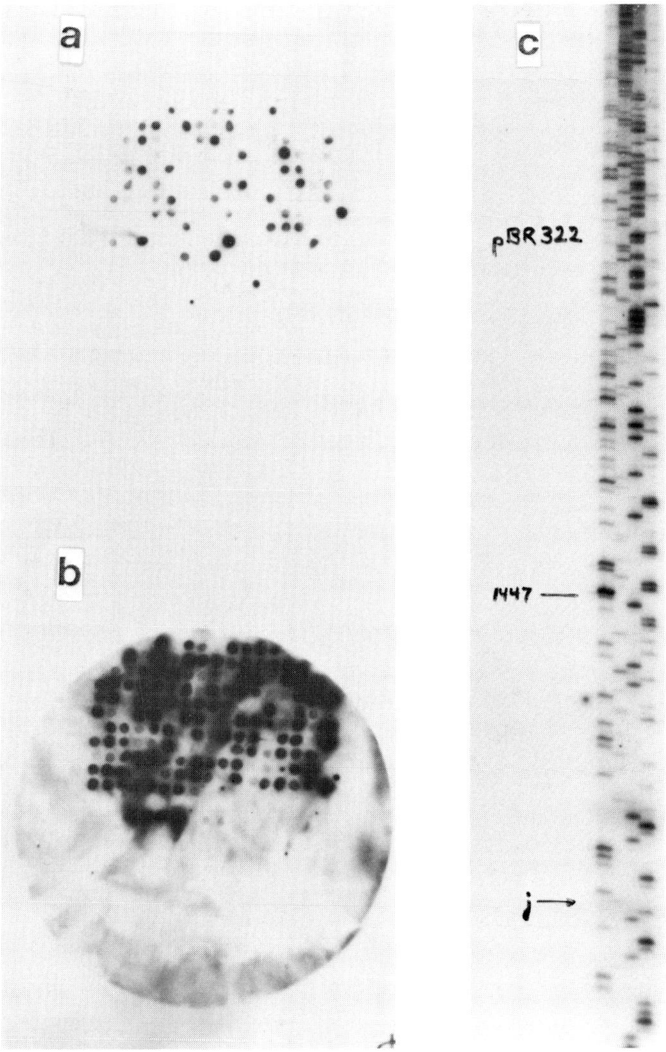

Fig. 2. Shotgun cloning of DNAase I fragments into M13mp7: a) Hybridization assay of recombinants with pBR322 as a probe; b) Hybridization assay of recombinants with insert as a probe; c) Sequence analysis of a random picked clone.

of 10 kb size or larger. To circumvent this the partially cleaved DNA is digested to completion with a second enzyme which recognizes sites of a higher specificity. Because of the higher specificity, the larger fragments of the partial digest are converted into smaller fragments. The second enzyme, however, has to produce the same type of ends as the first one to allow cloning of all fragments into the same cloning site. For instance, the enzyme Xho II produces the same ends as Sau 3A, but recognizes a nucleotide sequence that occurs less frequently (Table I).

An important feature of this procedure is to use fragments from many families. It may be difficult to determine the number of templates of a particular fragment family which should be sequenced before switching to another fragment family. The uniqueness of a nucleotide sequence has to be explored during the sequencing procedure itself. It may be the best to sequence a given fragment family as long as new regions can be discovered readily, and then to switch to the next one.

VI. THE PLASMID VECTOR IN THE SHOTGUN SEQUENCING APPROACH

Once a larger piece of DNA has been chosen to be sequenced it may be useful to transfer it from the lambda vector to a plasmid vector; larger amounts of the insert for sequencing can be obtained because of the higher insert to vector ratio. A possible next step is to separate the inserted DNA from its vector. Cross contamination, however, can still occur. In addition, in order to obtain complete restriction fragment families the ends of the purified fragment may have to be ligated to themselves to ensure that ends are included.

An alternative approach is to obtain the restriction fragment families from the entire plasmid. Hybridization techniques can then be used to sort out the appropriate templates (Fig. 2).

ACKNOWLEDGMENTS

We are grateful to Robert Swanson and the Minnesota Experiment Station for supporting this project. We thank Richard Gelinas, Steve Anderson and Shirley Halling for their helpful discussions.

REFERENCES

Ehrlich, S. D., Bertazzoni, U. and Bernardi, G. (1973) Eur. J. Biochem. 40, 143-147.

Heidecker, G., Messing, J. and Gronenborn, B. (1980) Gene 10, 69.

Gronenborn, B. and Messing, J. (1978) Nature 272, 375.

Legerski, R. H., Hodnett, J. L. and Gray, H. B. (1978) Nucl. Acids Res. 5, 1445.

Maniatis, T., Hardison, R. C., Lacy, E., Lauer, J., O'Connell, C., Quon, D., Sim, G. K. and Efstradiadis, A. (1978) Cell 15, 687.

Maxam, A. M. and Gilbert, W. (1977) Proc. Natl. Acad. Sci. USA 74, 560.

Messing, J., Gronenborn, B., Müller-Hill, B. and Hofschneider, P. H. (1977) Proc. Natl. Acad. Sci. USA 74, 3642.

Messing, J., Crea, R. and Seeburg, P. H. (1981) Nucl. Acids Res. 9, 309.

Sanger, F., Nicklen, S. and Coulson, A. R. (1977) Proc. Natl. Acad. Sci. USA 74, 5463.

Sanger, F. and Coulson, A. R. (1978) FEBS Lett. 87, 107.

Seeburg, P. H., Shine, J., Martial, J. A., Baxter, J. D. and Goodman, H. M. (1977) Nature 270, 486-494.

Staden, R. (1979) Nucl. Acids Res. 6, 2601-2610.

Staudenbauer, W. L., Kessler-Liebscher, B. E., Schneck, P. K., van Dorp, B. and Hofschneider, P. H. (1978) In "Single-Stranded DNA Phages" (D. T. Denhardt, D. H. Dressler and D. S. Ray, eds.), pp. 369-378. Cold Spring Harbor Laboratory, Cold Spring Harbor, New York.

Sternberg, N., Tiemeier, D. and Enquist, L. (1977) Gene 1, 255.

IN VITRO CONSTRUCTION OF SPECIFIC MUTANTS

Michael Smith
Shirley Gillam

Department of Biochemistry
Faculty of Medicine
University of British Columbia
Vancouver, B.C., Canada

I. ABSTRACT

The available methods for construction of mutants *in vitro* are summarized and the roles of the different strategies in systematic analysis of genetic functions are discussed.

II. INTRODUCTION

The combination of fine structural genetics and DNA sequence determination has been a dominant partnership in establishing precise definition of the role of DNA in bacteriophage and bacterial biology. Typical examples include the definition of sequences involved in the initiation and control of RNA synthesis (1,2) and the definition of reading frames for encoded proteins (3,4,5). In eukaryote systems, such detailed analyses have not generally been available; amongst the few examples of detailed and specific correlations are those for the yeast genes encoding iso-1-cytochrome c (6,7) and the tyrosine-inserting suppressor tRNA, SUP-4o (8,9,10). Recently, a remedy for the lack of fine structural genetic maps for eukaryote genomes and for deficiencies in prokaryote maps has been provided by the technological triad of restriction endonuclease dissection of DNA, recombinant DNA molecular cloning and rapid DNA sequence determination, all of which

DEVELOPMENTAL BIOLOGY USING PURIFIED GENES

ISBN 0-12-137420-3

were developed in the past decade. It is now possible to precisely modify DNA sequences *in vitro*, prior to studies on the genetic functions of the modified DNA, carried out either *in vitro* or *in vivo*. A variety of strategies have been developed, ranging from production of random deletions to specific changes of individual nucleotides. This new technology has been termed site-specific mutagenesis, site-directed mutagenesis, engineered mutagenesis, the production of constructed mutants and reverse genetics. Although it is more cumbersome, the phrase "*in vitro* mutant construction" is the most accurate description of the procedures. This article will describe the various strategies which are available for *in vitro* mutant construction. It will show how these various strategies complement one another in the information they provide about DNA function; there is no uniquely best method, since the best method for a given DNA depends on the questions being asked and the information already available.

III. DISCUSSION

A. Deletion Mutants

The possibility of cleaving small genomes with restriction endonucleases and religating the DNA after removal of specific fragments provided the first and most simple route to *in vitro* construction of mutants (11). This strategy was used, in concert with genetic complementation to define the positions of genes on the physical map of the SV40 genome (11). This basic strategy for physical mapping of genetic functions, using both *in vivo* and *in vitro* functional tests, continues to be a powerful approach for definition of the position of coding and regulatory functions. Because of its power and simplicity this method has been used very extensively in recent years and it consequently is not possible to provide a comprehensive listing. Interesting recent examples include definitions of the boundaries of eukaryote tRNA genes (12,13) and the functional elements of eukaryote gene promoters (14,15).

An important adjunct to this method and to all other *in vitro* mutant constructions is the cloning of the target DNA in a DNA vector which will replicate in a host cell where the mutated DNA is non-essential (16). Apart from permitting the isolation of a completely defective mutant

DNA, cloning also allows the purification of individual mutants even though the method of mutant construction produces a whole family of molecules with different changes.

Construction of deletion mutants by removal of specific fragments, whilst being a powerful technique, is limited by the random distribution of suitable pairs of restriction endonuclease cleavage sites. Consequently a number of strategies have been developed for producing deletion mutants which require only one restriction endonuclease cleavage or which are completely independent of such cleavages. Often these methods produce families of mutants as a result of deletion of progressively longer segments of DNA. Purification of each mutant by molecular cloning is, in these cases, an essential prelude to functional tests.

The simplest method for creating deletion mutants, centred on one restriction endonuclease cleavage, is one which is available with small circular DNA genomes. The double-stranded DNA is linearized by cleavage at a single site. After 5'-exonuclease treatment, the DNA can be directly used to transform or transfect host cells wherein it recircularizes after nucleolytic shortening of the ends of the molecule. A family of mutants with different sizes of DNA deletion is produced (17,18). The nucleolytic deletion and recircularization can be carried out in vitro prior to transformation of the host organism. A varient of this procedure, which facilitates subsequent manipulation of the mutant region, involves the insertion of an oligodeoxyribonucleotide duplex linker containing a specific restriction endonuclease cleavage site (19). This procedure, in that it provides a family of mutants, is especially useful in establishing the boundaries of genetic functions and the specific locations of key short genetic elements. Examples include definition of the promoter region of the Xenopus 5S RNA gene (19,20) and the different functional regions of the promoter of the iso-1-cytochrome c gene of yeast (21).

Methods for generating deletion mutants independent of restriction endonuclease cleavages have been developed. The principle behind these methods is the formation of unit length linear DNA from circular DNA by limited digestion with a non-specific nuclease followed by shortening of the linear molecule and recircularization (18). Two variants which have been described recently are of particular interest. Both take advantage of the physical properties of superhelical circular DNA. In the first procedure, a single-stranded segment of DNA (obtained by strand separation of a restriction endonuclease generated fragment) is used to relax the superhelical DNA by formation of a D-loop. The loop is then endonucleotically degraded using the single-strand specific S1 enconuclease. Subsequently the nuclease produces linear molecules. After recircularization with DNA ligase, a family of short deletions is obtained, all derived from the region of the D-loop (22). The second method for producing deletions from superhelical DNA involves nuclease digestion in the presence of ethidium bromide. Under these conditions cleavage is limited to one single-strand hit per molecule (16). Restriction endonucleases or non-specific endonucleases can be used. Treatment with an exonuclease to produce a single-strand gap, followed by endonuclease S1 and then ligation generates a family of deletion mutants (16).

B. Insertion Mutants

The strategies for production of deletion mutants described above provide a powerful set of tools for construction of mutants which disrupt specific DNA functions. A formally related strategy is the introduction of a DNA insert to produce defective mutants (23). The attraction of this approach lies in the fact that the inserted fragment can contain a specific restriction endonuclease cleavage site. This facilitates physical mapping of the mutant sites relative to a marker restriction endonuclease site. The physical map can be correlated with the corresponding genetic map. The method has been developed for constructing mutants in circular DNAs. The circular DNA is linearized by non-specific double-stranded cleavage by DNase I in the presence of Mn^{2+}; the oligodeoxyribonucleotide duplex is added and the DNA is religated. Specific mutants are isolated after molecular cloning in appropriate host cells (23). The method has been used to map the gene of *E. coli* toxin, ST I (24) and to map the genetic functions of the mating-type loci, MATa and MATα of *S. cerevisiae* (25). If the restriction endonuclease site of the introduced segment is chosen to be unique to the DNA under investigation, then the site provides a very convenient entry for DNA sequence determination (25) and precise definition of the mutational change. In fact, construction of a family of such mutants provides a very convenient strategy for systematic DNA sequence determination. This approach has been applied in the determination of the sequences of MATa and MATα and of the coding sequence of Semliki Forest virus glycoproteins (25,26,27). The sequences show that three types of mutant are produced. As well as the type with the insert directly inserted into the circular DNA, mutants with an adjacent region deleted are also formed, due to two double strand cleavages being produced in the target DNA. The third group of mutants results from a staggered cut in the target DNA. As part of the construction procedure the protruding ends are filled in prior to addition of the DNA insert. This results in a direct repeat on either side of the insert (23). One interesting result from studies of such mutants is that, even in protein coding regions, the mutants may be phenotypically silent (25). This is because the reading frame of the coding region is preserved in these mutants even though DNA is inserted or deleted and the resultant protein is still functional. From this it is clear that it is important to obtain a number of different mutated DNAs in order to be sure of detecting a particular genetic function.

C. Point Mutants Within a Defined DNA Segment

The above procedures provide for the definition of the location and boundaries of genetic functions. In order to define the role of specific nucleotides, it is necessary to construct point mutants which affect the genetic function of interest. If this function is not precisely localized, then the best strategy is to construct a family of point changes and screen or select for defective functions. These can be characterized by DNA sequence determination. Several strategies for producing families of point mutants have been developed. These can be grouped into two categories; mutants produced by chemical mutagens and those produced by incorporation of a mutagenic base analog into the DNA sequence.

The first method to be developed for the production of localized point mutants used a chemical mutagen to induce changes in a specific DNA fragment isolated from bacteriophage ϕX174 RF DNA (28). The fragment was first treated with methoxyamine and then reintroduced into genomic DNA using the marker rescue technique. In this way temperature sensitive mutants in a specific region of ϕX174 gene A were obtained (28). Presumably, these result from CG → TA transition mutations.

A second approach, which also yields predominantly CG → TA transition mutations, involves bisulphite-catalysed deamination of C residues in single-stranded DNA (29). Localized regions of single-stranded DNA are produced in circular double-stranded DNAs by enzymatic nicking and gap production on via production of a D-loop (29,30,31).

An alternate method for localized chemical mutagenesis involves the production of a R-loop where the specific RNA probe contains covalently-linked alkylating agents which can crosslink to the complementary DNA strand (32). Unlike many strategies for specific mutant construction discussed in this article, this method has been applied to a linear DNA, that of bacteriophage T7. Since the alkylating agent is most likely to react with G residues and cause depurination it is likely to produce other changes in addition to CG → TA transitions. In its present form the method is restricted to cases where a pure transcript is available; however, it is adaptable for use with D-loop structures to modify non-transcribed regions as well as transcribed regions for which a specific transcript is not available (33).

Incorporation of a mutagenic nucleotide into DNA at a single-strand gap has proved to be a useful method for production of point mutants in localized regions of a circular DNA. This involves DNA polymerase catalysed filling of the gap using 4-hydroxydeoxycytidine-5'-triphosphate in place of one of the pyrimidine deoxynucleotide triphosphates (34). The nucleotide induces both CG → TA and TA → CG transition mutations. It would be interesting to see if 7-methyldeoxyguanosine-5' triphosphate or 3-methyldeoxyadenosine-5' triphosphate can be used as substrates since after incorporation they could easily be depurinated.

D. Specific Nucleotide Changes Introduced by Synthetic Oligodeoxyribonucleotides

It was suggested some years ago that the ideal mutagen is a polynucleotide or oligonucleotide (35). However, the development of such a method as a generally useful tool required a number of advances, principally more convenient methods for oligodeoxyribonucleotide synthesis and defined DNA sequences. Improvements in the methods for oligodeoxyribonucleotide synthesis have been dramatic (36) and the imminent availability of automated synthesis will make any oligodeoxyribonucleotide easily obtainable.

The progenitor of the use of synthetic oligodeoxyribonucleotides as mutagens was marker rescue, where a wild-type DNA fragment, annealed to a mutant genome, is used to rescue the defective function in vivo (37,38). This process is inefficient; rescue is usually effected with an efficiency of 1% at best (38) and the method does not work at all with fragments much shorter than thirty nucleotides (38). Presumably this is a consequence of nuclease degradation. Consequently a method for in vitro covalent integration of oligodeoxyribonucleotides was developed (39). This uses the mutagenic oligodeoxyribonucleotide as a primer for E. coli DNA polymerase I (Klenow fragment) with wild-type DNA as template. It is critical that the template is circular because this allows the full length copy of the template to be ligated to the 5'-end of the primer. What results is a covalent closed duplex which is mismatched only at the predetermined site defined by the oligodeoxyribonucleotide. Studies on the use of this method for single nucleotide changes in the genome of bacteriophage ϕX174 have shown that a specific mutant can be produced with very high efficiency with oligodeoxyribonucleotides about ten nucleotides in length; in the most efficient experiments 40% of the progeny are the desired mutant and yields greater than 5% are usually obtained (40). Even shorter oligodeoxyribonucleotides can be used under favorable conditions (40). The method has been used to induce transition mutations (39,40,41,42) transversion mutations, involving purine-purine or pyrimidine-pyrimidine mismatches in the synthetic intermediates (40,43) and a single nucleotide deletion (44). Point mutations in bacteriophage ϕX174 were studied because the genome is well characterized (45). In addition a study on a circular DNA replicating in E. coli is an ideal model for most of the specific mutations which are likely to be required, since most cloned genes are conveniently manipulated in small circular E. coli plasmid vectors. Extension of the methodology to recombinant DNA clones in a filamentous phage DNA vector (46) or in the plasmid pBR322 (47) has proved to be straightforward.

E. Genotypic Selection of Mutants

Usually mutant isolation requires a changed phenotype to allow screening or selection. However many desirable mutants will have no discernable phenotype. This may be because the change affects a nonessential sequence in the genome under study or because the change affects a cloned genetic function which is not expressed in the host cell, e.g. a higher eukaryote DNA clone in an *E. coli* vector. In this case genotypic selection or screening is an essential adjunct of the mutagenic method. This is possible with deletion mutants because appropriate digestion of mutant DNA with restriction endonucleases reveals that specific DNA sequences are missing. The oligodeoxyribonucleotide linker-insertion method also is amenable to restriction endonuclease analysis (23,25). Because the entry point for producing point mutants within a defined DNA segment is a single-strand gap at a restriction endonuclease cleavage, some of the mutants lack the cleavage site (30,34). In addition, some of these mutants can result in the production of a new cleavage site (30). Both the removal (44) and the production (46) of restriction endonuclease cleavage sites are useful adjuncts to the oligodeoxyribonucleotide mutagenesis methods.

A more general approach to genotypic selection is provided by the oligodeoxyribonucleotide mutagen method. This is a consequence of the considerably reduced thermal stability of an oligodeoxyribonucleotide duplex with a mismatch, relative to a perfectly paired duplex (48). One application of this property is a mutant DNA selection procedure (49). At an appropriate temperature, a mutating oligodeoxyribonucleotide is a more effective primer of DNA synthesis with its cognate DNA as template than with wild-type DNA as template. Thus, with a circular DNA, mutant DNA template can be converted to a closed circular duplex under conditions where a wild-type template remains single-stranded and therefore susceptible to a single-strand specific enconuclease. This approach has been used to obtain up to a thirty-fold enrichment of a desired DNA (49). In addition it has allowed the isolation, with essentially 100% efficiency, of a phenotypically silent mutant of bacteriophage ϕX174 (44).

It also ought to be possible to use a synthetic oligodeoxyribonucleotide to screen for a mutant genome (50,51) and this appraoch has been used successfully in the case of a 14 base-pair deletion (47).

F. Conclusions and Prospects

A number of strategies for constructing mutants in vitro have been discussed. These range from methods for constructing deletions to methods for constructing specific point changes. Experiments conducted to date directed at modifying origins of replication (30) transcription (14,15,20,21,22) and translation (34,44,52) control sequences and coding sequences (39,41,42,43) clearly demonstrate the power of the methods for extending the precision of our understanding of the functions of DNA sequences. It is clear that a spectrum of methods is essential for a systematic and complete attack on DNA function; there is no uniquely "best" strategy for in vitro mutant construction. Because of the variety of in vitro manipulations to which DNA can be subjected, coupled with the unique power of biological cloning of DNA molecules for purification, it is likely that a number of strategies for mutant production will be added to those described in this article, in the not-too-distant future. However, it is exciting to know that, even with the present methods, we can ask and answer very specific questions about the function of any nucletoide in a DNA sequence.

ACKNOWLEDGMENTS

Research in the authors' laboratory was supported by the Medical Research Council of Canada of which M.S. is a Career Investigator.

REFERENCES

1. Rosenberg, M. and Court, D. (1979) Ann. Rev. Genet. 13, 319.
2. Pribnow, D. (1979) Biol. Reg. Devel. (Goldberg, R.F., ed.) Plenum Press, New York, Vol. 1, 219.
3. Barrell, B.G., Air, G.M. and Hutchison, C.A. III (1976) Nature 264, 34.
4. Smith, M., Brown, N.L., Air, G.M., Barrell, B.G., Coulson, A.R., Hutchison, C.A. III and Sanger, F. (1977) Nature 265, 702.
5. Brown, N.L. and Smith, M. (1977) J. Mol. Biol. 116, 1.
6. Smith, M., Leung, D.W., Gillam, S. Astell, C.R., Montgomery, D.L. and Hall, B.D. (1979) Cell 16, 753.

7. Sherman, F., Stewart, J.W. and Schweingruber, A.M. (1980) Cell 20, 215.
8. Kurjan, J., Hall, B.D., Gillam, S. and Smith, M. (1980) Cell 20, 701.
9. Koski, R.A., Clarkson, S.G., Kurjan, J., Hall, B.D. and Smith, M. (1980) Cell 22, 415.
10. Koski, R., Clarkson, S., Kurjan, J., Hall, B.D., Gillam, S. and Smith, M. (1981) This volume.
11. Lai, C.-J. and Nathans, D. (1974) J. Mol. Biol. 89, 179.
12. Telford, J.L., Kressmann, A., Koski, R.A., Grosschedl, R., Muller, F., Clarkson, S.G. and Birnstiel, M.L. (1979) Proc. Nat. Acad. Sci. USA 76, 2590.
13. Sprague, K.U., Larson, D. and Morton, D. (1980) Cell 22, 171.
14. Corden, J., Wasylyk, B., Buchwalder, A., Sassone-Corsi, P., Kedinger, C. and Chambon, P. (1980) Science 209, 1406.
15. Grosschedl, R. and Birnstiel, M.L. (1980) Proc. Nat. Acad. Sci. USA 77, 7102.
16. Peden, K.W.C., Pipas, J.M., Pearson-White, S. and Nathans, D. (1980) Science 209, 1392.
17. Carbon, J., Shenk, T.E. and Berg, P. (1975) Proc. Nat. Acad. Sci. USA 72, 1392.
18. Cole, C.N., Landers, T., Goff, S.P., Manteuil-Brutlag, S. and Berg, P. (1977) J. Virol. 24, 277.
19. Sakonju, S., Bogenhagen, D.F. and Brown, D.D. (1980) Cell 19, 13.
20. Bogenhagen, D.F., Sakonju, S. and Brown, D.D. (1980) Cell 19, 27.
21. Faye, G., Leung, D.W., Tatchell, K., Hall, B.D. and Smith, M. (1981) Proc. Nat. Acad. Sci. USA, in press.
22. Green, C. and Tibbetts, C. (1980) Proc. Nat. Acad. Sci. USA 77, 2455.
23. Heffron, F., So, M. and McCarthy, B.J. (1978) Proc. Nat. Acad. Sci. USA 75, 6012.
24. So, M., and McCarthy, B.J. (1980) Proc. Nat. Acad. Sci. USA 77, 4011.
25. Tatchell, K., Nasmyth, K.A., Hall, B.D., Astell, C.A. and Smith, M. (1981) Cell, in press.
26. Astell, C.R., Ahlstrom-Jonasson, L., Smith, M., Tatchell, K., Naysmith, K.A. and Smith, M. (1981) In preparation.
27. Garoff, H., Frischauf, A.M., Simons, K., Lehrach, H. and Delius, H. (1980) Nature 288, 236.
28. Borrias, W.E., Wilschut, I.J.C., Vereijken, J.M., Weisbeek, P.J. and van Arkel, G.A. (1976) Virology 70, 195.

29. Shortle, D. and Nathans, D. (1978) Proc. Nat. Acad. Sci. USA 75, 2170.
30. Shortle, D., Pipas, J., Lazarowitz, S., DiMaio, D. and Nathans, D. (1979) Genetic Engineering, Principles and Methods (Setlow, J.K. and Hollaender, A., eds.) Plenum Press, New York, Vol. 1, 73.
31. Shortle, D., Koshland, D., Weinstock, G.M. and Botstein, D. (1980) Proc. Nat. Acad. Sci. USA 77, 5375.
32. Salganik, R.I., Dianow, G.L., Ovchinnikova, L.P., Vorinina, E.M., Kokoza, E.B. and Mazin, A.V. (1980) Proc. Nat. Acad. Sci. USA 77, 2796.
33. Salganik, R.I. (1981) Personal communication.
34. Weissmann, C., Nagata, S., Taniguchi, T., Weber, H. and Meyer, F. (1979) Genetic Engineering, Principles and Methods (Setlow, J.K. and Hollaender, A., eds.) Plenum Press, New York, Vol. 1, 133.
35. Lederberg, J. (1960) Science 131, 269.
36. Itakura, K. and Riggs, A.D. (1980) Science 209, 1401.
37. Weisbeek, P.J. and van de Pol, J.H. (1976) Biochim. Biophys. Acta 224, 328.
38. Hutchison, C.A. III and Edgell, M.H. (1971) J. Virol. 8, 181.
39. Hutchison, C.A. III, Phillips, S., Edgell, M.H., Gillam, S., Jahnke, P. and Smith, M. (1978) J. Biol. Chem. 253, 6551.
40. Gillam, S. and Smith, M. (1979) Gene 8, 81.
41. Razin, A., Hirose, T., Itakura, K. and Riggs, A.D. (1978) Proc. Nat. Acad. Sci. USA 75, 4268.
42. Bhanot, O.S., Khan, S.A. and Chambers, R.W. (1979) J. Biol. Chem. 254, 12684.
43. Gillam, S., Jahnke, P., Astell, C., Phillips, S., Hutchison, C.A. III and Smith, M. (1979) Nucl. Acids Res. 6, 2973.
44. Gillam, S., Astell, C.R. and Smith, M. (1980) Gene 12, 129.
45. Sanger, F., Coulson, A.R., Friedmann, T., Air, G.M., Barrell, B.G., Brown, N.L., Fiddes, J.C., Hutchison, C.A. III, Slocombe, P.M. and Smith, M. (1978) J. Mol. Biol. 125, 225.
46. Wasylyk, B., Derbyshire, R., Guy, A., Molko, D., Roget, A., Teoule, R. and Chambon, P. (1980) Proc. Nat. Acad. Sci. USA 77, 7024.
47. Wallace, R.B., Johnson, P.F., Tanaka, S., Schold, M., Itakura, K. and Abelson, J. (1980) Science 209, 1396.
48. Gillam, S., Waterman, K. and Smith, M. (1975) Nucl. Acids Res. 2, 625.
49. Gillam, S. and Smith, M. (1979) Gene 8, 99.

50. Wallace, R.B., Shaffer, J., Murphy, R.F., Bonner, J., Hirose, T. and Itakura, K. (1979) Nucl. Acids Res. 6, 3543.
51. Szostak, J.W., Stiles, J.I., Tye, B.-K., Chiu, P., Sherman, F. and Wu, R. (1979) Methods Enzymol. 68, 419.
52. Smith, M. and Gillam, S. (1981) Genetic Engineering, Principles and Methods (Setlow, J.K. and Hollaender, A., eds.) Plenum Press, New York, Vol. 3, 1.

USE OF SYNTHETIC OLIGODEOXYRIBONUCLEOTIDES FOR THE ISOLATION OF SPECIFIC CLONED DNA SEQUENCES[1]

Sidney V. Suggs
Tadaaki Hirose[2]
Tetsuo Miyake[3]
Eric H. Kawashima[4]
Merrie Jo Johnson
Keiichi Itakura
R. Bruce Wallace

Division of Biology,
Molecular Genetics Section
City of Hope Research Institute
Duarte, California

I. ABSTRACT

A technique for the isolation of specific cloned DNA sequences has been developed using synthetic oligodeoxyribonucleotides as hybridization probes. To study the hybridization properties of such probes, oligonucleotides complementary to bacteriophage ØX174 DNA and cloned rabbit β-globin DNA were used as model systems. Results with the model systems show that: (a) oligonucleotides hybridize at specific sites in DNA; (b) a single mismatched base pair in an oligonucleotide: polynucleotide duplex significantly destabilizes the duplex; and (c) mixtures of oligonucleotides can be used to screen

[1]Supported by NIH Postdoctoral Fellowship GM07591 (SVS) and NIH grants GM26391 (RBW) and GM 25658 (KI).
[2]Present address: Keio University, Tokyo Japan.
[3]Present address: Wakunaga Pharmaceutical Co., Ltd., Osaka, Japan.
[4]Present address: Biogen, S.A., Geneva, Switzerland.

ISBN 0-12-137420-3

cloned DNA for the isolation of specific sequences. We have applied this methodology to the isolation of cloned cDNA for human β2-microglobulin.

II. INTRODUCTION

It is often difficult to obtain naturally occurring nucleic acid probes for the isolation of specific cloned DNA sequences. As an alternative, we have developed techniques for using chemically synthesized oligodeoxyribonucleotides as probes for specific cloned DNAs. Our general approach is to synthesize a mixture of oligodeoxyribonucleotides whose sequences represent all possible codon combinations predicted from a small portion of the amino acid sequence for a particular protein. One of this mixture must be complementary to a region of DNA coding for the protein. Stringent hybridization criteria would then be used to select the single correct sequence from the mixture. As a preliminary investigation, we have chosen to study two model systems.

In the first system, we investigate the hybridization behavior of three oligonucleotides, 11, 14, and 17 bases long, to DNA from wild-type (wt) and *am*-3 bacteriophage ØX174. The three oligonucleotide sequences are complementary to wt DNA at the region encompassing the *am*-3 point mutation (1). Duplexes formed between the oligonucleotides and *am*-3 DNA contain a single mismatched base pair. This system represents a useful model for the study of the effect of mismatched base pairs on duplex formation and stability.

In the second system, we examine the hybridization of oligonucleotides to cloned rabbit β-globin DNA sequences. We show that a mixture of eight different 13-base long oligonucleotides, one of which is complementary to rabbit β-globin DNA, hybridizes specifically to β-globin DNA under conditions where duplexes with single base mismatches will not form. In addition, colony screening experiments with transformed cells containing either pBR322 or pBR322 into which rabbit β-globin cDNA has been cloned demonstrate the feasibility of using of mixed oligonucleotides as specific probes for screening recombinant clones.

Finally, we have applied the information gained from the model studies to the isolation of the cloned cDNA sequences for human β2-microglobulin (β2m). The amino acid sequence for this protein is known (2). Using the amino acid sequence for β2m and the genetic code, we synthesized sets of oligodeoxyribonucleotides that should be specific for β2m. The synthetic oligonucleotides were labeled and used as hybridization probes to identify the cloned cDNA for human β2m.

III. RESULTS

A. Thermal Stability of Oligonucleotide-0X174 DNA Duplexes

In order to study the hybridization of synthetic oligodeoxyribonucleotides to natural DNA, we synthesized three oligonucleotides of chain length 11, 14, and 17, which are complementary to the single stranded DNA (+ strand) of the wild-type (wt) bacteriophage ØX174.

The 11-mer and 14-mer are synthetic intermediates of the 17-mer. The 17-mer is complementary to nucleotides 575 through 591 in the linear sequence of ØX174 DNA reported by Sanger and co-workers (1) (Figure 1). These sequences represent useful models for the study of single base pair mismatch since duplexes formed with *am*-3 ØX174 DNA contain one A-C base pair (amber mutation is a G-A transition at position 587 of the DNA sequence [Figure 1]).

The hybridization of all 3 oligonucleotides to wt ØX174 DNA was quite efficient. Between 13 and 22% of the sites on the phage DNA molecules hybridize with the labeled oligomers (Table I). The stability of the oligonucleotide:wt DNA duplexes were examined by thermal denaturation as described in reference 4. Note that in Table I thermal denaturation is described by the parameter T_d, the temperature at which one half of the duplexes are dissociated. This parameter is used rather than Tm since the experiment does not allow direct measurement of Tm in a thermodynamically rigorous way. Also note that the expected T_d's for the perfectly paired duplexes are calculated from an empirically derived formula, based on our observations of the hybridization properties of a large number of oligodeoxyribonucleotides.

As expected, an increase in thermal stability is seen with an increase in duplex length (Table I). Compared to the wt DNA, hybridization of the three oligonucleotides to *am*-3 DNA is much less efficient. In fact, the level of hybridization of the 11-mer to *am*-3 DNA was barely above background and determination of an accurate T_d was not possible. It can be seen that the thermal stability of the 14-mer and 17-mer duplexes with *am*-3 DNA is much lower than that of the corresponding wt duplexes.

TABLE I. Effect of Mismatch on Hybridization of Oligonucleotide Probes to ØX174 DNA

Number of nucleotides in probe	DNA hybridized	Number of nucleotides in duplex	% Sites hybridized	Td[a] observed (C°)	Td expected[b] (C°)
11	wt	11	20.0	34	32
11	am3	10	<0.5	-	-
14	wt	14	13.6	41	42
14	am3	13	2.9	30	-
17	wt	17	22.6	54	54
17	am3	16	11.3	43	-

[a]*Td is the temperature at which one half of the duplexes are dissociated under the conditions of the experiment (4).*

[b]*Calculated by the equation:*
Td = 2°C x number of AT base pairs + 4°C x number of GC base pairs.

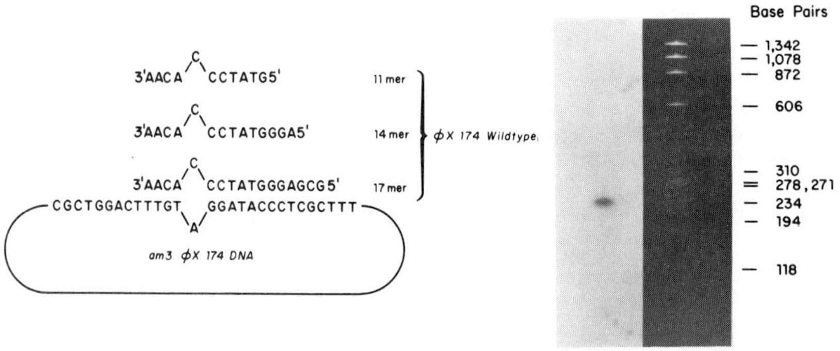

FIGURE 1. A representation of the mismatched duplexes formed between the three oligonucleotides and am-3 φχ174 RFI DNA, which was digested with HaeIII, *was subjected to electrophoresis on a 2% agarose gel and blotted onto nitrocellulose as described (3). The filter was hybridized to* ^{32}*P-labeled 14 mer at 12°C, washed at 12°C and autoradiographed. It can be seen that hybridization is to the 234 base pair long restriction fragment which contains the am-3 mutation at nucleotide 587 (1).*

B. Thermal Stability of Oligonucleotide-Globin DNA Duplexes

The model studies with cloned rabbit β-globin DNA were performed to determine the usefulness of mixtures of oligonucleotides in screening cloned DNAs. Table 2 shows a short amino acid sequence of rabbit β-globin (5) and the corresponding mRNA sequence as determined by Efstratiadis, *et al.* (6). Two oligodeoxyribonucleotides 14 bases long were synthesized, one perfectly complementary to the mRNA (RβG14A) and one with a single base change (T for C) at a position 3 nucleotides from the 5'-end (RβG14B). This latter oligonucleotide is meant to serve as a model probe forming a single mismatch. Since the base change is close to the end of the sequence and

TABLE II. Oligonucleotide Probes for Rabbit β-Globin DNA

		15	16	17	18	19	
Amino Acid Sequence		Trp	Gly	Lys	Val	Asn	
mRNA Sequence	5'	UGG	GGC	AAG	GTG	AAA	3'
Probe RβG14A	3'	ACC	CCG	TTC	CAC	TT	5'
Probe RβG14B	3'	ACC	CCG	TTC	CAT	TT	5'
Probe RβG13Mix	3'	CC	CCG	TTC	CAC	TT	5'
			A	T	T		

because it forms a G-T base pair, it was thought to be a good example of a mismatch with minimal destabilizing effect.

A mixture of eight different 13-base long oligonucleotides was also synthesized (RβG13Mix, Table II). This model mixture does not represent all possible combinations of anti-codon sequences (there are 32 possible sequences), but does contain the sequence complementary to the mRNA amongst the mixture of eight. This sequence was only 13-bases long (lacked 3'-deoxy-adenosine present in 14-mers) because 3'-deoxyadenosine is incompatible with the method of solid-phase synthesis used (7). The lack of a 3'-deoxyadenosine is not thought to have a significant effect on thermal stability.

To investigate the hybridization properties of the β-globin probes, labeled oligonucleotides were hybridized to blots prepared from *Eco*R1 digested RβG1 DNA (rabbit β-globin gene DNA cloned in bacteriophage Charon 4A (8)). When RβG1 DNA is digested with *Eco*R1 endonuclease, ten restriction fragments are produced. The portion of the gene complementary to the three probes is present in a 2.6 kilobase (kb) fragment (9). Figure 2 (A) shows the results of a hybridization of the ^{32}P labeled 14 mers to a blot of *Eco*R1 digested RβG1 DNA at 37°C in 0.9 M Na^+. It can be seen that virtually no hybridization of RβG14B to the 2.6 kb fragment is observed under these conditions while RβG14A hybrized well. Figure 2 (B) shows a competition experiment where ^{32}P-labeled RβG14A is hybridized in the presence or absence of a seven-fold excess of an unlabeled competitor, either RβG14A or RβG14B. It can be seen that only RβG14A effectively competes for hybridization of ^{32}P-labeled RβG14A to globin DNA. Figure 2 (C) shows the results of a blot hybridization with probe RβG13Mix. In this experiment, the 2.6 kb fragment is seen to hybridize with both RβG13Mix and RβG14A under conditions where RβG14B does not hybridize (Figure 2A). The specificity of hybridization of RβG13Mix to the globin DNA demonstrates the usefulness of oligonucleotides of mixed sequences as probes.

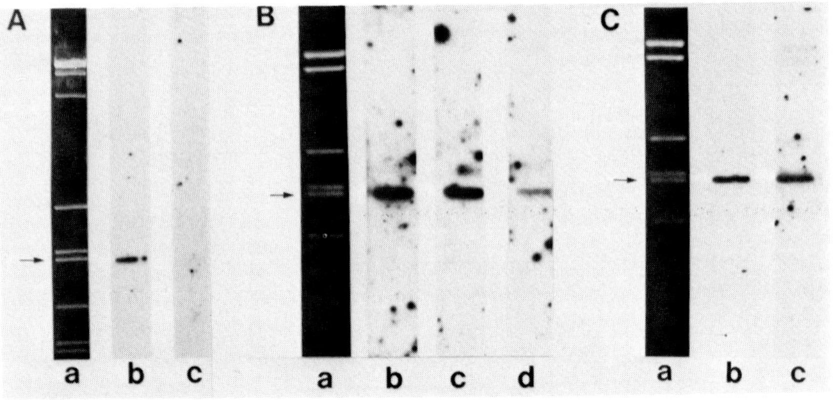

FIGURE 2. Hybridization of oligonucleotide probes to blots of RβG1 DNA. RβG1 DNA was digested with EcoRI *nuclease, the restriction fragments separated by electrophoresis on an agarose gel and transfered to nitrocellulose by blotting as described (3).*

A. Two lanes of such a blot were hybridized with ^{32}P-labeled RβG14A (lane b) or ^{32}P-labeled RβG14B(lane c) at 37°C as described (10). Lane a shows the ethidium bromide stained gel. The restriction map for EcoRI *sites has been determined (9). The 2.6 kb fragment (arrow) contains the portion of the globin gene complementary to the RβG14A probe.*

B. Three lanes of a blot were hybridized with ^{32}P-labeled RβG14A in the presence of a seven-fold molar excess of unlabeled RβG14B (lane c) or RβG14A (lane d) or in the absence of unlabeled probe (lane b). Lane a again shows the ethidium bromide stained gel. The specific activities of the two probes were 4.8 x 10^8 cpm/μg for RβG14A and 6.9 x 10^8 cpm/μg for RβG14B.

C. Two lanes of a blot were hybridized with ^{32}P-labeled RβG14A (lane b) or ^{32}P-labeled RβG13Mix (lane c). Lane a again shows the ethidium bromide stained gel. The specific activities of the probes were 4.8 x 10^8 cpm/μg for RβG14A and 6.9 x 10^8 cpm/μg for RβG13Mix.

C. Colony Hybridization Using the Rabbit β-globin Oligonucleotide Probes

The ultimate application of oligonucleotides of mixed sequences would be to use them as probes to screen recombinant clones for those which contain the desired sequence. In order to test the use of the mixed probe in colony screening, transformed cells which contained either pBR322 or pBR322 β-globin were grown and spotted onto a standard 10 cm diameter petri dish. Ten of fifty colonies contained globin DNA sequences. The colonies were grown, transferred to Whatman 540 paper, amplified and prepared for hybridization as described by Gergen, *et al.* (11). Filters were then hybridized with ^{32}P-labeled oligonucleotides. Figure 3 shows the results of such a screening. As expected, ^{32}P RβG14A clearly hybridizes to the ten globin DNA containing colonies and not to the others. ^{32}P RβG14B is not seen to hybridize in this exposure, but if the filters are exposed to film for a much longer time RβG14B also hybridizes to the ten globin DNA colonies. ^{32}P RβG13Mix is seen to hybridize specifically with the globin DNA colonies, albeit with a significantly higher background due to the

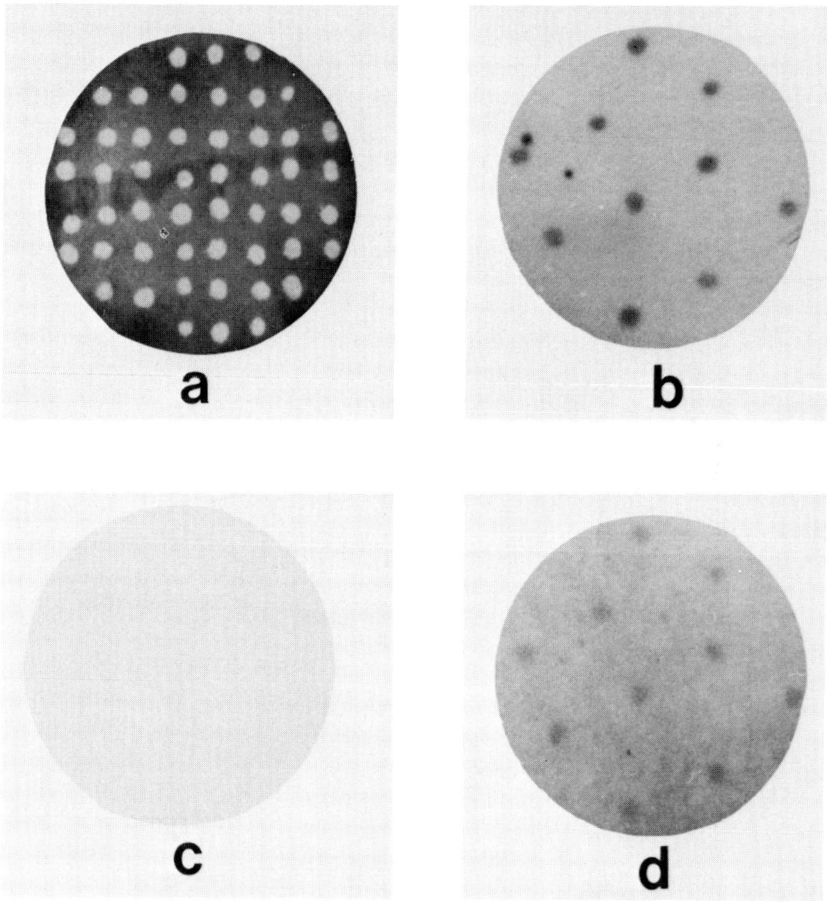

FIGURE 3. Hybridization of the oligonucleotide probes to globin and non-globin DNA containing colonies. Cells transformed with either pBR322 or pBR322 β-globin DNA were picked onto a 10 cm petri dish in an ordered array and grown overnight. Ten of the colonies contained globin DNA and 40 non-globin DNA. The colonies were lifted onto Whatman 540 filter discs, amplified with chloramphenicol and the filters prepared for hybridization as described (11). The filters were hybridized with ^{32}P-labeled RβG14A (b), RβG14B (c) or RβG13Mix (d) and washed and exposed to X-ray film as described (10). One of these filters was stained with ethidium bromide and photographed under ultraviolet light (a) showing the position of all of the colonies. The positive signal seen for the RβG14A and RβG13Mix coincide with the positions where the globin DNA-containing colonies were placed.

TABLE III. Oligonucleotide Probes for the Isolation of β2-Microglobulin

		95	96	97	98	99	
Amino Acid Sequence		Trp	Asp	Arg	Asp	Met	
Possible Codons	5'	UGG	GA^{U}_{C}	AG^{A}_{G} CGX^{a}	GA^{U}_{C}	AUG	3'
Probe β2m I	3'	ACC	CT^{A}_{G}	TC^{T}_{C}	CT^{A}_{G}	TAC	5'
Probe β2m II	3'	ACC	CT^{A}_{G}	GCX^{a}	CT^{A}_{G}	TAC	5'

[a]*X=A,C,G, or T(U).*

fact that eight-fold more labeled probe was present during the hybridization.

D. Isolation of a Bacterial Clone Containing β2-Microglobulin mRNA Sequences

In designing the probe sequences for β2m, we chose a region of the protein for which there are relatively few potential coding sequences. The region of amino acid sequence we used in designing the β2m-specific probes is shown in Table III. Amino acid residues 95-99 of β2m can be coded for by 24 possible sequences in the mRNA. We synthesized two sets of 15-mers corresponding to this region: β2mI is a mixture of 8 sequences and β2mII is a mixture of 16 sequences. The two sets of probes were synthesized as a mixture of sequences (10). Using polyA containing cytoplasmic RNA from a human lymphoblastoid cell line, double-stranded cDNA 500-800 base pairs in length was prepared (12) and inserted into the PstI site of the plasmid vector pBR322 by the standard G:C tailing method (13). The recombinant DNA was used to transform *E. coli* strain MC1061 by the Kushner procedure (14). 535 tetracycline-resistant clones were obtained and picked onto fresh plates in an ordered array. The bacterial clones were transferred to Whatman 540 filter paper, amplified with chloramphenicol and prepared for hybridization as described by Gergen, *et al.* (11). The filters were hybridized with ^{32}P-labeled oligonucleotide probes and washed as described (15). An autoradiogram of one of the five filters hybridized with the β2mII probe is shown in Figure 4. The amount of labeled probe hybridized to one of the clones is clearly greater than that hybridized to any of the other clones. Using the β2mI

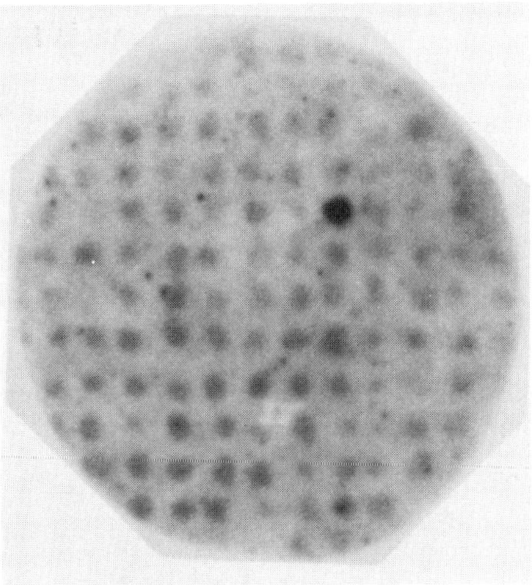

FIGURE 4. Hybridization of oligonucleotide probe β2mII to colonies transformed with G-tailed plasmid plus C-tailed cDNA. Bacterial cDNA clones were isolated and screened as described in the Materials and Methods Section. ^{32}P-Labeled β2mII probe was hybridized to the filters at 41°C overnight and washed with 0.9 M NaCl, 0.09 M sodium citrate at 41°C.

```
  -10 -9  -8  -7  -6  -5  -4  -3  -2  -1   1   2   3   4   5   6   7   8   9  10  11  12  13  14  15  16
  Ala Leu Leu Ser Leu Ser Gly Leu Glu Ala Ile Gln Arg Thr Pro Lys Ile Gln Val Tyr Ser Arg His Pro Ala Glu
C GCG CTA CTC TCT CTT TCT GGC CTT GAG GCT ATC CAG CGT ACT CCA AAG ATT CAG GTT TAC TCA CGT CAT CCA GCA GAG

17  18  19  20  21  22  23  24  25  26  27  28  29  30  31  32  33  34  35  36  37  38  39  40  41  42  43
Asn Gly Lys Ser Asn Phe Leu Asn Cys Tyr Val Ser Gly Phe His Pro Ser Asp Ile Glu Val Asp Leu Leu Lys Asn Gly
AAT GGA AAG TCA AAT TTC CTG AAT TGC TAT GTG TCT GGG TTT CAT CCA TCC GAC ATT GAA GTT GAC TTA CTG AAG AAT GGA

44  45  46  47  48  49  50  51  52  53  54  55  56  57  58  59  60  61  62  63  64  65  66  67  68  69  70
Glu Arg Ile Glu Lys Val Glu His Ser Asp Leu Ser Phe Ser Lys Asp Trp Ser Phe Tyr Leu Leu Tyr Tyr Thr Glu Phe
GAG AGA ATT GAA AAA GTG GAG CAT TCA GAC TTG TCT TTC AGC AAG GAC TGG TCT TTC TAT CTC TTG TAT TAT ... GAA TTC

71  72  73  74  75  76  77  78  79  80  81  82  83  84  85  86  87  88  89  90  91  92  93  94  95  96  97
Thr Pro Thr Glu Lys Asp Glu Tyr Ala Cys Arg Val Asn His Val Thr Leu Ser Gln Pro Lys Ile Val Lys Trp Asp Arg
ACC CCC ACT GAA AAA GAT GAG TAT GCC TGC CGT GTG AAC CAC GTG ACT TTG TCA CAG CCC AAG ATA GTT AAG TGG GAT CGA

98  99
Asp Met Stop          10         20         30         40         50         60         70         80         90
GAC ATG TAA GCAGCATCAT GGAGGTTGA AGATGCCGCA TTTGGATTGG ATGAATTCAA AATTCTGCTT GCTTGCTTTT TAATATTGAT ATGCTTATAC

        100        110        120        130        140        150        160        170        180        190
ACTTACACTT TATGCACAAA ATGTAGGGTT ATAATAATGT TAACATGGAC ATGATCTTCT TTATAATTCT ACTTTGAGTG CTGTCTCCAT GTTTGATGTA

        200        210
TCTGAGCAGG TTGCTCCACA GGTAGCT      3'
```

FIGURE 5. Nucleotide sequence for the cloned cDNA for human β2-microglobulin. The nucleotide sequence of the coding strand of the cloned cDNA is shown. Above the nucleotide sequence is the amino acid sequence of the protein, with the amino acid residues of the leader peptide indicated as negative numbers.

probe we could not distinguish specific hybridization to any clone (data not shown).

E. Nucleotide Sequence Analysis of the Cloned cDNA for β2-Microglobulin

The nucleotide sequence of the β2m plasmid DNA was determined using the base-specific cleavage reactions of Maxam and Gilbert (16). The nucleotide sequence of the cloned β2m cDNA is shown in Figure 5. The cloned cDNA includes the coding region for the protein, part of the coding region for the leader sequence, and a large portion of the 3'-untranslated region.

IV. DISCUSSION

From the two model studies utilizing oligonucleotides complementary to ØX174 DNA and rabbit β-globin DNA, it is clear that oligonucleotides are very specific probes when hybridization is performed under conditions where only perfectly base paired duplexes can form. This specificity allows the use of mixtures of oligonucleotides whose sequences represent all possible codon combinations for a short amino acid sequence as probes for the identification of recombinant clones containing DNA coding for that protein. The isolation of the human β2m cDNA clone from a collection of cDNA clones demonstrates the generality of this approach.

ACKNOWLEDGMENTS

We wish to thank Ting H. Huang for purification and sequence analysis of synthetic oligonucleotides and Monica Schold for excellent technical assistance.

REFERENCES

1. Sanger, F.; Air, G.M.; Barrell, B.G.; Brown, N.L.; Coulson, A.R.; Fiddes, J.C.; Hutchison, C.A.; Slocombe, P.M. and Smith, M. (1977) *Nature 264*, 687.
2. Cunningham, B.A.; Wang, J.L.; Berggard, I. and Peterson, P.A. (1973) *Biochemistry 12*, 4811.
3. Southern, E.M. (1975) *J. Mol. Biol. 98*, 503.

4. Wallace, R.B.; Shaffer, J.; Murphy, R.F.; Bonner, J.; Hirose, T. and Itakura, K. (1979) *Nucleic Acid Res. 6,* 3543.
5. Dayhoff, M.O. (1976) *Atlas of Protein Sequence and Structure, Vol. 5,* Supplement 2.
6. Efstratiadis, A.; Kafatos, F.C. and Maniatis, T. (1977) *Cell 10,* 571.
7. Miyoshi, K.; Huang, T. and Itakura, K. (1980) *Nucleic Acid Res. 8,* 5491.
8. Maniatis, T.; Hardisan, R.C.; Lacy, E.; Lauer, J.; O'Connell, C.; Quon, D.; Sim, G.K. and Efstratiadis, A. (1978) *Cell 15,* 687.
9. Lacy, E.; Hardisan, R.C.; Quon, D. and Maniatis, T. (1979) *Cell 18,* 1273.
10. Wallace, R.B.; Johnson, M.J.; Hirose, T.; Miyake, T.; Kawashima, E.H. and Itakura, K. (1981) *Nucleic Acid Res. 9,* 879.
11. Gergen, J.P.; Stern, R.H. and Wesink, P.C. (1979) *Nucleic Acid Res. 7,* 2115.
12. Goeddel, D.V.; Shepard, H.M.; Yelverton, E.; Leung, D. and Crea, R. (1980) *Nucleic Acid Res. 8,* 4057.
13. Chang, A.C.Y.; Nunberg, J.H.; Kaufman, R.J.; Ehrlich, H.A.; Schimke, R.T. and Cohen, S.N. (1978) *Nature 275,* 617.
14. Kushner, S.R. *in* "Genetic Engineering" (H.W. Boyer and S. Nicosia, eds.), p. 17. Elsevier, Amsterdam, (1978).
15. Suggs, S.V.; Wallace, R.B.; Hirose, T.; Kawashima, E.H. and Itakura, K. (1981), in preparation.
16. Maxam, A. and Gilbert, W. (1980) *Methods in Enzymology 65,* 499.

INDEX

D

E

F

G

T

U

V